NEUROEPIDEMIOLOGY IN TROPICAL HEALTH

NEUROEPIDEMIOLOGY IN TROPICAL HEALTH

Edited by

PIERRE-MARIE PREUX AND MICHEL DUMAS

ACADEMIC PRESS

An imprint of Elsevier

Academic Press is an imprint of Elsevier
125 London Wall, London EC2Y 5AS, United Kingdom
525 B Street, Suite 1800, San Diego, CA 92101-4495, United States
50 Hampshire Street, 5th Floor, Cambridge, MA 02139, United States
The Boulevard, Langford Lane, Kidlington, Oxford OX5 1GB, United Kingdom

Notices
Knowledge and best practice in this field are constantly changing. As new research and experience broaden our understanding, changes in research methods, professional practices, or medical treatment may become necessary.

Practitioners and researchers must always rely on their own experience and knowledge in evaluating and using any information, methods, compounds, or experiments described herein. In using such information or methods they should be mindful of their own safety and the safety of others, including parties for whom they have a professional responsibility.

To the fullest extent of the law, neither the Publisher nor the authors, contributors, or editors, assume any liability for any injury and/or damage to persons or property as a matter of products liability, negligence or otherwise, or from any use or operation of any methods, products, instructions, or ideas contained in the material herein.

Library of Congress Cataloging-in-Publication Data
A catalog record for this book is available from the Library of Congress

British Library Cataloguing-in-Publication Data
A catalogue record for this book is available from the British Library

ISBN: 978-0-12-804607-4

For Information on all Academic Press publications
visit our website at https://www.elsevier.com/books-and-journals

 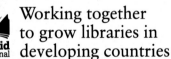 Working together
to grow libraries in
developing countries

www.elsevier.com • www.bookaid.org

Publisher: Nikki Levy
Acquisitions Editor: Natalie Farra
Editorial Project Manager: Kristi Anderson
Production Project Manager: Anusha Sambamoorthy
Cover Designer: Victoria Pearson

Typeset by MPS Limited, Chennai, India

Contents

PART I
FUNDAMENTAL CONCEPTS

PART II
TROPICAL NEUROEPIDEMIOLOGY: BETWEEN ENVIRONMENT AND GENETICS

PART III

TROPICAL NEUROEPIDEMIOLOGY BY LARGE AREAS OF THE WORLD

PART IV

FOCUS ON SPECIFIC NEUROLOGICAL SYNDROMES OR DISEASES IN TROPICAL AREAS

CONTENTS

21. Other Diseases: Traumatic Brain Injuries,
Tumors, and Multiple Sclerosis   297

MOUHAMADOU DIAGANA AND MICHEL DUMAS

List of Contributors

Thierry Adoukonou University of Parakou, Parakou, Benin

Daniel Ajzenberg University of Limoges, Inserm UMR 1094 NET, Limoges, France

Craig Barnett University of Melbourne, Melbourne, VIC, Australia

Farid Boumediene INSERM UMR 1094 NET, Limoges, France; University of Limoges, Limoges, France

Francisco J. Carod-Artal Raigmore Hospital, Inverness, United Kingdom; International University of Catalonia, Barcelona, Spain

Jean-Pierre Clément INSERM UMR 1094 NET, Limoges, France; University of Limoges, Limoges, France; Hospital Center Esquirol, Limoges, France

Philippe Couratier INSERM UMR 1094 NET, Limoges, France; University of Limoges, Limoges, France; CHU Limoges, Limoges, France

François Denis INSERM UMR 1092, Limoges, France; Laboratory of Bacteriology-Virology, CHU Limoges, Limoges, France

Jean-Claude Desport University Hospital of Limoges, Limoges, France; INSERM UMR 1094 NET, Limoges, France; University of Limoges, Limoges, France

Alain Dessein Aix Marseille University, Marseille, France

Hélia Dessein Aix Marseille University, Marseille, France

Mouhamadou Diagana Faculté de Médecine de Nouakchott, Nouakchott, Mauritania; University of Limoges, Limoges, France; INSERM UMR 1094 NET, Limoges

Catherine-Marie Dubreuil INSERM UMR 1094 NET, Limoges, France; University of Limoges, Limoges, France

Michel Dumas University of Limoges, Limoges, France; INSERM UMR 1094 NET, Limoges, France

Jean-François Faucher Limoges University Medical Center, Limoges, France

Philippe Fayemendy University Hospital of Limoges, Limoges, France; INSERM UMR 1094 NET, Limoges, France; University of Limoges, Limoges, France

Agnès Fleury National Autonomous University of Mexico, Mexico City, México

Hector H. Garcia Cayetano Heredia University, Lima, Peru; National Institute of Neurological Sciences, Lima, Peru

Maëlenn Guerchet King's College London, London, United Kingdom

Heber J. Hackembruch Neuroepidemiology Section, Institute of Neurology, School of Medicine, University of the Republic, Montevideo, Uruguay

Pascal Handschumacher Aix Marseille University, Marseille, France

Sébastien Hantz INSERM UMR 1092, Limoges, France; Laboratory of Bacteriology-Virology, CHU Limoges, Limoges, France

Dismand Houinato University of Abomey-Calavi, Cotonou, Bénin; University of Limoges, Limoges, France

Mohamad I. Idris University of Malaya Medical Centre, Kuala Lumpur, Malaysia

Pierre Jésus University Hospital of Limoges, Limoges, France; INSERM UMR 1094 NET, Limoges, France; University of Limoges, Limoges, France

Jacques Joubert University of Melbourne, Melbourne, VIC, Australia

Carlos N. Ketzoian Neuroepidemiology Section, Institute of Neurology, School of Medicine, University of the Republic, Montevideo, Uruguay

Dawit Kibru Bahir Dar University, Bahir Dar, Ethiopia

Philippe Lacroix INSERM UMR 1094 NET, Limoges, France

Annie Lannuzel University Hospital, Pointe-à-Pitre, Guadeloupe; Antilles University, Pointe-à-Pitre, Guadeloupe; Sorbonne University, Paris, France

Giancarlo Logroscino University of Bari "Aldo Moro", Bari, Italy; University of Bari "Aldo Moro", at "Pia Fondazione Cardinale G. Panico", Lecce, Italy

Laurent Magy CHU Limoges, Limoges, France

Benoît Marin INSERM UMR 1094 NET, Limoges, France; University of Limoges, Limoges, France; CHU Limoges, Limoges, France

Sandrine Marquet Aix Marseille University, Marseille, France

Stéphane Mathis CHU Bordeaux, Bordeaux, France

Athanase Millogo University of Limoges, Limoges, France; CHU, Bobo-Dioulasso, Burkina Faso

Charles R. Newton KEMRI/Wellcome Trust Research Programme, Kilifi, Kenya; University of Oxford, Oxford, United Kingdom

Edgard B. Ngoungou University of Health Sciences, Libreville, Gabon; University of Limoges, Limoges, France

Philippe Nubukpo INSERM UMR 1094 NET, Limoges, France; University of Limoges, Limoges, France; Hospital Center Esquirol, Limoges, France

Valerie S. Palmer Oregon Health & Science University, Portland, OR, United States

Abayubá Perna Neuroepidemiology Section, Institute of Neurology, School of Medicine, University of the Republic, Montevideo, Uruguay

Marie-Cécile Ploy UMR 1092 University of Limoges, Limoges, France

Pierre-Marie Preux Institute of Epidemiology and Tropical Neurology, University of Limoges, Limoges, France

Martin Prince King's College London, London, United Kingdom

Jacques Reis University of Strasbourg, Strasbourg, France

Peter S. Spencer Oregon Health & Science University, Portland, OR, United States

Chong T. Tan University of Malaya Medical Centre, Kuala Lumpur, Malaysia

Achille Tchalla University of Limoges, Limoges, France

Redda Tekle-Haimanot Addis Ababa University, Addis Ababa, Ethiopia

Jean-Michel Vallat CHU Limoges, Limoges, France

Andrea S. Winkler University of Oslo, Oslo, Norway; Technical University of Munich, Munich, Germany

Foreword

This book fills the existing void resulting from the lack of publications in two major areas of interests to neurologists worldwide: neuroepidemiology and tropical neurology. It provides substantial and updated information on the peculiar epidemiology and clinical manifestations of the diseases of the nervous system that occur predominantly or exclusively in the tropics.

A large majority of the inhabitants of the world live in the tropics, in regions situated along the Equator, extending 30 degrees latitude North and South between the Tropics of Cancer and Capricorn. Most of South America, Central America and the Caribbean islands, Africa, the Indian subcontinent, Southeast Asia, Papua New Guinea, and parts of Australia and the Pacific islands are situated in the tropics. The tropical climate is characterized by high daytime temperatures and abundant direct sunlight; days and nights of equal length, and absence of four seasons. Mountain ranges, jungles, and deserts provide wide variations in temperature and humidity, and a multitude of ecosystems for a large number of neurologic infections, poisonous animals, and neurotoxic plants capable of affecting the nervous system.

Although three-quarters of the total world population—or 5.55 billion of the current estimate of 7.4 billion people—live in the tropics, they consume a meager 6% of the worldwide food production. Not surprisingly, malnutrition affects more than 35% of the population in tropical Africa, about 25% of the people in India, and 5%–20% of Latin America and Caribbean populations. Hunger remains a widespread worldwide problem due to drought, famine, poverty, war, displacement of populations, and social unrest. Poverty is the best correlate of undernutrition; half of the world population lives on less than US$2 per day and 20% survive on less than US$1 per day. For this reason, hunger and neurologic problems linked with malnutrition and consumption of drought-resistant neurotoxic plants continue to occur since time immemorial. Contemporary epidemic outbreaks in the tropics, such as the Cuban blindness that affected over 50,000 patients in 1993–94 resulted from economic, political, nutritional, and toxic factors (Román, 2016). Research on this epidemic eventually led to better knowledge of the fragile equilibrium that exists between energy-providing nutrients and the failure of energy-dependent neurons of the optic nerve maculo-papillary bundle.

This book was planned and edited by the *Maître–Élève* team formed by two of the most prominent experts in the field, Profs. Michel Dumas and Pierre-Marie Preux. Honored in 2015 by the World Federation of Neurology with the *Gold Medal for Services to International Neurology*, Prof. Michel Dumas is the recognized creator of the specialty of Tropical Neuroepidemiology and Founding Director of the Institute of Neuroepidemiology and Tropical Neurology (IENT) which he launched in 1982 in Limoges, France. Prof. Dumas is a living legend: born in Bamako, Mali (Africa) in 1934, he became a neurologist specializing in tropical medicine under Prof. Henry Collomb in Dakar, Senegal, whom he succeeded as Chairman of Neurology. He trained innumerable generations of neurologists in Francophone Africa, the Pan-Arab nations, Latin America, and Asia. The result of Prof. Dumas' unfailing teaching and research activities in tropical neurology, parasitic diseases of the nervous system, and epidemiology was the eventual creation of one of the largest networks of tropical neurologists around the world. His disciple and co-editor, Prof. Pierre-Marie Preux MD PhD (Epidemiology and Public Health) is a clinical neurologist specialized in Epidemiology, Biostatistics, Clinical Research, and Tropical Medicine. Prof. Preux succeeded his mentor, Prof. Dumas, as Head of the IENT at the University of Limoges, where he also directs the INSERM UMR1094 Unit on Tropical Neuroepidemiology, and is Head of Clinical Research and Biostatistics at Limoges University Hospital. He has conducted research on numerous tropical diseases with emphasis on epilepsy throughout the tropical world including Africa, South East Asia, and Latin America. P.M. Preux is Professor of Epidemiology and Vice President for Research of the University of Limoges. He is an advisor to the World Health Organization (WHO) and to numerous research organizations in France and other countries. For this book, Preux and Dumas secured the collaboration of an international panel of distinguished experts that provided up-to-date information on their topics of expertise.

Many neurologists practicing in temperate areas of the world in North America and Europe may ask what is the need for a practicing neurologist to obtain information on the epidemiological distribution and the pattern of presentation of tropical neurological diseases. The answer is simple: neurologists around the world are now confronted with formerly "exotic" tropical diseases among travelers, foreign workers, and migrants. In these times of rapid air travel, the exponential increase in international travel and, in particular, ecologic tourism to the tropics, makes it far from exceptional to find neurologic cases of a tropical disease such as malaria in the midst of a heavy snowfall (Román, 2011). Moreover, global warming has recently extended the traditional ecological niche of many arthropod-borne diseases such as West Nile, Dengue, Chikungunya, and Zika. Despite claims to the contrary, global warming is changing the distribution of diseases formerly confined to the tropics.

Hospitals cater for the needs of millions of international business travelers and migrants from tropical countries. Tropical conditions mimic common ailments such as headache, seizures, or low-back pain. Isolated cases of parasitic tropical diseases may occur in recipients of blood transfusions or even organ donations from foreign-born asymptomatic carriers. Therefore, unless the neurologist obtains a history of travel, the correct diagnosis and treatment may be delayed with tragic consequences. Contemporary neurologists must be aware of the manifestations of diseases that are common in other regions of the world, as presented here.

Many tropical countries are underdeveloped producing only 15% of the world's net revenue. Health indicators such as life expectancy at birth, infantile mortality, and daily caloric intake, access to clean drinking water, literacy, and health expenditures are significantly lower in tropical countries than in industrial nations. These factors explain in part the neuroepidemiological pattern of tropical neurology resulting from problems of illiteracy, malnutrition, deficient sanitation, overcrowding, pollution with neurotoxins and other environmental pollutants, high human immunodeficiency virus (HIV) seroprevalence, as well as frequent waterborne and arthropod-transmitted infections of viral, bacterial, and parasitic origin. The two main determinants of neurological diseases in underdeveloped countries are lack of education and poor socioeconomic conditions (Toro et al., 1983).

Finally, the tropics continue to provide a fertile ground for research on neurological diseases. Two examples of novel neurological conditions recently reported in the tropics are the unexpected neurotropism of the Zika virus and the Nodding Syndrome described in Uganda, currently considered a post-measles brain disorder triggered by malnutrition (Spencer et al., 2016).

Zika virus is a striking example of a true tropical neurological disease producing a pandemic in tropical and temperate regions worldwide. Described 70 years ago, in 1947, in the Zika forest in Uganda, this *Flavivirus* was considered clinically unimportant given

that only 14 human symptomatic cases were reported between 1947 and 2006 (Smith et al., 2016). The first Zika case in the United States was reported in 2007 in Alaska, in a person infected during the epidemic outbreak in Yap Island in Micronesia. Local transmission of Zika in South America was reported in Brazil in 2015. Transmission was by the bite of mosquitoes *Aedes aegypti* and *Aedes albopictus*, who are also carriers of chikungunya and dengue viruses. Rapid spread of the virus throughout the Western Hemisphere resulted in over 650,000 reported cases. On February 1, 2016, the WHO declared Zika a Public Health Emergency of International Concern. In addition to mosquito bites, sexual transmission has been demonstrated and the virus has been isolated in blood, semen and maternal milk. Zika infection is manifested by fever, maculopapular rash, conjunctivitis and rare uveitis, arthralgias, muscle pain, and headache, but about 80% of infections are asymptomatic.

Although other Flaviviruses such as West Nile, yellow fever, Saint Louis encephalitis, dengue, and Japanese encephalitis are capable of producing neurological disease, the broad spectrum of neurological manifestations of Zika virus was unprecedented. Zika virus infection of pregnant women causes microcephaly probably because the virus infects pluripotent neural stem cells. Also, ocular macular abnormalities have been reported in association with microcephaly. Adults may develop Guillain—Barré syndrome (Parra et al., 2016), meningoencephalitis, and myelitis. Preventive measures to avoid mosquito bites and avoidance of unprotected sex are recommended when traveling to endemic areas. Currently there is no treatment available for Zika infection.

Gustavo C. Roman, MD DrHC

The Jack S. Blanton Presidential Chair for the Study of Neurological Disorders, Methodist Neurological Institute, Houston Methodist Hospital, Houston TX, USA

References

Parra B, Lizarazo J, Jiménez-Arango JA, et al. Guillain—Barré syndrome associated with Zika virus infection in Colombia. *N Engl J Med*. 2016;375:1513—1523.

Román GC. Parasitic diseases and malaria. *Continuum Am Acad Neurol*. 2011;17(1):113—133.

Román GC. *Cuban Blindness: Diary of a Mysterious Epidemic Neuropathy*. London: Academic Press; 2016.

Smith DE, Beckham JD, Tyler KL, Pastula DM. Zika virus disease for neurologists. *Neurol Clin Pract*. 2016;6:515—522.

Spencer PS, Mazumder R, Palmer VS, Lasarev ML, Stadnik RC, King P, et al. Environmental, dietary and case-control study of Nodding Syndrome in Uganda: a post-measles brain disorder triggered by malnutrition? *J Neurol Sci*. 2016;369:191—203.

Toro G, Román G, Navarro de Román LI. *Neurología Tropical: Aspectos Neuropatológicos de la Medicina Tropical*. Bogotá: Editorial Printer Colombiana; 1983.

Is Tropical Neurology Specific?

Only those countries situated between the Tropic of Cancer to the north and the Tropic of Capricorn to the south are strictly tropical, but for many healthcare professionals, the label also applies to other regions with similar environmental, cultural and socioeconomic characteristics. For our purposes concerning the specificity of tropical neurology, we consider these regions but take care not to stretch the concept to the point at which it becomes meaningless.

The specificity of tropical neurology was recognized by early neurologists working in these areas based on their environmental and cultural observations. Neurological conditions and their consequences are intimately linked to tropical environments, with outcomes that are often negative, but can also be positive. Furthermore, the understanding, interpretation and psychological experience of coping with disease, related to culture, are very different. Even before the many recent findings, neurologists working in tropical countries recognized the uniqueness of their observations.

The specificity is not apparent in the practice of neurology: neurological examination is the same everywhere, although, incidentally, the Babinski sign would probably not have been discovered by Babinski if he had lived in a tropical area where people often habitually walk barefoot, leading to the formation of a protective plantar rind but insensitive to stimulation.

The features characteristic of tropical medicine have been studied since the late 19th and early 20th centuries, resulting in an efflorescence of institutes, schools and scientific societies of tropical medicine in Hong Kong, London, Liverpool, Hamburg, Lisbon, Brussels, Puerto Rico, Manila, and Chicago.[1] Tropical medicine was an acknowledged specialty from very early on.

Tropical neurology emerged more gradually, and it was only in the second half of the 20th century that neurologists living in tropical regions shared ideas at medical meetings and through publications, making them aware of the existence of a specific neurological discipline in the tropical environment. This awareness has been particularly slow and gradual to emerge, as the number of neurologists working in relevant regions was very small, and findings have had to be submitted to western neurological societies. Western neurologists adhered to a singular conception of neurology, the one developed in the late 19th century and early 20th century. It was difficult to question the dogmas; it was said that if there were any differences, they could only be the result of inadequate observations or even a lack of knowledge of tropical doctors. It was only during the World Congress of Neurology held in Buenos Aires in 1961 under the leadership of Ludo Van Bogaert, President of the World Federation of Neurology, that the concept was recognized. Noshir Wadia, from Mumbai (India), Oscar Trelles, from Lima (Peru), Antonio Spina Franca, from São Paulo (Brazil), Henri Collomb, from Dakar (Senegal), and Gottlieb Monekosso, from Ibadan (Nigeria) all played crucial roles in communication. Thus, it became possible to create, within the World Federation of Neurology, the research group in Tropical Neurology that still exists today. This group contributed to the birth of neurological societies in many different tropical countries, enabling the emergence of genuine tropical neurology and recognition of its specificity.

Over the past two decades, a profound change in neurological practice in tropical countries has occurred. Progress continues apace; the emergence of more valid neuroepidemiological data has permitted a better understanding of the role of the environment and socioeconomic, ethnographic and cultural factors. In addition, advances in genetic and biological knowledge, and particularly steps forward in immunology, now allow an improved interpretation of disease, thus promoting prevention and treatment, principally in the field of infectious diseases, and reducing neurological complications.

Furthermore, neurological institutions comparable to those developed in the West have emerged in most tropical countries, mainly in the capital cities, and the numbers of doctors trained as neurologists is increasing. In addition, political and socioeconomic changes in tropical and subtropical countries are profound. But there are still huge disparities between and within countries, between urban and rural areas and among the social strata of the population.

All these innovations and their rapid evolution have led to a change in environmental conditions, such that the question arises whether the specificity of tropical neurology persists. Until we get a better understanding of the genetic issues, this specificity is primarily based on the following environmental factors.

First, *bad socioeconomic conditions* lead to poor sanitary conditions, lack of access to water, undernourishment, a cause of malnutrition, undernutrition or hypovitaminoses, disabilities of all kinds including neurological disorders such as nutritional neuropathies occurring mainly in adults, and learning difficulties often aggravated by iodine deficiency in young children.

The consequences of unfavorable socioeconomic conditions are many and varied given their multiplicity, but they are not the sole preserve of tropical regions, although they do predominate there. Such conditions have gradually declined in the West, but never completely disappeared. The changing world, the proliferation of conflicts and mass migrations for various reasons, show that all these medical issues, neurological in particular, are also subject to globalization. In the tropics, there is now a strong economic growth, even though it currently benefits only a small percentage of the population.

A second major characteristic of the tropical environment is *the high prevalence of infectious diseases.*

Some are ubiquitous, most being *bacterial infections,* among them meningitis. Although improvements in detection have led to better control, their prevalence remains significantly higher than in Western countries. Overall, we cannot consider them specific to the tropics, although for many reasons they still predominantly occur there; leprosy, causing neuritis, is a good example in this context.

Viral infectious diseases transmitted by arthropods are more specifically tropical. Climatic factors favor the development of conditions such as dengue, chikungunya or zika, which may lead to neurological complications. Climate change resulting in gradual warming of temperate countries, along with easier transport and migration, have favored the introduction of vectors and consequently the emergence of these disorders in very remote areas of tropics, where they were previously unknown. Again, this diminishes the specificity of tropical diseases and their neurological complications. Poliomyelitis and related paralysis is an ubiquitous condition *par excellence.* Despite efforts by the WHO, it persists in many tropical regions, despite being almost wiped out elsewhere. However, due to migration and poor prevention, in poorly controlled areas like northern Nigeria, it reappears outside the tropics as sporadic cases or small outbreaks and may even become endemic/epidemic. Once again, tropical specificity is lost. It may persist for other viral diseases such as Lassa fever, Marburg or Ebola, but for how long? The neurological complications of such diseases are often serious and may be deadly. However, they are exceptional and do not contribute to the emergence of a tropically specific neurology.

A large proportion of tropical infectious diseases with neurological complications are *parasitic infections.* Climatic and ecological conditions in the tropics promote hatching and development of vectors. No parasitosis is specifically neurological, but all can have such an effect via the invasion of the nervous system by parasite eggs or larva, which is the case for most parasites, via anoxic-type encephalopathic complications like cerebral malaria, or via allergic reaction or posttreatment as in lymphatic filariasis. Most of these parasitic diseases, such as cysticercosis, hydatidosis, coenurosis and distomatoses are ubiquitous but are declining in temperate regions due to the reduction in their transmission attributable to improved hygiene and pastoral activities. They are still predominant in the tropics, but they tend to diminish and disappear because of progress in prevention and therapy. Many parasitic infections have always been ubiquitous, including trichinosis, toxocariasis, telluric amoebiasis, alveolar echinococcosis, toxoplasmosis, and strongyloidiasis; they remain present everywhere even if some of them seem more frequent in tropical regions. Others are not totally ubiquitous but still have a high prevalence in tropical and subtropical regions. Because of the extent of their geographical distribution, they cannot be considered solely and specifically tropical. Examples include schistosomiasis, some distomatoses including paragominiasis, gnathostomiasis, and angiostrongyliasis, which still have a high prevalence in tropical and subtropical areas. Finally, there are few exclusively tropical parasitic diseases like African trypanosomiasis. Its vector, the tsetse fly cannot, for climatic and environmental reasons, develop and proliferate elsewhere than in the African tropics. The same is not true for American trypanosomiasis (Chagas disease) in South America, whose vector, the triatomine bug, extends far beyond the tropics.

It appears that changes in environmental conditions in the rest of the world have narrowed the concept of tropical specificity. Diseases such as high blood pressure, diabetes, cancer, their neurological complications, and dementias, currently greatly concern neurologists in temperate regions. They were not considered a priority in the tropics, but are becoming so. Their prevalence and their effects are already identical to, if not sometimes worse than in temperate regions. Again, there is nothing specific to the tropics, although some factors, particularly nutritional issues, may promote emergence and worsen severity.

All the findings briefly reported here show that neurology tends to standardize. Has tropical neurology therefore lost its specificity? It is possible to consider specificity as the unique result only of the environmental factors described above. However, that would be to ignore three facts of paramount importance:

- the role of specific cultural factors in the tropics;
- the role of sunshine, which may be disastrous but can also be beneficial; and
- the role of genetic factors.

CULTURAL ISSUES

Worldwide, culture guides the behavior of individuals who adapt and organize their own lives according to the cultural environment in which they operate and the beliefs they hold.

Culture is the set of activities, customs, intellectual productions, art, religious beliefs that define and distinguish a group. It is the nucleus of society and through it, the core of the individual. It is the source of life, well-being, harmony, and vitality; it is multi-faceted: artistic, musical, and culinary. It is expressed in customs that are preserved and perpetuated within the group. Culture is written in the genes of the group, giving individuals something to recognize and identify with. It is the crucible in which successive contributions of different generations are laid down like sediment. It enacts the group's laws and conventions. Culture is a way of thinking, acting and understanding the universe, offering strength and vitality. It is not static, but dynamic. It evolves, enriched by contributions from elsewhere. It directs the behavior of the group and the individual. It shapes those who identify with it and cannot live without it, enriching their thoughts and behaviors. It allows them to project into the future and defines their place in society. This way of being, existing and understanding existence, has gradually allowed the emergence of different cultures in all regions of the world, including and perhaps especially in tropical regions, for example native American, Aboriginal, Asian, African, and Arabian cultures.

Identifying the precise boundaries of a given culture is difficult because there are within each, many peculiarities, but there are also shared common denominators that serve to characterize and identify them.

The effect of culture on the experience of neurological disorders is particularly marked in the tropics. It is worth looking more closely at very common neurological disorders, headaches, epilepsy, and dementia, particularly Alzheimer's disease whose prevalence was considered lower in the tropics compared to temperate regions, although recent epidemiological studies suggest that at the same age, the prevalence is similar.[2]

Headaches have a very high prevalence, even more so because of psychosomatic headache in people with low-quality jobs who are obliged to adapt to another cultural environment, often of Western type. These individuals are deeply rooted in their culture and have difficulty integrating into another. They quickly become acculturated, searching for benchmarks they have lost, and headache becomes a way of life that persists as long as the person is unable to reconcile both cultural worlds.

Epilepsy in the tropics is not distinctive physiologically, neurobiologically or clinically. Its specificities are etiologic (infections, perinatal disorders) but these are not the most important. Indeed, tropical epilepsy is severe, often causing physical and social disability, and mortality remains high due to the lack of adequate medical facilities, shortage of doctors and neurologists, difficulty in obtaining medications, their poor distribution (distance to the dispensation facilities), high cost, and the need to compete with cheaper and sometimes adulterated medicines sold in markets. Epilepsy and other neurological conditions, even though frequent and debilitating, are not a priority for government officials; they receive far less attention than the endemics that poor countries still face. Yet the occurrence of epilepsy is very often the result of these endemic infectious diseases constituting a neurological complication, and requires considerable expenditure to be borne by families and governments. Such costs, of course, contribute to the economic deficit, and thereby exacerbate the vicious circle of poverty.

The specificity of tropical epilepsy is primarily cultural in origin. It is the consequence of the patient's representation in the traditional society in which he or she lives. As in former Western societies, the patient is still sometimes considered to be almost non-human. Epilepsy is experienced as a threat that must be protected against. It is considered transmissible through saliva, urine, feces, milk, and even breath and belching emitted during a seizure. The patient is supposed to have a strong power to create nuisance, to the point where no help is given should there be a seizure-related fall at home or even into a pool of water or an open fire. The social and cultural consequences of these beliefs, and the lack of knowledge of the disease, are obviously potentially catastrophic for the patient; at best, he may be simply pushed away and dismissed as being evil; at worst, he may suffer physical abuse. Psychoemotional stigma is constant and predominant. The patient is almost always marginalized to the point that in some societies customary funeral rites are no longer respected. Marginalization may occur at school, even if the teacher knows about epilepsy. The patient loses status and position in society; he cannot marry; very often he cannot even share the family meal. The result is great suffering for the patient and stigma. The first therapeutic recourse is usually traditional medicine, and only after several failures does the patient benefit from a modern treatment, prolonging the course of the disease. The traditional healer does, however, have a beneficial role in breaking the disease cycle and decreasing the distress of the patient and family. So, in these societies we should consider traditional medicine and Western-style medicine as complementary. It would be dangerous to oppose traditional approaches, especially when a lack of resources, personnel and medical facilities, necessitates non-drug remedies. Another consideration is that Western therapy for epilepsy must be taken daily which is culturally difficult to accept for the patient and family, again favoring the traditional therapist.[3]

The prevalence of *dementia* is still not sufficiently assessed and is discordant in tropical countries, due to numerous potential biases, low life expectancy, the dearth of high-quality epidemiological studies, and poor adaptation of tools to the populations of interest. Assessing the relationship between culture and dementia requires knowledge of the genetic heritage of each ethnic group with a distinct culture. Even then, it is often difficult to assess the respective roles of the cultural environment and genetic factors. Even subjects belonging to the same ethnic group, could lose or break away from the ancestral culture when living in a different cultural environment. For example, the prevalence of Alzheimer's disease is estimated at 8.2% among black people of Yoruba ancestry living in the USA in an American cultural environment, compared to only 2.3% among those born and living in their original African cultural environment.[4,5] Furthermore, assessment of the role of genetic factors is complicated by frequent interbreeding among populations.

Numerous studies have demonstrated that the apparently beneficial role of traditional cultural factors can be explained by the maintenance of the individual in his social group. People are valued because of their age and the status of ancestor confers power and wisdom. Older people acquire expert status, they are consulted, listened to and respected. They become the guardians and guarantors of the family group, and are often regarded as intermediaries between the living and the dead, becoming privileged interlocutors. The aged therefore remain helpful, efficient and productive, retaining their self-esteem rather than living in anticipation of death. They remain active members of their social group, perpetuating its culture. Adoption by traditional societies of modern habits tends to upset this psychosocial balance and favor the Western model characterized by the emergence of individualism. The

elderly person gradually loses her roots in history: her own history, that of the family, and that of the larger group in which it belongs. Access to the richness and wisdom characteristic of traditional societies is lost. Thus, the cultivation of traditional cultures could be a way of protecting people against dementia. In contrast, exposing traditional societies to modernity leads elderly people to lose their privileged status, and indirectly to decrease their intellectual functions. Traditional societies should therefore guard against the trend towards adopting Western ways.[6–9]

THE ROLE OF SUNSHINE

The only common denominator between all tropical regions, whether African, American, Australian or Asian, is sunshine. Ultraviolet B radiation has numerous effects. It can cause skin cancer, but is also known to have a beneficial impact on the immune system, disturbance of which is consistently seen in multiple sclerosis (MS). Many studies have reported a high prevalence of MS in cold or temperate regions, with a decline in the tropics and approaching the equator, regardless of the area. Ultraviolet acts first by depressing Langerhans' cells, antigen-presenting immunocompetent epidermal cells, and second by promoting the synthesis of vitamin D3, an essential vitamin widely recognized to be beneficial in various disorders. Sun exposure appears to protect against MS, the prevalence of which is very low in the tropics. This is consistent with proposed treatments aimed at depressing or regulating the immune system and thereby protecting genetically predisposed individuals.[10,11] Besides MS, the sun is now recognized as having a beneficial effect on cognitive function, particularly in the elderly.[12]

These protective and therapeutic effects of sunlight on neurological conditions contribute to constitute tropical neurological specificity. It is likely that in the distant future, it will be the main, if not the only, specificity that remains.

THE ROLE OF GENETIC FACTORS

It is currently difficult, if not impossible, to assess the roles of genetics and epigenetics in the determination of conditions encountered in tropical areas, for the following reasons:

• imperfect, if not absent, understanding of genetic factors and susceptibility to many diseases, particularly those rampant in the tropics;
• heterogeneity of the population; there is not one tropical population, but many, that spread around

the globe between the tropics. However, despite their heterogeneity these people have similarly colored skin and heightened melanocyte activity. Experience has also taught us that tropical populations probably have a propensity to make connective tissue, which results in a higher percentage of liver cirrhosis and keloid, but a low level of pseudoarthroses, and neurologically, more arachnoiditis, particularly spinal. These findings need to be confirmed, especially by epidemiological studies. It is not inconceivable that a common genetic specificity in these tropical populations may emerge one day, but right now it is impossible to say.

In conclusion, in view of all the above findings, it can be said that tropical neurology remains specific. This specificity may not exist much longer and will continue to evolve.

We hope that this book will incite readers to question this specificity and its evolution in the future.

Michel Dumas and Pierre-Marie Preux
University of Limoges, Limoges, France

References

1. Poser C, Dumas M, Spina Franca A. The history of tropical neurology. In: Shakir RA, Newman PK, Poser CM, eds. *Tropical Neurology*. London: Bailliere Division of W.B. Saunders Company Limited; 1996:1−4.

2. Guerchet M, M'belesso P, Mouanga AM, et al. Prevalence of dementia in elderly living in two cities of Central Africa: the EDAC survey. *Dement Geriatr Cogn Disord*. 2010;30:261−268.

3. Bruno E, Bartoloni A, Sofia V, et al. Sociocultural dimension of epilepsy: an anthropological study among Guaraní communities in Bolivia − an International League Against Epilepsy/International Bureau for Epilepsy/World Health Organization Global Campaign against Epilepsy regional project. *Epilepsy Behav*. 2011;22:346−351.

4. Hendrie HC, Osuntokun BO, Hall KS, et al. Prevalence of Alzheimer's disease and dementia in two communities: Nigerian Africans and African Americans. *Am J Psychiatry*. 1995;152:1485−1492.

5. Ogunniyi A, Baiyewu O, Gureje O, et al. Epidemiology of dementia in Nigeria: results from the Indianapolis-Ibadan study. *Eur J Neurol*. 2000;7:485−490.

6. Nubukpo P, Ouando JG, Darthout N, Clement JP. Démences et cultures. *Hommage à Yves Pélicier. Rev Fr Psychiatr Psychol Méd*. 2002;6:57−62.

7. Nubukpo P. *La culture peut-elle protéger de la démence ? Mémoire pour le DU de Gérontopsychiatrie*. Limoges: Université de Limoges; 1996.

8. Pollit PA. Dementia in old age: an anthropological perspective. *Psycho Med*. 1996;26:1061−1074.

9. Dumas M, Nubukpo P, Preux PM. Démences et cultures. *Bull Soc Pathol Exot*. 2010;103:194−195.

10. Dumas M, Jauberteau MO. The protective role of Langerhans' cells and sunlight in multiple sclerosis. *Med Hyp*. 2000;55:517−520.

11. Dumas M, Druet-Cabanac M, Preux PM, Jauberteau MO. Epidermal Langerhans' cells and sunlight are they protective factors in multiple sclerosis? *Rev Neurol*. 2000;156:172.

12. Miller JW, Harvey DJ, Beckett LA, et al. Vitamin D status and rates of cognitive decline in a multiethnic cohort of older adults. *JAMA Neurol*. 2015;72:1295−1303.

P A R T   I

FUNDAMENTAL CONCEPTS

CHAPTER

1

Methodological Challenges of Neuroepidemiological Studies in Low- and Middle-Income Countries

Farid Boumediene[1,2], *Benoît Marin*[1,2,3], *and Pierre-Marie Preux*[4]

[1]INSERM UMR 1094 NET, Limoges, France [2]University of Limoges, Limoges, France [3]CHU Limoges, Limoges, France [4]Institute of Epidemiology and Tropical Neurology, University of Limoges, Limoges, France

OUTLINE

1.1 INTRODUCTION

Most of the resources used to deal with neurological disorders around the world are spent in high-income countries. The medical demography (number of specialists) is highly correlated with the level of development of a country: most low- and middle-income countries (LMICs) have a very low number of neurologists, usually only in the cities (even sometimes only in the capital). Therefore, the departments of neurology are situated in the central hospital of these cities and most people with neurological disorders (PWND) do not receive appropriate care.[1] This is known, for example, in epilepsy as the treatment gap (TG).

Since the early 2000s, researchers have identified inadequate case ascertainment in a population, failure to diagnose neurological diseases and lack of appropriate care (which could include medical, surgical and/or social intervention) as the main reasons accounting for the TG. In addition, the TG can further be caused simply by a lack of treatment (unavailability of a drug in an area), as well as failure to continue treatment once started, due to sociocultural or economic factors. The latter is often referred to as the secondary TG.

Whatever the neurological disease concerned, there were only a few incidence studies of neurological disorders in LMICs. Prevalence studies used different study designs and are hardly comparable, usually identifying only main pathologies with the most explicit clinical signs. There were also very few unbiased case—control studies and natural history studies. Several studies mentioned the size of the TG but few analyzed it thoroughly.

Consistently applicable and valid epidemiological methods (descriptive, analytic and interventional) are needed in LMICs to generate comparable data on the burden of neurological disorders, outcome of interventions, risk factors and treatment. Systematic reviews and meta-analyses using such well-executed primary studies would be relevant to all involved in the care of neurological disorders. Epilepsy is probably the most studied neurological disorder and will be taken as a model in this chapter. It illustrates perfectly the various situations which could be transposed to the other neurological pathologies.

1.2 STUDY DESIGN AND FEASIBILITY IN THE FIELD

Standards of epidemiology are the same everywhere. It should not be considered that degraded methods are relevant for LMICs because of the difficulties in performing research. This is helpful neither for researchers, nor for the quality of research performed in those areas and is definitively not an ethical consideration.

Epidemiology can be divided in two branches: experimental and observational, depending on whether the exposure is assigned by the investigator. Observational studies can be either descriptive or analytic.

The main goal of descriptive epidemiology is to estimate the prevalence (number of subjects with a given characteristic—i.e., epilepsy—divided by the total number of subjects at risk in the underlying population) or rates (number of subjects developing a given characteristic—i.e., epilepsy (incidence rate) or a death (mortality rate)—divided by the total number subjects at risk in the underlying population for a given period

(person-time)). To be very clear, the term "prevalence rate" is totally wrong and should not be used.

Sample size should be sufficient to provide precise estimates of the 95% confidence interval. The design should include this estimation of the number of needed subjects. This of course should be carried out before starting the study. Even in a descriptive study, this calculation is relevant. For example, one could estimate the number of subjects to get an accurate precision of a proportion. Different types of calculations are available for other designs.

There should not be a confusion between this number (which is a key number for the study) and the feasibility to include this number. If the researchers can not include this minimal number of subjects, they should resign, or change their objective or design. Of course, the higher the number, the higher the challenges, because the area in which the inclusion will be performed will expand and the logistical constraints will increase. Another issue for the feasibility is the duration of the study. The period of inclusion should not be too long since the investigators will progressively resign with weariness.

Accurate sampling methods are of utmost importance to include a representative sample of the target population. In LMICs, it is not uncommon to lack recent and valid census data, and in this context, advanced sample methods (e.g., cluster sampling) can be used.

Descriptive epidemiology should not be considered as leading to low-impact studies (by comparison to experimental studies). Well-conducted original descriptive epidemiology might change the consideration of the scientific community's appreciation of certain health issues that were previously under-considered. For example, sustained efforts of researchers to estimate the high epilepsy burden in LMICs led the International League Against Epilepsy to launch a global campaign against epilepsy called "Out of the Shadows."

In terms of design, a cross-sectional study (a snapshot) will be performed to estimate a prevalence. Gold standard methodology in this case is a door-to-door case ascertainment that necessitates investigating the status of the disease in all the subjects present in the area on the day (or during the period) of investigation. This might be a huge challenge, especially for diseases with a low prevalence. In some cases, two-phase design can be used: (1) screening of all the subjects (to identify suspected cases), and then (2) in the suspected cases, confirmation of the diagnosis by the gold standard.

A longitudinal study should be performed to estimate rates. The major challenge in LMICs is the maintenance of a cohort for a relevant predefined period; that in some cases might be long (i.e., the follow-up of the subjects). This major difficulty in conducting cohorts is reflected in the field of epilepsy, by the fact that a recent meta-analysis identified 46 prevalence studies while only 8 incidence studies were available in sub-Saharan Africa.[2] It is also sometimes possible to reconstitute cohorts in the past, called "retrospective or historic cohorts."

Analytic epidemiology refers to studies aiming to identify associations between an exposure and an outcome. Two study designs can be used: case—control studies (leading to odds ratio estimation) and exposed/non-exposed cohort study (leading to relative risks estimation).

The case—control approach lies in the inclusion of cases (patients with the disease) and controls (patients proven to be without the disease). Exposure is then retrospectively assessed in both groups. The main methodological issues in case—control studies are: (1) the difficulty to recruit a sample of controls that provides a good picture (representative) of the subjects without the disease, and above all (2) the validity of exposure estimation (especially when no historical files are available and when a recall bias could be important). These two issues might lead to biased results. Given these limits, the level of evidence given by case—control studies is considered as low.

An exposed/non-exposed cohort study lies in the inclusion of exposed (subjects with the exposure at baseline) and non-exposed subjects (subjects proven to be without the exposure at baseline) who are followed for a relevant period, driven by the pathophysiological underlying mechanism and latency. As for the descriptive cohorts, the main issue is the follow-up of the subjects. Given that exposure is assessed at baseline and that the occurrence of the disease is followed-up, the level of evidence of analytic cohort studies is superior to case—control studies.

As a rule of thumb, case—control studies are best suited for a first evaluation of an association, when a long period of latency between exposure and the disease is expected, for rare diseases, for frequent exposure, and when the time constraints to deliver the answer is short. Conversely analytic cohort studies are best suited for providing a confirmation to a suspected association with highest level of evidence, for frequent disorders and rare exposures. At this stage, in the field of epilepsy for example, we face a lack of analytical studies. For sub-Saharan Africa, Ba-Diop et al. identified only eight analytic studies, only of case—control design.[2]

An experimental study consists of comparing an outcome in relation to the exposure to an intervention (drug, medical device, complex health intervention) that is assigned by the investigator. In practice, randomization is used to allocate the intervention. Results from experimental studies represent the highest level of evidence. The most famous design is the randomized two parallel arms control trial in which a sample of subjects is randomized to receive either the standard intervention (or a placebo, when none standard is available) or the experimental intervention (to be assessed). The main methodological components are individual randomization, blinding and intent-to-treat analysis that can lead to initial and sustained comparability between groups. When individual randomization is not possible (e.g., when the intervention includes the health system organization or when a contamination bias between subjects is possible), randomization can be performed on groups of subjects (clusters). The drawbacks of this approach are the increasing sample size and the complexity of statistical analysis.

A review dedicated to randomized controlled trials focused on neurological disorders in developing countries[3] pointed that (1) there has been an exponential evolution of the number of clinical trials in the recent years (64% of trials in between 2004 and 2014), (2) the Asian continent contributed significantly to the trials performed in LMICs (48%), followed by Africa (36%) and Latin America (16%), (3) there is a fairly good coverage of pathologic fields including non-communicable diseases, (4) also there is an increasing diversity of intervention types (therapeutic 72%; preventive 17%; rehabilitative 10.5%; diagnostic 0.5%) (5) besides there is a lack of early-phase trials (phases I and IIa), and (6) while the methodological quality of the trials has improved with time, there is still a tremendous need for improvement of some critical methodological issues. The development of clinical research units in LMICs is key for local researchers to be able to apply the highest standards in terms of methods, data-management and statistical analysis.

In some case, for ethical, political or logistical constraints, the randomization might be impossible, even cluster randomization. This is case for example when the intervention is complex (i.e., based on a modification of the organization). In this context, researchers can use quasi-experimental design based on the comparison of groups whose constitution is not random-based (e.g., choice of the investigator or location of the subject). Design can be "single difference impact estimates" which rely on the comparison between outcomes of a group with the intervention as compared to a control group without. This design is limited by the confusion driven by the possible lack of comparability between both groups at baseline. Another design consists of comparing the outcome before and after the intervention in a single group. This design is limited by the possibility for the outcome to evolve naturally or be impacted by any coincidental events. Another more robust design, that combines both previous approaches, is "difference-in-differences" which compares the changes in outcome over time between treatment and comparison groups to estimate the impact of the intervention. This design allows the initial difference between groups to be removed.

The absence of comparability between groups at baseline for known and unknown confounding factors can highly impact the results of quasi-experimental studies. Advanced statistical methods such as propensity score and instrumental variables are available to try to take this potential source of bias into account. While these approaches are useful, their application has been proved to be suboptimal in the literature.[4] Besides, residual confounding is difficult to exclude.

1.3 GENERAL CONTEXT

1.3.1 Geographical Difficulties

There are common characteristics in the LMIC environment, in particular in the tropical zone. First, of course, the climate is specific. There is often poor sanitation, which promotes the spread of infections. Environmental toxins are common, and malnutrition is frequent. Scarce resources also lead to poor access to care and greater inequality in access.

As an example, meta-analysis of prevalent studies of active epilepsy and lifetime epilepsy throughout the world found a higher prevalence in rural areas of LMICs, a lower one in urban parts of LMICs, and the lowest prevalence was found in higher-income countries (HICs).[5] The authors felt that access to healthcare was an important determinant of these results. This distribution, particularly explicit for epilepsy, could certainly be applied to many neurological diseases. As another example, population-based cross-sectional and case–control studies of active convulsive epilepsy were carried out in five centers in sub-Saharan Africa.[6] To reduce heterogeneity of findings, the authors used the same methods and definitions between regions. Interestingly, they found that heterogeneity could be accounted for by markers of birth trauma, exposure to a range of parasites and other factors including malnutrition.[6] The relationship between epilepsy and undernutrition is a complicated one. To ascertain which came first, one would need a long-term cohort study.[7]

There is inadequate public health sector and systems. There are many specific sociocultural characteristics that may by themselves be associated with neurological diseases. The population is young, with a low life-expectancy compounded by the HIV/AIDS pandemic. There is currently rapid and uncontrolled urban and suburban growth, again creating a mismatch in healthcare provision. Political factors, i.e., unstable governments, multiple levels of decision-making and the low priority given to health programs,

worsen the situation. Curative and preventive neurological health services are often very weak or lacking.

It is important for any investigator to assess this "big picture," not only from the point of view of designing a good study, but also to better place neurological disorders in the context of local difficulties. Carrying out epidemiological research in these settings is then very challenging for a variety of reasons—methodological, logistic, political, economic, ethical and often the low-perceived value of such works.

1.3.2 How to Move Forward?

Before any research, a review of literature is essential to draw up the state of the art of the problematic studied. However, studies of neurological disorders in the tropics are not numerous, and these gaps are amplified when one wishes to find original articles on the investigated country. Studies published in indexed journals are rare in the field of tropical neurology, and it is often difficult to obtain historical information (usually retrospective studies are impossible due to non-computerized medical records and lack of conceptual data models). The few existing studies are often difficult to compare because their methods are not standardized.[2,8]

Documentary research should go beyond indexed journals in which publications, which are very selective and do not always provide the detailed information sought. Medical or pharmaceutical theses in medicine or pharmacy, or PhD theses in public health (or in a specific discipline) represent an important source of detailed information. For over 30 years, the Institute of Neurological Epidemiology and Tropical Neurology has been supplying an original database on scientific productions dealing with neurological disorders, mainly on the African continent (http://www.unilim.fr/ient/base-bibliographique-de-l-ient/). There are also specific databases to be exploited depending on the pathologies investigated: for example, in epilepsy, it is pertinent to consult the Wan Fang (online) database which archives the Chinese and English journals that deal with studies conducted in mainland China (www.wanfangdata.com). It should be noted that journals with no impact factor (to date) are also emerging on the international scene to promote the results of original studies (e.g., Neurology Asia, http://www.neurology-asia.org/).

Finally, ministerial documents are often overlooked as they are a valuable source of information. The case of epilepsy in Lao PDR is explicit: when Tran et al. initiated the first on-site investigations in 2004, they did not find any published data.[9] However, through further research, they identified a survey conducted by the Ministry of Health in the mental health unit of the

Mahosot Hospital (located in the capital Vientiane). This survey was mentioned in a report stored in the Ministry.

1.3.3 Systematic Review and Meta-Analysis

There are a few systematic reviews and meta-analyses concerning some of the above-mentioned issues. Most of them have some or all the following problems: insufficient scrutiny of the existing literature, methodological differences in the included primary studies, (including differences in case ascertainment, various definitions or criteria of diagnosis) which could lead to heterogeneity, lack of standardization on age or sex, inclusion of samples that may not be representative of the general population.

The meta-analyses of incidence studies,[10] prevalence studies[5] and TG in epilepsy[11,12] have demonstrated a high degree of unexplained heterogeneity. This is due to the methodological differences between primary studies but also could be due to which studies to include or exclude from the meta-analysis. Problems with meta-analyses are greatest when there are few primary studies, as with incidence studies, and where there is wide variation between estimates of these studies.[10,13]

1.3.4 Socioeconomic and Sociocultural Factors

Consideration of sociocultural factors and economic status is fundamental to conducting investigations in LMICs. These two themes are discussed in detail in other chapters of this book, but the constraints and limitations that these two dimensions entail in the conduct of neuroepidemiological research programs should be mentioned.

Indeed, in so-called traditional societies, beliefs and perceptions about neurological diseases very often lead to a marked stigma because of the clinical expression of these pathologies. Thus, affected people generally do not want their neighbors or co-workers to know about their state of health. The consequences are dramatic because this directly constitutes a barrier to diagnosis and treatment seeking, especially in rural areas where anonymity is more difficult to preserve.

In addition, the economic difficulties are known to be associated with the treatment deficit, although this phenomenon still deserves specific research to be better understood. Indeed, sometimes, people seeking care in traditional systems afford much more than they would have to in a medical system.

Given the large share of the informal sector in LMICs, income information is often difficult to determine because it is irregular and a large majority of

people prefer not to disclose it. How to deal with the issues of education and socioeconomics is fundamental in understanding the difficulties of access to care. In 2012, Cooper et al. found that only 7% of poverty indicators were commonly measured by standardized validated methods.[14] Conversely, in 2012, Mbuba et al. showed that assessments based on socioeconomic status were very common for neglected diseases, but were much less frequent for neuroepidemiological investigations.[15]

It is therefore strongly recommended to involve in the research programs disciplines of human and social sciences such as ethnology, anthropology, sociology and of course health economics.

1.4 DIFFICULTIES IN THE AVAILABILITY AND MOBILIZATION OF DATA

1.4.1 Medical Data

Once epidemiological research in the tropics is initiated, the question of mobilizing information about known patients arises very quickly. In most LMIC countries, the health system is organized per a pyramidal system where the top is constituted by the central hospitals and the base by the primary health centers. The mobilization of useful information is very uneven depending on where you are in the pyramid: easier and often computerized in the upper part (central hospitals of the capital and provincial hospitals) and very difficult to access for in the lower part (district hospitals and primary health center). Although there is feedback from the bottom to top of the pyramid, the latter are generally aggregated (and therefore not very exploitable because they aim at quantitative assessments by groups of diseases for health surveillance or the supply of decentralized pharmacies). Regardless of the level of the pyramid where the research is located, useful information could be found in patient files in paper format. This could considerably impact the access and collection of data, computerization time, and supposed first that a safe storage of these files was ensured.

1.4.2 General Population Census

In High Income Countries, the general population censuses are carried out regularly and their protocols allow us to obtain detailed information, this is often not the case for the LMICs. These censuses are generally performed before (and motivated by) political elections and the modalities often lack rigor: a head of family is entitled to declare all the members of the household and the information provided can be approximate (e.g., on age). This low quality of population censuses is amplified by specific events (such as civil or ethnic wars) and large migratory flows (from rural to urban areas and from international migration). These migrant populations may be poorly accepted on their arrivals (usually ethnic and/or religious issues). Their marginalization often leads them to not attend medical and/or administrative structures. The negative consequences of this lack of information on the general population are particularly relevant in studies that are not door-to-door surveys. This problem is generally more pronounced in rural areas because of the low population density, lack of addresses, the majority of individuals working in the field and therefore being difficult to contact.[16]

1.4.3 Cartography and Geographical Information Systems

The integration of geographical information systems (GIS) into epidemiological studies is becoming more common. Indeed, the geo-epidemiological approach accompanies descriptive research as well as analytical research: the spatial organization of a territory is very often a key, notably for issues of accessibility to healthcare structures, or environmental exposures. Several free software programs exist and databases are easy to access (especially for the administrative divisions). The community of geomaticians promulgates the sharing of information and generally puts the GIS data resulting from their work at the disposal of the greatest number.

However, the manipulation of these tools and those that complement them (Global Positioning Systems, remote sensors, etc.) is complex, and in the most advanced approaches (spatial statistics), the researchers still lack training. This domain deserves to be developed (examples: clustering techniques, analysis of spatial interaction with the natural environment, analyses of distances to access to health centers, etc.). The cartographic language is universal and comprehensible by most of the actors. The map translates a convincing demonstration (spatial planning as an explanatory factor or as an answer to the problem), and local decision-makers are generally very sensitive.

1.5 STUDY DESIGN

1.5.1 Geographical and Logistical Challenges

The logistics of providing consumables and medical equipment in the field cause problems. These include transport across insecure regions and severe climatic conditions, which make maintaining equipment difficult. Rainy seasons should be avoided in limited-access rural areas. Another problem lies in the very low densities of population in certain regions of these areas, with dispersed villages, and then sometimes a long time walking to get few people included in a study.

1.5.2 Demographic and Follow-Up Issues

An additional problem in some remote rural areas of LMICs is the difficulty in determining the precise age of subjects, particularly if they are older. Indeed, official papers are often missing, and some may be falsified for various reasons. The knowledge of age is required to provide age-specific or age-adjusted rates. Different methods of age estimation have been developed and implemented, including the use of historical events to avoid such difficulties, for example in research on cognitive disorders.[17]

In epilepsy, age at onset is an important element for classifying seizures and exploring etiology. Without knowledge of the age at the first seizure, duration between first and second seizure and age at diagnosis, it is difficult to identify incident cases. An effort should be made (although not easy) to carry out case–control and outcome studies on incident rather than prevalent cases. If prevalent cases are used, it is not clear whether putative risk factors precede or follow onset of the disease. Prevalent cases represent a population of those who survive, thus resulting in an incidence–prevalence discrepancy.

The lack of death certificate data and autopsies make cause of death difficult to ascertain and verbal autopsies should not be relied upon. More outcome studies are imperative, particularly using incidence rather than prevalence cohorts.

1.6 CLINICAL AND REGULATORY ISSUES

1.6.1 International Ethic Rules

Ethical questions are extremely sensitive. The risk/benefit ratio should be accurately evaluated. There are still so far too few independent ethics committees in resource-poor countries. These committees have different functioning modalities, taking sometimes a very long time to give their advice. Their fees are variable, consisting sometimes of a percentage of the project funds, which is clearly not acceptable. Where no committee exists, health authorities should act, but they are considered to be less independent than a designated specific body. In addition, signed informed consent is often too complicated to obtain, and it is necessary to establish simpler consent protocols that the population can truly understand. Written and signed consent may be impossible because of illiteracy, but researchers can then seek oral consent and written evidence of comprehension of the included subject demonstrated through the signature of a witness.

The protocols of the studies, apart from obtaining a mandatory authorization in the concerned country, could be submitted also to an ethical committee in a developed country, if there is a collaborating institution located in such a country.

1.6.2 Suspected Cases Ascertainment

Identification of cases may also raise methodological challenges. Some disorders carry an important stigma. Above all, epilepsy is falsely associated with possession, or considered as a contagious disease.[18] This issue is also clear for neurological diseases whose clinical signs are not accepted by the general population. This can result in concealment of cases, even during door-to-door interviews, which remain the gold standard in neuroepidemiological approaches in LMICs.

In some studies, the head of the household is targeted and is interviewed on behalf of the entire household.[6] A study in Mexico found, however, that interviewing heads of households was inadequate and considerably underestimated the prevalence of epilepsy.[19] Multiple sources of information may be useful to complement door-to-door surveys as shown in a survey in Benin.[20] Information obtained through door-to-door interviews was supplemented with medical data from hospital records, clinics, pharmacies, etc., and by non-medical information from key informants, village chiefs, traditional healers and teachers. Of course, all the suspected cases identified by these sources were confirmed using the same definition. Using multiple sources of information allows the application of the capture—recapture method which is a statistical method evaluating the number of people not identified by all the sources of information. It is then possible to calculate an adjusted prevalence. In this study, the adjusted prevalence was much higher than the observed prevalence.[20]

In multi-phase studies, for confirming the good specificity of the screening test, rescreening a sample of a population will also help (e.g. specialist) to identify PWND missed in the initial survey. The false-negative proportion can be calculated to give a "maximum estimated prevalence."[21]

The same epidemiological assessment tools and definitions should be used throughout, despite diverse ethnicity, the existence of many different dialects, perceptions and sociocultural backgrounds. Hence, interviewers and interpreters need substantial training to ensure that the questions can be understood by the

population and responded to in the same manner. In addition, quality-control measures must be put in place to ensure the consistency of the procedures throughout the data-collection period. Questionnaires should be clearly adapted, validated, translated into the dialects used, and retranslated back into the language of origin to check their comprehensibility.[22–25]

It is particularly important that questionnaires are pretested for specificity and sensitivity in the same community prior to carrying out the pilot survey. To avoid impacting the targeted sample population, the testing of the questionnaire should not be performed with people residing in the area to be surveyed, nor should the pilot study be carried out among people to be included in the main study.[22] To test the questionnaire, it is necessary to have an adequate sample of people with and without the disease. Such testing is usually carried out in the setting of a hospital or clinic. However, it is difficult to ensure that the community on whom the questionnaire is being tested is similar to that in a distant rural area.

The first door-to-door screening protocols for neurological diseases were designed in the 1980s by World Health Organization[26] and the National Institutes of Health.[22] This protocol was for example used to carry out community-based surveys in China, Nigeria, and India.[23,27,28]

1.6.3 Diagnostic and Treatment

Methodological difficulties may also relate to diagnosis which is mainly clinical. An eye witness account is essential, e.g., for seizure occurrence. A cell phone can be used by the eye witness to record the seizure on video. Patients identified as positive on the screening questionnaire should be seen by a clinician as soon as possible but this is often difficult for logistic reasons. The clinician should ideally be an experimental neurologist or a specifically trained physician as it is important to make a correct diagnosis and to document the semiology to rule out negative cases and to specify certain criteria for diagnosis.

There is often a shortage of equipment and more importantly, appropriately trained people to use the material and interpret exams in the field. In epilepsy, video electroencephalograph (EEG) monitoring could be useful for electrophysiological and clinical correlation, but is rarely available and affordable in studies in LMICs. Both computed tomography (CT) scans and magnetic resonance imaging (MRI) are rarely available facilities and require substantial resources. If available, they are frequently undergoing maintenance or have broken down. Lack of clinical data, EEG and imaging data result in difficulty in providing an accurate diagnosis.

1.7 DESIGN-SPECIFIC ISSUES

1.7.1 Cases Census and Sample Studies

When a census has not previously been carried out (or realized recently), it can be done either before or during the main study. Two strategies can be adopted. One can carry out a door-to-door survey of an entire community or village or use a sampling approach. Among these approaches, the cluster sampling method is of interest in LMICs. It does not require prior enumeration of each person, but is based on a list of "clusters," which are groups of peoples (e.g., villages).[29] They are randomly selected and all their inhabitants will then be included in the survey. Thus, cluster surveys provide a geographical grouping of villages on which to focus the investigation. However, it is a challenge to be certain that all the inhabitants of a village have really been included. Also, cluster surveys require higher sample sizes than conventional surveys as the analysis must take the clusters into account. WHO simulations indicated that doubling the study population is required. Cluster sampling can also be used in urban areas using, for example, districts or parts of a town as clusters, but this practice could lead to biases since the clusters do not necessarily represent correctly the whole population. More complex sampling methods may then be used.[30]

1.7.2 Involvement of Institutions and Key Stakeholders

It is essential to involve key people in the country and in interest of the study, to develop a true partnership, a real co-operation in the sense of a joint operation. Repeated visits to the area targeted by the survey can be useful, as can pilot studies. In LMICs, lack of interest in neuroepidemiological studies leads to insufficient knowledge of the extent and impact of neurological diseases and hence low priority at all levels of policy making and public health. Many decision-making bodies encounter some difficulty in deciding on resource allocation.[31] The awareness of local authorities and population is important, and the protocol should include procedures which fully explain the course of events. At all levels of the health system—national, regional, district, and communal—it is necessary to obtain agreement on those surveys that have been conducted. This involvement is of utmost importance for the sustainability of the projects or interventions. The advocacy could lead to an integration of the recommendations in the policy or procedures of the national health systems. Ideally, there should be an integrated plan to improve knowledge of neurological epidemiology, which should then be incorporated into the national health programs.

1.7.3 Medical Examinations, Biological Specimens and Laboratory Tests

When biological exams are needed, the challenges increase exponentially. Storage and transportation of samples from remote areas are very difficult to realize in good conditions. The frozen chain could be broken. Availability of carbo-ice or liquid nitrogen is often low. Analyses should be realized in the country to avoid international transportation, submitted to more and more strict rules. National regulations are variable and complicate the process.

1.7.4 Monitoring and New Technologies

All epidemiological studies require a system of data collection, either on an ad hoc basis (observational study) or over a longer period (studies requiring follow-up and/or interventional studies). Multiple forms, which then must be gathered and linked, generate a monitoring that is greatly facilitated using digital tools. The paper forms gradually pass the relay to Personal Digital Assistants or smartphones and this has several advantages: automatic geolocation of the collected data, interactivity and synchronization of the information in a secure database, accessibility for all the actors of the program, etc. The use of these new technologies also has a significant advantage: as demonstrated by King et al., it significantly reduces the cost of research.[32] Moreover, the use of short messages (SMS) is also increasingly common for the collection and/or communication of information.[33]

1.8 VALORIZATION: SCIENTIFIC AND PUBLIC HEALTH ISSUES

1.8.1 International Acknowledgments of the Works Conducted in Tropical Areas

These methodological and logistical difficulties could result in the implementation of studies of relatively modest quality. Samples may be small, leading to a lack of power. The accuracy of diagnosis may be lower, as equipment is not available. Hence, these studies are more often published in local journals, which are often not indexed in major international databases. In the major international journals, it is sometimes more difficult to publish a study from an LMIC than similar work from the developed world although this problem seems to decrease. However, it is quite possible to publish the study in very good international journals if the study is truly original and adequately targeted and if the quality of the language is good. It is best to avoid dividing studies into multiple publications.

1.8.2 Advocacy—Policy and International Cooperation

Advocacy is defined by the act of pleading or arguing in favor of something, such as a cause, idea or policy, active support. In our case, a survey should try to help authorities to specify a policy in the studied domain. This document will give evidence of the appropriation of the results for the study by the stakeholders. This kind of valorization is a privileged ending of public health research, and should complement academic presentations of the results through papers and communications.

International partnership could enhance the project, with transfer of skills and experiences, from North to South, South to North or South to South. Efforts should be made to integrate projects in the development aid agenda, based on international organizations' priorities.

Research programs aimed at experimenting new interventional strategies for a better care of a disease must increase. Assessing the cost-effectiveness of a change or innovation in the management of people living with a disease (identification, diagnosis, prescription, adherence) is of the utmost importance in convincing governments to use it.

1.9 CONCLUSION

Challenges related to neuroepidemiological research in tropical countries should not be overlooked even if, overall, they diminish over time. The concepts and methods applied must be the same in LMICs as those used in developed world, even if their translation into practice may differ.

Neuroepidemiological research in these countries is still difficult, but feasible, and extremely useful to provide decision-makers with the knowledge they need to allocate limited resources in the best way to reduce TG. Genuine cooperation at all levels is necessary to conduct such research under appropriate conditions. The survey protocol should be particularly detailed and thorough. Specific epidemiological methodologies may be useful, and the collection of data must be standardized.

Neuroepidemiological research therefore requires perfect organization and the development of solutions specific to LMICs. Currently, 90% of health research addresses 10% of the world's health problems of the global population. This mismatch should not continue. Increased emphasis on neuroepidemiological studies in LMICs has a role to play in re-establishing a balance.

References

1. WHO; International Bureau for Epilepsy; International League Against Epilepsy. *Atlas: Epilepsy Care in the World 2005*. Geneva, Switzerland: World Health Organization; 2005.
2. Ba-Diop A, Marin B, Druet-Cabanac M, Ngoungou EB, Newton CR, Preux PM. Epidemiology, causes, and treatment of epilepsy in sub-Saharan Africa. *Lancet Neurol*. 2014;13:1029–1044.
3. Marin B, Agbota GC, Preux PM, Boumédiene F. Randomized trials in developing countries: different priorities and study design? *Front Neurol Neurosci*. 2016;39:136–146.
4. Shah BR, Laupacis A, Hux JE, Austin PC. Propensity score methods gave similar results to traditional regression modeling in observational studies: a systematic review. *J Clin Epidemiol*. 2005; 58:550–559.
5. Ngugi AK, Bottomley C, Kleinschmidt I, Sander JW, Newton CR. Estimation of the burden of active and life-time epilepsy: a meta-analytic approach. *Epilepsia*. 2010;51:883–890.
6. Ngugi AK, Bottomley C, Kleinschmidt I, et al. Prevalence of active convulsive epilepsy in Sub-Saharan Africa and associated risk factors: cross-sectional and case–control studies. *Lancet Neurol*. 2013;12:253–263.
7. Crepin S, Houinato D, Nawana B, Avode GD, Preux PM, Desport JC. Link between epilepsy and malnutrition in a rural area of Benin. *Epilepsia*. 2007;48:1926–1933.
8. Preux PM, Druet-Cabanac M. Epidemiology and aetiology of epilepsy in sub Saharan Africa. *Lancet Neurol*. 2005;4:21–31.
9. Tran DS, Odermatt P, Le TO, et al. Prevalence of epilepsy in a rural district of central Lao PDR. *Neuroepidemiology*. 2006;26:199–206.
10. Ngugi AK, Kariuki SM, Bottomley C, Kleinschmidt I, Sander JW, Newton CR. Incidence of epilepsy: a systematic review and meta-analysis. *Neurology*. 2011;77:1005–1012.
11. Mbuba CK, Ngugi AK, Newton CR, Carter JA. The epilepsy treatment gap in developing countries: a systematic review of the magnitude, causes, and intervention strategies. *Epilepsia*. 2008;49:1491–1503.
12. Meyer AC, Dua T, Ma J, Saxena S, Birbeck G. Global disparities in the epilepsy treatment gap: a systematic review. *Bull World Health Organ*. 2010;88:260–266.
13. Fiest KM, Pringsheim T, Patten SB, Svenson LW, Jette N. The role of systematic reviews and meta-analyses of incidence and prevalence studies in neuroepidemiology. *Neuroepidemiology*. 2014; 42:16–24.
14. Cooper S, Lund C, Kakuma R. The measurement of poverty in psychiatric epidemiology in LMICs: critical review and recommendations. *Soc Psychiatry Psychiatr Epidemiol*. 2012;47:1499–1516.
15. Mbuba CK, Ngugi AK, Fegan G, et al. Risk factors associated with the epilepsy treatment gap in Kilifi, Kenya: a cross-sectional study. *Lancet Neurol*. 2012;11:688–696.
16. Quet F, Odermatt P, Preux PM. Challenges of epidemiological research on epilepsy in resource-poor countries. *Neuroepidemiology*. 2008;30:3–5.
17. Paraiso MN, Houinato D, Guerchet M, et al. Validation of the use of historical events to estimate the age of subjects aged 65 years and over in Cotonou (Benin). *Neuroepidemiology*. 2010;35:12–16.
18. Rafael F, Houinato D, Nubukpo P, et al. Sociocultural and psychological features of perceived stigma reported by people with epilepsy in Benin. *Epilepsia*. 2010;51:1061–1068.

19. Quet F, Preux PM, Huerta M, et al. Determining the burden of neurological disorders in populations living in tropical areas. who would be questioned? Lessons from a Mexican rural community. *Neuroepidemiology*. 2011;36:194−203.

20. Debrock C, Preux PM, Houinato D, et al. Estimation of the prevalence of epilepsy in the Benin region of Zinvie using the capture-recapture method. *Int J Epidemiol*. 2000;29:330−335.

21. Placencia M, Sander JWAS, Shorvon SD, Ellison RH, Cascante SM. Validation of a screening questionnaire for the detection of epileptic seizures in epidemiological studies. *Brain*. 1992;115:783−794.

22. Schoenberg BS. Clinical neuroepidemiology in developing countries: neurology with few neurologists. *Neuroepidemiology*. 1982; 1:137−142.

23. Bharucha NE, Bharucha EP, Dastur HS. Pilot survey of the prevalence of neurologic disorder in the Parsi community Bombay. *Am J Prevent Med*. 1987;3:293−299.

24. Preux PM. Questionnaire in a study of *epilepsy* in tropical countries. *Bull Soc Pathol Exot*. 2000;93:276−278.

25. Diagana M, Preux PM, Tuillas M, Ould Hamady A, Druet-Cabanac M. Screening for epilepsy in tropical areas: validation of a questionnaire in Mauritania. *Bull Soc Pathol Exot*. 2006;99:103−107.

26. WHO. Research protocol for measuring the prevalence of neurological disorders in developing countries. *Neurosciences Programme*. Geneva, Switzerland: World Health Organization; 1982.

27. Li SC, Schoenberg BS, Wang CC, Cheng XM, Zhou SS, Bolis CL. Epidemiology of epilepsy in urban areas of the People's Republic of China. *Epilepsia*. 1985;26:391−394.

28. Osuntokun BO, Schoenberg BS, Nottidge VA, et al. Research protocol for measuring the prevalence of neurologic disorders in developing countries results of a pilot study in Nigeria. *Neuroepidemiology*. 1982;1:143−153.

29. Preux PM, Chea K, Chamroeun H, et al. First-ever, door-to-door cross-sectional representative study in Prey Veng province (Cambodia). *Epilepsia*. 2011;52:1382−1387.

30. Houinato D, Preux PM, Charriere B, et al. Interest of LQAS method in a survey of HTLV-I infection in Benin (West Africa). *J Clin Epidemiol*. 2002;55:192−196.

31. Dumas M, Preux PM. Epilepsy in tropical areas. *Bull Acad Natl Med*. 2008;192:949−960.

32. King JD, Buolamwini J, Cromwell EA, et al. A novel electronic data collection system for large-scale surveys of neglected tropical diseases. *PLoS One*. 2013;16:e74570.

33. Deglise C, Suggs LS, Odermatt P. SMS for disease control in developing countries: a systematic review of mobile health applications. *J Telemed Telecare*. 2012;18:273−281.

TROPICAL NEUROEPIDEMIOLOGY: BETWEEN ENVIRONMENT AND GENETICS

2

Sociocultural Factors

Philippe Nubukpo[1,2,3] and Catherine-Marie Dubreuil[1,2]

[1]INSERM UMR 1094 NET, Limoges, France [2]University of Limoges, Limoges, France
[3]Hospital Center Esquirol, Limoges, France

2.1 INTRODUCTION

Social and cultural factors are great determinants of healthcare all over the world, and especially in developing countries. The understanding, knowledge, attitude, and practices regarding a disease in a community are essential initial steps in developing strategies that aim at reducing the misconceptions and the social stigma associated with this condition.[1] This is very important in tropical health and especially in neuroepidemiology because brain diseases are amazing and mysterious for the common people in sub-Saharan Africa (SSA) and their explanatory model is sustained by magic—religious prevalent sociocultural representation and associated stigma. Epidemiologists working in SSA very often need "cultural interpreters" who come from human and social sciences areas.

Here we will first answer the modern legitimacy of collaboration between neuroepidemiology and anthropology in performing neuroepidemiological studies in SSA, then we will reflect on sociocultural factors in some neuropsychiatric disorders in SSA (psychiatric disorders, epilepsy, dementia).

2.2 TROPICAL NEUROEPIDEMIOLOGY AND HEALTH ANTHROPOLOGY: FOR EFFECTIVE COLLABORATION

This collective work, which queries the specificity of the neuroepidemiology in the so-called "tropical areas", is the opportunity to analyze aspects of collaboration now well established between this discipline in

DOI: http://dx.doi.org/10.1016/B978-0-12-804607-4.00002-2

the service of the public health and those that are often referred to by the generic term of "Social and Human Sciences" or SHS. Since the AIDS epidemic, these are regularly associated with operational research in public health.

Indisputable advances are, however, based on misunderstandings; contemporary forms of research situation of globalization of issues, are passing the questioning of the "disease" object to the object "global health", and require a broader approach.

2.2.1 The Expected and Assumed Collaborative Epidemiology/Anthropology

Humanities and Social Sciences, in contemporary forms of research situation, are considered as a whole, and to be relatively homogeneous: history, geography, social psychology, sociology, philosophy, psychology, cultural psychology, or psychiatry, are associated with the backward-thinking that they look alike and are potentially interchangeable (which would allow a historian to speak on psychology, a sociologist to speak on philosophy).

Anthropologists or ethnologists of health, yet tacitly accepting the amalgam they are submitted to, participate in the maintenance of misunderstanding or ignorance which their discipline is often the object of, both in scientific/scholarly communites, and the population and the public.

This promotes the confused and superficial use of their expertise, and leads to a loss of effectiveness of the programs which they are associated with. One speaks of anthropology applied to health when an anthropologist is solicited to help epidemiologists in their projects and field experiments. What is usually expected of him/her is the translation of local cultures.

If trust and the sincere expectation of improved outcomes "for the health of the people" are the foundations of the collaboration, which is carried out in mutual respect between the representatives of these domains, it does not avoid stereotypes. It consists of selective borrowing and rests on three supports: a method, a concept, and a mission.

As for the methodology, this is the very well known (but bad) methodology called "qualitative". It concerns almost exclusively the interviews, individual or collective. Used in a polymorphic way, it is not specifically anthropological, and by its success, in a simplified version, it participates in spreading the false idea that "making qualitative" means carrying out a few interviews. This fragmentary methodological mobilization is intended to remove obstacles, clarify practices, clarify areas of shadow or uncertainty (e.g., to explain why some figures do not give the expected answers,

why a population seems to not adhere to the use of allopathic drugs yet easy access, etc.).

Another concept which is misused from our view is the one of sociocultural representation.

2.2.2 A Concept: Almost Exclusive Use of the Sociocultural Representation Theory

This concept was developed by and for social psychology. Moscovici founded this field of reflection in the 1960s. With him we begin "to think that a same object can be associated with various representations, each of them emerging from social groups differentiated operating within the same company".[2] Since the 1990s, this concept has been used by many researchers in various disciplines. The concept of social representation designates a form of specific knowledge, the knowledge of common sense, whose contents show the operation of generative and functional processes to be socially marked. More broadly, it is a form of social thought.[3]

Sociocultural representation theory remains essentially the study of knowledge of common sense, especially from a psychosocial perspective. Popular at the international level, its popularity with many disciplines comes from its flexibility and its ability to cover all subjects: representation theory provides a conceptual framework of issues and varied approaches. Its global success continues unabated because it gives access to information about views and opinions, and thus helps to decipher the logic of social practices, to the point of passing for most of the research in "human and social sciences" (HSS), in social and medical anthropology.

It is the tool of those who belong to other disciplines; professionals refusing to engage in a long, complex research. This concept of sociocultural representation is not anthropology, and its use in this discipline is neither systematic nor unavoidable. It is perceived as a methodological tool that is accessible and easy to mobilize. Some ethnologists warn against its exclusive use, or its use attached to the study of the so-called local "traditional" cultures, that can no longer today be addressed without taking into account the context of globalization.[4]

A limitation of the use of the concept of sociocultural representation is that it is the declarative level, collected by questionnaires or interviews. The notion of sociocultural representations is adapted if the researcher wishes to obtain information on opinions.

In anthropology, we try to understand the logic underlying behavior: "It is a matter of questioning not the rationality of representations, but the rationality of behavior in relation to the social, cultural, and symbolic context in which they are to 'develop'".[5,6] The relationship between ethnology and public health, analyzed in an abundant literature, is described as problematic because of the fundamental imbalance which places the ethnologist in an asymmetrical relationship, faced with a discipline enjoying an unparalleled legitimacy.[7–10]

The model of medical health care consumption is now universal. Only access to care is not.

This creates "a multifaceted process of compression of space and time, corresponding to the intensification of worldwide social relations between remote places".[11]

"In the field of health these globalized expectations, prevented by structural violence of social inequalities, must continue to be documented. An analysis of the place occupied by health in what some call contemporary moral economies, is then required".[9] Despite the reservations set out above, studies exist and often result in a diagnosis of stigmatization of mental illness in SSA.

2.2.3 Health-Related Stigma

Greeks use the term "stigma" to describe marks on the body, which were put on people who had a morally detestable status. Stigma is an attribute that marks people as different and leads to devaluation.[12] Stigma can be visible (e.g., infirmity, skin color) or invisible (e.g., former user of psychiatry).

Health-related stigma is characterized by social disqualification of individuals who are identified as having the disease. According to the labeling theory, an individual bearing a stigma may accept and internalize what is projected onto him/her and act accordingly. Stigma is complex and multilayered. Individual consequences of stigma include reluctance to disclose the illness, shame, diminished self-esteem, and fear of exclusion. Families of people with a stigmatized health problem may also be affected by stigma or become perpetrators of stigma themselves.[13]

2.3 MENTAL DISORDERS AND TRADITIONAL BELIEFS IN SSA

In many African countries, the World Health Organization (WHO)'s plea for Primary Health Care (PHC) workers to deliver mental health services is yet to be heeded because of poor medical resources, a lack of equipment, and a predominance of traditional medicine with little or no collaboration with western biomedicine. For example, in Iganga and Jinja districts in Eastern Uganda, the prevalence of psychological distress in 400 patients over the age of 18 years attending traditional healers was 65.1%.[14]

Many authors have noted the significant similarities in beliefs relating to traditional religion and health in SSA cultures.[15,16] In SSA, there is a rich diversity of beliefs and a number of shared ideas about causality of mental health. We try to summarize here the most important traditional beliefs about mental illness in SSA in order to better understand mental health determinants in these countries.

2.3.1 Explanatory Models of Mental Illness in SSA

Explanatory models (EMs) are "notions about an episode of sickness and its treatment that are employed by all those engaged in the clinical process". EMs are formed from a variable cluster of cultural symbols, experiences and expectations associated with a particular category of illness. EMs reveal sickness labeling and cultural idioms for expressing the experiences of illness. EMs have been shown to influence health-seeking behavior and health service utilization.[3,17]

2.3.1.1 What are the Theories of Mankind in SSA?

Many African cultures do distinguish between the mind and body. The mind-soul is cited as residing in the head as well as in the heart or abdominal region.[18] It controls the body, is immortal and is the system of cognitive process that is thinking and perceiving; so mind is the deciding part of mankind.[19] In the AmaXhosa people of Southern Africa, the soul is considered to be the seat of feelings and resides in the blood and heart. The mind is located in the brain and is "the initiator of action required for health".[20] The Akan people of Ghana think that good health is dependent on a fine balance of three components: *onipadua* (the physical-mortal part), *sunsum* (personality), and *okra*, which is an intellectual non-personal life force.[21]

2.3.1.2 Causality of Mental Illness in SSA

A fundamental idea concerning causality is that the cause of any occurrence can be ascertained through divination, memory, reason and empirical judgment. Other beliefs must be considered while explaining mental illness in SSA; the idea of continuity between the living and dead postulates that after death, the spirit lives and maintains the wellbeing of its living descendants. Witchcraft, a widely held belief, continues to be seen as an important cause of illness and misfortune. The role of spirits is prominent. Unlike spiritual causation, the witch is often a living person with evil powers. The causal classifications of mental diseases, used are flexible and patient-dependent: one can distinguish natural causes (e.g., external environmental factors including climactic changes, infections, and substance abuse), inherited factors (mental illness can be inherited either due to spiritual or genetic factors), unnatural causes (the three-way model of supernatural forces: illness caused by other human beings, or illness caused by the behavior of an individual or his family).[16,17]

Traditional healers classified illness into several etiological categories, in order of importance: "God-given", witchcraft, curse, seasonality, hereditary, spirit possession, poisoning, broken taboos, infections, evil eye, and sorcery.

The Baganda in Ouganda, classify diseases of the brain into four groups: *eddalu* or violent madness, *ensimbu* or epilepsy, *obusiru* or foolishness which may be either congenital or acquired, *kantalooze* or dizziness. Two conditions similar to neurotic disorder are recognized: *emmeme etyemuka*, referring to a pounding of the heart with fright, and *emmeme egwa* which manifests as a general weakness of the body and a failure to eat.

Some authors describe other conditions of relevance to mental health: *amakiro* which afflicts women a few days postpartum, *akawango* characterized by a persistent headache, and *enjoka* referring to stomach aches.[22]

2.3.1.3 Psychosis and Neurosis

"Neurotic" symptoms are generally more difficult to identify cross-culturally. Though often not percieved to be mental disorders, they are still recognized by local people and traditional healers. In SSA, neurosis tends to be associated with somatic symptoms or diseases.

In Nigeria, Ebigbo[23] described a neurotic-like condition which associated several symptoms which appear to be unique including: heat and crawling sensation in the head, heat inside the body, neck stiffness and rigidity, feeling as if the tongue was trembling, a turning sensation in the head. It is close to the "brain fag syndrome" a disease with the mind going blank, thinking too much and the mind feeling boiling hot; this is a kind of depression with somatization which has been described in a number of SSA cultures for over 30 years.[23,24]

2.3.1.4 Stigma and Mental Illness

A study in Western Nigeria, within medical doctors ($n = 312$) from eight selected health institutions, exploring their knowledge and attitudes towards people with mental illness, showed a high prevalence of beliefs in supernatural causes. The mentally ill were perceived as dangerous and their prognosis perceived as poor.[25] A study in 164 students in Nigeria suggested that stigmatization of mental illness is also highly prevalent among Nigerian children.[26]

2.3.1.5 Managing Mental Illness: Traditional or Modern Medicine?

A survey in Ghana revealed many reasons for the appeal of traditional and faith healers in the provision of mental health care: cultural perceptions of mental disorders, the psychosocial support afforded by such healers, a number of barriers hindering collaboration, including human rights and safety concerns, skepticism around the effectiveness of "conventional" treatments were identified.[27]

In South Africa, a study examined traditional healers' EMs and treatment practices for psychotic and non-psychotic mental illnesses based on a focus group. Psychosis was the main exemplar of mental illness and was treated with traditional medicine. Non-psychotic illnesses were not viewed as mental illnesses at all. Traditional healers have incorporated into their traditional treatment practices modern ingredients that are potentially toxic.[28]

2.4 SOCIOCULTURAL REPRESENTATIONS AND STIGMA IN EPILEPSY

Epilepsy is a highly prevalent health problem in many developing countries. However, the proportion of people with epilepsy (PWE) who require treatment but are not receiving it, is very important. Uncontrolled seizures have serious consequences on morbidity and mortality rates.[29–32]

2.4.1 Explanatory Models of Epilepsy

For some authors,[33] "medical issues are really, semantic...". In traditional SSA, while talking about epilepsy, each cultural group focuses on one or many target symptoms. Very often, these are the symptoms of Generalized Tonic Clonic seizure (GTCS).

GTCS is in general very well recognized. However, other seizure types are hardly recognized or differentiated, with the exception of febrile convulsions.[34] The dramatic movements that characterize GTCS induce fear,[35,36] and bear many names and descriptions in communities. Epilepsy is frequently named by metaphor and periphrases. Two types of denomination can be identified: a symptomatic one and an etiologic one.

2.4.2 Symptomatic Denominations

Many denominations referring to the clinical semiology have been described.[37] That is the case for *kisenkiri* in Mossi, Burkina Faso, *sifosekuwa* in Swaziland ("disease that makes one fall"), *nwaa* ("throw someone to the ground") among the Bamilékés of Cameroon,[38] *kifafa* ("stiffness like half dead") among the Wapogoros of Tanzania,[39] *kobela ti makakou* ("the monkey disease") in Sango in Central African Republic.[36] Terms applied to GTCS in six Senegalese dialects signify agitation, fall or drop.[40]

Symptomatic denominations have also been reported in Ivory Coast,[41] in Togo[42] in Ethiopia[43] and in Mali.[44] In Bambara of Mali, the term *kirikirimasien* (disease that makes someone fall) used to designate a PWE is not pronounced. But a narrative description is used to talk about epilepsy, which focuses on physical and psychological semiology and context of seizure set-up.[44] In the same country, the Dogon considered the act of foaming at the mouth as the main symptom of the disease.[34] Sometimes the denomination of epilepsy in SSA underlined the circumstances of outcome of the disease and its suddenness.[41] In some cultural groups, as in Senegal, childhood epilepsy is identified.[45] To summarize, it is interesting to note similarities between very different cultural groups in naming epilepsy in SSA.

2.4.3 Etiological Denominations

In SSA, epilepsy is sometimes named by its supposed popular etiologies. Very often, in a magic–religious explanation model, epilepsy is considered as an evil spirit or supernatural disease. These beliefs are found in the whole SSA data published.[37,41] So, in Senegal PWE are considered as "one who is dead and born again", "one who sees the invisible", "vision of horror that makes him fall".[45]

A "spiritual" heredity concept as the origin of epilepsy usually appears in the name given to this disease in SSA.[34,38,46] In some cultural groups, epilepsy is named as "slave's disease or foreigner's disease"; for example, *heman baria* in Tigre, Ethiopia.[43] These etiological denominations very often reveal beliefs about the disease.

2.4.4 Beliefs About Epilepsy in SSA

Many beliefs about epilepsy coexist in SSA. The analysis of understanding models of epileptic seizure in different cultural groups showed the existence of major theories to explain the disease in traditional areas: magical, contagion, "biomedical".[47]

2.4.4.1 Magical Causes (Evil-Spirit, Possession, and Witchcraft)

Epilepsy is perceived as a discordance between the living and the dead; and is due to possession by an evil spirit. It is through seizures that a person gets into contact with the spiritual world. This supernatural hypothesis has been described in various versions in many studies.[33,46,48] Thus, the hypothesis of possession by an evil spirit (*shetani*) was evoked in Wapogoro (Bantu) of Tanzania[49] as well as in Wolof of Senegal and Cape Verde (Djin).[45] Beliefs in witchcraft were described in Ivory Coast,[41] and within the Bamileke of Cameroon.[38]

2.4.4.2 Contagion

Epilepsy is widely believed to be contagious, and contact with foam or urine from a seizing patient is thought to transmit the disease. Immediate contacts are always suspicious of contamination. Contact with breath, excrement, mother's milk, sperm, farts and excretions are believed to cause seizures.[50] The preparation of a dish in the same crockery or drinking in the same glass are also dreaded.[51]

Some particular circumstances could strongly suggest to PWE and their relatives that epilepsy is contagious; for example, when many relatives suffer from epilepsy due to genetically inherited reasons, or family cases of epilepsy caused by neurocystcercosis.[52]

In rural areas in Ethiopia, the frequency of beliefs on the contagion of the convulsive phase was underlined[43]; similar findings were made in Wolof and Sérère in Senegal[45] and in Togo.[53]

2.4.4.3 "Biomedical Model": Foreign Body Migration, Body Fluids-Bound, and Heredity

The organic nature of the disease is rarely mentioned, showing evidence of a great misunderstanding of the true nature of the disease in SSA.[40,54]

The "biomedical model" refers to all the beliefs that considered epilepsy as a consequence of the presence and migration of a foreign body in the patient's body, and which can block biological fluids.[38,46,47,51] We can also classify into this category all the beliefs in a kind of heredity, which is more likely religious and magical than truly biological.[43,54,55]

2.4.4.4 Epilepsy and Stigma

The social and cultural consequences of these beliefs and ignorance of the disease are extremely prejudicial for the patients. Overall, they lead to patient marginalization.

Epilepsy-related stigma is a complex phenomenon whose causes are not well known. However, today it is clear that sociocultural representations contribute to the rejection of PWE. Some authors even suggested that perceived stigma is less prevalent than enacted stigma in SSA.[1,56]

The exclusion from professional environments is frequent; but this is not highly underlined in the existing literature from SSA. This observation could be explained by the fact that formal employment or profession is rare in rural Africa. More studies must be conducted to specify the real occupations of PWE in SSA.[29,30,57]

PWE in SSA have high exposure to cognitive impairment, physical injuries, and disabilities. All these factors combined could exacerbate the stigma associated with epilepsy, which has long been described as a major burden of the disease.[1] In a study performed in Benin (West Africa) in PWE, about 68.7% reported feelings of stigma. Factors independently associated with feelings of stigma were experience of social isolation ($P < 0.001$), experience of marital problems ($P < 0.01$) and presence of anxiety disorder ($P < 0.01$). Social factors seem to be more influential than sociocultural representations of epilepsy. Stigma affected the help-seeking behavior contributing to the deficit of treatment.[13,58] Marginalization also comes from the family. Although PWE are almost never driven out of the house, they could be relegated to behind the house ("disease of behind the house"). The patient is never totally banished from the community. But he is tolerated, without much more of a social role.[40,41,59,60] When seizures occur and remain for a long time as a chronic disease, all the relatives of a PWE may be rejected.

The marital status of PWE is different from that of the general population. Patients have difficulty obtaining a bride or a groom.[61] In Cameroon, young women with epilepsy "are often married" to old men, and the usual rites of the marriage are bypassed as unnecessary.[38] In the Tamberma of Togo, a woman with epilepsy can be married without dowry.[62]

Schooling is difficult for patients; many children with epilepsy cannot go to school,[63] and many are stigmatized by their comrades.[55,64] The attitude of the teachers varies in different studies: there is either an overall positive attitude,[65] or a negative one.[66] PWE very often have a poor quality of life,[67,68] with serious consequences on their mental health. The results of surveys showed that depressive mood, often concomitant with anxiety disorders, represent the psychiatric co-morbidity frequently associated with epilepsy.[57,69] Epilepsy is very often regarded as an incurable disease (the marks of the disease as the ritual scarifications on the body, stigmatize the patient as a PWE, and are often considered as a criterion of incurability).[70] So the PWE are considered as deviants in a model of collective anguish and

social monitoring. However, in many cultural groups, exclusion is moderated by traditional assistance rules.[71] In the future, according to what happened in past centuries in developed countries, it can be expected that sociocultural representations of epilepsy in SSA will progress from a magical to a neurophysiological view.[54,55,71]

2.5 SOCIOCULTURAL REPRESENTATIONS AND STIGMA IN DEMENTIA

Dementia is among the most important public health problems, due to the increasing number of elderly people in the world.[72–74] In developed countries, the prevailing model of dementia is biomedical. Some authors criticize this tendency of modern medicine to "expand into almost all aspects of human existence", in particular aging.[75] The problem is different in low-income countries, where the prevalence of dementia is generally lower than in high-income countries.[76–81] Increasingly, studies concerned with the perception of dementia in these countries take account of sociocultural and socioeconomic situations of respondents.[82,83] It seems that the individual's social isolation, today's characteristic of the Western culture societies, and the frustrations borne by these individuals when they lose their social status, are factors favoring dementia.[73] Nowadays, the prevalence of dementia in traditional societies is at serious risk of increasing. The privileged status of the aged person unfortunately runs the risk of notwithstanding the modernity. Actually, lifestyles are changing, and like in the West, for economic reasons, young adults leave the countryside for urban areas. Worse, the outbreak of new diseases such as AIDS, that affect young adults, creates a mixture of fear and rejection of the aged persons who are still alive. They are suspected to have a witchcraft powers, or even cannibalism, which allow them to "eat the younger"; these beliefs bring people to treat them badly.

Faure et al.[84] performed a study to determine how dementia is perceived by different groups of people in urban areas of Central Africa: elderly people with dementia, cognitively impaired but not demented elderly, their relatives, and hospital caregivers.

During a prevalence survey (EDAC) among subjects aged 65 and over in Bangui (Central African Republic) and Brazzaville (Congo), 549 semi-directive interviews were carried out using the Explanatory Model Interview Catalogue. During the preparation of the investigation, there was no term equivalent to the biomedical concept of dementia in the Sango (Bangui), or Lingala or Kituba (Brazzaville) local languages. The term "disease of the elderly" was therefore used when conducting interviews.

Results showed that the biomedical concept of dementia was unknown other than among health professionals. However, the signs and consequences of the disease were well-recognized by the great majority, particularly in Bangui where 88.0% of individuals interviewed (all categories) reported they had already met or known a person with cognitive disabilities or behavioral troubles (57.5% in Brazzaville said the same).

2.5.1 Models of Understanding of Dementia

The results indicated that the majority of people interviewed perceived signs of dementia, in terms of more or less characteristic symptoms or the impact on daily life. Although dementia was not recognized as a medical state, dementia's signs and consequences were familiar to many elderly people and relatives interviewed. Traditional and magic—religious representations were not rejected by professional caregivers. Forgetfulness (77.8%—98.4%) was noted. Emotional and socioeconomic consequences (50.8%—82.7%) were commonly cited. Aging was seen as the principal cause, with ambivalent attitudes towards the elderly. In urban areas of Central Africa, people concerned with dementia have difficulties integrating biomedical representations that focus on biological aspects of the disease.

This confirmed what has already been shown,[77] for example, among the Yoruba who, as reported by Ineichen,[76] recognized elements of senility ("back to childhood", etc.) but did not assign to them a specific term.[82] A large proportion of elderly respondents identified themselves as affected by the "disease of the elderly", even if they associated other conditions with this notion. Sadness, anxiety, and concerns of the elderly were the subjects of significant attention. Their "socio-emotional concerns", were not inferior to "instrumental concerns".[85] The effects on the body of hard work carried out under the hot sun, were frequently implicated in the explanations of dementia. Nevertheless, anxiety or tension caused by the unusual behavior of people with dementia also encouraged beliefs in witchcraft and curses. Rasmussen et al.[86] made this observation among the Tuareg of Niger, who suspected people with dementia of having failed in their duty to say their daily prayers.

Studies on stigmatizing attitudes toward affected demented people have found that magico-religious beliefs were often associated with intergenerational conflicts.[87]

2.5.2 Social Tolerance of People With Dementia: Representations and Stigma

In our study,[84] the group with the lowest score for perceived stigma was the Dementia Persons (12.9 ± 6.6), in contrast to their relatives, whose score reached 14.3 ± 8.0. The score among health workers was higher, at 15.9 ± 5.4, and significantly different from that of all groups of relatives combined (13.5 ± 7.2; $P = 0.014$). The study revealed significant differences in the belief that older people with dementia would be less respected than others in the community (60.3% of health workers thought so, vs 36.3% of relatives, $P = 0.0003$); similar significant differences were also seen regarding whether dementia causes marital difficulties (74.6% vs 30.9%, $P < 0.0001$) and whether it leads to other health problems (79.4% vs 46.5%, $P < 0.0001$).

The narrative accounts provided by respondents indicate some intolerance, related to magico-religious beliefs, toward behavioral disturbances shown by people with dementia.

Social intolerance of dementia also resulted from the idea that older people become "useless" to society; "people do not respect them because they are unproductive", said an 81-year-old woman with cognitive impairment (CIEP) in Bangui. Nevertheless, there are limits to intolerance. Social support is important because it is based on traditional values, whether religious or simply deeply rooted in human society: the family ties between parents and children, marriage bonds, helping others, respect for older people who have given birth and brought up children, served their country and so on. "We must love and support our parents" said a cognitive impairment person's relative (CIEP-R, 57-year-old, Brazzaville). Similarly, one woman said, "If my husband became demented I would stay beside him until his death, because God has united us" (CIEP, 75-year-old, Bangui). "She is a senior citizen, so she needs to be treated like a human being. In my opinion, it is not good to separate her from the group" (dementia person's relative, 67-year-old, Bangui).

Among the Bantus' people in Central Africa, the aged persons have always been held as a sacred age group that used to be respected, or even held in awe. The aged persons used to be considered as the tradition guarantors. These aged persons, as a result of their closeness to the spirits of the protecting ancestors, could prevent witchcraft aggressions by the harmful spirits.

In most sub-Saharan systems of belief, people define themselves constantly and simultaneously with relationships to the actual, the imaginary, and the symbolic. It follows that the illness has always been of a supernatural origin. Consequently on the occasion of serious illnesses or death, it is a common practice to ask the healer about the supernatural origin of the illness or the death of a close person; thus it happens that the aged subject, wrongly accused in supposed witchcraft aggressions, acknowledges the accusations in the presence of the village oracle, insisting on his not being dignified and expressing his regrets; this can set the issue of associated depressive states. It is in this context that acts of violence are actually posed against the aged persons, thus transgressing customs and traditions that made them loved, protected, respected, and even feared persons because of their powers.

Studies on stigmatizing attitudes towards people with dementia have found that magic–religious beliefs are often associated with intergenerational conflicts. In extreme cases, abusive behavior may occur (abuse, stoning, beating). The problem of abuse or neglect of the elderly is not new.

Attitudes are changing, but perceptions are difficult to change at their core.[88] It is therefore essential to consider the representational universe of people affected by dementia before trying to introduce new representations that define a purely biomedical condition.

Representations are influenced by age, education level, and knowledge system. Cultural environment and closeness with the disease also seem to be influential variables. This also concerns stigma. Some studies showed that doctors and medical students endorsed stigmatizing attitudes towards mental illnesses and were especially prone to see patients as blameworthy.[25,88,89]

The preservation of a traditionally cultural milieu is a true defense phenomenon against the developed culture of the Western countries, where the individual, as soon as he/she loses his/her social status, is isolated.[73,90]

Knowledge of EMs of illness in SSA, should be used to conduct cross-cultural epidemiological studies which, while being culturally sensitive, are also comparable with other studies.

2.6 CONCLUSION

Health burst as the Central positive standard in many societies... The planet is largely medicalized and this is a little questionable... Health is the most accepted standard, alongside of merchant capitalism and the Western political democracy.[8]

The effects of globalization, essentially derived from Western models, are not applied as such in different societies; the medicines are more and more hybrids, mixed. There is a reinterpretation; that is why, in the

eyes of some health ethnologists, simply exploring local models as if they were free from this influence, participates in hypocrisy. The western global biomedical model should be seen as a cultural system among others, analyzing its influence and resisting its hegemony. However, the devices exported by public health abroad are little questioned. Whatever the outcome, EMs of illness in SSA, should be used to conduct cross-cultural epidemiological studies which, while being culturally sensitive, are also comparable with other studies. In many SSA societies, people with mental disorders frequently experience stigma, discrimination, violence and limited support, but views about illness are changing.

In addition to the rapid social changes in Africa, the biomedical approaches risk causing harm in a context where spirituality plays such an important role. Local caregivers are, from this perspective, the key elements in creating a health system that reconciles these multidimensional aspects.

Traditional societies will have to guard against the tendency to evolve towards a Western model, which no longer reserves for the aged person the place that used to be his/hers in traditional society. Right now, the African elite must integrate these new dimensions in national health policies, owing to dementia prevalence in these regions of the world.

References

1. Jacoby A, Snape D, Baker GA. Epilepsy and social identity: the stigma of a chronic neurological disorder. *Lancet Neurol.* 2005;4:171–178.
2. Moscovici S. *La psychanalyse, son image et son public.* Paris: Presses Universitaires de France; 1961:652 p.
3. Jodelet D. *Les représentations sociales.* Paris: Presses Universitaires de France; 1997:447 p.
4. Hours B. Pour une anthropologie post-culturelle de la santé. Bull Amades. 2003;56.
5. Fainzang S. Faire du nouveau avec de l'ancien, et un peu plus... pour penser les nouveaux objets en anthropologie de la santé. Anthropol Santé. 2010;1.
6. Falavigna A, Teles AR, Roth F, Velho MC, Roxo MR, Dal Bosco AL, et al. Awareness, attitudes and perceptions on epilepsy in Southern Brazil. *Arq Neuro-Psiquiatr.* 2007;65(4B):1186–1191.
7. Collignon R, Gruénais ME, Vidal L. *L'annonce de la séropositivité au VIH en Afrique. Psychopathologie Africaine.* vol. 26. Dakar: Société de psychopathologie et d'hygiène mentale de Dakar; 1994:149–291.
8. Hours B. D'un patrimoine (culturel) à l'autre (génétique). J Anthropol. 2002;88–89, 21–28.
9. Fassin D. La globalisation et la santé. Eléments pour une analyse anthropologique. In: Hours B, ed. *Systèmes et politiques de santé. De la santé publique à l'anthropologie.* Paris: Karthala; 2001:24–40.
10. Desclaux A, Sarrandon-Eck A. Introduction au dossier: l'éthique en anthropologie de la santé: conflits, pratiques, valeur heuristique. Ethnographiques.org [online]. No 17; November 2008.
11. Cunin E. *La globalisation de l'ethnicité? Autrepart. Revue des sciences sociales au Sud, 38.* Bondy/Paris: IRD/Colin; 2006:203 p.

12. Goffman E. Stigmate: Les usages sociaux des handicaps [Le Sens Commun Collection, Kihm A, Trans.]. Paris: Editions de Minuit; 1975 [Original work published 1963].

13. Weiss MG, Ramakrishna J. Stigma interventions and research for international health. *Lancet*. 2006;367:536–538.

14. Abbo C, Ekblad S, Waako P, Okello E, Muhwezi W, Musisi S. Psychological distress and associated factors among the attendees of traditional healing practices in Jinja and Iganga districts, Eastern Uganda: a cross-sectional study. *Int J Ment Health Syst*. 2008;2:16.

15. De Jong JT, Komproe IH, Van Ommeren M. Common mental disorders in post conflict settings. *Lancet*. 2003;361 (9375):2128–2130.

16. Dobricki M, Komproe IH, de Jong JT, Maercker A. Adjustment disorders after severe life-events in four postconflict settings. *Soc Psychiatry Psychiatr Epidemiol*. 2010;45(1):39–46.

17. Kleinman A. Anthropology and psychiatry. The role of culture in cross-cultural research on illness. *Br J Psychiatry*. 1987;151:447–454.

18. Patel V, Pereira J, Coutinho L, Fernandes R. Is the labelling of common mental disorders as psychiatric illness clinically useful in primary care? *Indian J Psychiatry*. 1997;39(3):239–246.

19. Mutambirwa J. Health problems in rural communities, Zimbabwe. *Soc Sci Med*. 1989;29(8):927–932.

20. Cheetham WS, Cheetham RJ. Concepts of mental illness amongst the rural Xhosa people in South Africa. *Aust N Z J Psychiatry*. 1976;10(1):39–45.

21. Osei Y, Brautigam W. Psychosomatic illness concept and psychotherapy among the Akan of Ghana. *Can J Psychiatry*. 1979;24 (5):451–457.

22. Orley J. Epilepsy in Uganda (rural). A study of eighty-three cases. *Afr J Med Sci*. 1970;1(2):155–160.

23. Ebigbo PO. Development of a culture specific (Nigeria) screening scale of somatic complaints indicating psychiatric disturbance. *Cult Med Psychiatry*. 1982;6(1):29–43.

24. Guinness EA. Social origins of the brain fag syndrome. *Br J Psychiatry Suppl*. 1992;(16):53–64.

25. Adewuya AO, Oguntade AA. Doctors' attitude towards people with mental illness in Western Nigeria. *Soc Psychiatry Psychiatr Epidemiol*. 2007;42(11):931–936.

26. Ronzoni P, Dogra N, Omigbodun O, Bella T, Atitola O. Stigmatization of mental illness among Nigerian schoolchildren. *Int J Soc Psychiatry*. 2010;56(5):507–514.

27. Ae-Ngibise K, Cooper S, Adiibokah E, Akpalu B, Lund C, Doku V, Mhapp Research Programme Consortium. "Whether you like it or not people with mental problems are going to go to them": a qualitative exploration into the widespread use of traditional and faith healers in the provision of mental health care in Ghana. *Int Rev Psychiatry*. 2010;22(6):558–567.

28. Sorsdahl KR, Flisher AJ, Wilson Z, Stein DJ. Explanatory models of mental disorders and treatment practices among traditional healers in Mpumulanga, South Africa. *Afr J Psychiatry*. 2010;13 (4):284–290.

29. Preux PM, Druet-Cabanac M. Epidemiology and aetiology of epilepsy in sub-Saharan Africa. *Lancet Neurol*. 2005;4(1):21–31.

30. Ba-Diop A, Marin B, Druet-Cabanac M, Ngoungou EB, Newton CR, Preux PM. Epidemiology, causes, and treatment of epilepsy in sub-Saharan Africa. *Lancet Neurol*. 2014;13(10):1029–1044.

31. Birbeck GL, Kalichi EMN. The functional status of people with epilepsy in rural sub-Saharan Africa. *J Neurol Sci*. 2003;209:65–68.

32. Birbeck G, Chomba E, Atadzhanov M, Mbewe E, Haworth A. The social and economic impact of epilepsy in Zambia: a cross-sectional study. *Lancet Neurol*. 2007;6:39–44.

33. Uchoa E, Corin E, Bibeau G, Koumare B. Représentations culturelles et disqualification sociale. L'épilepsie dans trois groupes ethniques au Mali. *Psychopathol Afr*. 1993;25:33–57.

34. Miletto G. Vues traditionnelles sur l'épilepsie chez les dogons. *Méd Trop*. 1981;41:291–296.

35. Colomb H, Ayat H, Dumas M. L'épilepsie, maladie sociale au Sénégal. *Bull Soc Méd Afr Noire Lgue Fr*. 1968;13:925–932.

36. Bernet-Bernady P, Tabo A, Druet-Cabanac M, et al. L'épilepsie et son vécu au nord-ouest de la République Centrafricaine. *Méd Trop*. 1997;57:407–411.

37. Arborio S, Jaffre Y, Farnarier G, Doumbo O, Dozon JP. Etude du kirikirimasien (épilepsie) au Mali : dimensions étiologique et nosographique. *Méd Trop*. 1999;59:176–180.

38. Nkwi PN, Ndonko FT. The epileptic among the bamilékés of Maham in the Nde division, West Province of Cameroon. *Cult Med Psychiatry*. 1989;13:437–448.

39. Jilek-Aall L. Morbus Sacer in Africa: some religious aspects of epilepsy in traditional cultures. *Epilepsia*. 1999;40:382–386.

40. Karfo K, Kere M, Gueye M, Ndiaye IP. Aspects socio-culturels de l'épilepsie grand mal en milieu dakarois : enquête sur les connaissances, attitudes et pratiques. *Dakar Méd*. 1993;38:139–145.

41. Arborio S, Dozon JP. La dimension socioculturelle de l'épilepsie (Kirikirimasien) en milieu rural Bambara (Mali). *Bull Soc Pathol Exot*. 2000;93:241–246.

42. Amani N, Durand G, Delafosse RCJ. Incidence des données culturelles dans la prise en charge des épileptiques en Afrique noire. *Nervure*. 1995;7:47–51.

43. Anani TK, Hegbe KM, Balogou KA, Mbella Mam E, Grunitzky EK. Aspects culturels des épilepsies au Togo. Communication orale (ATW11) au Congrès de Neurologie Tropicale. Limoges, France; 21 and 23 September 1994.

44. Tekle-Haimanot R, Abebe M, Forsgren L, Gebre-Mariam A, Heijbel J, Holmgren G, et al. Attitude of rural people in central Ethiopia toward epilepsy. *Soc Sci Med*. 1991;32:203–209.

45. Adotevi F, Stephany J. Représentations culturelles de l'épilepsie au Sénégal (région du Cap-Vert et du Fleuve). *Méd Trop*. 1981;41:283–288.

46. Nubukpo P, Preux PM, Clément JP. Représentations socioculturelles de l'épilepsie en Afrique Noire. *Ann Psychiatr*. 2001;16:219–227.

47. Anderman LF. Epilepsy in developing countries. *Transcult Psychiatr Res Rev*. 1995;32:351–384.

48. Pilard M, Brosset C, Junod A. Les représentations sociales et culturelles de l'épilepsie. *Méd Afr Noire*. 1992;39:652–657.

49. Jilek-Aall L, Rwiza HT. Prognosis of epilepsy in a rural African community: a 30-year follow-up of 164 patients in an outpatient clinic in rural Tanzania. *Epilepsia*. 1992;33:645–650.

50. Rwiza HT, Matuja WBP, Kilonzo GP, Haule J, Mbena P, Mwang'ombola R, et al. Knowledge, attitude, and practice toward epilepsy among rural Tanzanian residents. *Epilepsia*. 1993;34:1017–1023.

51. Awaritefe A. Epilepsy: the myth of contagious disease. *Cult Med Psychiatr*. 1989;13:449–456.

52. Grunitzky EK, Dumas M, Dabis F, Deniau M, Kassankogno Y, Belo M, et al. Cysticercose cérébrale et épilepsie au Togo. *Afr J Neurol Sci*. 1989;8:17–19.

53. Nubukpo P, Preux PM, Clément JP, et al. Etude comparée des représentations socioculturelles des épilepsies en France (Limousin) et en Afrique (Togo et Bénin). *Méd Trop*. 2003;63:143–150.

54. Nyame PK, Biritwum RB. Epilepsy: knowledge, attitude and practice in literate urban population, Accra, Ghana. *West Afr J Med*. 1997;16:139–145.

55. Millogo A, Traoré ED. Études des connaissances et des attitudes en matière d'épilepsie en milieu scolaire à Bobo-dioulasso (Burkina-Faso). *Epilepsies*. 2001;13:103—107.

56. Reis R, Meinardi H. ILAE/WHO "Out of the shadows Campaign" Stigma: does the flag identify the cargo? *Epilepsy Behav*. 2002;3:S33—S37.

57. Nubukpo P, Preux PM, Houinato D, et al. Psychosocial issues in people with epilepsy in Togo and Benin (West Africa) I. Anxiety and depression measured using Goldberg's scale. *Epilepsy Behav*. 2004;5(5):722—727.

58. Rafael F, Houinato D, Nubukpo P, Dubreuil CM, Tran DS, Odermatt P, et al. Sociocultural and psychological features of perceived stigma reported by people with epilepsy in Benin. *Epilepsia*. 2010;51(6):1061—1068.

59. Mc Queen A, Swartz L. Reports of epilepsy in a rural south african village. *Soc Sci Med*. 1995;40:859—865.

60. Nerwell ED, Vyungimana F, Geerts S, Van Herckoven I, Tsang WCV, Engels D. Prevalence of cysticecosis in epileptics and members of their families in Burundi. *Trans Roy Soc Trop Med Hyg*. 1997;91:389—391.

61. Farnarier G, Moubeka-Mounguengui M, Kouna P, Assengone-Zeh Y, Gueye L. Epilepsies dans les pays tropicaux en voie de développement: étude de quelques indicateurs de santé. *Epilepsies*. 1996;8:189—213.

62. Nubukpo P, Grunitzky EK, Pélissolo A, Radji A, Preux PM, Clément JP. Epilepsie et personnalité chez les tamberma du Togo à partir d'une étude en population générale à l'aide du TCI de Cloninger. *Encephale*. 2006;32(6 Pt 1):1019—1022.

63. Ndiaye IP, Ndiaye M, Tap D. Sociocultural aspects of epilepsy in Africa. *Prog Clin Biol Res*. 1983;124:345—351.

64. Matuja WBP, Rwiza HT. Knowledge, attitude and practice (KAP) towards epilepsy in secondary school students in Tanzania. *Centr Afr J Med*. 1994;40:13—18.

65. Mielke J, Adamolekun B, Ball D, Mundada T. Knowledge and attitudes of teachers towards epilepsy in Zimbabwe. *Acta Neurol Scand*. 1997;96:133—137.

66. Danesi MA. Epilepsy and the secondary schools in Nigeria. *Trop Geogr Med*. 1994;46:S25—S27.

67. Cramer JA, Perrine K, Devinsky O, Bryant-Comstock L, Meador K, Hermann B. Development and cross-cultural translations of 31-item quality of life in epilepsy inventory. *Epilepsia*. 1998;39:81—88.

68. Nubukpo P, Clément JP, Houinato D, et al. Psychosocial issues in people with epilepsy in Togo and Benin (West Africa) II: quality of life measured using the QOLIE-31 scale. *Epilepsy Behav*. 2004;5(5):728—734.

69. Trimble MR, Krishnamoorthy ES. Neuropsychiatric disorders in epilepsy: some transcultural Issues. *Epilepsia*. 2003;44:21—24.

70. Osakwe C, Otte WM, Alo C. Epilepsy prevalence, potential causes and social beliefs in Ebonyi State and Benue State, Nigeria. *Epilepsy Res*. 2014;108(2):316—326.

71. Andriantseheno LM, Rakotoarivony MC. Aspects socioculturels de l'épilepsie chez le Malagasy : étude C.A.P. en population générale à Antananarivo, Madagascar. *Bull Soc Pathol Exot*. 2000;93:247—250.

72. WHO. *Dementia in Later Life: Research and Action: Report of a WHO Scientific Group on Senile Dementia*. Geneva, Switzerland: WHO; 1986.

73. WHO. *Towards an International Consensus on Policy for Long-Term Care of the Ageing*. Geneva, Switzerland: WHO/Milibank Memorial Fund; 2000.

74. WHO. *Ageing and intellectual disabilities: improving longevity and promoting healthy ageing*. Summative report. WHO/MSD/HPS/MDP/003, Geneva, Switzerland; 2000.

75. Goodwin JS. Geriatrics and the limits of modern medicine. *NEJM*. 1999;340(16):1283—1285.

76. Ineichen B. The epidemiology of dementia in Africa: a review. *Soc Sci Med*. 2000;50:1673—1677.

77. Pollitt PA. Dementia in old age: an anthropological perspective. *Psychol Med*. 1996;26:1061—1074.

78. Herbert CP. Cultural aspects of dementia. *Can J Neurol Sci*. 2001;28(Suppl 1):S77—S82.

79. Hendrie HC, Murrell J, Gao S, Unverzagt FW, Ogunniyi A, Hall KS. International studies in dementia with particular emphasis on populations of African origin. *Alzheimer Dis Assoc Disord*. 2006;20:S42—S46.

80. Guerchet M, Houinato D, Paraiso MN, Von Ahsen N, Nubukpo P, Otto M, et al. Cognitive impairment and dementia in elderly people living in rural Benin, West Africa. *Dement Geriatr Cogn Disord*. 2009;27(1):34—41.

81. Guerchet M, M'belesso P, Mouanga A, et al. Prevalence of dementia in elderly living in two cities of Central Africa: the EDAC survey. *Dement Geriatr Cogn Disord*. 2010;30(3):261—268.

82. Enjolras F. Incidence du pronostic sur la construction des modèles explicatifs de la maladie d'Alzheimer à l'île de la Réunion. [Impact of the prognostic into the construction of disease's explanatory models: the Alzheimer's disease in l'île de la Réunion]. *Sci Soc Santé*. 2005;23(3):69—94.

83. Blay SL, Toledo Pisa Peluso E. Public stigma: the community's tolerance of Alzheimer disease. *Am J Geriatr Psychiatry*. 2010;18(2):163—171.

84. Faure-Delage A, Mouanga AM, M'belesso P, et al. Socio-cultural perceptions and representations of dementia in Brazzaville, Republic of Congo: the EDAC survey. *Dement Geriatr Cogn Dis Extra*. 2012;2:84—96.

85. Hulko W. From "not a big deal" to "hellish": experiences of older people with dementia. *J Aging Stud*. 2009;23:131—144.

86. Rasmussen SJ. *The Poetics and Politics of Tuareg Aging: Life-Course and Personal Destiny in Niger*. Dekalb: Northern Illinois University Press; 1997.

87. WHO. *La maltraitance des personnes âgées. Chapitre 5 [Abuse of the elderly. Chapter 5]*. OMS/WHO Rapport mondial sur la violence et la santé [OMS/WHO: World Report on Violence and Health]. Geneva, Switzerland: OMS; 2002:139—162.

88. Kohl FS. *Les représentations sociales de la schizophrénie. [Social Representations of Schizophrenia]*. Paris: Masson; 2006:210 p.

89. Fernando SM, Deane FP, Mc Leod HJ. Sri Lankan doctors' and medical undergraduates' attitudes towards mental illness. *Soc Psychiatry Psychiatr Epidemiol*. 2010;45(7):733—739.

90. Ndamba-Bandzouzi B, Nubukpo P, Mouanga AM, M'belesso P, Tognide M, Tabo A, et al. Violence and witchcraft accusations against older people in Central and Western Africa: toward a new status for the older individuals? *Int J Geriatr Psychiatry*. 2014;29(5):546—547.

Further Reading

Moliner P, Guimelli C. *Les représentations sociales. Fondements historiques et développements récents*. Grenoble: Presses Universitaires de Grenoble; 2015:139 p.

3

Climatic Factors Under the Tropics

Jacques Reis[1], Pascal Handschumacher[2], Valerie S. Palmer[3],
and Peter S. Spencer[3]

[1]University of Strasbourg, Strasbourg, France [2]Aix Marseille University, Marseille, France
[3]Oregon Health & Science University, Portland, OR, United States

3.1 INTRODUCTION

This chapter examines the impact of climatic factors on health and disease in tropical zones. The tropics are the regions surrounding the Equator, delimited in latitude by the Tropic of Cancer in the Northern Hemisphere and the Tropic of Capricorn in the Southern Hemisphere. We examine the physical determinants that define the tropics and their direct and indirect roles in the maintenance of health and induction of disease. We discuss the influence of geographical features on ecosystems and living organisms, interactions that can promote the onset of disease (with cases studies) in different ecosystems, and anthropogenic changes and their impact on the ecosystem and, consequently, on human health. These physical constraints of the climate interfere largely with sociocultural and economical issues (see dedicated chapters), creating specific epidemiological conditions under the tropics.

3.2 A PHYSICAL GEOGRAPHIC APPROACH: THE CLIMATE AND ITS CHARACTERISTICS

The climate system is driven by one energy source: the sun. Three-quarters of this energy is absorbed (by the atmosphere and the Earth's surface), the remaining quarter is reflected back into space.

The solar radiations over planet Earth have a heterogeneous distribution, dependent notably on Earth rotation, latitude, altitude, and surface (land vs water). The non-uniformity of the Earth also contributes to heterogeneous energy distribution, for example, the irregular distribution of continents, topography with mountain barriers, and large water surfaces (oceans, seas). These physical features cause heating differences that generate circulations of air and water masses, which allow heat exchanges between the Earth's zones. In general, tropical zones (air, oceans, land) receive more solar radiation than the poles. The main characteristic of a tropical climate is that the daytime temperature is constantly over 18°C.[1]

3.2.1 The Role of the Inter-Tropical Convergence Zone

One of the main determinants of tropical weather and climates is the Inter-Tropical Convergence Zone (ITCZ), "the location where northeast winds in the Northern Hemisphere converge with the southeast winds from the Southern Hemisphere". ITCZ is a low-pressure zone characterized by a line of cumulus clouds in the tropics.[2,3]

At the Equator, the ITCZ causes significant rainfall throughout the year. Sometimes, there are two rainy seasons, a long and a short one. Some areas, such as parts of the Amazon Basin in South America, receive 3 or more meters (9 ft) of rain per year. Temperatures range from 25 to 35°C, with the hottest months only 2−3°C higher than those of cooler months. Constant temperature and high humidity are linked with luxuriant vegetation and a great biodiversity of flora, fauna and microbial organisms.

At locations progressively distant from the Equator, seasons become progressively more distinct (i.e., rainy season and dry season); these are linked with seasonal variations in the position of the ITCZ. For example, in Sahelian regions, there is a cold dry season (October−November to March), followed by a hot dry season (April to June) followed by a short rainy season with low rainfall (200−400 mm/year). The dry season is characterized by a continental trade wind blowing from the Northeast (Harmattan), which often brings much dust.

3.2.2 Types of Tropical Climate

Between the Equator and the Tropics, dependent mostly on temperature variability, humidity and precipitation, as well as winds, there are gradients leading from tropical moist to tropical dry. Three main types are described: the Tropical Rainforest or Equatorial, the Tropical Monsoon, and the Tropical Wet and Dry or Savannah.[3]

At the limit of the tropical area, the hot and arid climatic type is characterized by a huge temperature variation (from a daily temperature of 45°C to nocturnal frost), a scarcity of precipitation (mostly absent) and desertic vegetation. The Sahara Desert in northern Africa receives only 2−10 cm of rain per year, and the Chilean Atacama Desert is famous for its usual dryness. As the vegetal soil's cover is reduced, the permanent wind can erode the earth's surface and increase airborne particulate matter (PM), originating sometimes in major dust storms.

3.3 A BIOGEOGRAPHICAL APPROACH: HOW CLIMATE'S FEATURES DETERMINE NATURAL BIOMES AND ECOSYSTEMS

Functional biogeography brings the link between "the distribution of species and ecosystems across space and time and of the underlying biotic and abiotic factors, mechanisms, and processes".[4]

Biomes are characterized by distinctive plant and animal species that are maintained under regional climatic conditions. Several continental biomes exist in the tropics, including rainforests (Amazonia, Congo), dry deciduous forests (Madagascar, Deccan, Thailand), spiny forests (Madagascar), grasslands and deserts.[5]

Tropical rainforests, thought to be the oldest biome on Earth, comprise only 40% of the world's tropical forests and only 20% of the world's total forests. Flora are organized into four strata depending on the sunlight's access: giant trees constitute the emergent layer. The canopy layer above, formed by the trees' crown and often covered with other plants and tied together with woody vines (lianas), is home to 90% of the organisms found in the rainforest. The understory (under-canopy and shrub layer) allows the growth of young trees and plants that tolerate low light. The forest floor receives less than 2% of the sunlight. In most tropical rainforests, the soils are relatively poor in nutrients, apart from recent volcanic soils.[6,7] The plant-available soil phosphorus and the dry-season intensity are, in Panama, the strongest predictors for the distribution of more than half of tree species.[8]

Many forested areas in the tropics are not rainforests. Forests that receive irregular rainfall (monsoons followed by a dry season) are moist deciduous forests. Trees in these forests may drop their leaves in the dry season.[7]

Arid zones are characterized by excessive heat and inadequate, variable precipitation and scarce vegetation. In the deserts, vegetation is virtually absent, whereas semi-desertic vegetation includes a mixture of grasses, herbs, and small, short trees and shrubs.[9] The majority of grasslands (among the 26% of total land area) are located in tropical low-income countries, such as the savannahs of Sahelian countries and shrublands and llanos in South America. Grasses, often 3–6 ft tall at maturity, are the dominant vegetation, with shrubs or bush associated with scarce drought- and fire-resistant trees. Grasses are particularly important for the resident populations since they provide the feed base for grazing livestock.[10] Desertification and vulnerability of water resources, including freshwater,[11,12] are great concerns in arid zones.

3.3.1 The Marine Biome

Tropical marine biomes also require consideration.[13] Marine regions cover about three-quarters of the Earth's surface and represent the largest of all ecosystems: they include oceans, coral reefs, and estuaries.[14] The seas have an enormous capacity to store heat.[15] Over 1.3 billion people live on tropical coasts, primarily in low- and middle-income countries.[16] Marine

food chains that support seafood species have been used by human societies for food since earliest times, e.g., in the Red Sea.[17]

Tropical marginal seas (TMSs) are natural subregions of tropical oceans containing notably three key ecosystems: coral reefs and emergent atolls, deep benthic systems, and pelagic biomes.[18] In the oceans, two other zones are described: the intertidal and abyssal zones. Coral reefs are widely distributed in warm shallow waters, in the Atlantic Ocean (Wider Caribbean) and the Indo-Pacific (from East Africa and the Red Sea to the Central Pacific Ocean).[15] These tropical reefs shelter one-quarter to one-third of all marine species. This diversity is mainly concentrated in the central Indo-Pacific (the "Coral Triangle") and decreases with increasing distance from the Indo-Australian archipelago.[19]

Tropical costal zones constitute an important ecosystem for humans since they serve as "food factories". The highest primary productivity and the richest fisheries are found within the Exclusive Economic Zones (EEZs), narrow strips of 200 nmi/370 km from coastlines.[17] As shown for the South American coastlines, the influence of the ocean currents, with the phenomenon of upwelling (Humboldt current) and the high environmental variability caused by the El Nino Southern Oscillation (ENSO) and La Nina Southern Oscillation (LNSO), have important roles in the biodiversity of the oceans and coastal lines.[20]

Some tropical biomes support the highest biodiversity in the world (e.g., tropical forests and coral reefs); others have the lowest rates of biodiversity, like the arid zones.[21] "Tropical ecosystems support a diversity of species and ecological processes that are unparalleled anywhere else on Earth".[22] The relation of the species' richness and a latitudinal gradient in geographical range is called the Rapoport's rule after Eduardo H. Rapoport.[23] However, other factors are also involved, notably geographical (longitude, elevation and depth) and environmental conditions (topography and aridity). One of the mechanisms of generation of this biodiversity is clearly the climate, particularly in the tropics.[24]

3.3.2 Microbial Diversity and Biogeography

The essential role of microbes in ecosystem homeostasis has been recognized recently with the emergence of a new field of Microbial Ecology.[25,26] Besides their great abundance (Earth hosts more than an estimated 10^{30} microbial cells), microorganisms are also immensely diverse and constitute about 60% of the Earth's biomass. Bacteria and Archaea have a central role in ecosystem processes. In the soil, they participate in water purification, soil fertility, decomposition and catalysis, and are involved in the carbon (C), nitrogen (N), sulfur (S), and phosphorus (P) cycles, providing nutrients (N and P) and storage of carbon (up to half of carbon in living organisms).[27] Their diversity changes with distance, season, climate, soil texture and other environmental parameters. In arid and semi-arid soils (Israel), the microbial biogeography is determined more by specific environmental factors (moisture, organic matter, and silt/clay content) than dispersal limitation (geographic distances and spatial distribution patterns). As everywhere in the Biosphere, these microorganisms interact constantly with humans, in pathogenic or beneficial ways.[28] We discuss microorganisms in desert dust later.

3.4 BIOLOGY: ADAPTATION AND ACCLIMATION TO TROPICAL CLIMATES

A fundamental issue in biology is the way that climate influences the physiology and evolution of organisms, either directly or indirectly. Water availability, temperature, but also radiation (especially ultraviolet radiation (UVR)), are typical direct tropical climatic constraints. Indirect influences on health include those related to nutrition, which is heavily dependent on the type of plant and animal species able to thrive in tropical climates. These, together with the anatomical and physiological adaptations of human to life in the tropics, such as skin pigmentation, are important drivers of ecosystem and human health.

3.4.1 Thermal Tolerance

Thermal tolerance affects plants, viruses,[29] insects[30,31] and marine and terrestrial ectotherms (reptiles).[32,33] Janzen, in the 1960s, noted that tropical species are exposed to low annual variations in ambient temperature; he concluded that selection had favored organisms with a narrow physiological tolerance to temperature. This leads to smaller areas of biological distribution and increases the turnover of species. Thus, any perturbation can endanger these tropical species because their physiological adaptation is limited.[32,34]

Seasonality, which involves a greater climatic tolerance and a spatial heterogeneity, also plays a critical role in thermal adaptation.[35] Another consequence is the capacity of organisms (e.g., vectors and reservoirs) to spread to higher latitudes. Ability for thermal adaptation is key. To survive at higher latitudes, organisms must be able to withstand greater temporal variability of climate relative to that at lower latitudes.[31]

3.4.2 Water Availability

Rodents from arid and semi-arid habitats live under conditions where the spatial and temporal availability of free water is limited or scarce. For example, South American desert rodents possess structural as well as physiological systems for water conservation.[36]

Tropical plants are stressed by a dynamic network of interacting stressors, such as availability of water, CO_2, light and nutrients, temperature and salinity. Several strategies have allowed them to adapt to a wide ecophysiological variety of habitats; e.g., some employ a method of carbon fixation adapted to arid conditions; other plants have developed effective physiological (nocturnal stomatal opening for CO_2 uptake and daytime closure of stomata) and biochemical plasticity (crassulacean acid metabolism plants fix CO_2 nocturnally in the dark period) allowing a high rate of responsiveness by readily reversible variations in performance.[37]

3.4.3 UVR and Airborne Propagation of Neurotropic Viruses

Weather patterns also affect the distribution of viral organisms and the risk for certain human viral diseases. One example is Varicella Zoster Virus (VZV) because it is mainly transmitted via virus-contaminated air. The associated human disease (chickenpox) is a highly contagious acute disorder that may reappear as herpes zoster if the dormant virus in the nervous system is later reactivated. Whereas, in the tropics, primary acute VZV infection occurs in later childhood, in temperate zones most infection occurs before leaving school. Therefore, adults are more susceptible to infection in tropical countries (30%−50%) compared to those in temperate countries (5%−10%). Thus VZV infection is subject to climatic modulation. Rice[38] suggests that UVR is the climatic factor responsible for the geographical differences in transmission. Furthermore, he hypothesizes that UVR has been involved in the co-evolution of VZV as humans migrated out of Africa, with loss of the selective advantage of resistance to UVR in the tropics, where UVR is at its highest level. This hypothesis is debated. Different genotypes of VZV segregate geographically into tropical and temperate areas. The UVRs do not explain the distribution of VZV genotypes in different tropical and temperate regions of Mexico.[39] Nevertheless, the seasonality of varicella in Perth, Western Australia, which peaks during August−September (Australian Spring when both UVR and temperature are relatively low) and a significant association between VZV infection and UVR were confirmed, as well as the role of the temperature in the VZV transmission.[40]

Human biology, notably skin pigmentation, also shows adaptation to climate and UV radiation. Many genes (MC1R, MATP (SLC45A2), SLC24A5, TYR, DCT, OCA2, KITLG, SLC24A4, and IRF4) are involved in melanogenesis, such that pigmentation-associated gene variants are specific to either Europeans or East Asians.[41] Skin protects against the deleterious effects of UVR and allows photosynthesis of UVR B-dependent vitamin D3. There is a strong correlation between human skin pigmentation and latitude, i.e., UVR levels. Living under high UVR (near the Equator), our ancestors had a rich protective eumelanin skin. Positive selection during evolution led to elimination of the MC1R locus polymorphism and to a continuous purifying selection acting on the same locus. When *Homo sapiens* dispersed out of the tropics, the goals changed. To adapt to low and highly seasonal UVB conditions, the need to maximize cutaneous biosynthesis of pre-vitamin D3 prevailed and drove a depigmentation selection.[42,43] The role of vitamin D3 has already been illustrated, notably in the pathogenesis of multiple sclerosis.[44] The deleterious impact (survival and/or reproductive fitness), on which the selection pressure was exercised, is still debated: effect on folate metabolism? Skin cancer?[42,43,45]

The skin is the first stage of the immune, innate and humoral response. External perturbations, such as UVR, can modify this response and lead to an immune-suppressive effect. Cytokines made by UV-irradiated keratinocytes play an essential role. Keratinocyte-derived interleukin (IL)-10 is responsible for the systemic impairment of antigen-presenting cell function and the UV-induced suppression of delayed-type hypersensitivity.[46]

3.4.4 Thermoregulation in Man

Although Man is a poikilotherm, thermoregulation is a challenge in the tropics. Skin keratinization aids in controlling the transepidermal modulation of water loss. Keratinization and epidermal differentiation genes are under accelerated evolution in the human lineage, driven by environmental selection pressure.[47]

3.4.5 Nutrition

Plants adapted to climatic extremes, such as severe drought and water-logging, comprise an important source of food for many in tropical climes. The same plants may be eaten in temperate regions but consumption patterns differ. In tropical regions subject to severe drought, most food and feed plants become depleted, resulting in disease and death of cattle and other sources of animal protein. Food dependency on one or more environmentally tolerant plants increases progressively with time, leading to malnutrition and disease. Since many plant species contain chemicals with toxic or neurotoxic potential, those in the tropics who must rely on individual plant species both for subsistence and emergency food, are susceptible to plant-specific neurological disease. Outbreaks of such diseases (e.g., lathyrism, cassavism) are almost always restricted to impoverished rural communities that depend on locally available sources for food and feed, are subject to weather extremes, and at risk for plant losses from these and other causes, such as viral and fungal infection. Toxic human neurological disease may affect such communities seasonally when crop yields are low, food stores have diminished, and protein malnutrition is evolving. By contrast, in temperate climates, where the same plants may be consumed throughout the year, populations remain relatively well nourished and are spared from neurotoxicity because they typically consume a mixed diet of animals and plants, such that the threshold for toxicity for any individual plant species is never exceeded.[48]

The most important example is cassava (*Manihot esculenta*), the tuber and leaves of which are eaten for their carbohydrate and protein content, respectively. The plant is eaten without detectable illness by hundreds of millions of people in the tropics, subtropics and beyond. However, in certain impoverished communities of sub-Saharan Africa and India (inner Kerala) that must subsist on cassava as their sole or major dietary source, outbreaks of irreversible upper motor neuron disease in the form of spastic paraparesis are not infrequent. Local names for cassavism include *konzo* and *mantakassa*, from the Democratic Republic of Congo and Mozambique, respectively. While the culpable neurotoxin has yet to be pinpointed, cassava is protein-poor and contains cyanogenic glucosides that produce cyanide, thiocyanate and cyanate in the consumer.[49]

Another example of great historical importance is grasspea (*Lathyrus sativus*), a remarkably environmentally tolerant and nutritious legume that, with prolonged dietary dependency, results in a neurotoxic upper motor neuron disease (lathyrism). While grasspea in small quantities is widely eaten as a tasty snack in affluent Bangladeshi households, those who are impoverished are heavily dependent on grasspea and develop varying degrees of spastic paraparesis. Today, lathyrism is largely confined to the Ethiopian highlands. The culpable neurotoxin is a plant-specific amino acid, beta-*N*-oxalylamino-L-alanine (L-BOAA), that acts as an excitant via a specific class of neuronal glutamate receptors. The same receptor has been implicated in cassavism, which might explain the comparable neurologic outcome.[50,51]

A third example is the cycad (*Cycas* spp.), an exceptionally poisonous largely tropical plant that is resistant to extreme environmental conditions, including drought, fire and cyclones. Nevertheless, humans and animals have used cycads as a source of food and feed especially after tropical cyclones when less hardy plants are destroyed. Animals that eat untreated plant components (leaves, seed) develop a chronic and perhaps progressive neuromuscular disease. Humans that detoxify cycad seed (Australian aborigines) or sago (Ryukyu islanders, Japan) through extensive water leaching are spared illness, while others at high risk for amyotrophic lateral sclerosis and Parkinsonism—dementia complex have used raw cycad seed for medicine or incompletely detoxified seed for food.[52]

Possible independent influences of dietary habits (nutrition) and chemical environment (related to climate and biomes) have been demonstrated. Lactase persistence in adulthood is a heritable condition providing a physiological advantage, the capacity to digest lactose contained in fresh milk. This enzymatic shift represents the best-known adaptation related to diet. A similar shift has been demonstrated for the enzyme, arylamine *N*-acetyltransferase 2 (NAT2) involved in acetylation (a well-known pharmacogenetic trait), although the environmental causative factor (if any) driving its evolution is as yet unknown. Investigations in the African population of the Sahelian belt showed a clear difference for the activity of NAT2 between nomadic pastoralists and hunter-gatherers (slow acetylators) versus agriculturalists or food-producing populations (sedentary farmers).[53]

3.5 DOES THE TROPICAL CLIMATE FAVOR PATHOGENS AND INFECTIONS IN HUMANS?

Galiana[54] proposed the holistic Ecosystem Screening Approach to understand pathogen-associated microorganisms affecting host disease. Actually, pathogens challenge all kinds of living organisms: plants, animals, and humans. The questions to be addressed are numerous.

3.5.1 Biogeography

The distribution of human-associated pathogens seems to follow Rapoport's rule: a study showed a positive correlation between range, size and latitude for 290 human pathogenic species.[55] Latitudinal gradient, nested species pattern, and Rapoport's rule, are the factors that produce the observed geographical distribution of human pathogenic species.[55] However, there is no tropical effect for parasite species. Kamiya[56] has shown that the rich diversity of parasite species

across animal, plant, and fungal hosts does not correlate with, and even runs counter to, Rapoport's rule. Parasite diversity tends to be greater as one moves further from the Equator.[56]

3.5.2 Role of Environmental Conditions

"The life cycles and transmission of many infectious agents—including those causing disease in humans, agricultural systems and free-living animals and plants—are inextricably tied to climate".[57] The life cycle of parasites is influenced by environmental factors. For example, the development, survival, distribution, and migratory behavior of free-living helminth larvae on pasture are primarily weather-related. Eggs hatch and develop more readily at higher temperatures, and optimal temperature allows for larval activity and thus motility. Moisture affects motility and must be present to prevent desiccation and death of developing larvae. Rainfall favors larval dispersal. As larvae migrate deep into the soil, the soil type has a major effect on their ability to migrate.[58] Endohelminths (mainly trematodes) additionally are influenced by temperature, water salinity, pH, oxygen content, water mineral content (hardness), light (linked with depth/water pressure), UVRs, and desiccation.[59]

3.5.3 Infections of Humans

The impact of climatic conditions on human pathogens depends on several features: free persistence outside the host (parasites), transmission via a biological vector or a non-biological physical vehicle (water, soil, etc.) and, commonly, involvement of natural reservoirs (mice, rodents, small mammals, deer, birds, fish, zooplankton).[57,60] Based on their natural host and transmissibility, epidemiologists classify infectious diseases into anthroponoses (i.e., human to animal) or zoonoses (animal to human). In some cases, a pathogen can spread outside its normal zoonotic cycle and affect humans, e.g., Rift Valley fever, when floods trigger *Culex* mosquitoes to feed on both infected ungulate hosts and then on humans.

Climatic factors act on different targets: the life cycle of the pathogen, its hosts and reservoirs. For example, the rates of replication, development, transmission, and mortality of a pathogen and many vectors depend on temperature and humidity. Both pathogens (protozoa, bacteria, viruses, etc.) and their associated vector organisms (snails, mosquitoes, ticks, sandflies, fleas and flies, etc.) live in a limited range of climatic conditions, the "climate envelope", which allows continuation of their natural life cycle. Besides temperature, climatic and geographic factors, important modulating factors include: precipitation, sea level elevation, wind and duration of sunlight, estuary flow, and water salinity in marshes. Patz[60] reviewed extensively the impact of these different factors on the life cycles of pathogens. An increasing concern addresses the ecosystems where the pathogens (Ecosystem Screening Approach), vectors and reservoirs reside. Ecophysiological approaches must consider multiple host species and parasite developmental stages that have a different sensitivity to climatic factors.[57]

3.5.4 Seasonality Under the Tropics

The seasonality of infections, first described by Hippocrates in *Air*, is well known.[60] However, the topic is still debated and mysterious, mainly because of its complexity (climatic factors but also runoff, increased snowmelt, floods).[61–63] Complexity is related to the numerous interacting factors and their variability and strength in the environment in which they occur. Different clusters of seasonal determinants have been studied: human activities; host susceptibility (related to the endocrine-immune systems variability); vitamin D levels; melatonin; epithelia's mucosal integrity and pathogen survival and transmissibility (directly under the influence of climatic factors that define seasons, temperature, humidity, and precipitation) depending of course on their type of environment (e.g., sewage, aerosol, droplets, etc.), changes in vector abundance, and natural reservoirs.[61,62,64] Seasonal patterns in parasitism have been shown: the long dry season may limit development and survival of parasite stages in the environment and, as a result, host contact and parasite transmission. Seasonal increased disease risk with increasing temperature and/or rainfall occurs for Japanese encephalitis, Chikungunya, West Nile viruses, Rift Valley fever, yellow fever and other mosquito-borne viral diseases. The ENSO is responsible for fluctuations of numerous arthropod-borne viral diseases, including Rift Valley fever, Ross River virus infection and dengue (Fisman, 2012). The Noumea study (New Caledonia) showed that "the epidemic dynamics of dengue were essentially driven by climate during the last forty years. Specific conditions based on maximal temperature and relative humidity thresholds were determinant in outbreaks occurrence". As dengue fever affects at least 500,000 patients leading to 25,000 deaths yearly, the forecast operational model of outbreaks would be helpful in term of Public Health.[65]

3.5.5 The Tropical Origin of Infectious Diseases

In the Tropics, seasonality is mainly responsible for population dynamics of vectors, hosts and sometimes, pathogenic agents. For example, malaria, even though it reached its maximum extent in the world during the so-called "Little Ice Age", is mainly transmitted in tropical areas and during the rainy season. These climatic conditions allow us to distinguish two major areas of transmission: stable and unstable. Stable malaria transmission areas are located in equatorial and rainy regions of the tropics while unstable transmission areas are located in dry parts of the tropics.[66] These degrees of stability are determined following the duration of the transmission season and have a crucial importance in degrees of disease severity. In the stable areas, immunological mechanisms will, after some time, lead to a protection due to permanent exposure whereas seasonal transmission will lead each year to loss of immunity between rainy seasons and most frequent severe forms of the disease.

On the contrary, very wet conditions and high clay soils can be unfavorable environmental conditions for the larva of the Tsetse fly because they permit development of their predators, such as termites (clay soils) or fungi (wet conditions).[67]

Seasonality can also be the reflection of human activities associated with climate characteristics. In the Casamance region in southern Senegal, agriculture uses excessive phytosanitary products because of exposure to numerous pests. In 2005, the Bignona Department was confronted with an alarming digestive and neurological syndrome characterized by high lethality (58%). The syndrome affected three villages, especially in the rainy season. The study showed a link between the syndrome and the use of pesticides or the presence of stocks treated by these products in the houses.[68]

Another example of the seasonal interplay between human activity and an environmental agent is the "acute encephalitis syndrome" that seasonally affects children in northern regions of Bangladesh, India, and Vietnam in communities where lychee fruit is cultivated. While this often-fatal disease has been attributed to an unknown virus, pesticide, fruit coloring, or heat stroke, the actual cause is most probably an amino acid in lychee fruit pulp that blocks gluconeogenesis in the consumer. Hungry, poorly nourished children who eat the more toxic unripe lychee fruit develop an acute illness (seizures, coma, death) resulting from severe hypoglycemia that is readily treatable with dextrose infusion. A similar acute hypoglycemic encephalopathy occurs in West African and Jamaican children who consume unripe ackee fruit (*Blighia sapida*), which contains hypoglycin A, the higher homolog of the hypoglycemic agent α-(methylenecyclopropyl) glycine in *Litchi sinensis*.[69,70]

3.5.6 Fungal Proliferation Under the Tropics

Climatic conditions also affect the concentration of mycotoxins, including neurotoxic mycotoxins, in food.

Warmer weather, heat waves, increased precipitation and drought will have various region-dependent impacts on mycotoxin production. Overall mycotoxin production will vary with temperature, rainfall and crop production. In China, there is a differential seasonal incidence of moldy sugarcane poisoning, which results in an acute encephalopathy, with putaminal necrosis and persistent dystonia in survivors.[71]

Plants are susceptible to infection by fungi that produce secondary metabolites (mycotoxins) with neurotoxic potential in humans and the animals on which they depend for food and transport. Climatic conditions modulate fungal infection and growth both before and after harvest, and during storage and transport. Again, the same organism may flourish in both temperate and tropical climes, but exposure may be controlled by human intervention in the former but not the latter. For example, the ergot-generating fungus *Claviceps purpurae*, food contamination of which can cause extremity gangrene, dystonic and convulsive disorders, was once widespread in Europe and Russia (last major outbreak in 1938) but is now reported only in Ethiopia. Other mycotoxins with immunosuppressive potential, such as aflatoxin, are generally found in higher concentrations in food materials grown in tropical climes. Outbreaks of food-associated aflatoxicosis resulting in acute hepatic encephalopathy have occurred in Malaysia.[72]

Whether and how exposure to these mycotoxins increases risk for other infectious diseases has not been addressed. A recent environmental, nutritional and case control study of Nodding Syndrome, a pediatric epileptic disorder associated with infection with the nematode *Onchocerca volvulus*, found an association with prior measles infection and food dependency on moldy maize prior to onset of head nodding. The authors suggested the nematode infection was secondary to measles immunosuppression, while Nodding Syndrome may be a post-measles disorder akin to subacute sclerosing panencephalitis in which central nervous system measles virus activation is triggered by heavy dietary exposure to immunosuppressive mycotoxins.[73]

3.5.7 Algal Blooms and Seafood Poisoning

Harmful algal blooms (HABs) are increasing in frequency and intensity worldwide, perhaps in association with climate change. Toxic blue—green algae flourish in warm waters, increase their growth rate with higher levels of carbon dioxide, and thrive on nutrients in water run-off associated with heavy rainfalls.[74] Cyanobacteria, which are found in fresh, estuarine and marine waters, are composed of a variety of non-toxic and toxic strains, the latter releasing a range of toxic and neurotoxic substances into water, including microcystins, cylindrospermopsins, anatoxins and saxitoxins, the latter being responsible for Paralytic Shellfish Poisoning. Other algal neurotoxins known to cause human neurological illness include domoic acid, maitotoxin and ciguatoxin. Ciguatoxin, which accumulates to toxic levels in large predatory reef fish, is one of the most common causes of human neurotoxic illness (ciguatera food poisoning) worldwide. Additionally, most genera of cyanobacteria elaborate beta-N-methylamino-L-alanine (L-BMAA), a low-potency neurotoxin of interest in relation to the presence of the amino acid in cycads linked to western Pacific amyotrophic lateral sclerosis and Parkinsonism—dementia complex.

3.6 WEATHER EVENTS AND EXTREME EFFECTS OF CLIMATE

Extreme weather events, including heat waves, droughts and floods, dust winds and tropical cyclones, have occurred since time immemorial, even if climate change is increasing their occurrence. These different events can be mild or catastrophic depending on the climatic zones and the biomes involved.

In the tropics, ocean-atmospheric phenomena (sea surface temperature anomalies associated with large-scale air circulation changes) are involved in many extreme weather events. The Pacific Ocean with its dominant mode of climate variability, the El Niño-Southern Oscillation (ENSO), increases the risk of droughts, floods and tropical cyclones and infection risk from certain vector- and rodent-borne diseases (e.g., malaria and dengue).[75] In Australia, the "Big Dry" (2003-2012) was related to the Indian Ocean Dipole (IOD), the mode of variability in the tropical Indian Ocean; it is also involved in rainfalls in countries surrounding the tropical Indian Ocean.[76] Disasters can affect the biomes and destroy ecosystems (e.g., the 1997—98 ENSO-related drought in Sabah, Borneo).[77] The material consequences from fires, dwelling collapse, landslide, and mud torrent are supplemented by human displacement, traumatic injury, drowning, food shortages and famine, poisoning of water supplies and food, chemical dissemination, waterborne diseases[78] and increased infection.[79] The impact of these events can be evaluated by different indices, such as the Climate Change Vulnerability Index (CCVI)[80] and the Germanwatch Global Climate Risk Index (CRI).[81] These also take into account socioeconomic factors, such as poor development, poor sanitation, lack of preparedness, etc. The health issues are in the neurological field and include: injuries and head trauma, anxiety, and emotional stress (especially in the elderly), and post-traumatic stress disorder.[82,83]

Heat waves will become more frequent, more intense, and longer-lasting with the advance of global warming triggered by climate change. Exposure to extreme heat can have lethal consequences ranging from heat rash and heat cramps to heat exhaustion and heat stroke. Those at great risk are the elderly, infants and children, those with chronic illness, and city dwellers who reside in dwellings that lack air conditioning. Body temperature that rise to 105°F or more can result in delirium, convulsions, coma and death. Only scarce epidemiological data are available in tropical countries even if heatwaves occur more and more. During the European heat wave in 2003, an estimated 35,000 people died from stroke, heart attack, lung disease, and other causes exacerbated by heat.[84] In Pakistan in 2005, temperatures as high as 49°C (120°F) caused the deaths of about 2000 people from dehydration and heat stroke, mostly in Sindh province and its capital city, Karachi. The heat wave also claimed the lives of zoo animals and countless agricultural livestock. The event followed a separate heat wave in neighboring India that killed 2500 people in May 2015.[85]

Desert winds aerosolize tons of soil-derived dust, notably in the tropical deserts of Africa and Asia. At a regional level, sand winds are responsible for the high concentrations of particulate matter (PM) in deserts. Desert dust is seasonally or constantly injected into the atmosphere and winds facilitate its transoceanic and transcontinental dispersal. The estimated annual quantity of desert dust subject to regional or global airborne migration is 0.5–5.0 billion tons. As dust and airborne PM serve as the vehicles for pathogens, they pose environmental challenges that extend well beyond tropical zones.[86,87] Dust clouds allow pathogens to be disseminated beyond their usual geographical range and thereby pose a wider risk to human health.[88] Bacterial abundance is correlated with the dust events, with >10-fold higher concentrations on severe dust days. Airborne microorganisms impact indigenous microbial communities.[89]

3.7 A HEALTH GEOGRAPHICAL APPROACH: CASE STUDIES OF CLIMATIC FEATURES THAT MODULATE TROPICAL DISEASES

3.7.1 Mainly Climatic Determinants

3.7.1.1 Epidemic Meningococcal Meningitis

Every year, West African countries within the Sahelo-Sudanian band are afflicted with major meningococcal meningitis (MCM) disease outbreaks, which affect up to 200,000 people, mainly young children, in one of the world's poorest regions. The timing of the epidemic year, the dry months of February to May and the spatial distribution of disease cases throughout the "Meningitis Belt", strongly indicate a close linkage between the life cycle of the causative MCM agent (the throat bacterium *Neisseria meningitides*) and climate variability. While dry throats are speculated to increase risk for bacterial infection and transmission, the mechanisms responsible for the observed patterns are still not clearly identified.[90] The strong seasonality could be due to changes in temperature, humidity, and dust. The amount of dust is particularly high in this part of the world thanks to the Harmattan, a strong wind that comes in from the northeast. The Harmattan picks up dust as it blows over desert regions like the Bodélé Depression, a dried-up lakebed in central Chad that is the largest dust source on Earth. The resulting dust storms are so thick that they can block out sunlight for several days.[91]

In Mali, an approach based on the construction of an index reflecting the movement of the lower layers of the atmosphere, has illustrated the temporal correspondence between the evolution of the Harmattan and the occurrence of cases of MCM.[90]

However, a study conducted at the local level in west central Senegal, which compared 6 years (3 epidemic and 3 non-epidemic years) showed that the occurrence of sandstorms, haze, and blowing sands appeared not to correlate with the emergence of epidemic episodes. Only a decrease of the relative humidity (<30%) during a continuous period over 10 days distinguished MCM-epidemic and non-epidemic years significantly.[92] The humidity decrease seems to be necessary but certainly not sufficient to trigger epidemic MCM, as two non-epidemic years include a succession of several days with a relative humidity below 30%. The relatively short duration of the study, coupled with the failure to take account of the immunological status of the population, force us to remain cautious about the findings. However, it appears that the southward spread of desert should amplify the occurrence of phenomena associated with the Harmattan; a dynamics survey of MCM over long period is needed.

3.7.2 Complex Relations Between Climate, Soil, and Human Activities

3.7.2.1 Climate Impact on Tropical Endemic Diseases Vectors Ecology: Case Studies of Malaria and Human African Trypanosomiasis in Africa

Human African trypanosomiasis (HAT) or sleeping sickness is caused by a flagellated protozoan parasite of the genus *Trypanosoma*: the species *T. gambiense* occurs in West Africa and *T. rhodesiense* in East Africa. Transmission to humans (and animals) is due to a vector, a fly of the genus *Glossina*. The Sudan-Guinea equatorial climate zone favors the ecosystem where this vector lives and acts. Endemic HAT has undergone a remarkable evolution during the 20th century:s its historical spots were drastically reduced due to a combination of colonial health policies, population growth and pressure on forest areas and urbanization effects. However, sleeping sickness remains at a low level, especially in the Guinea mangrove coast and western Ivory Coast. Moreover, HAT has taken advantage of the multiple conflicts and political instability that occur in the continent to re-emerge, for example in the Democratic Republic of Congo, Angola, Uganda, Central African Republic, Cameroon, Sudan, or Ivory Coast.[93,94]

The climate change appears here more or less restrictive in contrast to anthropogenic space transformation and territories health management, which appear to be crucial determinants. This can also be observed with malaria, whose present distribution area (characterized by stability or endemic and instability or epidemic zones) might spread northward, following the hypothesis of nowadays.

Malaria and its often deadly meningeal syndromes are mainly due to *Plasmodium falciparum*, which is mostly prevalent in Africa (between the Sahelian and the equatorial zones) but also affects Southeast Asia and Latin America. However, historically, this disease was not only rampant in the Mediterranean but also spread over North Europe and North America.[95] The maximum extension of this disease northwards happened at the end of the Little Ice Age for example, indigenous cases were diagnosed in Canada or Scandinavia during the 19th century! The decline and disappearance of malaria arose from environmental sanitation and the widespread use of quinine curative treatment. Improvement in health systems delivered the coup de grâce, thereby allowing eradication in malaria areas of Europe and North America. The present global warming could lead to an extension of the malaria-endemic area to the countries of the North only if health systems collapse. Of course, individual sporadic cases might still occur and be detected; thus, the major risk determinant in the countries of the North is not the climate but rather political, economic, and social pressures. The main malarial risk linked with climate change concerns the endemic areas of malaria in the countries of the South by extending transmission seasons and modifying altitudinal limits of disease extension.[96]

3.7.2.2 *The Amazon Basin, Soil Chemical Composition, Soil—Air Interface, Climate Dependency, and Environmental (Surface Water) Contamination by Mercury*

Human poisoning by methylmercury (MetHg), the highly bioavailable form of mercury that readily crosses the blood—brain and placental barriers, is characterized by neurological signs in adults and neurodevelopmental effects in children. Thus, the populations at risk are mainly young children and pregnant women.[97]

Although gold mining and panning are often involved in the mercury pollution of rivers (because of the use of this metal to amalgamate gold), it is not the only cause of pollution. A significant part comes from natural causes (earth's crust degassing, erosion and landslip) or other human activities (stubble-burning, farming on steep slopes) contributing to pollute the environment. Natural mercury sources provide for 60% of the mercury carried by the Amazon basin's rivers in Bolivia[98]. These processes are likely to lead to an environmental contamination only in mercury-containing soil areas. However, it is this pollution process that is the most efficient in the rainy tropics and equatorial areas where chemical processes producing oxysoils, characterized by high mercury content[99-101] can occur. But not all exposed populations in affected areas are contaminated in the same way. Studies have shown there is a differential exposure in human communities linked with the availability of alternative resources. Thus, access to commercial activities allowed the diversification of their food supply and their food consumption modification. Two activities have been identified: selling of their own agricultural products in surrounding town markets and illegal extraction and trade of wood in remote areas far from regulatory centers. However, the main food of recently settled Indian populations remains potentially mercury-contaminated fish. Thus, on a small scale, a theoretical equal exposure to mercury leads to variations based on human activities, social behavior, commercial links, and ethnicity.[102]

In French Guiana, where exposure to mercury is mainly due to gold extraction, a study showed concordant results with the importance of diet and food consumption. The factors contributing the most to explaining the high level of contamination of a population of 500 individuals consulting in 13 health centers were linked to consumption of fresh fish and livers from game, more then residence in the proximity of a gold-mining community.[103] But each rainy season may increase pollution of the lowerlands and the rivers because of the process of deforestation followed by land abandonment contributing to increasing human population exposure.[104]

3.8 THE "BOOMERANG EFFECT" UNDER THE TROPICS: HEALTH CONSEQUENCES OF ANTHROPOGENIC CHANGES

Human activity has imposed many changes on planet Earth. Many ecosystems have been modified, for example by the introduction of agriculture, industrialization, and the creation of cities. While these anthropogenic activities have dramatically supported population growth, they have also had adverse environmental effects on human health. Here we examine some of these "boomerang effects" and their causes.

3.8.1 The Bangladesh Story: Persistent Dryness, Safe Drinking Water Availability and Deep-Water Wells

Perhaps the greatest manmade environmental disaster is traceable to the goal of reducing outbreaks of infectious diarrhea among impoverished Bangladeshi people dependent on bacteria-laden well water. In the 1980s and 1990s, a third of the country's wells were replaced by tube wells drawing on deep aquifers that yielded water contaminated with arsenic.[105,106] Half of the new wells had an arsenic concentration of 50 mg/L, far above the World Health Organization (WHO) acceptable standard (<10 mg/L). This exposed 35−77 million people to daily doses of arsenic, a proven cause of skin lesions, cancers and neurological effects. While the goal of preventing waterborne infectious disease was laudable, the resulting chronic waterborne intoxication illustrates why diverse expertise must be considered before such far-reaching decisions are made. Unfortunately, tube-wells have sometimes failed to protect against gastrointestinal diseases in Bangladesh, despite regular use of tube-well water for drinking.[107]

3.8.2 Agriculture Under the Tropics

Climate has an important influence on soil characteristics and vegetation and thus determines an area's suitability for agriculture, farming practice, and the type of crops that can be grown. Climates affect agriculture in four different ways: by solar radiation, temperature, precipitation and wind. Indirectly, climate influences agriculture by its effects on soil formation. The diverse types of tropical climate correspond with great variation to the agricultural potential of different parts of the tropics, named agro-ecological zones. "Three worlds of the tropics" can be distinguished in Africa: the humid tropics or rainforest zone, the subhumid tropics or savannah zone and the semiarid or Sahel zone, which extends into subtropical latitudes.

Even if all the types of tropical agriculture depend on atmospheric conditions, climatic and especially rain variability make tropical agriculture often a risky business, and many famines in the tropics are related to drought. As crops react differently to drought stress, selecting the right crop diminishes the risk of dryland farming. In dry and seasonal dry regions, water shortage and low soil fertility are the most important constraints on crop production.

Global projections describe continued but geographically disproportionate expansion of the human population based on differential regional fertility rates, such that almost 25% of humans will reside in Africa by 2050. By 2025, the Food and Agriculture Organization predicts that 480 million people in Africa could be living in areas with very scarce water, and that as climatic conditions deteriorate, 600,000 km^2 of land currently classed as moderately constrained will become severely limited. Global warming is predicted to have a general negative effect on plant growth due to the damaging effect of high temperatures on plant development, with predictable results for the availability of animal feed and human food. The increasing threat of climatological extremes, including very high temperatures, might lead to catastrophic reduction of crop productivity, widespread famine, and heavy dependence on environmentally tolerant plants such as cassava and grasspea (*vide supra*).[108,109]

3.8.3 Perturbation of the Global Nutrient Cycles

Along with climate change, nutrient enrichment of water bodies, caused by the widespread use and run-off of fertilizers, is one of the most profound changes in the Earth's ecosystems. Acting in synergy with the climate, this excess of nutrients, phosphorus and nitrogen favors the growth and proliferation of many organisms involved in the emergence of human (vector-borne infections) and wildlife diseases (HABs, coral reef diseases).[110]

3.8.4 Climate in Tropical Cities

More than the half of the world's population is now urban and many of the rapidly expanding cities are located in the tropicsnotably sub-Saharan Africa (e.g., Lagos, Kinshasa, Kampala).[111,112] Cities have some particular climatic features due to several factors: the urban solar energy's cycle; the effect of urbanism (heat circulation); evapotranspiration and anthropogenic heat sources. There is a clear temperature difference (up to 10°C) between cities and surrounding rural areas, a gradient termed the urban heat island

(UHI).[113] UHI intensities are generally lower in the tropics as compared to those of comparable cities in temperate zones. There is a double fluctuation in temperature associated with diurnal changes and seasonal variation.[114,115] Cities induce health effects related directly to the temperature (UHI) and to the related production of ozone. The relation between outdoor temperature and mortality risk for many cities is well established.[113]

3.9 CONCLUSIONS

Climate scientists have shown an expansion over the past few decades of the tropics both northward and southward, thus confirming several climate models. While there is uncertainty about the causes and mechanisms, the forecast is a widening of tropical climes in the future, notably related to climate change. This will cause fundamental shifts in ecosystems and in human settlement.[115,116] There is also a huge change in demography (population increase) in many countries, notably in Africa, which is associated with a constant urbanization. Climate change in these conditions could have major negative consequences.

Diverse expertise must be combined to analyze climate trends and thereby predict impacts on biological systems, including human health. There is precedence in veterinary health when, in the 1970s, the science of Ecopathology emerged to address the biological, physical, human, and economic causal elements of disease among livestock. That heuristic concept led to a search for previously unexplored issues, including habitat, hygiene, herd movements, climate and husbandry.[117] Relevant here is the "One Health" approach that seeks to address the emergence of human disease from wildlife and livestock populations in multiple regions of the world. "More than 60% of human infectious diseases are caused by pathogens shared with wild or domestic animals."[118] The "Ecosystem approaches to health" or "Ecohealth" extends "One Health" in an ecological approach to the interactions between ecosystems, society and health of animals and humans. "One World One Health" became a protected trademark in 2009![119] Although there are some nuances between these concepts,[120] their importance is huge, with major practical applications in public health management[121,122] and in facing the health consequences of climate change.[123] Of course, a holistic approach that addresses the whole biological complexity of "Life on Earth" is needed to address the manifold causes of neurological disease in the tropics and beyond.

References

1. Ritter ME. *The physical environment: an introduction to physical geography.* <www.earthonlinemedia.com/ebooks/>; 2006. Accessed 27.07.16.
2. National Weather Service. *Inter-tropical convergence zone.* <www.srh.noaa.gov/jetstream/tropics/itcz.html>. Accessed 31.11.16.
3. The British Geographer. *Tropical revolving storms: Cuba 2008.* <http://thebritishgeographer.weebly.com/the-climate-of-tropical-regions.html>. Accessed 31.11.16.
4. Violle C, Reich PB, Pacala SW, Enquist BJ, Kattgei J. The emergence and promise of functional biogeography. *PNAS.* 2014;111:13690—13696.
5. Osborne PL. *Tropical Ecosystems and Ecological Concepts.* 2nd ed. Cambridge: Cambridge University Press; 2012.
6. Internet Geography. *Tropical rainforest.* <www.geography.learnontheinternet.co.uk/topics/rainforest.html#structure>; 2015. Accessed 27.07.16.
7. Conservatory of Flowers. *Tropical ecosystems,* update 07/2014. <www.conservatoryofflowers.org/sites/default/files/Tropical%20Ecosystem.pdf>; 2014. Accessed 27.07.16.
8. Condit R, Bettina MJ, Engelbrecht BMJ, Delicia Pino D, Pérez R, Turner BL. Species distributions in response to individual soil nutrients and seasonal drought across a community of tropical trees. *PNAS.* 2013;110:5064—5068.
9. FAO. *Arid Zone Forestry: A Guide for Field Technicians.* Rome: FAO; 1989. <www.fao.org/docrep/t0122e/t0122e03.htm>. Accessed 27.07.16
10. Boval M, Dixon RM. The importance of grasslands for animal production and other functions: a review on management and methodological progress in the tropics. *Animals.* 2012;6:748—762.
11. Al-Kalbani MS, Martin F, Price MF, Abahussain A, Ahmed M, O'Higgins T. Vulnerability assessment of environmental and climate change impacts on water resources in Al Jabal Al Akhdar, Sultanate of Oman. *Water.* 2014;6:3118—3135.
12. Gain AK, Giupponi C, Renaud FG. Climate change adaptation and vulnerability assessment of water resources systems in developing countries: a generalized framework and a feasibility study in Bangladesh. *Water.* 2012;4:345—366.
13. Corlett RT. Where are the subtropics? *Biotropica.* 2013;0:1—3, The Association for Tropical Biology and Conservation.
14. University of California Museum of Paleontology. *The marine biome.* <www.ucmp.berkeley.edu/exhibits/biomes/marine.php>. Accessed 27.07.16.
15. UNEP. *Vital Water Graphics: an Overview of the State of the World's Fresh and Marine Waters.* 2nd ed. Nairobi, Kenya: UNEP; 2008. <http://www.unep.org/dewa/vitalwater/>. Accessed 27.07.16.
16. Sale PF, Agardy T, Ainsworth CH, Feist BE, Bell JD, Christie P, et al. Transforming management of tropical coastal seas to cope with challenges of the 21st century. *Mar Pollut Bull.* 2014;85:8—23.
17. Price ARG. The marine food chain in relation to biodiversity. *Scientific World Journal.* 2001;1:579—587.
18. McKinnon AD, Williams A, Young J, Ceccarelli D, Dunstan P, Brewin RJ, et al. Tropical marginal seas: priority regions for managing marine biodiversity and ecosystem function. *Ann Rev Mar Sci.* 2014;6:415—437.
19. Plaisance L, Caley MJ, Brainard RE, Knowlton N. The diversity of coral reefs: what are we missing? *PLoS One.* 2011;6:e25026. Available from: http://dx.doi.org/10.1371/journal.pone.0025026.
20. Miloslavich P, Klein E, Dıaz JM, Hernandez CE, Bigatti G, Campos L. Marine biodiversity in the Atlantic and Pacific coasts of South America: knowledge and gaps. *PLoS One.* 2011;6:e14631. Available from: http://dx.doi.org/10.1371/journal.pone.0014631.

21. Gaston KJ. Global patterns in biodiversity. *Nature*. 2000;405: 220–227.

22. Bawa KS, Kress WJ, Nadkarni NM, Lele S. Beyond paradise—meeting the challenges in tropical biology in the 21st century. *Biotropica*. 2004;36:437–446.

23. Stevens GC. The latitudinal gradient in geographical range: how so many species coexist in the tropics. *Am Nat*. 1989;133:240–256.

24. Erwin DH. Climate as a driver of evolutionary change. *Curr Biol*. 2009;19:R575–R583.

25. Guerrero R. Microbial ecology comes of age. *Int Microbiol*. 2002;5:157–159. Available from: http://dx.doi.org/10.1007/s10123-002-0093-9.

26. Escalante AE, Pajares S. The coming of age of microbial ecology. In: Benítez M, Miramontes O, Valiente-Banuet A, eds. *Frontiers in Ecology, Evolution and Complexity*. Mexico City: CopIt-arXives; 2014:1–12.

27. Allison SD, Martiny JBH. Resistance, resilience, and redundancy in microbial communities. *PNAS*. 2008;105:11512–11519.

28. Womack AM, Bohannan BJM, Green JL. Biodiversity and biogeography of the atmosphere. *Philos Trans R Soc B*. 2010;365: 3645–3653.

29. Knies JL, Kingsolver JG, Burch CL. Hotter is better and broader: thermal sensitivity of fitness in a population of bacteriophages. *Am Nat*. 2009;173:419–430.

30. Frazier MR, Huey RB, Berrigan D. Thermodynamics constrains the evolution of insect population growth rates: "warmer is better". *Am Nat*. 2006;168:512–520.

31. Addo-Bediako A, Chown SL, Gaston KJ. Thermal tolerance, climatic variability and latitude. *Proc R Soc Lond B*. 2000;267:739–745.

32. Ghalambor CK, Huey RB, Martin PR, Tewksbury JJ, Wangy G. Are mountain passes higher in the tropics? Janzen's hypothesis revisited. *Integr Comp Biol*. 2006;46:5–17.

33. Sunday JM, Bates AE, Dulvy NK. Thermal tolerance and the global redistribution of animals. *Nat Clim Change*. 2012;2:686–690.

34. Rozner H. Survival of the flexible. *Nature*. 2013;494:22–23.

35. Bonebrake TC, Deutsch CA. Climate heterogeneity modulates impact of warming on tropical insects. *Ecology*. 2012;93:449–455.

36. Bozinovic F, Gallardo P. The water economy of South American desert rodents: from integrative to molecular physiological ecology. *Comp Biochem Physiol C Toxicol Pharmacol*. 2006;142:163–172.

37. Lüttge U. Ability of crassulacean acid metabolism plants to overcome interacting stresses in tropical environments. *AoB Plants*. 2010;2010:plq005. Available from: http://dx.doi.org/10.1093/aobpla/plq005.

38. Rice PS. Ultra-violet radiation is responsible for the differences in global epidemiology of chickenpox and the evolution of varicella-zoster virus as man migrated out of Africa. *Virol J*. 2011;8:189–195. Available from: http://dx.doi.org/10.1186/1743-422X-8-189.

39. Vaughan G, Rodríguez-Castillo A, Cruz-Rivera MY, Ruiz-Tovar K, Ramírez-González JE, Rivera-Osorio P. Is ultra-violet radiation the main force shaping molecular evolution of varicella-zoster virus? *Virol J*. 2011;8:370–373. Available from: http://dx.doi.org/10.1186/1743-422X-8-370.

40. Korostil IA, Regan DG. Varicella-Zoster virus in Perth, Western Australia: seasonality and reactivation. *PLoS One*. 2016;11:e0151319. Available from: http://dx.doi.org/10.1371/journal.pone.0151319.

41. Jeong C, Di Rienzo A. Adaptations to local environments in modern human populations. *Curr Opin Genet Dev*. 2014;29:1–8.

42. Jablonski NG, Chaplin G. Human skin pigmentation as an adaptation to UV radiation. *PNAS*. 2010;107:8962–8968.

43. Jablonski NG, Chaplin G. Human skin pigmentation, migration and disease susceptibility. *Philos Trans R Soc B*. 2012;367:785–792.

44. Dumas M, Preux PM. Is tropical neurology specific? In: Preux PM, Dumas M, eds. *Neuroepidemiology in Tropical Health*. 1st ed. Amsterdam: Elsevier/Academic Press; 2016:1–10.

45. Greaves M. Was skin cancer a selective force for black pigmentation in early hominin evolution? *Proc Biol Sci*. 2014;281:20132955. Available from: http://dx.doi.org/10.1098/rspb.2013.2955.

46. Ullrich SE. Does exposure to UV radiation induce a shift to a Th-2-like immune reaction? *Photochem Photobiol*. 1996;64:254–258.

47. Gautam P, Chaurasia A, Bhattacharya A, Grover R, Indian Genome Variation Consortium R, Mukerji M. Population diversity and adaptive evolution in keratinization genes: impact of environment in shaping skin phenotypes. *Mol Biol Evol*. 2014;32: 555–573.

48. Spencer PS, Ludolph AC, Kisby GE. Neurologic diseases associated with use of plant components with toxic potential. *Environ Res*. 1993;62:106–113.

49. Tshala-Katumbay DD, Spencer PS. Toxic disorders of the upper motor neuron system. In: Eisen A, Shaw P, eds. *Handbook of Clinical Neurology: Motor Neuron Disorders and Related Diseases*. Vol. 82. Edinburgh: Elsevier; 2007:353–372.

50. Spencer PS. Lathyrism. In: Vinken PJ, Bruyn GW, Klawans HL, eds. *Handbook of Clinical Neurology*, Part 2. Vol. 21. Amsterdam: Elsevier Science Publishers; 1995:1–20.

51. Reis J, Spencer PS. Lathyrism. In: Chopra J, Sawhney MS, eds. *Neurology in Tropics*. 2nd ed. New Delhi: Elsevier India; 2015: 369–378.

52. Spencer PS, Gardner E, Palmer VS, Kisby GE. Environmental neurotoxins linked to a prototypical neurodegenerative disease. In: Aschner M, Costa L, eds. *Environmental Factors in Neurodevelopment and Neurodegenerative Disorders*. New York: Elsevier; 2015:212–237.

53. Podgorná E, Issa Diallo I, Vangenot C, Sanchez-Mazas A, Sabbagh A, Černý V. Variation in NAT2 acetylation phenotypes is associated with differences in food producing subsistence modes and ecoregions in Africa. *BMC Evol Biol*. 2015;15:263. Available from: http://dx.doi.org/10.1186/s12862-015-0543-6.

54. Galiana E, Marais A, Mura C, Industri B, Arbiol G, Ponchet M. Ecosystem screening approach for pathogen-associated microorganisms affecting host disease. *Appl Environ Microbiol*. 2011;77: 6069–6075.

55. Guernier V, Guégan JF. May Rapoport's rule apply to human associated pathogens? *Ecohealth*. 2009;6:509–521. Available from: http://dx.doi.org/10.1007/s10393-010-0290-5.

56. Kamiya T, O'Dwyer K, Nakagawa S, Poulin R. What determines species richness of parasitic organisms? A meta-analysis across animal, plant and fungal hosts. *Biol Rev Camb Philos Soc*. 2014;89:123–134. Available from: http://dx.doi.org/10.1111/brv.12046.

57. Altizer S, Ostfeld RS, Johnson PTJ, Kutz S, Harvell CD. Climate change and infectious diseases: from evidence to a predictive framework. *Science*. 2013;341:514–519. Available from: http://dx.doi.org/10.1126/science.1239401.

58. Stromberg BE. Environmental factors influencing transmission. *Vet Parasitol*. 1997;72:247–264.

59. Pietrock M, Marcoglies DJ. Free-living endohelminth stages: at the mercy of environmental conditions. *Trends Parasitol*. 2003;19: 293–299.

60. Patz JA, Githeko AK, McCarty JP, Hussein S, Confalonieri U, deWet N. Climate change and infectious diseases. In: McMichael AJ, Campbell-Lendrum DH, Corvalan CF, Ebi KL, Scheraga JD, Woodward A, eds. *Climate Change and Human Health: Risks and Responses*. Geneva, Switzerland: WHO; 2003:103–132.

61. Grassly NC, Fraser C. Seasonal infectious disease epidemiology. *Proc Biol Sci*. 2006;273:2541–2550. Available from: http://dx.doi.org/10.1098/rspb.2006.3604.

62. Naumova EN. Mystery of seasonality: getting the rhythm of nature. *J Public Health Policy*. 2006;27:2–12.

63. Fisman D. Seasonality of viral infections: mechanisms and unknowns. *Clin Microbiol Infect.* 2012;18:946–954.

64. Fares A. Factors influencing the seasonal patterns of infectious diseases. *Int J Prev Med.* 2013;4:128–132.

65. Descloux E, Mangeas M, Menkes CE, Lengaigne M, Leroy A, Tehei T. Climate-based models for understanding and forecasting Dengue epidemics. *PLoS Negl Trop Dis.* 2012;6:e1470. Available from: http://dx.doi.org/10.1371/journal.pntd.0001470.

66. Mouchet J, Carnevale P, Coosemans M, Julvez J, Manuin S, Richard-Lenoble D. *Biodiversité du Paludisme Dans le Monde.* Montrouge, France: John Libbey Eurotext; 2004.

67. Handschumacher P, Schwartz D. Do pedo-epidemiological system exists? In: Landa ER, Feller C, eds. *Soil and Culture.* New York: Springer, IRD; 2010:355–368.

68. Touré K, Coly M, Toure D, Fall M, Sarr MD, Diouf A. Investigation of death cases by pesticides poisonning in a rural community, Bignona, Senegal. *Epidemiol.* 2011;1:105. doi:10.4172/2161-1165.1000105. <http://www.omicsonline.org/investigation-of-death-cases-by-pesticides-poisonning-in-a-rural-community-bignona-senegal-2161-1165.1000105.php?aid=2193>. Accessed 06.11.16.

69. Anonymous. Toxic hypoglycemic syndrome – Jamaica, 1989–1991. *MMWR.* 1992;41:53–55.

70. Spencer PS, Palmer VS, Mazumder R. Probable toxic cause for suspected lychee-linked viral encephalitis [letter]. *Emerg Infect Dis.* 2015;21:904–905. Available from: http://dx.doi.org/10.3201/eid2105.141650.

71. Ming L. Moldy sugarcane poisoning—a case report with a brief review. *J Toxicol Clin Toxicol.* 1995;33:363–367.

72. Lye MS, Ghazali AA, Mohan J, Alwin N, Nair RC. An outbreak of acute hepatic encephalopathy due to severe aflatoxicosis in Malaysia. *Am J Trop Med Hyg.* 1995;53:68–72.

73. Spencer PS, Mazumder R, Palmer VS, Lasarev MR, Stadnik RC, King P. Environmental, dietary and case-control study of Nodding Syndrome in Uganda: a post-measles brain disorder triggered by malnutrition? *J Neurol Sci.* 2016;369:191–203. Available from: http://dx.doi.org/10.1016/j.jns.2016.08.023.

74. EPA USA. *Nutrient pollution.* <www.epa.gov/nutrientpollution/climate-change-and-harmful-algal-blooms>. Accessed 06.11.16.

75. Kovats RS. El Nino and human health. *Bull World Health Organ.* 2000;9:1127–1135.

76. Ummenhofer CC, England MH, McIntosh PC, Meyers GA, Pook MJ, Risbey JS. What causes southeast Australia's worst droughts? *Geophys Res Lett.* 2009;36:L04706. Available from: http://dx.doi.org/10.1029/2008GL036801.

77. Walsh RP, Newbery DM. The ecoclimatology of Danum, Sabah, in the context of the world's rainforest regions, with particular reference to dry periods and their impact. *Philos Trans R Soc Lond B Biol Sci.* 1999;354:1869–1883.

78. Davies GI, McIver L, Kim Y, Hashizume M, Iddings S, Chan V. Water-borne diseases and extreme weather events in Cambodia: review of impacts and implications of climate change. *Int J Environ Res Public Health.* 2015;12:191–213. Available from: http://dx.doi.org/10.3390/ijerph120100191.

79. Doocy S, Dick A, Daniels A, Kirsch TD. The human impact of tropical cyclones: a historical review of events 1980–2009 and systematic literature review. *PLoS Curr.* 2013;16:5. Available from: http://dx.doi.org/10.1371/currents.dis.2664354a5571512063ed29d25ffbce74.

80. Verisk Maplecroft. *Climate Change Vulnerability Index (CCVI).* <https://maplecroft.com/about/news/ccvi.html>. Accessed 30.07.16.

81. Kreft S, Eckstein D, Dorsch L, Fischer L. *Global Climate Risk Index 2016 Who Suffers Most From Extreme Weather Events? Weather-Related Loss Events in 2014 and 1995 to 2014.* Bonn, Germany: Germanwatch; 2015. <www.germanwatch.org/en/cri>. Accessed 27.07.16

82. Stanke C, Kerac M, Prudhomme C, Medlock J, Murray V. Health effects of drought: a systematic review of the evidence. *PLoS Curr.* 2013;5:pii. Available from: http://dx.doi.org/10.1371/currents.dis.7a2cee9e980f91ad7697b570bcc4b004.

83. Ahern M, Kovats RS, Wilkinson P, Few R, Matthies F. Global health impacts of floods: epidemiologic evidence. *Epidemiol Rev.* 2005;1:36–46. Available from: http://dx.doi.org/10.1093/epirev/mxi004.

84. McGregor GR, Bessemoulin P, Ebi K, Menne B. *Heatwaves and Health: Guidance on Warning-System Development, WMO-No. 1142.* Geneva, Switzerland: World Meteorological Organization and World Health Organization; 2015.

85. Wikipedia. *2015 Pakistan heat wave.* <https://en.wikipedia.org/wiki/2015_Pakistan_heat_wave>. Accessed 06.11.16.

86. Díaz J, Tobías A, Linares C. Saharan dust and association between particulate matter and case-specific mortality: a case-crossover analysis in Madrid (Spain). *Environ Health.* 2012;11:11. Available from: http://dx.doi.org/10.1186/1476-069X-11-11.

87. Goudie AS. Desert dust and human health disorders. *Environ Int.* 2014;63:101–113. Available from: http://dx.doi.org/10.1016/j.envint.2013.10.011.

88. Griffith DW. Atmospheric movement of microorganisms in clouds of desert dust and implications for human health. *Clin Microbiol Rev.* 2007;20:459–477. Available from: http://dx.doi.org/10.1128/CMR.00039-06.

89. Yamaguchi N, Park J, Kodama M, Ichijo T, Baba T, Nasu M. Changes in the airborne bacterial community in outdoor environments following Asian dust events. *Microbes Environ.* 2014;29:82–88. Available from: http://dx.doi.org/10.1264/jsme2.ME13080.

90. Sultan B, Labadi K, Guegan JF, Janicot S. Climate drives the meningitis epidemics onset in West Africa. *PLoS Med.* 2005;2:e6.

91. Shirber M. *Climate conditions help forecast meningitis outbreaks;* March 24, 2014. <http://climate.nasa.gov/news/1054/climate-conditions-help-forecast-meningitis-outbreaks/>.

92. Mbaye I, Handschumacher P, Chippaux J-P, Diallo A, Ndione JA, Paul P. Influence du climat sur les épidémies de méningite à méningocoque à Niakhar (Sénégal) de 1998 à 2000 et recherche d'indicateurs opérationnels en santé publique. *Environ Risques Santé.* 2004;3:219–226.

93. Berrang-Ford L, Jamie L, Breau S. Conflict and human African trypanosomyasis. *Soc Sci Med.* 2011;72:398–407.

94. Courtin F, Jamonneau V, Duvallet G, Camara M, Kaba D, Solano P. Un siècle de "trypano" en Afrique de l'Ouest. Communication affichée lors des journées du centenaire de la SPE. *Bull Soc Pathol Exot.* 2008;101:287–289.

95. Hay SI, Guerra CA, Tatem AJ, Noor AM, Snow RW. The global distribution and population at risk of malaria: past, present and future. *Lancet Infect Dis.* 2004;4:327–336. doi:10.1016/S1473-3099(04)01043-6.

96. Yamana TK, Bomblies A, Eltahir EAB. Climate change unlikely to increase malaria burden in West Africa. *Nat Clim Change.* 2016;6:1009–1015.

97. Marques RC, Bernardi JVE, Dórea JG, Leão RS, Malm O. Mercury transfer during pregnancy and breastfeeding: hair mercury concentrations as biomarker. *Biol Trace Elem Res.* 2013;154:326–332.

98. Maurice-Bourgoin L., Aalto R., Guyot JL, 2002, Sediment-associated mercury distribution within a major Amazon tributary: century-scale contamination history and importance of flood plain accumulation, *The Structure, Function and Management Implications of Fluvial Sedimentary Systems* (Proceedings of an international symposium, Alice Springs, Australia, September 2002). IAI IS Publ. no. 276, pp.161-168.

99. Lechler PJ, Miller JR, Lacerda LD, Vinson D, Bonzongo JC, Lyons WB. Elevated mercury concentrations in soils, sediments, water, and fish of the Madeira River basin, Brazilian Amazon: a function of natural enrichments? *Sci Total Environ.* 2000;260:87–96.

100. Roulet M, Lucotte M, Saint-Aubin A, Tran S, Rhéault I, Farella N. The geochemistry of Hg in Central Amazonian soils developed on the Alter-do-Chao formation of the lower Tapajos river valley, Para state, Brazil. *Sci Total Environ.* 1998;223:1–24.

101. Do Valle CM, Santana GP, Augusti R, Egreja Filho FB, Windmöller CC. Speciation and quantification of mercury in Oxisol, Ultisol, and Spodosol from Amazon (Manaus, Brazil). *Chemosphere.* 2005;58:779–792.

102. Tschirhart C, Handschumacher P, Laffly D, Bénéfice E. Resource management, networks and spatial contrasts in human mercury contamination along the Rio Beni (Bolivian Amazon).. *Hum Ecol.* 2012;40:511–523.

103. Cordier S, Grasmick C, Pasquier-Passelaigue M, Mandereau L, Weber JP, Jouan M. Mercury exposure in French Guiana: levels and determinants. *Arch Environ Health.* 1998;53:299–303.

104. Wantzen K, Mol JH. Soil erosion from agriculture and mining: a threat to tropical stream ecosystem. *Agriculture.* 2013;3:660–683.

105. Yu WH, Harvey CM, Harvey CF. Arsenic in groundwater in Bangladesh: a geostatistical and epidemiological framework for evaluating health effects and potential remedies. *Water Resour Res.* 2003;39:1146. Available from: http://dx.doi.org/10.1029/2002WR001327.

106. Chakraborti D, Rahman MM, Mukherjee A, Alauddind M, Hassane M, Dutta RH. Groundwater arsenic contamination in Bangladesh—21 Years of research. *J Trace Elem Med Biol.* 2015;31:237–248.

107. Islam MS, Siddika A, Khan MN, Goldar MM, Sadique MA, Kabir ANMH. Microbiological analysis of tube-well water in a rural area of Bangladesh. *Appl Environ Microbiol.* 2001;67:3328–3330.

108. Uni-Goettingen. *Tropical Agroecosystem Function SS08 U2&U11—climatic factors & permanent cropping with annuals;* 4–23. <www.uni-goettingen.de/de/363593.html>. Accessed 27.07.16.

109. Ramirez-Villegas J, Thornton PK. *Climate change impacts on African crop production.* CCAFS working paper no. 119. Copenhagen, Denmark: Research Program on Climate Change, Agriculture and Food Security. <www.ccafs.cgiar.org>; 2015. Accessed 27.07.16.

110. Johnson PTJ, Townsend AR, Cleveland CC, Glibert PM, Howarth RW, McKenzie VJ, et al. Linking environmental nutrient enrichment and disease emergence in humans and wildlife. *Ecol Appl.* 2010;20:16–29.

111. Butsch C, Sakdapolrak P, Saravanan VS. Urban health in India. *Int Asienforum.* 2012;43:13–32.

112. Roth M. Review of urban climate research in (sub) tropical regions. *Int J Climatol.* 2007;27:1859–1873. Available from: http://dx.doi.org/10.1002/joc.1591.

113. Rydin Y, Bleahu A, Davies M, Dávila JD, Friel S, De Grandis G, et al. Shaping cities for health: complexity and the planning of

urban environments in the 21st century. *Lancet.* 2012;379:2079–2108.

114. Reiner Jr RC, Smith DL, Gething PW. Climate change, urbanization and disease: summer in the city. *Trans R Soc Trop Med Hyg.* 2015;109:171–172. Available from: http://dx.doi.org/10.1093/trstmh/tru194.

115. Seidel DJ, Fu Q, Rander WJ, Reichler TJ. Widening of the tropical belt in a changing climate. *Nat Geosci.* 2008;1:21–24. Available from: http://dx.doi.org/10.1038/ngeo.2007.38.

116. Lu J, Deser C, Reichler T. Cause of the widening of the tropical belt since 1958. *Geophys Res Lett.* 2009;36:L03803. Available from: http://dx.doi.org/10.1029/2008GL036076.

117. Anonymous. Ecopathology: the influence of environment on disease. *Spore.* 1994;49:6.

118. Karesh WB, Dobson A, Lloyd-Smith JO, Lubroth J, Dixon MA, Bennett M, et al. Ecology of zoonoses: natural and unnatural histories. *Lancet.* 2012;380:1936–1945.

119. Zinsstag J, Schelling E, Waltner-Toews D, Tanner M. From "one medicine" to "one health" and systemic approaches to health and well-being. *Prev Vet Med.* 2011;101:148–156.

120. Roger F, Caron A, Morand S, Pedrono M, de Garine-Wichatitsky M, Chevalier V, et al. One health and ecohealth: the same wine in different bottles? *Infect Ecol Epidemiol.* 2016;6:1–4. Available from: http://dx.doi.org/10.3402/iee.v6.30978.

121. Asakura T, Mallee H, Tomokawa S, Moji K, Kobayashi J. The ecosystem approach to health is a promising strategy in international development: lessons from Japan and Laos. *Globalization Health.* 2015;11:1–8. Available from: http://dx.doi.org/10.1186/s12992-015-0093-0.

122. Leung Z, Middleton D, Morrison K. One Health and Ecohealth in Ontario: a qualitative study exploring how holistic and integrative approaches are shaping public health practice in Ontario. *Public Health.* 2012;12:358. :<http://www.biomedcentral.com/1471-2458/12/358>

123. Patz JA, Hahn MB. Climate change and human health: a One Health approach. *Curr Top Microbiol Immunol.* 2013;366:141–171. Available from: http://dx.doi.org/10.1007/82_2012_274.

Further Reading

Courtin F, Jamonneau V, Duvallet G, Garcia A, Coulibaly B, Doumenge JP, Cuny G, Solano P. Sleeping sickness in West Africa (1906–2006): changes in spatial repartition and lessons from the past. *Trop Med Int Health.* 2008;13:334–344.

Greaves M. Response to Jablonski and Chaplin. *Proc Biol Sci.* 2014;281:20140940. Available from: http://dx.doi.org/10.1098/rspb.2014.0940.

Mbaye I, Sy MD, Handschumacher P. *Contraintes biologiques, gestion agricole et risque éco-toxicologique dans le district de Bignona (Région de Ziguinchor/Sénégal): Pourquoi une vulnérabilité hétérogène? Natures Tropicales, enjeux et perspectives.* Bordeaux: Presses Universitaires; 2012:57–68.

4

Economic Aspects

Achille Tchalla

University of Limoges, Limoges, France

OUTLINE

4.1 INTRODUCTION

Developing countries and especially Africa are faced with the increased attention to issues of cost and efficiency, prompted by the pervasive scarcity of resources relative to health needs and demands, driven by factors such as the HIV pandemic, aging populations, the development of innovative but often expensive technologies and also by the heightened knowledge and expectations of healthcare consumers. Developing countries are faced with chronic neurological diseases such as epilepsy and cerebral malaria, but must also prepare for the risk of emergence or resurgence of other diseases such as dementia or vectors such as dengue, yellow fever, chikungunya and human African trypanosomiasis with neurological forms. Despite the proven effectiveness of various strategies that are in place, the economic aspects are crucial in the implementation of strategies and especially accessibility to new therapies for a significant reduction in the incidence and severity of neurological pathologies. This accessibility faces economic considerations.

Economic considerations have assumed an increasingly prominent role in the planning, management and evaluation of health systems, ranging from the design of ways to pay providers or to improve access to care for households, to the definition of essential packages for insurance, to decisions about whether or not to include new medicines on hospital, state or national formularies.[1,2] The basic economic aspects of health systems, individual income and representations are linked to neurological diseases such as epilepsy and dementia.

4.2 HEALTH SYSTEM FINANCING: KEY CHALLENGES

4.2.1 The Changing Landscape of Disease and Injury

Globally, great improvements in health have been observed over the last few decades, as evidenced by increased life expectancy and reduced child mortality in the vast majority of countries. Such improvements can be attributed partly to an increase in living standards, improvements in the availability of environmental conditions such as water and sanitation, and more recently, to advances in medical technologies and services.[3,4]

DOI: http://dx.doi.org/10.1016/B978-0-12-804607-4.00004-6

But of late there have also been some notable and significant reverses. For example, average life expectancy in a worrying number of countries in Sub-Saharan Africa (SSA) has actually declined since 1990 while adult male mortality rates have increased in parts of some countries.[3] These developments have different causes, such as economic decline and increases in HIV, alcohol abuse and its associated chronic disease in Africa.[5,6]

However, in all regions of the world, non-communicable diseases are becoming increasingly important, leaving poor countries with double the burden of financing high burdens of infectious diseases while at the same time tackling the escalating toll of cancer, cardiovascular disease, diabetes and road traffic injuries.[7] Now that treatment is substantially prolonging life, the "lifestyle"-related non-communicable diseases, plus HIV, require relatively expensive, long-term care, increasing the need to find more resources for health.

4.2.2 Health Spending: Inefficient or Insufficient?

From an economic perspective, it is always tempting to point out that if resources were more carefully targeted towards the priority of the health needs of the population, more benefit could be obtained from existing levels of expenditure. While the search for such improvements is certainly needed, it is important to distinguish between the relative efficiency with which resources are allocated versus the absolute quantity of resources available in the first place. Of the 192 countries that are members of the World Health Organization (WHO), and for which data are available, 39 spent < US $25 per capita on health in 2004, and 60 spent < US $50.[8] This includes funding from all sources; external and national donors, governments, firms and household out-of-pocket contributions. This level of funding is not sufficient to ensure universal access to even a minimum level of basic health services, estimated to cost somewhere between US $35 and US $50 depending on the country.[9,10] In most countries, health expenditures have been rising more rapidly than national income, to the extent that the world's richest countries are finding that, in spite of cost containment measures, some are now devoting well in excess of 10% of their entire production to pay for health-related services and goods.[8] Tax- or social health insurance-funded systems have been unable to find the funds required to ensure universal access to all the interventions that have the possibility of improving health or extending life, resulting in apparent resource scarcity even in rich countries. The first key issue of health financing policy, therefore, is to raise sufficient funds for health, an issue that is important in all countries but critical in the poorest parts of the world.[11]

4.2.3 Composition of Health Spending

Raising funds is important, but not sufficient. The two other questions relate to how the funds are raised and how they are used. In terms of the former, out-of-pocket payments made by households directly to providers are regressive forms of health financing; they penalize those least able to afford care and ensure that some people who need services do not seek them. Moreover, they lead in many cases to health spending levels that have been labeled as "catastrophic" because they cause households to reallocate their budgets away from other essential needs such as education, food and housing. In some cases, the simple act of seeking care can even push households under the poverty line because of the need to pay for services. To protect people against the financial risks involved in falling ill and seeking care, there is now solid agreement that it is important to increase the extent to which most low- and middle-income countries rely on pre-payment mechanisms to finance health, via insurance, taxation or a mix, rather than relying on user charges.[11] Pooled funds can then be used to allow people to access services when they need them, spreading the financial risks of ill health across the population. How this is achieved when many people work in the informal sector is a major issue capturing the attention of both the countries involved and the international community.[12]

4.3 ECONOMICAL ASPECTS RELATED TO CHRONIC NEUROLOGICAL DISEASES

4.3.1 Epilepsy

Regarding epilepsy for example, in 1990, the WHO determined that on average the cost of the anti-epileptic drug phenobarbitone, which can be used alone to treat seizures in a substantial proportion of epilepsy and is on the WHO list of essential drugs, could be as little as US $5 a year per patient. The report of the World Bank, *Investing in Health*,[2] indicates that in 1990 epilepsy accounted for nearly 1% of the disease burden in the world. Epilepsy commonly affects young people in their most productive years and is often a preventable cause of unemployment.

Epilepsy is a cosmopolitan affection of multifactorial origin, which now represents a major public health problem. Evaluation of this condition from both epidemiologic and economic viewpoints has been the focus of increasing interest in developed countries.[13] In developing countries, it poses a problem in view of its medical, cultural, and economic consequences. However, it remains a condition that may not always be recognized and whose costs are largely unknown.[14] Economic analyses in developing countries are rare. The impact of the disease will depend on its age at onset, its natural history, its prevalence, and the possibility of treatment. On average, the prevalence in Africa is 15/1000. However, the prevalence varies markedly from country to country.[15] In these regions, people with epilepsy are often dependent on their families and sometimes are excluded from the socioeconomic life of the country. This marginalization constitutes an important loss of productivity, as young people make up 70% of cases, and exclusion poses problems for the medical evaluation of these patients. It also hampers economic analysis. The WHO has estimated that 80% of active persons with epilepsy receive no treatment or inadequate treatment.[16]

Traditionally, economic analysis is based on estimation of direct and indirect costs.[17] The absence of a reliable system of medical information complicates the methods of cost estimation in developing countries. These methods are based on two techniques:[18] a top-down technique, which estimates aggregates of nationwide costs, and a bottom-up method, which records a volume of medical goods and services with secondary cost estimation. Depending on the methods of calculation, average costs may vary by a factor of two.[19] However, numerous studies conducted in developed countries have been the subject of a review.[20] Some of these studies are referred,[21,22] but it should be borne in mind that such results cannot be readily compared with those from developing countries, in view of the differences in patient management between countries. However, it emerges from these studies in different countries that the proportion of health expenses devoted to epilepsy ranges from 0.12% to 1.12% of gross domestic product (GDP). Marked differences are found between countries in the direct costs involved, with the percentage of hospital costs ranging from 19% to 82%. Comparisons were made only with respect to the direct costs, as there is so much variation in the calculation of indirect costs. In developing countries, the direct costs of epilepsy have been evaluated in Indonesia, Latin America, and India.[23,24] The direct costs in these countries are approximately one-quarter of the total cost of the pathology. Even in countries with low GDPs, it is worth estimating the impact of this type of pathology to define public health policy and quantify the real needs of the population. Few accurate estimates are available of the costs of epilepsy in Africa. A simple treatment may lead to a considerable reduction in healthcare costs, especially in the

indirect costs of the pathology. We have no explanation for this result. One major hypothesis could be that epilepsy in treated patients was more severe; another one could be that treated patients more often went to hospital, leading to an increase of the number of days of family life disrupted. This last measure may not be a good reflection of indirect costs in treated patients.[25]

In Lao, the availability and quality of drugs are also critical factors for effective management and control of epilepsy. The availability of antiepileptic drugs (AEDs) particularly of phenobarbital, is restricted to higher-category pharmacies and within those it is rather limited. To meet the costs of AEDs in this setting is a major challenge for people with epilepsy. However, the quality of the available phenobarbital is rather satisfactory.[26]

4.3.2 Dementia and Cognitive Disorders

Current data from developing countries suggest that age-adjusted dementia prevalence estimates in 65-year-olds are high (≥5%) in certain Asian and Latin American countries, but consistently low (1%–3%) in India and SSA; Alzheimer's disease accounts for 60% whereas vascular dementia accounts for ~30% of the prevalence.[27] Dementia is associated with increased mortality risk. These results highlight the need for targeted health policies and strategies for dementia care in SSA.[28] The impact of dementia could be understood at three inter-related levels: (1) the person with dementia, who experiences ill health, disability, impaired quality of life and reduced life expectancy; (2) the family and friends of the person with dementia, who, in all world regions, are the cornerstone of the system of care and support; (3) wider society, which, either directly through government expenditure or in other ways, incurs the cost of providing health and social care and the opportunity cost of lost productivity. Other social impacts may be harder to quantify, but no less real.

There is no study evaluating the economic impact of dementia in low- or middle-income countries; one approach to evaluate the economicals aspects may be the global burden approach.

4.3.2.1 The Global Burden of Disease Approach

One approach for assessing the impact of dementia, and comparing it with other health conditions, is to use the Global Burden of Disease (GBD) estimates. The key indicator is the disability adjusted life year (DALY), a composite measure of disease burden calculated as the sum of years lived with disability (YLD) and years of life lost (YLL). Thus, the DALY summarizes the effects of diseases, both on the quantity (premature mortality) and quality of life (disability). These effects are summed across estimated numbers of affected individuals to express the regional and global impact of disease. The effect of living for 1 year with disability depends upon the disability weight attached to the health condition concerned. In a wide international consensus consultation for the GBD report, disability from dementia was accorded a higher disability weight (0.67) than that for almost any other condition, with the exception of severe developmental disorders.[29] This weight signified that each year lived with dementia entails the loss of two-thirds of one DALY.

4.3.2.2 Limitations of the GBD Approach

Concerns have been expressed, in general, regarding the use of GBD estimates to determine allocation of resources. An important critique is that such decisions should be based not on burden alone, but on potential to reduce burden through the scaling up of interventions that are cost-effective.[30,31] A counter argument for conditions such as dementia, where no such interventions yet exist, would be that the size of the burden should be an important factor in determining research spending into new treatments, and that diagnostic and supportive services are required to meet the need arising from the burden.

4.3.2.3 Alternative Approaches to Understanding the Economic Impact of Dementia

The most important critique of the GBD estimates is that these fail, in important ways, to capture the true impact of different chronic diseases upon disability, needs for care, and attendant societal costs.[32] This limitation is most evident for older people, among whom most of these needs arise, and for conditions such as dementia, vision and hearing loss and musculoskeletal disorders, where most of the impact comes from disability rather than associated mortality.[33] This was already a problem for the WHO GBD estimates,[34] and has been greatly exacerbated with the shift to the Institute for Health Metrics and Evaluation (IHME) system with its new weights, and its focus upon "health loss". One approach would be simply to stop using currently formulated DALYs (or the YLD component) when assessing the impact of disabling conditions.[32] The IHME burden of disease estimates are highly discrepant with findings from studies of the directly measured disability, dependence, and cost associated with chronic diseases, which provide a very different picture regarding the societal impact of dementia relative to other non-communicable conditions. The "evidence test" is much more important than the "common sense" test referred to earlier. The relevant evidence has been reviewed previously in the World Alzheimer Report 2009 and the World Alzheimer Report 2013,[35,36] and is updated and summarized briefly here:

1. Dementia and cognitive impairment are by far the leading chronic disease contributors to disability, and, particularly, needs for care (dependence) among older people. While older people can often cope well and remain reasonably independent even with marked physical disability, the onset of cognitive impairment quickly compromises their ability to carry out complex but essential tasks and, later, their basic personal care needs. The need for support from a caregiver often starts early in the dementia journey, intensifies as the illness progresses over time, and continues until death.[36]

2. For low- and middle-income countries, the population-based surveys carried out by the 10/66 Dementia Research Group have shown clearly that disorders of the brain and mind (dementia, stroke and depression) make the largest independent contribution, both in terms of the strength of the association and the "population attributable prevalence fraction" (PAPF), to disability[34] and dependence.[37] Dementia makes the dominant contribution, particularly to needs for care.

3. Dementia is typically associated with a particular intensity of needs for care, exceeding the demands associated with other conditions. For example, caregivers of people with dementia were more likely than caregivers of people with other conditions to be required to provide help with getting in and out of bed (54% vs 42%), dressing (40% vs 31%), toileting (32% vs 26%), bathing (31% vs 23%) managing incontinence (31% vs 16%) and feeding (31% vs 14%).[36] These findings were confirmed in reports from the 10/66 Dementia Research Group; among those needing care, those with dementia stood out as being more disabled, as needing more care (particularly support with core activities of daily living), and as being more likely to have paid caregivers; dementia caregivers also experienced more strain than caregivers of those with other health conditions.[38,39]

4. Another proxy indicator of the relevance of dementia to dependence is the extent to which older people with dementia use different types of care services that reflect the increasing levels of needs for care, and the extent to which they are over-represented among older users of those services.

5. Moving into a care home is generally a marker of particularly high needs for care, although other factors can be involved. But this way is not yet overcome in low- and middle-income countries.

6. Therefore, the current and future costs of long-term care will be driven to a very large extent by the coming epidemic of dementia.[36] Our success in designing and implementing successful strategies for the prevention of dementia,[40] and in identifying treatments that can alter the course of the disease will be important determinants of future health and social care costs, currently rising inexorably in the context of population ageing.

7. The enormous global societal costs of dementia were estimated in the World Alzheimer Report 2010, and these estimates have been updated to 2015. There have been relatively few attempts to compare dementia costs with those of other chronic diseases. Few of these studies take into account comorbidity, and estimate the independent or "attributable" costs of dementia. For all countries other than India, the attributable cost of dementia exceeded that of other conditions (depression, hypertension, diabetes, ischemic heart disease, and stroke). Medical care costs for dementia were negligible, reflecting limited access to services, but dementia costs dominated for social care, informal care, and paid home care.

4.4 CONCLUSION

Economic evaluation is sometimes perceived to be an end in itself rather than a tool that is part of an approach to overall health system financing. Accordingly, a number of central components of health financing are used to consider the role and value of cost effectiveness in the planning, management or monitoring of health systems. The two key roles of health financing systems are to raise sufficient funds for health and to do so in a way that allows people to access services without the risk of financial catastrophe or impoverishment. We have highlighted that the GBD estimates fail to reflect the societal impact of dementia, relative to other chronic diseases and, as such, cannot be considered to be a reliable tool for prioritization for research, prevention, and health or social care among older people.

On a final note, economic evaluation focuses on only one outcome, population health. There are many other outcomes people also care about; inequalities in health outcomes, utilization of services, responsiveness and fairness of financing. Therefore, the results of economic evaluation cannot be used to set priorities by themselves but should be introduced into the policy debate to be considered along with the impact of different policy and intervention mixes on other outcomes.

Further studies are required in Africa to provide more comparative data. The search for simple indicators for the evaluation of indirect costs such as the number of days disrupted or the impact on family life should continue. They are more relevant to the notion of human capital, which is more representative of these low-income populations in which few patients are salaried workers.

References

1. Drummond M, Jonsson B, Rutten F. The role of economic evaluation in the pricing and reimbursement of medicines. *Health Policy*. 1997;40:199−215.
2. World Bank. *World Development Report 1993: Investing in Health*. New York: Oxford University Press; 1993.
3. World Health Organization. *World Health Report 1999: Making a Difference*. Geneva, Switzerland: WHO; 1999.
4. McMichael AJ, McKee M, Shkolnikov V, Valkonen T. Mortality trends and setbacks: global convergence or divergence? *Lancet*. 2004;363:1155−1159.
5. Stover J, Way P. Projecting the impact of AIDS on mortality. *AIDS*. 1998;12(Suppl. 1):S29−S39.
6. Andoh SY, Umezaki M, Nakamura K, et al. Correlation between national income, HIV/AIDS and political status and mortalities in African countries. *Public Health*. 2006;120:624−633.
7. World Health Organization. *Preventing Chronic Diseases: A Vital Investment*. Geneva, Switzerland: WHO; 2006.
8. World Health Organization. *National Health Accounts (NHA) website*. <www.who.int/nha>.
9. World Health Organization. *World Health Report 2000. Health Systems: Improving Performance*. Geneva, Switzerland: WHO; 2000.
10. World Health Organization. *Macroeconomics and health: investing in health for economic development*. Report of the Commission on Macroeconomics and Health. Geneva, Switzerland: WHO; 2001.
11. Chisholm D, Evans DB. Economic evaluation in health: saving money or improving care? *J Med Econ*. 2007;10:325−337.
12. Carrin G, James C. Social health insurance: key factors affecting the transition towards universal coverage. *Int Soc Secur Rev*. 2005;58: 45−64.
13. Begley CE, Annegers JF, Lairson DR, et al. Cost of epilepsy in the United States: a model based on incidence and prognosis. *Epilepsia*. 1994;35:1230−1243.
14. Pachlatko C. The relevance of health economics to epilepsy care. *Epilepsia*. 1999;40:3−7.
15. Preux PM, Tiemagni F, Fodzo L, et al. Antiepileptic therapies in the Mifi Province in Cameroon. *Epilepsia*. 2000;41:432−439.
16. Shorvon SD, Farmer PJ. Epilepsy in developing countries: a review of epidemiological, sociocultural, and treatment aspects. *Epilepsia*. 1988;29:S36−S54.
17. Drummond M, Stoddart G, Torrance G. *Methods for Economic Evaluation of Health Care Programmes*. Oxford: Oxford University Press; 1990.
18. Begley CE, Annegers JF, Lairson DR, et al. Estimating the cost of epilepsy. *Epilepsia*. 1999;40:S8−S13.
19. Halpern M, Rentz A, Murray M. Cost of illness of epilepsy in the US: comparison of patient-based and population-based estimates. *Neuroepidemiology*. 2000;19:87−99.
20. Kotsopoulos IA, Evers SM, Ament AJ, et al. Estimating the costs of epilepsy: an international comparison of epilepsy cost studies. *Epilepsia*. 2001;42:634−640.
21. Murray MI, Halpern MT, Leppik IE. Cost of refractory epilepsy in adults in the USA. *Epilepsy Res*. 1996;23:139−148.

22. Banks G, Regan K, Beran R. The prevalence and directs costs of epilepsy in Australia. In: Beran RG, ed. *Cost of Epilepsy: Proceedings of the 20th International Epilepsy Congress.* Wehr/Baden: Ciba-Geigy Verlag; 1995:39−48.

23. Galdames D, Erazo S. Cost and requirement of new antiepileptic drugs in an urban hospital in Santiago, Chile. *Epilepsia.* 2001;42 (Suppl 2):26.

24. Chandra B. Economic aspects of epilepsy in Indonesia. In: Beran RG, ed. *Cost of Epilepsy: Proceedings of the 20th International Epilepsy Congress.* Wehr/Baden: Ciba-Geigy Verlag; 1995:75−82.

25. Nsengiyumva G, Druet-Cabanac M, Nzisabira L, Preux PM, Vergnenègre A. Economic evaluation of epilepsy in Kiremba (Burundi): a case−control study. *Epilepsia.* 2004;45(6):673−677.

26. Odermatt P, Ly S, Simmala C, et al. Availability and costs of antiepileptic drugs and quality of phenobarbital in Vientiane municipality, Lao PDR. *Neuroepidemiology.* 2007;28(3):169−174.

27. Kalaria RN, Maestre GE, Arizaga R, et al. Alzheimer's disease and vascular dementia in developing countries: prevalence, management, and risk factors. *Lancet Neurol.* 2008;7(9):812−826.

28. Samba H, Guerchet M, Ndamba-Bandzouzi B, et al. Dementia-associated mortality and its predictors among older adults in sub-Saharan Africa: results from a 2-year follow-up in Congo (the EPIDEMCA-FU study). *Age Ageing.* 2016;45(5):681−687.

29. World Health Organization. *Global Burden of Disease 2004 Update: Disability Weights for Diseases and Conditions.* Geneva, Switzerland: WHO; 2004.

30. Williams A. Calculating the global burden of disease: time for a strategic reappraisal? *Health Econ.* 1999;8(1):1−8.

31. World Economic Forum and World Health Organization. *From Burden to "Best Buys": Reducing the Economic Impact of Non-Communicable Diseases in Low- and Middle-Income Countries.* Cologny/Geneva, Switzerland: World Economic Forum; 2011.

32. Grosse SD, Lollar DJ, Campbell VA, Chamie M. Disability and disability-adjusted life years: not the same. *Public Health Rep.* 2009;124(2):197−202.

33. Prince MJ, Wu F, Guo Y, Gutierrez Robledo LM, O'Donnell M, Sullivan R, et al. The burden of disease in older people and implications for health policy and practice. *Lancet.* 2015;385(9967): 549−562.

34. Sousa RM, Ferri CP, Acosta D, Albanese E, Guerra M, Huang Y, et al. Contribution of chronic diseases to disability in elderly people in countries with low and middle incomes: a 10/66 Dementia Research Group population-based survey. *Lancet.* 2009;374(9704): 1821−1830.

35. Alzheimer's Disease International. *World Alzheimer Report 2009.* London: Alzheimer's Disease International; 2009.

36. Prince M, Prina M, Guerchet M. *World Alzheimer Report 2013. Journey of Caring: An Analysis of Long-Term Care for Dementia.* London: Alzheimer's Disease International; 2013.

37. Sousa RM, Ferri CP, Acosta D, Guerra M, Huang Y, Ks J, et al. The contribution of chronic diseases to the prevalence of dependence among older people in Latin America, China and India: a 10/66 Dementia Research Group population-based survey. *BMC Geriatr.* 2010;10(1):53.

38. Liu Z, Albanese E, Li S, Huang Y, Ferri CP, Yan F, et al. Chronic disease prevalence and care among the elderly in urban and rural Beijing, China: a 10/66 Dementia Research Group cross sectional survey. *BMC Public Health.* 2009;9:394.

39. Acosta D, Rottbeck R, Rodriguez G, Ferri CP, Prince MJ. The epidemiology of dependency among urban-dwelling older people in the Dominican Republic; a cross-sectional survey. *BMC Public Health.* 2008;8(1):285.

40. World Health Organization. *Dementia: A Public Health Priority.* Geneva, Switzerland: WHO; 2012.

CHAPTER

5

Tropical Epidemiology: Nutritional Factors*

Jean-Claude Desport[1,2,3], Pierre Jésus[1,2,3], and Philippe Fayemendy[1,2,3]

[1]University Hospital of Limoges, Limoges, France [2]INSERM UMR 1094 NET, Limoges, France
[3]University of Limoges, Limoges, France

5.1 INTRODUCTION

The nutritional status of neurological patients is often altered, and the major risk is malnutrition. Malnutrition, but also obesity, is detected by examinations with several interesting characteristics: they are robust, simple, low cost, and consequently easy to use in tropical countries.

In practice, malnutrition is the result of either an insufficient alimentary intake, or an excess of energy losses, or a combination of these two processes (Fig. 5.1). It is fundamental to find and, when possible, to treat these causes. They can differ from one neurological disease to another, but a number of them are shared by several pathologies.

Among many consequences, malnutrition increases the risk of infection, favors a loss of autonomy and a decrease of muscular capacity, reduces the healing possibilities, alters the digestive and the psychological status, and can be accompanied by bone fragility (Fig. 5.1).[1] Death may also be a consequence of malnutrition.

Obesity is an excess of fat mass, due to insufficient physical activity and/or an excess of energy intake, compared to the needs. Obesity may also be a problem concerning neurological patients, even in tropical countries. Its prevalence is lower in emerging countries than in occidental ones, but nutritional transition is in progress, and excess weight is now developing worldwide. Its consequences are numerous, including

*Conflict of interest: none.

47

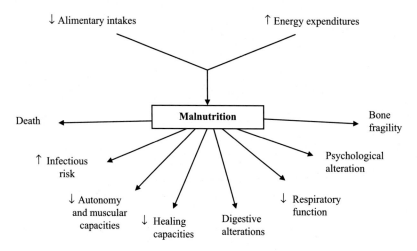

FIGURE 5.1 Causes and consequences of malnutrition.

diabetes, cardiovascular complications, loss of autonomy and rheumatologic problems, psychological problems, increase of risk of cancer, etc.[2]

In neurological tropical epidemiology, the nutritional data at our disposal are scarce, but especially concern epilepsy and dementia. A large field of research remains to be explored.

5.2 THE TOOLS FOR POINTING OUT MALNUTRITION AND OBESITY AND THEIR THRESHOLDS

Classically, only one criterion is enough for the diagnosis of malnutrition, and the most used measurement for defining obesity is body mass index (BMI).

5.2.1 Body Mass Index

BMI is calculated by dividing the weight in kilogram by the square of the height in square meter. For the World Health Organization (WHO), in adults, malnutrition is present when BMI is <18.5, and obesity when BMI >30 kg/m². [3] For malnutrition, other limits are proposed for the elderly in occidental countries (like in France BMI <21 when age >70), but they are not used in tropical countries. BMI calculation seems easy, but one needs to know the true weight and the true height of patients, and not declarative values, too often respectively under- or over-evaluated in normal people. In tropical countries, for patients in which the height cannot be measured in an upright position, we have no formula to assess the height from knee height as in occidental countries.[4] A proposed WHO formula uses the half-arm span (HAS), but its validation is uncertain. HAS is the distance from the middle of the sternal notch to the tip of the middle finger with the

arm held out horizontally to the side. The formula is: height (m) = 0.73 × (2 × HAS (m)) − 0.43.[3]

For BMI, a frequently noted bias is hyperhydration with edemas, consequently normalizing or majoring the weight, even if a severe malnutrition is present. Of course, edemas may be linked to a disease other than malnutrition, such as heart, renal or liver failure, or due to troubles of venous or lymphatic drainage. Nevertheless, it may be too a kwashiorkor (from Ghanean words meaning "red-skin child," because of associated dermatitis or rashes sometimes noted). Kwashiorkor is a form of malnutrition caused by a severe protein-energetic insufficiency and an inflammation or an infection, with hypoalbuminemia and troubles of the microvascular status allowing a transfer of water in interstitium.[5] It is important to detect these edematous forms (which may be associated in children with a big liver, and for which the prognosis is bad), in children as well as in adults.

5.2.2 Unintentional Loss of Weight

It is a criterion of increasing importance for detecting malnutrition. However, it is not proposed by the WHO, probably because it must contain at least two weights, and if possible the weights must be measured and not declarative. The calculation formula for loss of weight is: ((usual weight − actual weight)/usual weight) × 100. Malnutrition is present if the weight loss is >5% compared to usual weight or compared to the weight 3 months before, or >10% compared to the weight 6 months before.[6] However, these thresholds may vary from one recommendation or one neurological disease to another.

5.2.3 Main Other Anthopometric Assessments

Body compartments can be easily assessed by measurements performed on arms.[7] One hypothesis is that skinfolds, especially triceps skinfold (TSF) measured at the back part of arm at the mid-point between the tip of the shoulder and the tip of the elbow (olecranon process and the acromium), represents total body fat mass. Conversely, the mid-arm muscle circumference (MAMC), given by the formula MAMC = brachial circumference (cm, measured by a tape at the same place as TSF) − 0.314 × TSF (mm), could represent lean body mass. A malnutrition could be present if MAMC <5th percentile of the distribution in occidental countries.[8] For the WHO, it is possible to use more simply the mid-upper arm circumference (MUAC), that is the circumference of the left upper arm, and malnutrition is possible when MUAC is <24 cm for men and <23 cm for women,[9] but the validation of this index is questionable again.

These indices are easily and rapidly obtained, with a low cost. Another particularity is that they do not take into account the weight, and consequently are more interesting in the case of edemas.

5.2.4 Albuminemia

Albumin is a protein synthesized by the liver in an adequate quantity if nutritional intakes, and specially protein intakes, are well-fitted to the requirements. In case of insufficient intakes, hypoalbunemia appears. Below 35 g/L, according to the standard of the laboratory, there is malnutrition.[10] However, there are numerous biases, possibly inducing a false hypoalbuminemia, like an excess of hydration, an inflammatory associated syndrome, a liver failure, the presence of an albuminuria or of severe diarrheas, etc. Consequently, albuminemia is an indicative criterion, but for a correct interpretation, a clinical examination needs to be conducted, and if possible a biological assessment of inflammatory and hepato-renal status.

5.2.5 Dietician Survey

It does not allow a diagnosis of malnutrition, but is an alarm factor. Moreover, on a visual analogical scale from 0 to 10, a daily alimentary consumption <25% of proposed meals seems to be associated with an increase of mortality.[11] The dietician survey guides largely the care, if oral intakes are allowed, in order to propose meals enriched, fitted to tolerated textures and to patients' preferences.

5.3 THE MAIN CAUSES OF MALNUTRITION DURING NEUROLOGICAL DISEASES AND THEIR TREATMENT

5.3.1 Disabilities

Disabilities are of great importance. These difficulties of daily life, in connection with the neurological disease, may lead to reduced food intake. It is easily understood that for a dementia patient or a victim of stroke, or for a serious epileptic patient, getting food by going to the market or to a store can be very difficult or impossible, along with as preparing meals, using cutlery or bringing food to the mouth, or sometimes even recognizing the food.[12–14] The person is dependent not only on equipment, but also specifically on unpaid services. These volunteers may experience too much fatigue, and it is rare in tropical countries that families can appeal to specialized structures adapted for dependents. Therefore, the family environment, but also the entourage care sector must be informed of the importance of a sufficient supply of patients and eventually be trained and helped.

5.3.2 Anorexia

Anorexia is multifactorial. Thus, when dementia,[14] but also at a Parkinson's disease[15] or Lou Gehrig's disease (amyotrophic lateral sclerosis or ALS) is advanced,[12] central lesions may explain the loss of appetite. Respiratory disorders, often related to neurological damage with swallowing disorders and infectious problems, generate anorexia. Digestive disorders, especially constipation, very common in neurological diseases, are also appetite suppressants,[16] as well as poor glucose tolerance in the first days after stroke[17] or degradation of the psychological state regardless of the disease. The support involves the correction of respiratory or infectious diseases, the treatment of constipation and hyperglycemia, and psychological help if needed.

5.3.3 Swallowing Disorders

Swallowing disorders are very common, often linked to central nerve damage[14,18] but can also be favored by facial deformities in cases of neuromuscular disease, or facial weakness after stroke. Neurological disorders are readily risk factors of swallowing troubles. The question is whether these disorders are related to liquid, solid or mixed food intake. A decision flow chart is useful (Fig. 5.2). At least it is possible to use DePippo's test,[19] and perhaps, depending on availability, an ear, nose and throat (ENT) endoscopy or radiovideoscopy, with an adaptation of textures and head positions for the disability.

5.3.4 Disorders of Salivation

Disorders of salivation are rarely reported by patients but can cause serious difficulties for food intake and social life.[16] These are true hypersialorrhea of central origin, translated into a drooling, or

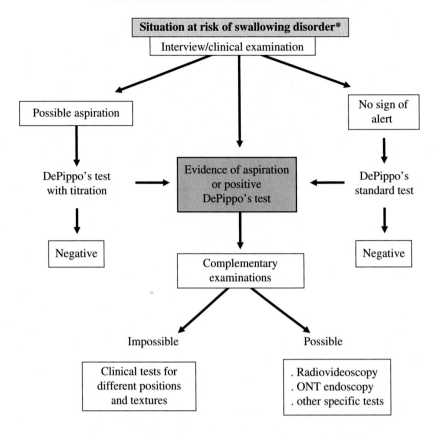

*Stroke, neuromuscular and neurodegenerative disorders are at risk situations

FIGURE 5.2   Decisional tree before a situation at risk of swallowing disorders.

false hypersialorrhea due to difficulty in swallowing and/or holding food and saliva in the mouth, for example in connection with facial hypotonia after stroke. The treatments use the drug atropine. Abnormal secretion of very thick and sticky saliva is rare but very disabling for food; the treatment is beta-blocker medication.

5.3.5 The Increase in Energy Expenditure

There may be an increase in energy expenditure related to physical activity, if there is a hypertonia, abnormal movements, ambulation, respectively, as in Parkinson's disease, Huntington's disease or some dementia.[14,20] It can also be a pathological increase in resting energy expenditure in the 8 days after a stroke, during an infection, or over 50% of cases during ALS.[21] The treatment is then based almost exclusively on drugs, or a restriction on movement of patients in case of ambulation. In Parkinson's disease, it is important to ensure that levodopa is taken at least 20–30 minutes before meals, both to limit hypertonia, and to promote efficiency of this drug, which is in competition with food proteins to their digestive absorption and cerebral passageway.[22]

5.3.6 The Role of Therapeutics

Finally, some specific drugs can play a role. And cholinesterase inhibitors used in dementia may promote anorexia, nausea and vomiting, constipation, especially during the first month of treatment.[14] Levodopa may also be accompanied by anorexia and digestive disorders.[23] For epilepsy, several drugs support a weight loss, such as topiramate, felbamate, zonisamide, and stiripentol. The traditional emetogenic or purgative treatments can also be involved in the development of malnutrition.[24]

Before malnutrition develops due to an impossibility of feeding sufficiently by the mouth, the recourse to enteral nutrition in theory is advised. However, for economic reasons, this technique is very seldom routinely available in the tropical countries.

Obesity care needs, if possible, a multidisciplinary approach, involving a physician, a dietician and a psychologist. As well as for malnutrition, the financial possibilities of the patients are a problem, but also the availability of particular professionals.

5.4 NUTRITION AND EPILEPSY IN TROPICAL COUNTRIES

5.4.1 Malnutrition or Obesity as Possible Consequences of Epilepsy

A primary relationship between epilepsy and nutritional status has been suspected for a long time, epilepsy being incriminated in the genesis of either malnutrition or obesity.

Malnutrition can be attributed to antiepileptic drugs with anorexic effects[25] (Table 5.1), but also to food taboos (depending on the area in the world), and depression due to stigma.[26] Moreover, in the particular case of drug resistance epilepsy, patients can have

TABLE 5.1 Effects of Antiepileptic Drugs on Weight[26]

Weight gain	No effect	Weight loss
Carbamazepine	Lamotrigine	Felbamate
Oxcarbamazepine	Levetiracetam	Topiramate
Gabapentine	Phenytoine	Zonisamide
Pregabaline	Thiagabine	Stiripentol
Sodium valproate	Lacosamide	
Vigabatrin		

other neurological disabilities like chewing problems, behavior problems or cerebral palsy, for example.[26]

Conversely, it is notable that physical activity in people with epilepsy (PWE) can be limited by their treatments, and that some treatments may increase appetite (Table 5.1), favoring an excess of weight.[27,28]

A limited number of studies concerning nutrition and epilepsy are available in developing countries (Table 5.2).[32–34] They suggest first that there is a real more marked prevalence of malnutrition in PWE than in control patients. The sometimes divergent data could be explained by methodological differences or lack of power. However, only in one case-control study with enough power were malnutrition and epilepsy clearly linked.[31] In an Ethiopian study by Vaid et al., authors noted that they excluded patients with significant neurological disability, to minimize the effect of feeding difficulties.[30] Consequently, a number of malnourished PWE might not have been taken into account. A second point is that the Ethiopian population explored was globally often malnourished (20.5% in PWE and 23.5% in controls), emphasizing the huge burden of feeding problems in this country. Nevertheless, authors concluded that PWE had evidence of stunting and disproportionate skeletal growth, raising the possibility of a link between early malnutrition and epilepsy.

A recent Brazilian study showed different results, without any malnutrition, but a high prevalence of obesity in PWE, 2.5-times more common in the sample than in a healthy population of a similar age (respectively, 37.5% and 14.8%).[29] This weight excess could be due to elevated carbohydrate and protein intakes

TABLE 5.2 Studies on Nutrition and Epilepsy Performed in Humans in Developing Countries

Place (first author/year)	Study design	Population	Age of subjects (years)	Results
Brazil (De Azevedo Fernandez, 2015)[29]	Cross-sectional study	72 PWE	46.3 ± 12.4	No malnourished patients; 29.5% overweight and 37.5% obesity; no link between alimentary intake, nutritional status and seizure control
Ethiopia (Vaid, 2012)[30]	Case-control study	112 PWE, 149 controls	18–45	BMI similar, but lower MUAC ($p = 0.01$) and FFM ($p = 0.04$) in PWE. No difference between patients with or without seizure control
Benin (Crepin et al., 2007)[31]	Case-control study	131 PWE, 262 controls	All ages	Risk of malnutrition higher for PWE (OR = 2.9, $p = 0.0006$); MUAC linked with epilepsy (OR = 0.7, $p = 0.002$)
Benin (Nkwetngam Ndam, 2004)[32]	Case-control study	39 PWE, 39 controls	All ages	No link between malnutrition and epilepsy ($p > 0.05$, lack of power)
India (Hackett et al., 1997)[33]	Case-control study	26 PWE, 1146 controls	8–12	Mean BMI lower in PWE (OR = 0.74; $p = 0.023$)
India (Pal, 1999)[34]	Case-control study	61 PWE, 59 controls	2–18	No link between BMI and epilepsy (OR = 31.3; $p > 0.05$, lack of power)

BMI, body mass index; FFM, fat free mass; MUAC, mid-upper arm circumference; OR, odds ratio; PWE, people with epilepsy.

among PWE (approximately two- and four-times higher than recommended, respectively), without a clear explanation. Obesity in PWE was a problem already noted in occidental countries, such as Canada with a prevalence of 19.1% for PWE versus 15.4% in the general population.[27] It is consistent with studies in the United States,[28] and could be linked with a lower physical activity than people without epilepsy, and/or with the use of anticonvulsant medications or psychotropic drugs that can stimulate appetite and cause sedation and lethargy.[27,28] As well in these occidental populations as probably in Brazilian PWE, deleterious complications of obesity are possible.[28]

According to another point of view, antiepileptic treatments may also play a role in developing micronutrient deficiencies. A recent review of 60 articles, including 40 on epileptic patients, showed that there was in the treated epilepsy patients versus untreated a serum deficiency of zinc, copper and magnesium suggesting a link between treatments and metabolism of trace elements.[35] However, it was not possible to say whether the deficits were the cause or the consequence of epilepsy. Concerning vitamin D, a study in Malaysia found 22.5% of epileptic patients with deficiency, in spite of the large sun exposure in this country. In multivariate analysis, there was an increased risk of deficiency in the case of polytherapy (>1 treatment; odds ratio (OR): 2.16, 95% confidence interval (CI): 1.16−4.36), and an interest for a vitamin supplementation was suggested.[36]

5.4.2 Epilepsy Favored by Malnutrition

The second relationship is between malnutrition and epilepsy, because malnutrition is responsible for protein-energy, micronutrient and immuno deficiencies. Mechanisms implicated are decrease of seizure threshold, altered neurotransmission or increase of sensibility to infectious diseases as neurotropic diseases.[26]

In animals, neuro-anatomical studies have provided evidence for many years that malnutrition, and particularly protein insufficiency may affect the seizure threshold, perhaps through alterations of cholinergic, GABAergic, serotoninergic and glutaminergic transmission,[26] but these findings are not always confirmed in malnourished humans.[26]

People with malnutrition are more vulnerable to infections that may lead to epilepsy, such as neurotropic viruses or cysticercosis.[37]

Some antiepileptic drugs can change metabolism or absorption of micronutrients, and consequently favor deficiencies possibly decreasing the level of epileptogenesis.[38] Regarding micronutrients, a

number of them have been involved, including calcium, sodium, magnesium, phosphorus, vitamins B1, B6, B12, D, zinc, selenium, copper with various mechanisms[24,26,35]

A double-blind Iranian study of vitamin E supplementation, whose antioxidant effects are known was conducted on 32 epilepsy patients, versus 33 controls.[39] It showed improved antioxidant status and a reduction in seizure frequency of 50% in vitamin E, versus 12.1% in the placebo group ($p < 0.001$). However, like in Saghazadeh's study, it was not possible to say whether the deficits were the cause or the consequence of epilepsy.[35]

At least after the first seizure, it is probably important to recommend, if possible, to assess the micronutrient status of patients.

In developed countries, it is unlikely that malnutrition can induce epileptic seizures, whereas in emerging countries the causal relationship is more likely. However, the relationship "epilepsy → malnutrition" is possible in any part of the world.

5.5 NUTRITION AND DEMENTIA IN TROPICAL COUNTRIES

Alzheimer's disease (AD), degenerative dementia, is the most common neurodegenerative disease in the world. Dementia leads to multiple cognitive deficits, with memory impairment, aphasia, apraxia, agnosia or disturbance in executive functions. These alterations can affect the nutritional status of the person with dementia and therefore the evolution of the dementia. Furthermore the nutritional status and food intake (vitamins, trace elements, antioxidants, polyunsaturated fatty acid, alcohol, etc.) may have an impact on the prevention or on the onset of dementia.

5.5.1 Nutritional Status

The alteration of nutritional status is related to atrophy of the medial temporal cortex, resulting in the occurrence of cognitive disorders, behavioral disorders, anorexia, and may also increase physical activity. The alteration of higher functions with apraxia and/or agnosia leads to a loss of autonomy in different food processes and therefore increases the risk of food intake reduction.[40−43] Behavior disorders with agitation or perambulatory and/or eating disorders with refusal, opposition may also disrupt food.[44] All of this can therefore lead to weight loss then to a state of malnutrition.

Several studies in tropical areas found a high prevalence of malnutrition in elderly with dementia

TABLE 5.3 Studies on Malnutrition and Dementia Performed in Humans in Tropical Countries

Place (first author/year)	Study design	Population (*n*/age)	Dementia/MCI (%)	Malnutrition (%)
CAR/Congo (Pilleron, 2015)[45]	Cross-sectional study	2002/over 65 years	Dementia: 8.4% in CAR; 6.9% in Congo	In dementia: 65.6% in CAR; 52.3% in Congo (BMI <18.5 kg/m^2)
			MCI: 7.2% in CAR; 6.1% in Congo	In MCI: 30.5% in CAR; 42.0%, in Congo (BMI <18.5 kg/m^2)
				Association with dementia OR = 2.3; $p < 0.001$ in CAR
				OR = 2.2; $p = 0.09$ in Congo
CAR/Congo (De Rouvray, 2014; Guerchet, 2012)[46,47]	Cross-sectional study	1016/over 65 years	Dementia: 8.1% in CAR; 6.7% in Congo	34.7% (BMI <18.5 kg/m^2)
				Association with dementia OR = 1.77; $p = 0.042$
Bangladesh (Palmer, 2014)[48]	Cross-sectional study	850/over 60 years	Dementia: 11.5%	Association with dementia OR: 5.9; $p = 0.02$
Brazil (Marino, 2013)[49]	Cross-sectional study	36/over 54 years	Dementia: 100%	11.1% (MNA <17)
Brazil (Galesi, 2013)[50]	Cross-sectional study	150/over 60 years	Dementia: 48.0%	20.7% (BMI <22 kg/m^2)
Egypt (Khater, 2011)[51]	Cross-sectional study	120/over 60 years	MCI: 38.3%	17.4% (MNA <17)
Nigeria (Ochayi, 2006)[52]		280/over 65 years	Dementia: 6.4%	13.0% (BMI <18.5 kg/m^2)
				Association with dementia OR = 3.5; $p = 0.02$

BMI, body mass index; CAR, Central African Republic; MCI, mild cognitive impairment; MNA, mini nutritional assessment; OR, odds ratio.

(Table 5.3). Indeed, in Africa, in Nigeria, Ochayi et al. found a prevalence of malnutrition (BMI < 18.5 kg/m^2) of 13.0% in people with dementia (PWD).[52] De Rouvray et al. found 34.7% of malnutrition (BMI < 18.5 kg/m^2) in PWD at home in urban areas of Central African Republic (CAR) and the Republic of the Congo (Congo).[46] Pilleron et al. confirmed these data, using several malnutrition criteria, BMI (<18.5 kg/m^2), brachial circumference (BC < 24 cm) and MUAMC (MUAMC < 5th percentile): malnutrition prevalence was of 65.6%, 43.9%, and 30.8% respectively in CAR and of 52.3%, 38.0%, and 28.0%, respectively in Congo.[45] This study also provided data on the nutritional status of people with mild cognitive impairment (MCI) at home. The prevalence of malnutrition in MCI was from 10.0% to 30.5% in CAR and from 32.0% to 42.0% in the Congo based on these different nutritional criteria. An Egyptian study found 17.4% of malnutrition according to the mini-nutritional assessment (MNA < 17) in MCI patients in institutions.[51]

In other tropical areas such as Brazil, Galesi et al. found 20.7% of malnutrition (BMI < 22 kg/m^2) in patients with dementia in institutions.[50] Marino et al. found a prevalence of malnutrition of 11.1% with MNA (MNA < 17) in patients with dementia at home in Brazil.[49]

5.5.2 Nutritional Factors Associated With Dementia

5.5.2.1 Malnutrition and Weight Loss

In Nigeria, Ochayi et al. found that a BMI < 18.5 kg/m^2 was positively associated with dementia (OR = 3.5 (95% CI: 1.2–9.9)).[52] Guerchet et al. in the EDAC study in CAR and Congo showed similar results (OR = 1.77 (95% CI: 1.02–3.06); $p = 0.042$).[47] Pilleron et al. in the EPIDEMCA study in Central Africa showed that BMI < 18.5 kg/m^2, BC < 24 cm and MUAMC < 5th percentile were positively associated with the presence of dementia in CAR.[45] However, in Congo only the MUAMC was associated with MCI but not with dementia. MUAMC appear to be more relevant than BMI or BC. Indeed, it reflects the fat free mass (FFM) whereas BMI and BC take into account the FFM and fat mass indiscriminately. The association of cognitive disorders with the MUAMC is consistent with the fact that FFM is correlated with brain volume and cognitive status.[53] In addition, the percentile evaluation of the MUAMC depends on age, unlike BMI in adults (according to WHO criteria) and BC, which probably allows better assessment of nutritional status according to age.[8] In Bangladesh, Palmer et al. also found a strong association between malnutrition and dementia (OR: 5.9 (95% CI: 1.3–26.3); $p = 0.02$).[48] All these studies were cross-sectional studies, therefore we cannot make a causal link between the presence of dementia and malnutrition. However, in the longitudinal study of EPIDEMCA (EPIDEMCA-FU) after a follow-up of 2 years in Congo, PWD had 2.5-times higher mortality risk (hazard ratio = 2.53 (95% CI: 1.42–4.49); $p = 0.001$),[54] but malnutrition was not a risk factor of death in PWD in this study. On another aspect of the nutritional status, controlled weight loss in obesity could be protective of cognitive status. Indeed, the Brazilian study of Horie et al. found that intentional weight loss over 12 months in obese elderly with MCI improved their cognitive status in global cognition, verbal memory, language, and executive function.[55]

5.5.2.2 Food Consumption

In Central Africa in the study of De Rouvray et al., PWD usually did not eat in the same room as the rest of the family, but it was not a factor associated with malnutrition.[46] However, isolation is usually associated with an increased risk of malnutrition in the elderly.[56] PWD also consume more often simple food such as cookies, easy to eat with fingers. For Suominen et al. the "snacking" tends to reduce the risk of malnutrition.[57]

De Rouvray et al. also found that PWD comsumed less fruit and fish than people without cognitive disorders. Foods containing antioxidants may have a role in the genesis and evolution of dementia.[58–60] For Cordoso et al. in Brazil, fruit consumption was also positively associated with memory ($\beta = 0.14$; $p < 0.001$). Indeed, consumption of antioxidant vitamins (C, E) could reduce the risk of dementia.[61–63] However, the review of Loef et al. found a protective role of a high consumption of vegetables but no association with fruit consumption.[64]

Pilleron et al. found in rural areas of CAR, that a low oilseed consumption is associated with the presence of cognitive impairment (OR = 2.8 (95% CI: 1.02–7.70); $p = 0.046$; for dementia; OR = 3.67 (95% CI: 1.37–9.85); $p = 0.01$; for MCI).[65] Indeed, peanut oil is used in rural areas and contains 32.0% of polyunsaturated fatty acids (PUFAs)[66,67] In Brazil, Baierle et al. found in 45 patients that cognitive functions were positively correlated with serum levels of 24:1 n-9 fatty acid, docosahexaenoic acid (DHA) and total n-3 PUFAs.[68] Inversely saturated fatty acids (SFAs) and n-6/n-3 ratio were negatively correlated with cognitive functions in this study. PUFAs limit the oxidative stress and could have a protective role in dementia and AD.[69–71] In occidental countries a high intake of n-3 PUFAs was associated with a decreased risk of

MCI.[72] In contrast, SFAs could be associated with an increased risk of cognitive disorders.[73]

In Brazil, Cardoso et al. found that selenium intakes were significantly lower in patients with MCI and dementia compared to controls without cognitive impairment.[74,75] Serum and erythrocyte levels of selenium were also significantly decreased in these patients. Indeed, deficiency of this antioxidant trace element might play a role in cognitive decline.[76]

In CAR, alcohol consumption in the general population is negatively associated with dementia (OR = 0.34 (95% CI: 0.14–0.83); $p = 0.018$). However, no association between cognitive impairment and food and alcohol in Congo was found.[65] The literature found a positive association between moderate alcohol consumption and decreased risk of degenerative dementia.[77,78]

In another tropical area, an Indonesian study noted that high soy consumption (rich in phytoestrogens) such as tofu was associated with worse memory ($\beta = -0.18$; $p < 0.01$) but tempe (a fermented whole soybean product) was independently related to better memory ($\beta = 0.12$; $p < 0.05$), particularly in people over 68 years.[79]

Current studies in tropical areas although mainly cross-sectional studies and few in number find data close to occidental countries. Weight loss and malnutrition are associated with cognitive impairment. Weight loss could appear presymptomatically as in occidental countries.[80–83] Also food consumption (PUFAs, selenium, alcohol, phytoestrogen) seems related to the evolution of the cognitive status.

References

1. Stratton RJ, Green CJ, Elia M. *Consequences of disease-related malnutrition. Disease-Related Malnutrition: An Evidence-Based Approach to Treatment*. Wallingford, UK: CABI Publishing; 2003:113–155.
2. Apovian C. Obesity: definition, comorbidities, causes, and burden. *Am J Manag Care*. 2016;22(7 Suppl):s176–s185.
3. World Health Organization (WHO). *Management of Severe Malnutrition: A Manual for Physicians and Other Senior Health Workers*. Geneva, Switzerland: WHO; 1999. Available from: <http://www.who.int/nutrition/publications/severemalnutrition/9241545119/en/>.
4. Chumlea WC, Roche AF, Steinbaugh ML. Estimating stature from knee height for persons 60 to 90 years of age. *J Am Geriatr Soc*. 1985;33(2):116–120.
5. Torum B, Chew F. *Protein-energy malnutrition. Modern Nutrition in Health and Disease*. 9th ed. Baltimore, MD: Williams & Wilkins; 1999:963–988.
6. Melchior J, Hanachi M, Hankard R. *Méthodes d'évaluation de l'état et du risque nutritionnel. Traité de Nutrition Clinique*. Paris: SFNEP, Knoé; 2016:647–668.
7. Hammond K. *Assessment: Dietary and Clinical Data. Krause's Food & Nutrition Therapy*. St Louis, MO: Saunders Elsevier; 2008:383–410.
8. Frisancho AR. New norms of upper limb fat and muscle areas for assessment of nutritional status. *Am J Clin Nutr*. 1981;34(11):2540–2545.

9. World Health Organization (WHO). Physical status: the use and interpretation of anthropometry. Report of a WHO Expert Committee. *World Health Organ Tech Rep Ser*. 1995;854:1–452.

10. Demarest-Litchford M. *Assessment: laboratory data. Krause's Food & Nutrition Therapy*. St Louis, MO: Saunders Elsevier; 2008:411–431.

11. Hiesmayr M, Schindler K, Pernicka E, Schuh C, Schoeniger-Hekele A, Bauer P, et al. Decreased food intake is a risk factor for mortality in hospitalised patients: the NutritionDay survey 2006. *Clin Nutr*. 2009;28(5):484–491.

12. Golaszewski A. Nutrition throughout the course of ALS. *NeuroRehabilitation*. 2007;22(6):431–434.

13. Bouteloup C, Ferrier A. Nutrition et accident vasculaire cérébral. *Nutr Clin Metab*. 2011;25(4):217–226.

14. Secher M, Gilette-Guyonnet S, Nourashémi F. Nutrition et maladie d'Alzheimer. *Nutr Clin Metab*. 2011;25:227–232.

15. Sheard JM, Ash S, Silburn PA, Kerr GK. Prevalence of malnutrition in Parkinson's disease: a systematic review. *Nutr Rev*. 2011;69(9):520–532.

16. Desport J, Maillot F. Nutrition et sclérose latérale amyotrophique (SLA). *Nutr Clin Metab*. 2002;16:91–96.

17. Radermecker RP, Scheen AJ. Management of blood glucose in patients with stroke. *Diabetes Metab*. 2010;36:S94–S99.

18. Desport J-C, Jésus P, Fayemendy P, De Rouvray C, Salle J-Y. Évaluation et prise en charge des troubles de la déglutition. *Nutr Clin Metab*. 2011;25(4):247–254.

19. DePippo KL, Holas MA, Reding MJ. Validation of the 3-oz water swallow test for aspiration following stroke. *Arch Neurol*. 1992;49(12):1259–1261.

20. Markus HS, Cox M, Tomkins AM. Raised resting energy expenditure in Parkinson's disease and its relationship to muscle rigidity. *Clin Sci (Lond)*. 1992;83(2):199–204.

21. Bouteloup C, Desport J-C, Clavelou P, Guy N, Derumeaux-Burel H, Ferrier A, et al. Hypermetabolism in ALS patients: an early and persistent phenomenon. *J Neurol*. 2009;256(8):1236–1242.

22. Cereda E, Barichella M, Pedrolli C, Pezzoli G. Low-protein and protein-redistribution diets for Parkinson's disease patients with motor fluctuations: a systematic review. *Mov Disord*. 2010;25(13):2021–2034.

23. Barichella M, Marczewska A, De Notaris R, Vairo A, Baldo C, Mauri A, et al. Special low-protein foods ameliorate postprandial off in patients with advanced Parkinson's disease. *Mov Disord*. 2006;21(10):1682–1687.

24. Crepin S, Godet B, Chassain B, Preux P-M, Desport J-C. Malnutrition and epilepsy: a two-way relationship. *Clin Nutr*. 2009;28(3):219–225.

25. Ben-Menachem E. Weight issues for people with epilepsy: a review. *Epilepsia*. 2007;48:42–45.

26. Crepin S, Godet B, Preux P, Desport J. Arguments for a relationship between malnutrition and epilepsy. *Handbook of Behavior, Food and Nutrition*. New York: Springer New York; 2011:2329–2342.

27. Hinnell C, Williams J, Metcalfe A, Patten SB, Parker R, Wiebe S, et al. Health status and health-related behaviors in epilepsy compared to other chronic conditions: a national population-based study. *Epilepsia*. 2010;51(5):853–861.

28. Kobau R, DiIorio CA, Price PH, Thurman DJ, Martin LM, Ridings DL, et al. Prevalence of epilepsy and health status of adults with epilepsy in Georgia and Tennessee: Behavioral Risk Factor Surveillance System, 2002. *Epilepsy Behav*. 2004;5(3):358–366.

29. De Azevedo Fernandez R, Corrêa C, Muxfeld Bianchim M, Schweigert Perry ID. Anthropometric profile and nutritional intake in patients with epilepsy. *Nutr Hosp*. 2015;32(2):817–822.

30. Vaid N, Fekadu S, Alemu S, Dessie A, Wabe G, Phillips DIW, et al. Epilepsy, poverty and early under-nutrition in rural Ethiopia. *Seizure*. 2012;21(9):734–739.

31. Crepin S, Houinato D, Nawana B, Avode GD, Preux P-M, Desport J-C. Link between epilepsy and malnutrition in a rural area of Benin. *Epilepsia*. 2007;48(10):1926–1933.

32. Nkwetngam Ndam M. Epilepsie et statut nutritionnel dans la commune de Djidja, Département du Zou au Bénin [Thèse]. Bénin: Université d'Abomey-Calavi; 2004:74p.

33. Hackett R, Hackett L, Bhakta P. The prevalence and associated factors of epilepsy in children in Calicut District, Kerala, India. *Acta Pædiatrica*. 1997;86(11):1257–1260.

34. Pal DK. Methodologic issues in assessing risk factors for epilepsy in an epidemiologic study in India. *Neurology*. 1999;53(9):2058–2063.

35. Saghazadeh A, Mahmoudi M, Meysamie A, Gharedaghi M, Zamponi GW, Rezaei N. Possible role of trace elements in epilepsy and febrile seizures: a meta-analysis. *Nutr Rev*. 2015;73(11):760–779.

36. Fong CY, Kong AN, Poh BK, Mohamed AR, Khoo TB, Ng RL, et al. Vitamin D deficiency and its risk factors in Malaysian children with epilepsy. *Epilepsia*. 2016;57(8):1271–1279.

37. Preux P-M, Druet-Cabanac M. Epidemiology and aetiology of epilepsy in sub-Saharan Africa. *Lancet Neurol*. 2005;4(1):21–31.

38. Soltani D, Ghaffar Pour M, Tafakhori A, Sarraf P, Bitarafan S. Nutritional aspects of treatment in epileptic patients. *Iran J Child Neurol*. 2016;10(3):1–12.

39. Mehvari J, Motlagh F, Najafi M, Ghazvini MA, Naeini A, Zare M. Effects of vitamin E on seizure frequency, electroencephalogram findings, and oxidative stress status of refractory epileptic patients. *Adv Biomed Res*. 2016;5(1):36.

40. Gillette-Guyonnet S, Nourhashémi F, Andrieu S, de Glisezinski I, Ousset PJ, Rivière D, et al. Weight loss in Alzheimer disease. *Am J Clin Nutr*. 2000;71(2):637s–642s.

41. Romatet S, Belmin J. La perte de poids dans la maladie d'Alzheimer. *Rev Gériatrie*. 2002;27(7):587–596.

42. Wang P-N, Yang C-L, Lin K-N, Chen W-T, Chwang L-C, Liu H-C. Weight loss, nutritional status and physical activity in patients with Alzheimer's disease. *J Neurol*. 2004;251(3):314–320.

43. Rullier L, Lagarde A, Bouisson J, Bergua V, Barberger-Gateau P. Nutritional status of community-dwelling older people with dementia: associations with individual and family caregivers' characteristics. *Int J Geriatr Psychiatry*. 2013;28(6):580–588.

44. Greenwood CE, Tam C, Chan M, Young KWH, Binns MA, van Reekum R. Behavioral disturbances, not cognitive deterioration, are associated with altered food selection in seniors with Alzheimer's disease. *J Gerontol A Biol Sci Med Sci*. 2005;60(4):499–505.

45. Pilleron S, Jésus P, Desport J-C, Mbelesso P, Ndamba-Bandzouzi B, Clément J-P, et al. Association between mild cognitive impairment and dementia and undernutrition among elderly people in Central Africa: some results from the EPIDEMCA (Epidemiology of Dementia in Central Africa) programme. *Br J Nutr*. 2015;114(02):306–315.

46. De Rouvray C, Jésus P, Guerchet M, Fayemendy P, Mouanga AM, Mbelesso P, et al. The nutritional status of older people with and without dementia living in an urban setting in Central Africa: the EDAC study. *J Nutr Health Aging*. 2014;18(10):868–875.

47. Guerchet M, Mouanga AM, M'belesso P, Tabo A, Bandzouzi B, Paraïso MN, et al. Factors associated with dementia among elderly people living in two cities in Central Africa: the EDAC multicenter study. *J Alzheimers Dis*. 2012;29(1):15–24.

48. Palmer K, Kabir ZN, Ahmed T, Hamadani JD, Cornelius C, Kivipelto M, et al. Prevalence of dementia and factors associated with dementia in rural Bangladesh: data from a cross-sectional, population-based study. *Int Psychogeriatr*. 2014;26(11):1905–1915.

49. Marino LV, Ramos LFA, de O, Chiarello PG. Nutritional status according to the stages of Alzheimer's disease. *Aging Clin Exp Res*. 2014;27(4):507−513.

50. Galesi LF, Leandro-Merhi VA, de Oliveira MRM. Association between indicators of dementia and nutritional status in institutionalised older people. *Int J Older People Nurs*. 2013;8(3):236−243.

51. Khater MS, Abouelezz NF. Nutritional status in older adults with mild cognitive impairment living in elderly homes in Cairo, Egypt. *J Nutr Health Aging*. 2011;15(2):104−108.

52. Ochayi B, Thacher TD. Risk factors for dementia in central Nigeria. *Aging Ment Health*. 2006;10(6):616−620.

53. Burns JM, Johnson DK, Watts A, Swerdlow RH, Brooks WM. Reduced lean mass in early Alzheimer disease and its association with brain atrophy. *Arch Neurol*. 2010;67(4):428−433.

54. Samba H, Guerchet M, Ndamba-Bandzouzi B, Mbelesso P, Lacroix P, Dartigues J-F, et al. Dementia-associated mortality and its predictors among older adults in sub-Saharan Africa: results from a 2-year follow-up in Congo (the EPIDEMCA-FU study). *Age Ageing*. 2016;45(5):681−687.

55. Horie NC, Serrao VT, Simon SS, Gascon MRP, dos Santos AX, Zambone MA, et al. Cognitive effects of intentional weight loss in elderly obese individuals with mild cognitive impairment. *J Clin Endocrinol Metab*. 2015;101(3):1104−1112.

56. Ramic E, Pranjic N, Batic-Mujanovic O, Karic E, Alibasic E, Alic A. The effect of loneliness on malnutrition in elderly population. *Med Arh*. 2011;65(2):92−95.

57. Suominen M, Muurinen S, Routasalo P, Soini H, Suur-Uski I, Peiponen A, et al. Malnutrition and associated factors among aged residents in all nursing homes in Helsinki. *Eur J Clin Nutr*. 2005;59(4):578−583.

58. Von Arnim CAF, Herbolsheimer F, Nikolaus T, Peter R, Biesalski HK, Ludolph AC, et al. Dietary antioxidants and dementia in a population-based case-control study among older people in South Germany. *J Alzheimers Dis*. 2012;31(4):717−724.

59. Craggs L, Kalaria R. Revisiting dietary antioxidants, neurodegeneration and dementia. *Neuroreport*. 2011;22(1):1−3.

60. Polidori MC, Schulz R-J. Nutritional contributions to dementia prevention: main issues on antioxidant micronutrients. *Genes Nutr*. 2014;9(2):1−11.

61. Morris M, Evans DA, Bienias JL, Tangney CC, Bennett DA, Aggarwal N, et al. Dietary intake of antioxidant nutrients and the risk of incident alzheimer disease in a biracial community study. *JAMA*. 2002;287(24):3230−3237.

62. Engelhart MJ, Geerlings MI, Ruitenberg A, van Swieten JC, Hofman A, Witteman JC, et al. Dietary intake of antioxidants and risk of alzheimer disease. *JAMA*. 2002;287(24):3223−3229.

63. Devore EE, Grodstein F, van Rooij FA, Hofman A, Stampfer MJ, Witteman JC, et al. Dietary antioxidants and long-term risk of dementia. *Arch Neurol*. 2010;67(7):819−825.

64. Loef M, Walach H. Fruit, vegetables and prevention of cognitive decline or dementia: a systematic review of cohort studies. *J Nutr Health Aging*. 2012;16(7):626−630.

65. Pilleron S, Desport J-C, Jésus P, Mbelesso P, Ndamba-Bandzouzi B, Dartigues J-F, et al. Diet, alcohol consumption and cognitive disorders in Central Africa: a study from the EPIDEMCA program. *J Nutr Health Aging*. 2015;19(6):657−667.

66. Guerchet M. Démences en Afrique Subsaharienne: outils, prévalence et facteurs de risque [Thèse]. Limoges: Université de Limoges; 2010:183p.

67. Prince M. Methodological issues for population-based research into dementia in developing countries. A position paper from the 10/66 Dementia Research Group. *Int J Geriatr Psychiatry*. 2000;15(1):21−30.

68. Baierle M, Vencato P, Oldenburg L, Bordignon S, Zibetti M, Trentini C, et al. Fatty acid status and its relationship to cognitive decline and homocysteine levels in the elderly. *Nutrients*. 2014;6(9):3624−3640.

69. Barberger-Gateau P, Samieri C, Féart C, Plourde M. Dietary omega 3 polyunsaturated fatty acids and Alzheimer's disease: interaction with apolipoprotein E genotype. *Curr Alzheimer Res*. 2011;8(5):479−491.

70. Lin P-Y, Chiu C-C, Huang S-Y, Su K-P. A meta-analytic review of polyunsaturated fatty acid compositions in dementia. *J Clin Psychiatry*. 2012;73(9):1245−1254.

71. Shinto L, Quinn J, Montine T, Dodge HH, Woodward W, Baldauf-Wagner S, et al. A randomized placebo-controlled pilot trial of omega-3 fatty acids and alpha lipoic acid in Alzheimer's disease. *J Alzheimers Dis*. 2014;38(1):111−120.

72. Roberts RO, Cerhan JR, Geda YE, Knopman DS, Cha RH, Christianson TJH, et al. Polyunsaturated fatty acids and reduced odds of MCI: the Mayo Clinic study of aging. *J Alzheimers Dis*. 2010;21(3):853−865.

73. Soto ME, Secher M, Gillette-Guyonnet S, van Kan GA, Andrieu S, Nourhashemi F, et al. Weight loss and rapid cognitive decline in community-dwelling patients with Alzheimer's disease. *J Alzheimers Dis*. 2012;28(3):647−654.

74. Cardoso BR, Ong TP, Jacob-Filho W, Jaluul O, Freitas MI, Cozzolino SM. Nutritional status of selenium in Alzheimer's disease patients. *Br J Nutr*. 2010;103(06):803−806.

75. Cardoso BR, Silva Bandeira V, Jacob-Filho W, Franciscato Cozzolino SM. Selenium status in elderly: relation to cognitive decline. *J Trace Elem Med Biol*. 2014;28(4):422−426.

76. Berr C, Balansard B, Arnaud J, Roussel A-M, Alpérovitch A, EVA Study Group. Cognitive decline is associated with systemic oxidative stress: the EVA study. *J Am Geriatr Soc*. 2000;48(10):1285−1291.

77. Peters R, Peters J, Warner J, Beckett N, Bulpitt C. Alcohol, dementia and cognitive decline in the elderly: a systematic review. *Age Ageing*. 2008;37(5):505−512.

78. Anstey KJ, Mack HA, Cherbuin N. Alcohol consumption as a risk factor for dementia and cognitive decline: meta-analysis of prospective studies. *Am J Geriatr Psychiatry*. 2009;17(7):542−555.

79. Hogervorst E, Sadjimim T, Yesufu A, Kreager P, Rahardjo TB. High tofu intake is associated with worse memory in elderly Indonesian men and women. *Dement Geriatr Cogn Disord*. 2008;26(1):50−57.

80. Barrett-Connor E, Edelstein S, Corey-Bloom J, Wiederholt W. Weight loss precedes dementia in community-dwelling older adults. *J Nutr Health Aging*. 1998;2(2):113−114.

81. Stewart R, Masaki K, Xue Q, Peila R, Petrovitch H, White L, et al. A 32-year prospective study of change in body weight and incident dementia: the honolulu-asia aging study. *Arch Neurol*. 2005;62(1):55−60.

82. Johnson DK, Wilkins CH, Morris JC. Accelerated weight loss may precede diagnosis in alzheimer disease. *Arch Neurol*. 2006;63(9):1312−1317.

83. Knopman DS, Edland SD, Cha RH, Petersen RC, Rocca WA. Incident dementia in women is preceded by weight loss by at least a decade. *Neurology*. 2007;69(8):739−746.

CHAPTER

6

Genetics of Infections and Diseases Caused by Human Parasites Affecting the Central Nervous System

Alain Dessein[1], Agnès Fleury[2], Hélia Dessein[1], and Sandrine Marquet[1]

[1]Aix Marseille University, Marseille, France [2]National Autonomous University of Mexico, Mexico City, México

6.1 INTRODUCTION

Human genetics has progressed considerably over the last 20 years, with most diseases shown to have significant genetic determinants. Conceptual and technical progress has led to partial elucidation of the control exerted by genetic variants on disease development. Human genetics is now widely used to analyze the mechanisms of diseases and is contributing to the development of personalized medicine and the identification of druggable targets. The impact of genetic variants on disease development was first analyzed in diseases with monogenic determinism caused by rare variants with large effects (high risk). The recent advent of high-throughput DNA sequencing and genotyping, and the availability of large cohorts of patients have made it possible to analyze the determinism of complex diseases dependent on a large number of rare (frequency <5%) and common alleles with modest effects. Here, we summarize the most recent and convincing findings concerning the genetic control of human infections with parasites of the central nervous system (CNS). Some of these diseases affect only

certain populations, particularly in Africa, but various genetic variants that were selected and reached equilibrium in ancestral African populations have been transported out of Africa and into Europe, Asia and America, by migrating human populations, on several occasions. These variants may modulate human susceptibility to other diseases of the CNS.

The most prevalent parasites affecting the CNS are the helminths *Taenia* (neurocysticercosis) and *Schistosoma* (neuroschistosomiasis), and the protozoans *Toxoplasma* (toxoplasmosis), *Trypanosoma* (sleeping sickness) and *Plasmodium* (cerebral malaria, CM). The observation of a markedly higher frequency of certain genetic variants of hemoglobin in regions of high malaria transmission provided the first strong evidence for an impact of host genetics on the outcome of parasite infections. Many studies have since investigated the genetics of susceptibility to malaria. However, analyses of the genetic determinism of human susceptibility to other parasites affecting the CNS, such as *Trypanosoma* and *Toxoplasma*, lag well behind those on malaria, principally due to difficulties in evaluating exposure to protozoan parasites and their various clones and variants.

DOI: http://dx.doi.org/10.1016/B978-0-12-804607-4.00006-X

Nevertheless, the very wide spectrum of symptoms observed in parasitic infections, from asymptomatic infection to lethal disease, with many intermediate clinical phenotypes, cannot be explained solely by differences in parasite virulence or exposure, because some endemic populations are highly exposed to many strains of the same parasite.

The heritability of parasitic diseases has been evaluated in various ways, the simplest and most direct of which is to look for correlations of disease between blood relatives (especially siblings), including monozygotic and dizygotic twins in particular. Differences in the prevalence of clinical phenotypes between ethnic groups may also suggest some degree of heritability of the phenotype studied. Analyses of this type have been performed in several well-documented studies on twins,[1–4] families[5] and ethnic groups[6–11] infected with *Plasmodium* and *Toxoplasma*. Strong familial correlations have been reported in subjects with severe schistosomiasis, and genetic modeling has identified the gene models best fitting the familial segregation of particular diseases. These models have been used to identify loci containing susceptibility variants. However, in most infectious diseases, association studies (AS) are the strategy most widely used to identify the allelic variants contributing to disease. AS test for a higher or lower frequency of genetic variants in patients than in healthy subjects. Cases and controls must be independent (not blood relatives). Such case–control studies are easy and cheap to perform, but they are often biased by cohort recruitment problems, because a good knowledge of the structure of the population and of family relationships is required to avoid population admixture and the selection of blood relatives. However, several methods, such as the Transmission Disequilibrium Test (TdT) method, which assesses allele transmission from heterozygous parents to affected children (see studies on *Toxoplasma* and malaria) can circumvent these problems. Furthermore, some methods can detect recruitment problems and take them into account in the analysis. Regardless of the method used, it is vital to replicate genetic associations in independent populations. The human genome contains millions of genetic polymorphisms, hundreds affecting each gene. It would therefore be too expensive to test all the genetic variants of each candidate gene. Fortunately, genetic polymorphisms can be grouped into correlation groups ($r^2 > 0.8$; or linkage disequilibrium groups), and testing can be limited to the allele (Tag allele) best correlated with all polymorphisms in the group, to the identification or exclusion of associations of any of the correlated alleles with disease.

In this chapter, we discuss the principal results obtained in studies of the genetic factors governing the development of neurocysticercosis (*Taenia solium*), toxoplasmosis (*Toxoplasma gondii*), and CM (*Plasmodium falciparum*). There has been little exploration of the genetics of neuroschistosomiasis and of amebiasis and these diseases will not, therefore, be discussed here.

6.2 NEUROCYSTICERCOSIS

Neurocysticercosis (NCC) is caused by cysts of *T. solium* accumulating in the human CNS. Humans infected with adult worms (through the ingestion of contaminated, undercooked pork meat) shed eggs into the environment. On ingestion, these eggs produce larvae (cysticerci) that perforate the gastro-intestinal tract and migrate to all types of tissues, including those of the brain. The parasite may even evolve between two closely related vertebrate hosts, increasing the fitness of the parasite and its adaptation to its hosts, humans in particular. Epileptic seizures are the most frequent clinical manifestation of brain invasion by the cysticerci, but NCC may also manifest as an increase in intracranial pressure, ischemic cerebrovascular disease, dementia and signs of spinal root/cord compression.[12–14] Combinations of two or more symptoms are common and depend on various factors, such as lesion number and location in the CNS, the developmental stage of the parasite and, most importantly, the intensity of the host inflammatory reaction.[15,16] NCC is thought to be responsible for between one-third and one-half of all epilepsy cases in the developing world. Pigs are the principal reservoir of the disease and, in poor populations in which pigs live among humans and can feed anywhere, epilepsy is frequent in certain families, suggesting a key role for genetic factors in disease development. This view is also supported by studies on pigs showing a familial clustering of *T. solium* NCC.[17] There has been some suggestion that NCC is controlled by major genes in human populations[18] as in human schistosome infections.[19–21] The difficulty of performing brain evaluations on a large number of individuals preclude, however, genetic studies on large cohorts. As described above, disease is associated with inflammation around brain cysts.[22–24] Subjects with non-inflammatory cysts in the brain are more likely to be asymptomatic than those with inflammatory cysts. Unsurprisingly, given the expected high level of parasite fitness of a parasite that cycles between two mammal hosts that are very closed, asymptomatic subjects account for up to 50% of subjects with cysticerci in the brain. Immunological studies in human populations have revealed the production of higher levels of pro-inflammatory cytokines, such as interferon (IFN)-gamma, interleukin (IL)-1beta, and tumor necrosis factor (TNF), and the presence of soluble cell adhesion molecules (sICAMs) in subjects with symptomatic NCC than in those with asymptomatic NCC. Genes encoding proteins promoting, terminating or modulating inflammation have been tested as candidates gene in AS. The glutathione S-transferase (GST) enzymes protect against the tissue damage caused by reactive oxygen species (ROS). Singh and coworkers[25] showed, in 75 symptomatic and 75 asymptomatic subjects, that those with deletions affecting the GSTM1 and GSTT1 genes were more likely to have symptomatic NCC. Deletions in the GSTM1 and GSTT1 genes were associated with lower levels of activity for the encoded enzymes. This findings has not yet been confirmed by an independent group. When a parasite invades a vertebrate host, the innate immunity of the host is triggered by pattern-recognition receptors, such as Toll-like receptors (TLRs). TLR stimulation leads to the induction of an inflammatory response and the development of acquired immunity. TLRs have also been implicated in the development of CNS inflammation and neurodegeneration. TLR4 serves as the receptor of LPS but it may also be activated by other types of lipid. *Plasmodium falciparum* produces, which can also activate TLR4. Two *TLR4* single nucleotide polymorphisms (SNPs) are co-expressed with high penetrance in African populations: *TLR4* 896A>G (corresponding to an Asp299Gly substitution mutation, SNP identification number: rs4986790) and *TLR4* 1196C>T (corresponding to a Thr399Ile mutation, SNP 4986791). Asp299Gly and Thr399Ile alter TLR-4 signaling by affecting TLR4 folding, cell surface expression and interactions with other regulatory proteins downstream in the regulatory pathway[26−28] inflammatory cytokines. These mutations were associated with hyporesponsiveness to LPS.[29] V.K Paliwar and coworkers[30] investigated these polymorphisms in 58 symptomatic subjects with epileptic seizures and positive serological results and 82 subjects with asymptomatic NCC from India. The number of *T. solium* cysts in the brain was similar in these two groups, but patients with symptoms had larger numbers of degenerating cysts, consistent with the view that disease is associated with cyst destruction by the host immune response. The Asp299Gly and Thr399 Ile substitutions were found to be associated with symptomatic NCC, in comparisons with asymptomatic patients and 150 healthy controls. Moreover, no difference in allelic variant frequency was found between asymptomatic cases and the healthy controls. These results suggest that these *TLR4* alleles are associated with neurological disease rather than with parasite control. The association of both these SNPs with disease is expected, because these two alleles are in total linkage disequilibrium

(LD = 1). The association of the second SNP with disease did not increase the strength of the association. In a more recent study, Singh and coworkers[31] investigated these polymorphisms in 143 Indian patients with epileptic seizures caused by solitary (single cyst) *Cysticercus granuloma* (SCG) evaluated by computed tomography. Edema and inflammation were not among the inclusion criteria, and patients with persistent intracranial pressure or evidence of progressive neurologic deficits were excluded. The controls were 134 healthy subjects matched with the patients for age and sex. The authors reported a convincing association of *TLR4* 1196C>T and a suggestive association of the *TLR4* A>G with SCG (in spite of LD = 1 since $r^2 = 0.79$ between these two SNPs). Moreover, SCG persisted 6 months after treatment in patients with both heterozygous genotypes. Seizure recurrence was also associated with both heterozygous genotypes, providing a link between cyst persistence and seizures. Together, these results indicate that *TLR4* rs4986790 and rs4986791 are associated with symptomatic NCC in Indian patients. These results now require confirmation in non-Asian populations.

Subjects with symptomatic NCC have high serum sICAM-1 concentrations.[32] ICAM-1 is a cell adhesion molecule involved in leukocyte activation, cell adhesion and migration to inflammatory sites. It may regulate inflammation, particularly at the blood−brain barrier. *ICAM1* SNP rs458 affects the binding of ICAM-1 to endothelial cells, macrophages and leukocytes. ICAM-1 variants have been shown to be associated with susceptibility to inflammatory diseases, including malaria. The ICAM-1 SNP 458 allelic variant was also found to be associated with symptomatic NCC, in a comparison to 75 symptomatic and 75 asymptomatic studies, whereas no association with NCC per se was found when NCC patients were compared with healthy subjects.[33]

Again, these results require confirmation in independent studies on larger samples, and an analysis of SNPs in tight linkage disequilibrium with these markers should be carried out to rule out the possibility of these SNPs being responsible for the association.

6.3 TOXOPLASMOSIS

Toxoplasma gondii is an obligate intracellular coccidian parasite that infects humans and a large number of other mammals and even birds. During acute infection, the infected macrophages disseminate the parasite throughout the host. In immunocompetent human adults and children (after the neonatal stage), toxoplasmosis is usually benign or asymptomatic, and acute infection progresses to the chronic stage without overt signs of disease. By contrast, congenital infections can be severe. They occur only if the mother suffers acute infection during pregnancy. One-third of mothers acquiring the infection during pregnancy pass it on to the fetus. Congenital infection can cause abortion or premature birth. The most severe clinical manifestations are retinochoroiditis (toxoplasmic retinochoroiditis, TR), psychomotor disturbances, intracerebral calcification, hydrocephaly, and microcephaly. Symptoms are present at birth in 20% of affected infants, and they develop by adolescence in 82%. Toxoplasmic encephalitis is also a common complication of AIDS associated with the reactivation of latent infection. *Toxoplama gondii*, which multiplies in the macrophage, stimulates the production of high levels of proinflammatory cytokines, which are thought to play a crucial role in innate immunity to *Toxoplasma* and in neurological disorders. More specifically, as reviewed by Fabiani[34] the T-cell/IFN-γ stimulates degradation of tryptophan to *N*-formylkynurenine by the immune system generating kynurenine (KIN) and 3-hydroxykynurenine, which are thought to play a role in pathophysiology of schizophrenia.[35] Administration of L-kynurenine induces a depressive-like behavior in mice.[35] Kynurenic acid (KYNA) is likely involved in the development of schizophrenia.[36] *Toxoplama gondii* activation of astrocytes, results in a high synthesis of KYNA. KYNA in the brain is able to inhibit the NMDA receptors for glutamate and the α7-nicotinic acid receptors for acetylcholine.[37] Thus, the inhibition exerted by KYNA on NMDA or α7nAch receptors, causes a reduction in glutamatergic and nicotinergic neurotransmission and could cause cognitive and sensory impairment.

IL-1β is required for the optimal polarization of IFN-γ-producing T cells; IL-1bβ is a key proinflammatory cytokine; it is generated from an inactive precursor by the inflammasomes, the role of which in immunity to *Toxoplasma* has been investigated in outstanding studies on both animals[38] and humans.[39–41] Nucleotide-binding oligomerization domain (NOD)-like receptors (NLRs), the principal components of inflammasomes, are intracellular immune receptors with leucine-rich repeats (LRRs) and a nucleotide-binding domain (NBD) (for a review see Refs. [42,43]). The LRR domains play a role in autoregulation, and the recognition of pathogen-associated molecular patterns (PAMPs), and/or protein–protein interactions. NLRs have different N-terminal domains. All the NLRs in one large group (NLRPs or NALPs) have an N-terminal pyrin domain (PYD); all those in another group containing the NOD–containing proteins 1 and 2 (NOD1, NOD2) have an N-terminal caspase recruitment domain (CARD). In response to activators, members of the NLR family assemble into multimolecular complexes that activate caspases. Activated caspase-1 controls the maturation of the cytokines of the IL-1 family. The multimolecular NLR complexes formed upon activation are referred as to "inflammasomes" because they activate inflammatory caspases. Caspase-1 cleaves other members of the IL-1 family, such as pro-IL-18 and IL-33. IL-1 is required for the optimal polarization of IFN-γ-producing T cells directly required for the killing of intracellular pathogens. Under certain circumstances, cytokine release by inflammasomes occurs at the same time as a caspase-1-induced type of inflammatory cell death, pyroptosis. Pyroptosis is characterized by caspase-1-dependent DNA fragmentation and pore formation, leading to cellular lysis.

Studies in rats identified a locus (Toxo-1) controlling susceptibility to toxoplasmosis.[44] Genetic analysis narrowed the locus involved down to a region containing 29 genes including the NOD-like receptor NLRP1a/caspase-1, which was identified as one of the best candidates. Functional studies demonstrated that the Toxo1-resistant locus was associated with the ability of macrophages to limit *T. gondii* growth and to induce the death of infected macrophages.[38] Parasite-induced cell death in macrophages displayed pyroptosis-like features and was dependent on caspase-1 activation and IL1-beta secretion.[38]

NALP1 belongs to the NLR family of proteins that form inflammasomes. Witola and coworkers[39] demonstrated that allelic variants of NALP1 were associated with greater susceptibility to congenital toxoplasmosis in humans. They also used RNA interference to knock down the level of NALP1 expression in a monocytic cell line. NALP1 silencing attenuated the progression of *T. gondii* infection and accelerated host cell death. Monocytic lines in which NALP1 expression was knocked down displayed no upregulation of IL-1β, IL-18, and IL-12 on infection with *T. gondii*. The NALP1 inflammasome is, therefore, critical for innate immune responses to *T. gondii* infection and pathogenesis. Similarly, Dutra and coworkers[40] reported an association ($p = 0.04$) between TR and polymorphisms of the NOD2 gene. They found no difference in IFN-γ production (which is highly dependent on IL-12) between patients with symptomatic disease and asymptomatic individuals. However, interleukin 17A (IL-17A) levels were higher in patients with symptomatic disease, suggesting a possible effect of NOD2 on IL-17A production by IFN-γ-negative TH17 cells in human toxoplasmosis. Other studies, including that by Hsu and coworkers,[45] have shown that challenging macrophages with muramyl dipeptide induces the formation of NOD2, NALP1 and caspase-1 complexes, resulting in the induction of IL-1β in a NOD2- and caspase-1-dependent manner.

During the acute phase of inflammation following tissue injury, the ATP and UTP released from apoptotic cells signal through P2 purinergic receptors, to recruit monocytes, dentritic cells (DCs) and neutrophils. In inflamed tissues, the ATP released from damaged cells binds to low-affinity P2X7Rs on macrophages, activates the inflammasome and stimulates the secretion of IL-1β. Blackwell and coworkers[41] investigated the possible role of one of these receptors (P2X7 encoded by *P2RX7*) in susceptibility to congenital toxoplasmosis, by studying polymorphisms of *P2RX7*. They reported an association ($p = 0.015$) between clinical toxoplasmosis and the rs1718119 SNP (1068T>C; Thr348Ala). An analysis of clinical subgroups found no association with hydrocephalus, but associations with retinal disease and brain calcifications were detected. Moreover, the association with TR ($p = 0.002$) was replicated in a family-based study (60 families; 68 affected offspring) in Brazil.[41]

Given the results of these studies, polymorphisms of IL-1β might have been expected to modulate susceptibility to ocular toxoplasmosis and/or congenital toxoplasmosis. Two studies[46,47] reported associations between IL1β polymorphisms and TR or congenital toxoplasmosis. However, these associations were weak and require confirmation in additional studies. TNF is a proinflammatory cytokine that is thought to have an early protective effect in *T. gondii* infections since it increases the microbicidal activity of the macrophages and stimulates INF-γ production by neurokinin cells. Work in mice showed that variants of TNF expressing high levels of TNF mRNA are resistant to toxoplasmic encephalitis in mice. Suzuki and coworkers,[48] however, suggested that another gene in the TNF genetic region accounted for the association; finally Cordeiro and coworkers[49] tested the functional TNF (−308G>A) polymorphisms in TR and found no association.

IL-10 is a cytokine with strong anti-inflammatory effects that can be recruited by the inflammasomes. IL-10 has been implicated in the modulation of TR. Cordeiro and coworkers[50] analyzed the association of SNP in IL10 promoter (−1082 (G>A)) with TR in 100 patients and 100 controls (with anti-toxoplasma IgG, without ocular disease). Previous studies had shown that −1082 AA was associated with lower IL-10 production. The authors found association between TR and the AA genotype supporting the view that IL-10 protects against TR. This result was not entirely expected since a previous study[51] on susceptibility to another protozoan parasite (*Leishmania braziliensis*) had shown that this same genotype (−1082 AA) reduced the risk of severe lesions in *L. braziliensis*-infected subjects (TdT on 148 trios), indicating that increasing IL-10 inhibited anti-Leishmania immunity. This illustrates that Il-10 may have the opposite role on disease development depending on which arm of immunity—clinical or sterile—has more crucial effects.

IL-6 is produced in large amounts in human and experimental toxoplasmosis. IL-1β and TNF are major activators of IL6 expression. IL-6 has a broad effect on cells in and out of the immune system (see review Ref. [52]). IL-6 has pro- and anti-inflammatory properties depending on the tissue context. While IL-6 has a protective role in many infections, it also plays a key role in maintaining chronic, deleterious inflammation. Transgenic mice with increased expression of IL-6 develop various disorders, including neurological disease when IL-6 is overexpressed in the CNS. Injection of antibody against IL-6 reduces Toxoplasma cyst numbers and reduces brain inflammation. Moreover, IL-6 inhibits IFN-γ-mediated Toxoplasma killing in vitro.[53] Other studies have shown, however, that mice deficient in IL-6 exhibit high numbers of brain cysts and die of toxoplasmic encephalitis, these mice also develop more severe inflammation in the retina and vitreous humor. A G-to-C mutation in the 5′ flaking region of the gene was shown to lower levels of IL6 expression.[54] This allele was associated with protection against inflammatory disease like early-onset juvenile arthritis. The −174 C allele has been associated with an increased risk of congenital toxoplasmosis[47] and with a higher occurrence of toxoplasmic retinochoroiditis.[55] These results support the conclusion that IL-6 has protective effects against *T. gondii* congenital infection and against TR.

6.4 CEREBRAL MALARIA

CM and severe anemia (SA) are the most severe clinical complications of *P. falciparum* infection. Malaria occurs mostly in young children, killing about half a million African children per year. CM is a reversible encephalopathy characterized by seizures and a loss of consciousness. SA results from a combination of enhanced red blood cell destruction and inefficient red blood cell production. Little is known about the pathophysiology of CM in humans, but it is thought to involve the sequestration of parasites in the small blood vessels of the brain and the deregulation of key immune system elements. Disease outcome is influenced by several factors, including environmental and host factors that may play an important role in disease severity. It has recently been suggested that host nutritional status can affect the outcome of CM in mice through metabolic effects on host immune response and the parasite.[56] Dietary restriction (DR) seems to prevent neurological disease without affecting parasite growth in experimental CM, by reducing parasite accumulation in the brain and increasing parasite clearance from the spleen. The authors showed that leptin, a cytokine secreted by adipocytes upon exposure to pro-inflammatory stimuli, such as the malaria parasite, was involved in pathological extracellular matrix (ECM) features and that DR protected against disease by decreasing leptin production,[56] thereby decreasing mTORC1 activity in T cells. There is currently no evidence for a host genetic factor common to CM and neurodegenerative diseases, but it has been suggested that epilepsy is a consequence of CM in areas of endemic malaria.[57−60]

Studies of genetic susceptibility to severe malaria in humans have a long history, starting with the identification of the sickle hemoglobin variant as a major resistance factor. However, despite the large number of studies, the full impact of human genetics on resistance to the disease remains largely unknown. Several genetic mutations have been shown to be associated with resistance/susceptibility to disease, but they can explain only a small proportion of cases. There are, therefore, almost certainly other unknown gene mutations. For this reason, new approaches have been developed in recent years, based on the use of the most recent technological advances to identify new susceptibility genes.

Several malaria phenotypes, including CM, severe anemia and respiratory distress, have been considered in genetic studies. These phenotypes were evaluated separately or in the same cohort, to evaluate the possible association between certain polymorphisms and susceptibility/resistance to severe disease. Most of the studies carried out were AS, few studies to date have been performed with a family-based cohort (TdT), to avoid the selection of the control group and the introduction of biases due to population stratification. The genes found to be associated with severe disease are summarized in Table 6.1. The most convincing results obtained are those demonstrating the protective role of hemoglobinopathies, with sickle-cell disease (HbS mutation of the alphaglobin gene) potentially protecting against severe clinical forms of malaria in different populations (Table 6.1). These findings build on the observations reported by Allison,[61] highlighting a similarity between the geographic distribution of this hemoglobinopathy (with a high frequency of this mutation, despite its deleterious nature) and regions of endemic *P. falciparum* malaria. Since the first demonstration of an association between HbS and protection against malaria 50 years ago, several studies have confirmed this result in genetically different populations (Table 6.1), suggesting that *P. falciparum* may exert a strong selection pressure in areas of endemic disease.

Other genes encoding proteins involved in red blood cell physiology, such as HBA, ABO, G6PD, SLC4A1, and HP, have been shown to be associated with severe disease (Table 6.1). A number of AS have also detected associations between severe malaria and candidate genes involved in both the sequestration of *P. falciparum* in blood vessels within the brain (CD36, ICAM1, and CR1) and immune responses (Table 6.1). Several recent studies have shown mutations of the gene encoding the endothelial protein C receptor (EPCR) may be involved in malaria pathogenesis,[62] with lower levels of this receptor in patients with CM.[63,64] One polymorphism (rs867186) has been associated with protection against severe malaria.[65,66]

The possible association of some polymorphisms with severe disease has been studied in detail, but often with conflicting results or the demonstration of an absence of association. These conflicting results suggest several hypotheses including epistatic interactions, as described for the common variant of haptoglobin and alpha+ thalassemia.[65,67]

Technological advances have recently made it possible to carry out genome-wide association studies (GWAS) to identify susceptibility genes without the need for prior hypotheses concerning gene function. The first such study was performed by Jallow et al.,[68] who genotyped 500,000 SNPs in a large discovery cohort including 2500 children from Gambia (Table 6.2) and then performed a replication study on 3400 children. The cases were subjects admitted to hospital with severe malaria (82% with CM, 30% with severe malarial anemia, and 11% with respiratory distress). Surprisingly, this study was only partly successful. It did not lead to the identification of any new genes associated with severe malaria, but it did confirm the protective role of HbS. This problem may have resulted from strong phenotypic heterogeneity, with overlapping clinical entities, due to the considerable population stratification reported by the authors and differences in haplotype structure between ethnic groups that greatly attenuated the signals of association. These results raise questions about how best to conduct effective GWAS in Africa. Three years after this initial study, the second genome-wide association study identified two new loci associated with resistance to severe malaria—ATP2B4 and MARVELD3 (Table 6.2)[69]—and confirmed previous reports on the protective effects of the sickle-cell trait and blood group O. The study included 1325 cases of severe anemia or CM and 828 unaffected controls from Ghana. The combination of genotyping and genome-wide imputation resulted in a total of 5,010,634 SNPs being available for further analysis. Replication experiments were performed in an additional 1320 cases and 2222 controls from the same population. These four

TABLE 6.1 Genetic Associations Reported in Malaria Studies

Locus	Phenotype	Study location	Reference
HBB	SM	Nigeria	Gilles et al. (1967)
(HbS)	SM	Gambia	Hill et al. (1991)
HBB	SM, SA	Kenya	Aidoo et al. (2002)
(HbC)	SM	Ghana	Mockenhaupt (2004a)
	SM	Gambia	Ackerman et al. (2005)
	SM	Mali	Toure et al. (2012)
	SM	Tanzania	Manjurano et al. (2012)
	SM	Cameroon	Apinjoh et al.
	SM	Kenya	Atkinson et al. (2014)
	SM	Gambia	Shah et al. (2016)
	SM	Ghana	Mockenhaupt (2004)
	SM	Mali	Agarwal et al. (2000)
HBA	SM	Papua New Guinea	Allen et al. (1997)
	SM	Ghana	Mockenhaupt (2004)
	SM	Kenya	Williams et al. (2005)
	SM	Kenya	Atkinson et al. (2014)
ABO	SM	Zimbabwe	Fischer et al. (1998)
	SM	Gabon	Lell et al. (1999)
	SM	Mali	Toure et al. (2012)
	SM	Gambia, Kenya, Malawi	Rowe et al. (2007)
	SM		Fry et al. (2008)
G6PD	SM	Nigeria	Gilles et al. (1967)
	SM	Gambia and Kenya	Ruwende et al. (1995)
	SM	Tanzania	Manjurano et al. (2012)
	CM	Gambia	Shah et al. (2016)
SLC4A1 (band 3)	CM	Papua New Guinea	Allen et al. (1999)
	CM	Papua New Guinea	Genton et al. (1995)
HP	SM	Kenya	Atkinson et al. (2014)
	CM	Sudan	Elagib et al. (1998)
CD36	SM	Kenya	Pain et al. (2001)
	CM	Thailand	Omi et al. (2003)
ICAM1	CM	Kenya	Fernandez-Reyes et al. (1997)
	SM	Gabon	Kun et al. (1999)
	SM	Vietnam	Dunstan et al. (2012)
CR1	CM	Thailand	Teeranaipong et al. (2008)
	CM	India	Panda et al. (2012)
EPCR	SM	Thailand	Naka et al. (2014)
	SM	Uganda	Shabani et al. (2016)

(Continued)

TABLE 6.1 (Continued)

Locus	Phenotype	Study location	Reference
NOS2	FCM	Gambia	Burgner et al. (1998)
	SM	Gabon	Kun et al. (1998)
	CM	Gambia	Burgner et al. (2003)
	SM	Thailand	Ohashi et al. (2002)
HLA-B	SM	Gambia	Hill et al. (1991)
HLA-DR	SM	Gambia	Hill et al. (1991)
HLA-A	CM	Mali	Lyke et al. (2011)
TNF	CM	Gambia	McGuire et al. (1994)
	FM	Gambia	McGuire et al. (1994)
	SM	Thailand	Knight et al. (1999)
	CM	Thailand	Ubalee et al. (2001)
	CM	Vietnam	Hananantachai (2007)
	SM	Nigeria	Dunstan et al. (2012)
	SM		Olaniyan et al. (2016)
IL1A	SM	Vietnam	Dunstan et al. (2012)
IFNG	CM	Mali	Cabantous et al. (2005)
	SM	Saudi Arabia	Nasr et al. (2014)
IL12B	CM(death)	Tanzania	Morahan et al. (2002)
	CM	Mali	Marquet et al. (2008)
	CM	Thailand	Naka et al. (2009)
IL4	SM	Mali	Cabantous et al. (2009)
IL17F	CM	Mali and Nigeria	Marquet et al. (2016)
IL17RC	SM	Vietnam	Dunstan et al. (2012)
IL13	SM	Thailand	Naka et al. (2009)
	SM	Vietnam	Dunstan et al. (2012)
CD40L	SM	Gambia	Sabeti et al. (2002)
IFNGR1	CM	Gambia	Koch et al. (2002)
TLR4	SM	Ghana	Mockenhaupt et al. (2006)
TLR1	SM	Melanesia	Manning et al. (2016)
MBL2	SM	Gabon	Luty et al. (1998)
	SM	Gabon	Boldt et al. (2006)
	SM	Ghana	Holmberg et al. (2008)
	SM	India	Jha et al. (2014)
FOXO3A	SM	Gabon	Nguetse et al. (2015)
ABCA1	CM	India	Sahu et al. (2013)
RNASE3	CM	Ghana	Adu et al. (2011)
ADAMTS13	CM	Thailand	Kraisin et al. (2011)

ABCA1, ATP-binding cassette transporter 1; ABO, blood group ABO system transferase; ADAMTS13, Willebrand factor-cleaving protease; CD36, CD36 molecule; CD40L, CD40 ligand; CM, cerebral malaria; CR1, complement component 3b/4b receptor 1; EPCR, endothelium protein receptor; FCM, fatal cerebral malaria; FM, fatal malaria; FOXO3A, forkhead box protein 3; G6PD, glucose-6-phosphate dehydrogenase; HBB/HBA, HBB/HBA gene encoding the hemoglobin subunit; HLA, human leukocyte antigen; HP, haptoglobin; ICAM1, intercellular adhesion molecule 1; IFNG, interferon-γ; IL, interleukin; IL17FRc, IL-17 receptor C; MBL2, mannose binding lectin 2; NOS2, nitric oxide synthase 2; RNASE3, ribonuclease A family member 13; SA, severe anemia; SLC4A1, solute carrier family 4 member 1; SM, severe malaria; TLR1,4, Toll like receptor 1,4; TNF, tumor necrosis factor.

TABLE 6.2 Malaria: Loci Identified by Association Studies on Large Case—Control Cohorts

Phenotypes	Populations	Type of study	Discovery study	Replication study	Gene	Reference
SM: CM, SA, RD	Gambia	GWAS	1,060 SM 1,500 Co	1,087 SM 2,376 Co	HBB (HbS)	*Nat. Genetics* [68]
SM: CM, SA	Ghana	GWAS	1,325 SM 828 Co	1,320 SM 2,222 Co	ATP2B4 MARVELD3 ABO HBB (HbS)	*Nature* [69]
SM: CM, SA, CM + SA	12 locations Africa, Asia Oceania	Candidate genes	11,890 SM 17,441 Co	None	HBB (HbS) ABO ATP2B4 G6PD CD40L	*Nat. Genetics* [70]
SM	Burkina, Cameroon, Gambia, Ghana, Malawi, Mali, Tanzania	GWAS	5633 SM 5919 Co	13,946 SM/ Co	HBB (HbS) ABO ATP2B4 FREM3 GYPE, GYPB, GYPA	*Nature* [71]

ATP2B4, plasma membrane Ca-transporting ATPase 4; CM, cerebral malaria; Co, control; DR, respiratory distress; FREM3, FRAS1-related extracellular matrix 3; GWAS, genome-wide association study; GYPE, A, B, glycophorin E, A, B; MARVELD3, MARVEL domain-containing protein 3; SA, severe anemia; SM, severe malaria.

signal SNPs were then tested in an independent replication group constituted from data provided by the MalariaGEN Network and including Gambian children with severe malaria.[68] The same risk alleles were identified, with similar ORs. The ATP2B4 gene encodes the plasma membrane Ca^{2+}-ATPase type 4 (PMCA4), a major erythrocyte Ca^{2+} pump.[72] The product of MARVELD3 is part of the tight-junction structures of epithelial and vascular endothelial cells.[73]

Genes previously associated with severe malaria in case—control studies were evaluated in a large multi-center case—control study of 11,890 cases with severe malaria and 17,441 controls from 12 locations in Africa, Asia and Oceania.[70,74] Multicenter studies involve larger numbers of subjects, but may also result in the introduction of considerable genotypic and phenotypic heterogeneity. A Bayesian framework for analysis has been proposed, to overcome the problems associated with heterogeneity across different populations. In total, 27 genes were tested, and there was strong evidence of association ($p < 10^{-4}$) for SNPs in five genes (ABO, ATP2B4, CD40LG, G6PD, and HBB), and no strong evidence of interaction between these genes. HbS (rs334) and ABO (rs8176719) were confirmed to be associated with resistance to severe malaria, with reduced risks of about 90% and 30%, respectively.[70] This study indicated that G6PD deficiency (rs1050828) was associated with opposite effects on different severe complications of *P. falciparum* infection, conferring a higher risk of severe malarial anemia but a lower risk of CM. This finding suggests that the evolutionary origins of

this genetic disorder are more complex than initially thought. The ATP2B4 gene, previously identified in a genome-wide association study,[69] was also associated with resistance/susceptibility to severe malaria in this multicenter study, with two SNPs (rs10900585, rs55868763) identified as increasing the risk and two other SNPs (rs4951074, rs1541255) found to decrease it. For CD40LG, the SNP rs3092945 was associated with a lower risk of severe malaria when data for all the sites were included, but the results obtained were not consistent between sites. One key finding is that most previously reported candidate gene associations are not clearly replicated, even using statistical methods taking heterogeneity across populations into account. In 2015, a multicenter genome-wide association study was performed on 5633 children with severe malaria and 5919 controls from Gambia, Kenya, and Malawi.[71] In total, 2.5 million SNPs were used and appropriate statistical approaches were developed to take heterogeneity between study populations into account. Several previously reported GWAS signals, such as HBB, ABO, and ATP2B4 have been replicated,[68,69,75] and new candidate loci have been identified.[71] The strongest signal of association was obtained for rs184895969, which is located between the FREM3 gene and a cluster of three glycophorin genes (GYPE, GYPB, and GYPA) involved in the invasion of erythrocytes by *P. falciparum*. A replication study was performed to evaluate the candidate SNPs in a case—control sample including 13,946 individuals from Burkina Faso, Cameroon, Gambia, Ghana, Malawi, Mali, and Tanzania. An analysis of the combined 25,498 samples provided convincing evidence of association with rs186873296, which is located close to rs184895969, with which it is in strong LD ($r^2 > 0.9$). The minor allele (non-ancestral) reduced the risk of severe malaria by about 40%. The protective allele at rs18495969 was associated with higher levels of GYPA transcription, as reported in public gene expression data for HapMap.[76] However, the complex pattern of variation in the glycophorin gene cluster makes it difficult to identify the causal variant with any degree of certainty. The loci reaching significance according to the conventional criteria for genome-wide association ($p < 5 \times 10^{-8}$) correspond to genes playing a key role in erythrocytes, the primary host cell of *P. falciparum*, or located close to such genes. The FREM3/GYPE region coincides with a locus thought to have been subject to ancient balancing selection, identified by an analysis of haplotype sharing between humans and chimpanzees. This finding suggests that large, multisite genetic studies may be of scientific interest but that such studies at present are challenging in African populations.

The Hapmap Project[77] and the 1000 Genomes Project[78] have greatly enhanced our understanding of genetic variation in general, but the characterization of African populations remains limited. The African Genome Variation Project (AGVP) was recently developed to investigate the genetic diversity of African populations in detail, which should facilitate studies of human malaria susceptibility. The AGVP is an international collaboration involving the use of both dense genotypes from 1481 individuals from 18 African populations (including two populations from the 1000 Genomes Project) and whole genome sequences (WGS) from 320 individuals from across Sub-Saharan Africa (SSA).[79] Evidence has been obtained for historically complex and regionally different admixtures with multiple hunter-gatherer and Eurasian populations across SSA, and new loci under positive selection have been identified for malaria resistance. This project should provide a useful data resource for researchers performing genetic studies in African populations.

References

1. Couvreur J, Desmonts G, Girre JY. Congenital toxoplasmosis in twins: a series of 14 pairs of twins: absence of infection in one twin in two pairs. *J Pediatr*. 1976;89:235−240.
2. Jepson AP, Banya WA, Sisay-Joof F, Hassan-King M, Bennett S, Whittle HC. Genetic regulation of fever in *Plasmodium falciparum* malaria in Gambian twin children. *J Infect Dis*. 1995;172:316−319.
3. Jepson A, Sisay-Joof F, Banya W, et al. Genetic linkage of mild malaria to the major histocompatibility complex in Gambian children: study of affected sibling pairs. *BMJ*. 1997;315:96−97.
4. Peyron F, Ateba AB, Wallon M, et al. Congenital toxoplasmosis in twins: a report of fourteen consecutive cases and a comparison with published data. *Pediatr Infect Dis J*. 2003;22:695−701.
5. Ranque S, Safeukui I, Poudiougou B, et al. Familial aggregation of cerebral malaria and severe malarial anemia. *J Infect Dis*. 2005;191:799−804.
6. Arama C, Maiga B, Dolo A, et al. Ethnic differences in susceptibility to malaria: what have we learned from immuno-epidemiological studies in West Africa? *Acta Trop*. 2015;146:152−156.
7. Cherif M, Amoako-Sakyi D, Dolo A, et al. Distribution of FcγR gene polymorphisms among two sympatric populations in Mali: differing allele frequencies, associations with malariometric indices and implications for genetic susceptibility to malaria. *Malar J*. 2016;15:29.
8. Dolo A, Modiano D, Maiga B, et al. Difference in susceptibility to malaria between two sympatric ethnic groups in Mali. *Am J Trop Med Hyg*. 2005;72:243−248.
9. Israelsson E, Maiga B, Kearsley S, et al. Cytokine gene haplotypes with a potential effect on susceptibility to malaria in sympatric ethnic groups in Mali. *Infect Genet Evol*. 2011;11:1608−1615.
10. Maiga B, Dolo A, Touré O, et al. Human candidate polymorphisms in sympatric ethnic groups differing in malaria susceptibility in Mali. *PLoS One*. 2013;8:e75675.
11. Blocher J, Schmutzhard E, Wilkins PP, et al. A cross-sectional study of people with epilepsy and neurocysticercosis in Tanzania: clinical characteristics and diagnostic approaches. *PLoS Negl Trop Dis*. 2011;5:e1185.
12. Pittella JE. Neurocysticercosis. *Brain Pathol*. 1997;7:681−693.
13. Mahale RR, Mehta A, Rangasetty S. Extraparenchymal (racemose) neurocysticercosis and its multitude manifestations: a comprehensive review. *J Clin Neurol*. 2015;11:203−211.
14. Carabin H, Ndimubanzi PC, Budke CM, et al. Clinical manifestations associated with neurocysticercosis: a systematic review. *PLoS Negl Trop Dis*. 2011;5:e1152.
15. Fleury A, Cardenas G, Adalid-Peralta L, Fragoso G, Sciutto E. Immunopathology in *Taenia solium* neurocysticercosis. *Parasite Immunol*. 2016;38:147−157.
16. Fleury A, Escobar A, Fragoso G, Sciutto E, Larralde C. Clinical heterogeneity of human neurocysticercosis results from complex interactions among parasite, host and environmental factors. *Trans R Soc Trop Med Hyg*. 2010;104:243−250.
17. Sciutto E, Martínez JJ, Huerta M, et al. Familial clustering of *Taenia solium* cysticercosis in the rural pigs of Mexico: hints of genetic determinants in innate and acquired resistance to infection. *Vet Parasitol*. 2003;116:223−229.
18. Fleury A, Gomez T, Alvarez I, et al. High prevalence of calcified silent neurocysticercosis in a rural village of Mexico. *Neuroepidemiology*. 2003;22:139−145.
19. Abel L, Demenais F, Prata A, Souza AE, Dessein A. Evidence for the segregation of a major gene in human susceptibility/resistance to infection by Schistosoma mansoni. *Am J Hum Genet*. 1991;48:959−970.
20. Rodrigues V, Abel L, Piper K, Dessein AJ. Segregation analysis indicates a major gene in the control of interleukin-5 production in humans infected with Schistosoma mansoni. *Am J Hum Genet*. 1996;59:453−461.
21. Dessein AJ, Hillaire D, Elwali NE, et al. Severe hepatic fibrosis in Schistosoma mansoni infection is controlled by a major locus that is closely linked to the interferon-gamma receptor gene. *Am J Hum Genet*. 1999;65:709−721.
22. Chavarría A, Roger B, Fragoso G, et al. TH2 profile in asymptomatic *Taenia solium* human neurocysticercosis. *Microbes Infect*. 2003;5:1109−1115.
23. Chavarría A, Fleury A, García E, Márquez C, Fragoso G, Sciutto E. Relationship between the clinical heterogeneity of neurocysticercosis and the immune-inflammatory profiles. *Clin Immunol*. 2005;116:271−278.

24. Sáenz B, Fleury A, Chavarría A, et al. Neurocysticercosis: local and systemic immune-inflammatory features related to severity. *Med Microbiol Immunol.* 2012;201:73−80.

25. Singh A, Prasad KN, Singh AK, et al. Human glutathione s-transferase enzyme gene polymorphisms and their association with neurocysticercosis. *Mol Neurobiol.* 2016.

26. Ohto U, Yamakawa N, Akashi-Takamura S, Miyake K, Shimizu T. Structural analyses of human Toll-like receptor 4 polymorphisms D299G and T399I. *J Biol Chem.* 2012;287:40611−40617.

27. Yamakawa N, Ohto U, Akashi-Takamura S, et al. Human TLR4 polymorphism D299G/T399I alters TLR4/MD-2 conformation and response to a weak ligand monophosphoryl lipid A. *Int Immunol.* 2013;25:45−52.

28. Figueroa L, Xiong Y, Song C, Piao W, Vogel SN, Medvedev AE. The Asp299Gly polymorphism alters TLR4 signaling by interfering with recruitment of MyD88 and TRIF. *J Immunol.* 2012;188:4506−4515.

29. Smirnova I, Mann N, Dols A, et al. Assay of locus-specific genetic load implicates rare Toll-like receptor 4 mutations in meningococcal susceptibility. *Proc Natl Acad Sci USA.* 2003;100:6075−6080.

30. Verma A, Prasad KN, Gupta RK, et al. Toll-like receptor 4 polymorphism and its association with symptomatic neurocysticercosis. *J Infect Dis.* 2010;202:1219−1225.

31. Singh A, Garg RK, Jain A, et al. Toll like receptor-4 gene polymorphisms in patients with solitary *Cysticercus granuloma*. *J Neurol Sci.* 2015;355:180−185.

32. Prasad A, Prasad KN, Gupta RK, Pradhan S. Increased expression of ICAM-1 among symptomatic neurocysticercosis. *J Neuroimmunol.* 2009;206:118−120.

33. Singh A, Singh AK, Singh SK, Paliwal VK, Gupta RK, Prasad KN. Association of ICAM-1 K469E polymorphism with neurocysticercosis. *J Neuroimmunol.* 2014;276:166−171.

34. Fabiani S, Pinto B, Bonuccelli U, Bruschi F. Neurobiological studies on the relationship between toxoplasmosis and neuropsychiatric diseases. *J Neurol Sci.* 2015;351:3−8.

35. Hinze-Selch D, Däubener W, Eggert L, Erdag S, Stoltenberg R, Wilms S. A controlled prospective study of *Toxoplasma gondii* infection in individuals with schizophrenia: beyond seroprevalence. *Schizophr Bull.* 2007;33:782−788.

36. Schwarcz R, Hunter CA. *Toxoplasma gondii* and schizophrenia: linkage through astrocyte-derived kynurenic acid? *Schizophr Bull.* 2007;33:652−653.

37. Miüller N, Schwarz MJ. The immunological basis of glutamatergic disturbance in schizophrenia: towards an integrated view. *J Neural Transm Suppl.* 2007;(72):269−280.

38. Cavailles P, Flori P, Papapietro O, et al. A highly conserved Toxo1 haplotype directs resistance to toxoplasmosis and its associated caspase-1 dependent killing of parasite and host macrophage. *PLoS Pathog.* 2014;10:e1004005.

39. Witola WH, Mui E, Hargrave A, et al. NALP1 influences susceptibility to human congenital toxoplasmosis, proinflammatory cytokine response, and fate of *Toxoplasma gondii*-infected monocytic cells. *Infect Immun.* 2011;79:756−766.

40. Dutra MS, Béla SR, Peixoto-Rangel AL, et al. Association of a NOD2 gene polymorphism and T-helper 17 cells with presumed ocular toxoplasmosis. *J Infect Dis.* 2013;207:152−163.

41. Jamieson SE, Peixoto-Rangel AL, Hargrave AC, et al. Evidence for associations between the purinergic receptor P2X(7) (P2RX7) and toxoplasmosis. *Genes Immun.* 2010;11:374−383.

42. Stutz A, Golenbock DT, Latz E. Inflammasomes: too big to miss. *J Clin Invest.* 2009;119:3502−3511.

43. Latz E, Xiao TS, Stutz A. Activation and regulation of the inflammasomes. *Nat Rev Immunol.* 2013;13:397−411.

44. Cavaillès P, Sergent V, Bisanz C, et al. The rat Toxo1 locus directs toxoplasmosis outcome and controls parasite proliferation and spreading by macrophage-dependent mechanisms. *Proc Natl Acad Sci USA.* 2006;103:744−749.

45. Hsu LC, Ali SR, McGillivray S, et al. A NOD2-NALP1 complex mediates caspase-1-dependent IL-1beta secretion in response to Bacillus anthracis infection and muramyl dipeptide. *Proc Natl Acad Sci USA.* 2008;105:7803−7808.

46. Cordeiro CA, Moreira PR, Costa GC, et al. Interleukin-1 gene polymorphisms and toxoplasmic retinochoroiditis. *Mol Vis.* 2008;14:1845−1849.

47. Wujcicka W, Gaj Z, Wilczyński J, Nowakowska D. Contribution of IL6 -174 G>C and IL1B +3954 C>T polymorphisms to congenital infection with *Toxoplasma gondii*. *Eur J Clin Microbiol Infect Dis.* 2015;34:2287−2294.

48. Suzuki Y, Joh K, Orellana MA, Conley FK, Remington JS. A gene(s) within the H-2D region determines the development of toxoplasmic encephalitis in mice. *Immunology.* 1991;74:732−739.

49. Cordeiro CA, Moreira PR, Costa GC, et al. TNF-alpha gene polymorphism (-308G/A) and toxoplasmic retinochoroiditis. *Br J Ophthalmol.* 2008;92:986−988.

50. Cordeiro CA, Moreira PR, Andrade MS, et al. Interleukin-10 gene polymorphism (-1082G/A) is associated with toxoplasmic retinochoroiditis. *Invest Ophthalmol Vis Sci.* 2008;49:1979−1982.

51. Salhi A, Rodrigues V, Santoro F, et al. Immunological and genetic evidence for a crucial role of IL-10 in cutaneous lesions in humans infected with *Leishmania braziliensis*. *J Immunol.* 2008;180:6139−6148.

52. Hunter CA, Jones SA. IL-6 as a keystone cytokine in health and disease. *Nat Immunol.* 2015;16:448−457.

53. Beaman MH, Hunter CA, Remington JS. Enhancement of intracellular replication of *Toxoplasma gondii* by IL-6. Interactions with IFN-gamma and TNF-alpha. *J Immunol.* 1994;153:4583−4587.

54. Fishman D, Faulds G, Jeffery R, et al. The effect of novel polymorphisms in the interleukin-6 (IL-6) gene on IL-6 transcription and plasma IL-6 levels, and an association with systemic-onset juvenile chronic arthritis. *J Clin Invest.* 1998;102:1369−1376.

55. Cordeiro CA, Moreira PR, Bessa TF, et al. Interleukin-6 gene polymorphism (-174G/C) is associated with toxoplasmic retinochoroiditis. *Acta Ophthalmol.* 2013;91:e311−e314.

56. Mejia P, Treviño-Villarreal JH, Hine C, et al. Dietary restriction protects against experimental cerebral malaria via leptin modulation and T-cell mTORC1 suppression. *Nat Commun.* 2015;6:6050.

57. Ngoungou EB, Dulac O, Poudiougou B, et al. Epilepsy as a consequence of cerebral malaria in area in which malaria is endemic in Mali, West Africa. *Epilepsia.* 2006;47:873−879.

58. Ngoungou EB, Koko J, Druet-Cabanac M, et al. Cerebral malaria and sequelar epilepsy: first matched case-control study in Gabon. *Epilepsia.* 2006;47:2147−2153.

59. Dulac O. Cerebral malaria and epilepsy. *Lancet Neurol.* 2010;9:1144−1145.

60. Christensen SS, Eslick GD. Cerebral malaria as a risk factor for the development of epilepsy and other long-term neurological conditions: a meta-analysis. *Trans R Soc Trop Med Hyg.* 2015;109:233−238.

61. Allison AC. The distribution of the sickle-cell trait in East Africa and elsewhere, and its apparent relationship to the incidence of subtertian malaria. *Trans R Soc Trop Med Hyg.* 1954;48:312−318.

62. Turner L, Lavstsen T, Berger SS, et al. Severe malaria is associated with parasite binding to endothelial protein C receptor. *Nature.* 2013;498:502−505.

63. Moxon CA, Wassmer SC, Milner DA, et al. Loss of endothelial protein C receptors links coagulation and inflammation to parasite sequestration in cerebral malaria in African children. *Blood.* 2013;122:842−851.

64. van der Poll T. The endothelial protein C receptor and malaria. *Blood*. 2013;122:624–625.

65. Naka I, Patarapotikul J, Hananantachai H, Imai H, Ohashi J. Association of the endothelial protein C receptor (PROCR) rs867186-G allele with protection from severe malaria. *Malar J*. 2014;13:105.

66. Shabani E, Opoka RO, Bangirana P, et al. The endothelial protein C receptor rs867186-GG genotype is associated with increased soluble EPCR and could mediate protection against severe malaria. *Sci Rep*. 2016;6:27084.

67. Atkinson SH, Uyoga SM, Nyatichi E, et al. Epistasis between the haptoglobin common variant and α + thalassemia influences risk of severe malaria in Kenyan children. *Blood*. 2014; 123:2008–2016.

68. Jallow M, Teo YY, Small KS, et al. Genome-wide and fine-resolution association analysis of malaria in West Africa. *Nat Genet*. 2009;41:657–665.

69. Timmann C, Thye T, Vens M, et al. Genome-wide association study indicates two novel resistance loci for severe malaria. *Nature*. 2012;489:443–446.

70. Network MGE. Reappraisal of known malaria resistance loci in a large multicenter study. *Nat Genet*. 2014;46:1197–1204.

71. Band G, Rockett KA, Spencer CC, Kwiatkowski DP, Network MGE. A novel locus of resistance to severe malaria in a region of ancient balancing selection. *Nature*. 2015;526:253–257.

72. Stauffer TP, Guerini D, Carafoli E. Tissue distribution of the four gene products of the plasma membrane Ca^{2+} pump. A study using specific antibodies. *J Biol Chem*. 1995;270: 12184–12190.

73. Steed E, Rodrigues NT, Balda MS, Matter K. Identification of MarvelD3 as a tight junction-associated transmembrane protein of the occludin family. *BMC Cell Biol*. 2009;10:95.

74. Network MGE. A global network for investigating the genomic epidemiology of malaria. *Nature*. 2008;456:732–737.

75. Band G, Le QS, Jostins L, et al. Imputation-based meta-analysis of severe malaria in three African populations. *PLoS Genet*. 2013;9:e1003509.

76. Stranger BE, Montgomery SB, Dimas AS, et al. Patterns of cis regulatory variation in diverse human populations. *PLoS Genet*. 2012;8:e1002639.

77. Consortium IH. The International HapMap Project. *Nature*. 2003;426:789–796.

78. Abecasis GR, Auton A, Brooks LD, et al. An integrated map of genetic variation from 1,092 human genomes. *Nature*. 2012;491:56–65.

79. Gurdasani D, Carstensen T, Tekola-Ayele F, et al. The African Genome Variation Project shapes medical genetics in Africa. *Nature*. 2015;517:327–332.

TROPICAL NEUROEPIDEMIOLOGY BY LARGE AREAS OF THE WORLD

CHAPTER

7

Asia

Mohamad I. Idris and Chong T. Tan

University of Malaya Medical Centre, Kuala Lumpur, Malaysia

7.1 INTRODUCTION

The term "Asia" was originally used by the ancient Greeks to refer to civilizations east of their empire. Modern day Asia makes up 30% of the world's landmass. It can be divided into geopolitical regions to West, Central, South, Southeast and East Asia. The population in Asia was estimated to be 4.4 billion in 2016, or about 60% of the global population. There are 48 countries in Asia, including China and India, the two most populous countries on earth. Economically, about 5% are of high income, the rest are of middle or low income. Asia is also very diverse in terms of its ethnic composition, culture, language, religion, politico-social system and environment. Asia as a whole has enjoyed high economic growth rates in recent years, with rapid socioeconomic changes in many of the countries. Given the tremendous varieties and changes outlined above, it is hardly surprising that there is considerable heterogeneity in the epidemiology of neurological conditions in Asia. As such, the aim of this chapter is twofold: firstly, to give an overview of the available epidemiological data on various common neurological conditions in Asia, and the recent trends; and secondly, to highlight the aspects that may be more unique in the region. This is to be followed by a discussion on networking, education, and other aspects of the provision of neurological care.

7.2 STROKE

7.2.1 Incidence, Prevalence, and Mortality

The Global Burden of Disease (GBD) 2013 study estimated that in 2013, there were almost 25.7 million stroke survivors, 10.3 million new strokes, and 6.5 million deaths from stroke worldwide. The burden of stroke in Asia is correspondingly high, as it accounts for almost two-thirds of the total mortality due to stroke in the world. However, in general, many of the Asian countries do not have good epidemiological data on stroke. There are also considerable differences in the methodology of many of the stroke studies from which the data are obtained.

The incidence of stroke in Asia varied considerably in one systematic review, from 67/100,000 per year in Malaysia to 329/100,000 per year in Taiwan, hovering between 100/100,000 per year and 200/100,000 per year in most of the other Asian countries.[1] Another systematic review of stroke in South Asia demonstrated how the age-adjusted incidence rate varied from 145/100,000 per year to 262/100,000 per year.[2] For comparison, elsewhere in the world, the total crude stroke incidence rates have been reported to be between 112/100,000 and 223/100,000 person-years in developed countries and between 73/100,000 and 165/100,000 person-years in low- to middle-income countries.[3]

Data on the prevalence of stroke in Asia also demonstrate wide variability ranging from 260/100,000 to 719/100,000 people in China to 4800/100,000 in Pakistan.[1] In developed countries, the age-adjusted prevalence of stroke in 2013 was 577.6/100,000 people for ischemic strokes, and 128.3/100,000 people for hemorrhagic strokes.[4]

In terms of the type of strokes that are occurring in Asia, ischemic stroke accounts for between 61% and 90% of stroke cases, with the remaining proportion being due to hemorrhagic strokes.[1]

There are important trends shown in certain countries. One of the best epidemiological evidences for stroke in Asia comes from the Hisayama stroke registry in Japan.[5] Studies on this cohort of patients have shown that the incidences of ischemic and hemorrhagic stroke have both declined over time in Japan, reflecting the overall improvement in healthcare, not only in Japan but also in other developed countries in Asia.

Conversely, in China, as demonstrated in the Sino-MONICA-Beijing Project, the incidence rate of hemorrhagic stroke declined by 1.7% but the incidence rate of ischemic stroke increased by 8.7% annually on average between 1984 and 2004.[6] This suggests that in developing countries, as the economy continues to prosper, there is a trend towards a higher incidence of stroke, especially ischemic stroke. Possible reasons for this are that some of the risk factors for atherosclerosis including dietary fat intake, mean serum cholesterol level, diabetes prevalence and obesity prevalence were found to be increasing in trend in China. At the same time, better control of hypertension may have led to a reduction in hemorrhagic stroke in the country.

In contrast, resource-poor countries often report a high proportion of hemorrhagic stroke likely due to suboptimal control of vascular risk factors. In Bangladesh, for example, the proportion of hemorrhagic stroke is particularly high, accounting for up to 39% of stroke cases.[7]

Compared to the developing countries, people in the developing world experience a high mortality rate following stroke. This is also true in Asia. This is demonstrated in a study from Kolkata, India, in which the 30-day case-fatality rate was 41.08%.[8]

7.2.2 Stroke Subtypes

Ischemic strokes can be further categorized according to the Trial of Org 10172 in Acute Stroke Treatment (TOAST) classification: large artery atherosclerosis, small-vessel occlusion, cardioembolism, stroke of other determined etiology, and stroke of undetermined etiology. This is of particular relevance in Asia because intracranial atherosclerosis has been found to be more prevalent in Asian patients, whereas extracranial atherosclerosis is more commonly found in patients from western countries.[9] Many reasons have been postulated for this finding, including: hypercholesterolemia being less common in Asians, advanced and untreated hypertension being more common in Asians, interethnic differences in intracranial vascular tortuosity, as well as genetic differences such as a more prevalent ring finger protein (RNF) 213 variant that could lead to increased vascular fragility.

7.2.3 Young Stroke

Young stroke is another important issue in Asia. The definition of young stroke varies, with many reports using a cutoff point of 50 or 55 years or less for their study populations. For example, the mean age at the onset of stroke is 59 years in Pakistan, 60 years in China, and 63 years in India, in contrast to Western countries where it is 68 years in America and 71 years in Italy.[10] This could be partly due to a generally younger population among many of the Asian countries. There is evidence also to suggest that stroke in Asia may occur earlier, as compared to the Western countries. A comparative study between two centers in Kuala Lumpur, Malaysia and Melbourne, Australia demonstrated that large-vessel atherosclerosis and small-vessel occlusion were significantly more common in young Malaysians with ischemic stroke, suggesting premature atherosclerosis, whereas the Australians have more in the determined etiologies category, especially vascular dissection.[11]

7.2.4 Other Causes of Stroke

Apart from the usual causes of stroke outlined above, a small but important proportion of stroke cases in Asia can be attributed to other diseases. A prime example is Moyamoya disease, a chronic occlusive cerebrovascular disease of unknown etiology that is relatively more common in East Asian countries such as Korea and Japan.[12] A study in 2008 showed an incidence rate of 1/100,000 per year and a prevalence rate of 9.1/100,000 people for Moyamoya disease in Korea.[13] Stroke secondary to tuberculous meningitis (TBM) has been reported to occur in up to two-thirds of the cases.[14] Nasopharyngeal carcinoma (NPC) is particularly prevalent among southern Chinese, Malays and Filipinos. Post-radiation extracranial carotid stenosis is common among NPC survivals. Young NPC patients had a higher risk of stroke compared to the general population.[15] Systemic lupus erythematosus[16] has high prevalence among the Chinese and Thai population, and rheumatic heart disease[17] is more common in the resource-poor countries. Both are associated with complication of stroke.

7.3 EPILEPSY

7.3.1 Prevalence and Incidence

The lifetime prevalence of epilepsy in Asia is reported to be between 1.5/1000 and 14/1000, with a median of 6/1000.[18] This is similar to the prevalence of epilepsy in developed countries in the West usually quoted at around 5/1000. However, it is lower than some developing countries in other regions such as in sub-Saharan Africa and Latin America, where the prevalence is estimated to be 15/1000 and 18/1000, respectively. The prevalence of epilepsy in Asia should also be viewed in the context of high mortality reported from some parts of rural Asia.[19]

The incidence of epilepsy in Asia is estimated to be between 28.8/100,000 and 60/100,000 person-years. This is similar to the rate reported in developed countries, where the age-adjusted incidence is estimated to be between 24/100,000 and 53/100,000 person-years.[18]

In developed countries, the incidence of epilepsy has been reported to have two age peaks distributed in a U-shaped curve: in childhood and in the elderly. While the reports of high incidence in childhood in Asia is common, there have not been many studies demonstrating the high incidence in the elderly among Asians.[20] This is probably due to the generally younger age of the various study populations.

As the Asian population is very diverse, with large variations in its ethnic composition as well as culture, and religion, it is interesting to note that in Singapore, which has a multi-ethnic population of Chinese, Malay and Indian, there have been reports of higher prevalence of epilepsy among the Chinese and Indians as compared to the Malays.[21,22]

7.3.2 Classification and Causes

The classification of epilepsy has gone through many changes over the years, with the main types being idiopathic (genetic), symptomatic and cryptogenic (unknown).[23] The proportion of the various types reported in Asia are: 4%–42% idiopathic (genetic), 22%–53% symptomatic, and 13%–60% cryptogenic (unknown). In our own prospective series of newly diagnosed epilepsy from Kuala Lumpur, about a third are idiopathic (genetic), a fifth are symptomatic, and half are cryptogenic (unknown) which is similar to the proportions reported elsewhere.[24] As for the syndromic classification of epilepsy, a report of 336 Singaporean children also showed similar prevalence of the various syndromes as elsewhere, with benign epilepsy with centro-temporal spikes (BECTS) being the most common.[22]

Variations exist in the cause of epilepsy and seizures in the different populations. In general, the common causes of symptomatic epilepsy listed include head injury, birth trauma, stroke, and central nervous system (CNS) infections such as neurocysticercosis, and Japanese encephalitis (JE).

As for traumatic brain injury, many Asian countries are enjoying robust economic growth with rapid increases in the use of motor vehicles together with a rise in road accidents. As an example of the increasing trend for traumatic brain injury from traffic accidents, between 2006 and 2010, the Ministry of Health in Vietnam reported 15,000–18,000 deaths each year from road traffic injuries, compared to 6394 deaths in 1998.[25] Serious brain injury per vehicle is also much more common in many of the Asian countries, especially when compared to developed countries. For example, according to the World Health Organization (WHO), road fatalities per 100,000 motor vehicles in Myanmar in 2010 were 50-times that of the United Kingdom or Norway. Traumatic brain injury is thus an emerging cause of seizures and epilepsy in Asia.

In terms of CNS infections, the specific causes vary according to geography. Neurocysticercosis is an important cause of seizures and epilepsy in regions that are endemic. Nepal, India, Bali, Papua New Guinea, Sulawesi, as well as parts of Vietnam and China are regions in Asia that have high prevalence of neurocysticercosis. Another important CNS infection in Asia is JE, in which 65% of patients have acute symptomatic seizures, and 13% go on to develop epilepsy.[26]

As for the contribution of genetics to epilepsy, it is important to note the fairly common practice of consanguineous marriage especially among the Indians and some of the Muslim population. Consanguineous marriage has been found to increase the risk of idiopathic and cryptogenic epilepsy.[27]

7.3.3 Mortality

As for mortality of epilepsy, there are some reports of high mortality particularly from the rural Asia. For example, a case-fatality rate of 90.9/1000 person-years has been reported from a rural district in Lao PDR.[28] A standardized mortality ratio exceeding 23 was also reported among those between the ages of 15 and 29 years in rural China.[19] In contrast, the standardized mortality ratio for epilepsy in most developed countries is two to threefold.[18]

7.3.4 Psychosocial Burden

There are a number of studies indicating that the psychosocial burden secondary to epilepsy is particularly high in Asia. Firstly, the name of epilepsy in Chinese is associated with madness. Because of the influence of Chinese medicine, the name of epilepsy in most parts of East and Southeast Asia is also commonly derogatory, often associated with madness and animals.[29] There are a number of knowledge, attitude and practice studies conducted in many countries in Asia, indicating a significant level of stigma, even in countries that are economically developed.[30] The high psychosocial burden is also reflected in the high proportion of those who are unemployed, in poverty, who remain single, and with lower levels of education among epilepsy patients as compared to the general population.[31]

7.3.5 Treatment Gap

The "treatment gap" is the proportion of epilepsy patients who are not receiving treatment by modern medicine. In most of the Asian countries, the treatment gap is between 50% and 80%. In some rural areas, for example in Lao PDR, it is higher at more than 90%.[18] However, even when a "basic" level of medical care is available, a large proportion of Asian epilepsy patients are still inadequately managed, in what constitutes a "care gap" or a "management gap". For example, it is estimated that currently only about 180 cases of epilepsy surgeries are being performed yearly in Southeast Asia, a region with a population of 620 million people. By comparison, 300 epilepsy surgeries are being performed yearly in South Korea, with a population of only 50 million.

7.3.6 Steven–Johnson Syndrome Secondary to Carbamazepine

Another relevant issue in the management of epilepsy in Asia is the association between Steven–Johnson syndrome (SJS) and carbamazepine. Strong association between the human leukocyte antigen (HLA) allele B*1502 and SJS was first reported in the Han Chinese in 2004. Since then, similar associations have been found among the various Southeast Asian ethnic groups, including Thais, Malays, and Vietnamese.[32] This suggests that in populations where the HLA-B*1502 allele is more prevalent, the HLA status should be determined prior to prescription of carbamazepine.

7.4 HEADACHE

7.4.1 Prevalence of Primary Headache

Headache is a common neurological symptom and in most cases is caused by one of the primary headache disorders like migraine or tension-type headache (TTH). In this section, headache will be defined by the International Classification of Headache Disorders (ICHD), first or second edition.[33,34] The 1-year prevalence of migraine globally is thought to be around 10%.[35] Early studies conducted in Asia, however, seemed to suggest that the prevalence of migraine in Asia was lower. Two studies conducted in China in 1985 and 1990 showed that a 1-year prevalence of migraine ranging from 0.63% to 1.5%.[36,37] Likewise, a survey performed in Bangalore, India between 1993 and 1995 reported a 1-year prevalence of headache disorders of 1.1%.[38] All this changed when studies started to adopt the ICHD criteria for various headache disorders. In a recent review, population-based studies of migraine in Asia generally showed a 1-year prevalence of migraine that ranged from 8.4% to 12.5%, comparable to that in North America (8.5%−14.7%) and Europe (9.6%−24.6%).[39]

In terms of gender, migraine was found to be more common in women in Asia, with a gender ratio of 1.2−4.0:1. A peak age of 20−60 years was demonstrated in Asian women, although the peak age for men was more variable amongst the different countries.[39]

In terms of migraine with aura, a study conducted in Malaysia reported a lower proportion of cases, accounting for 10.6% of the people with migraine,[40] as compared to the usually cited figure of around one-third of all migraine cases. A study carried out in Zimbabwe also reported that there were no cases of migraine with aura found, suggesting ethnic variation in the prevalence of migraine with aura.[41]

Reports on the epidemiology of TTH vary partly according to whether they refer to episodic TTH (ETTH), chronic TTH (CTTH), overall TTH (ETTH + CTTH), or probable TTH. Overall, the prevalence of TTH in Asia ranged between 10.8% and 33.3%.[39] In terms of gender, females are more commonly affected, with a gender ratio of 1.7−2:1. The peak age of TTH varied between countries, with Malaysia and Hong Kong reporting younger peak ages between 16 and 35 years and 25−34 years, compared to Korean studies showing peak ages between 50 and 59 years.

The prevalence of chronic daily headache, ranging between 1.0% and 3.9%, is lower than the global average.[42] Chronic daily headache is the old term for chronic TTH, as mentioned in the ICHD first edition.

7.4.2 Headache in the Elderly

Studies among the elderly in Asia show that headache continues to be a significant symptom in the older age group. A community-based study in Taiwan reported that the prevalence of migraine, but not TTH, declines in the elderly.[43] Overall, two-thirds of the elderly reported that there was no change in their headache when compared to "10 years ago". An urban primary care clinic population in Malaysia reported that TTH was the commonest subtype of headache among the elderly, and chronic daily headache was more common in the elderly than in the younger patients; whereas migraine without aura was more common among young adults.[44]

7.4.3 Cultural Influence on Concepts and Management

The cultural influence on concepts, perceived precipitants and management of headache is shown in a community-based study performed in Kuala Lumpur, Malaysia and Melbourne, Australia.[45] Of the triggering factors for headache, significantly more Malaysians attributed their headaches to heat, exposure to sun and change of weather, whereas more Australian subjects attributed it to glare. As for food, significantly more Malaysian subjects associated headache with fried food, mutton, and "heaty food". As for coping, significantly more Malaysians resorted to drinking lots of water and taking "cooling" food as remedies for headache. The perceived precipitants, food, and management were consistent with the common cultural belief among Malaysians that "heatiness" causes headache. "Heaty food", "cooling food", and "heatiness" are all Chinese medicine concepts.

7.5 NEUROINFLAMMATION

7.5.1 Multiple Sclerosis

Multiple sclerosis (MS) is an inflammatory demyelinating disease of the CNS that usually affects young adults. According to the Atlas of MS 2013, the estimated number of people with MS worldwide was 2.3 million in 2013, with the global median prevalence estimated at 33/100,000 in the same year.[46]

Table 7.1 summarizes the result of a recent systemic review of the epidemiology of MS across Asia.[47] As shown, the prevalence of MS in most parts of Asia is low, at less than 5/100,000 population. The exceptions are the Middle Eastern countries with Arab, Turkish and Iranian populations. The female-to-male ratio is 3.6:1. The mean age at onset is about 30 years. Most of the studies included in the review are based on multi-center, hospital-based studies, with few nationwide epidemiological studies performed. The diagnoses are mostly based on McDonald's criteria.

TABLE 7.1 Prevalence, Mean Age at Onset and Gender Ratio of Multiple Sclerosis in Asia (data from Eskandarieh et al)[47]

Country	Prevalence (per 100,000 population)	Mean age at onset (years)	Gender ratio (F:M)
Japan	0.8–16.2	28.3–33	1.1–3.38:1
China	1.39	32.6–46.4	1.2–9:1
Hong Kong	0.77–4.8	25.9–31.8	2.9–9.6:1
Korea	2.4–3.6	25.8–38.4	1.26–3.2:1
Taiwan	0.84–2.96	30–35.6	2.5–5:1
Malaysia	2	28.6–31	5–6.6:1
Thailand	2	30.4–33	4–6.2:1
India	0.17–8.35	26.3–38.3	0.7–2.1:1
Pakistan	–	27–32	1.45–1.5:1
Iran	12.94–85.8	27.1–34.25	1.8–3.8:1

In Japan, there is evidence pointing to an increase in the prevalence of MS in recent years. A study conducted in 2004 had already demonstrated a fourfold increase in the number of patients with MS compared to the year 1972,[48] and this number continued to increase, from 7.7/100,000 to in 2003 to 13.1/100,000 in 2006[49] and 16.2/100,000 in 2011,[50] suggesting a change in lifestyle may be a factor in the prevalence of the disease.

According to the Atlas of MS 2013, there is no prevalence data on MS in some of the Asian countries with large populations, such as Bangladesh, Philippines, Vietnam, and Myanmar. However, based on feedback from physicians practicing there, these countries appear to also fall into the low prevalence category for MS prevalence.

There have been enormous advances in the availability of disease modifying treatment of MS. However, these treatments are mostly out of reach for the majority of MS patients in Asia, whose population consists mainly of people in the middle- or low-income group, and the patients often have to pay for these treatments out of pocket. Thus, access to treatment is an important issue in the care of MS patients in Asia.

7.5.2 Neuromyelitis Optica

A nationwide survey conducted in Japan in 2012 showed a prevalence of neuromyelitis optica (NMO) of 3.65/100,000 people.[51] This is about a quarter of the prevalence of MS in Japan; i.e., 16.2/100,000 people, as mentioned above. Likewise, a study conducted in urban Mangalore, South India demonstrated a prevalence of 2.6/100,000 in that region.[52]

One aspect of NMO or NMO spectrum disorder (NMOSD) that is unique in Asia is the relatively high proportion of NMO/NMOSD amongst all the inflammatory demyelinating diseases (IDD) of the CNS in many countries in the region. NMO makes up 13.7% of IDD cases in India[53] and 39.3% in Thailand.[54] Ethnicity seems to be important in determining the ratio between NMO/NMOSD and MS, as shown by a study in Malaysia where it was 63% in Chinese, 41% in Malays, and 20% in Indians.[55] As NMO/NMOSD is only separated from MS after the discovery of the aquaporin-4 antibody (AQP4) in 2005,[56] the older estimation of MS prevalence and other descriptions of MS often include patients with NMO/NMOSD.

7.6 NEUROMUSCULAR DISORDERS

7.6.1 Amyotrophic Lateral Sclerosis

The prevalence of amyotrophic lateral sclerosis (ALS) in Asia is 0.95–9.9/100,000 population.[57] These data are derived from single-center or multicenter cohort studies, rather than population-based studies. The estimated incidence of ALS in Western population-based studies is 2.16/100,000 person-years. There are few incidence studies of ALS in Asia. The studies in Japan and South Korea show incidence rates that are similar to the Western cohort, 2.2/100,000 and 1.78/100,000 person-years. Lower incidences have been estimated in the Chinese population in Taiwan and Hong Kong. The reasons for this difference are uncertain.

The mean age at the onset of disease is 46 years in India, 52 years in China, 58 years in South Korea, and 61 years in Japan. This probably partly reflects the differences in the mean age of the general population in these countries. Males are more likely to be affected than females in Asia, with a gender ratio between 1.5 and 1.7:1. In the Caucasian population, the C9orf72 genetic mutation is known to account for greater than 40% of familial and 5%–20% of sporadic ALS. Reports from Asian cohorts suggest that this mutation occur less frequently. On the other hand, mutations in superoxide dismutase 1 (SOD1) have been found to be the most common cause of familiar and sporadic ALS.

While ALS is still the most common form of motor neuron disorder in Asia, there are motor neuron syndromes that are peculiar to the region. Juvenile muscular atrophy of distal upper extremity or Hirayama disease is characterized by muscle atrophy of the hand and forearm,[58] and rarely also involving the distal lower limb.[59] Reports have almost exclusively arisen from Asian countries including China, India, Japan, Malaysia, and Singapore. There is a preponderance of males and the onset is typically in adolescence or young adulthood, reaching a plateau between 5 and 10 years.[60] Madras motor neuron disease is a young-onset progressive disease characterized by atrophy and weakness of the limbs, multiple cranial nerve palsies particularly the 7th, 9th to 12th nerves, and sensorineural hearing loss, reported from southern part of India.[61] The ALS—parkinsonism—dementia complex has been described in the Kii peninsula of Japan and Guam island in the Pacific. There is ALS pathology together with neurofibrillary tangles in the brain.[62]

7.6.2 Guillain—Barré Syndrome

Since the eradication of polio, Guillain—Barré Syndrome (GBS) has become the most common cause of acute flaccid paralysis worldwide. Several variants exist including acute inflammatory demyelinating polyneuropathy (AIDP), acute motor axonal neuropathy (AMAN), acute motor and sensory axonal neuropathy (AMSAN), and Miller Fisher syndrome.[63] The incidence and prevalence of GBS overall is similar worldwide, but the subtypes of GBS vary across geographical regions. In Europe and North America, AIDP is the most common variant (accounting for up to 90% of GBS cases), whereas in Asia there is a relatively higher frequency of AMAN.[64,65] The geographic differences in demography and phenotype likely relate to differences in exposure to specific infectious agents or to host-related or environmental factors. The initial reports of AMAN from China demonstrated a seasonal distribution with peaks in the summer months, associated with *Campylobacter jejuni* infection.[66] Large studies in northern China, Japan, and Pakistan show that axonal GBS constitutes 30%—47% of GBS cases.[67—69] In a study from Bangladesh where *C. jejuni* is endemic, AMAN is seen in 92% of cases.[70] AMAN has been associated with the presence of anti-ganglioside antibodies.[71] Molecular mimicry of human gangliosides by the *C. jejuni* lipo-oligosaccharide has since been established as a trigger to the immunological pathogenesis associated

with AMAN.[72] Miller Fisher syndrome is also more commonly seen in GBS patients from Eastern Asia, accounting for up to 20% of patients in Taiwan in one study, and 25% of patients in Japan.[73,74]

7.6.3 Myasthenia Gravis

The prevalence of myasthenia gravis (MG) worldwide is 7.7/100,000 population, and incidence is 0.53/100,000 person-years.[73] The prevalence of MG in Asia is largely consistent with these rates, though there are reports suggesting a higher occurrence of ocular MG in Asian countries.

In Taiwan, a study using the National Health Insurance Research Database found that the prevalence of MG increased from 8.4/100,000 in 2000 to 14.0/100,000 in 2007, with an average annual incidence rate of 2.1/100,000 per year.[75] A study from Korea looked at Health Insurance Review and Assessment (HIRA) data from 2010 to 2014 and found that the prevalence of MG increased from 10.42/100,000 in 2010 to 12.99/100,000 in 2014, with an average incidence of 0.69 cases per 100,000 person-years. Another study found a standardized prevalence rates of 9.67/100,000 and 10.66/100,000 in 2010 and 2011, respectively, and a standardized incidence rate of 2.44/100,000 person-years. The peak age at onset was in the elderly population between 60 and 69 years.[76,77]

A study looking at 391 unselected MG patients in Hubei province in China found that 50% of the patients were children, with a peak age at onset of 5—10 years. Interestingly, 75% of the children, and 28% of adults had ocular MG.[78] There have been other studies in the region reporting a higher incidence of ocular MG in Chinese cohorts. In one study from Singapore, 90% of MG patients were of Chinese decent and 55% of their patient cohort had ocular MG.[79] These findings were comparable to data from Taiwan[80] and Hong Kong.[81]

7.6.4 Muscle Diseases

There have been recent advances in the diagnosis and classification of idiopathic inflammatory myositis (IIM). A recent study based on a large number of muscle biopsy samples in Japan showed that out of 460 patients with IIM, 177 (39%) had immune-mediated necrotizing myopathy (IMNM), 73 (16%) inclusion body myositis (IBM), 56 (12%) dermatomyositis, 51 (11%) anti-synthetase syndrome, 19 (4%) polymyositis, and 84 (18%) non-specific myositis.[82] As for the autoantibody profiles of the 387 IIM patients excluding IBM, anti-signal recognition particle (anti-SRP) (18%), anti-aminoacyl-transfer-ribonucleic-synthetase (anti-ARS) (13%), and anti-3-hydroxy-3-methylglutaryl-coenzyme

A reductase (anti-HMGCR) (12%) were the most common antibodies found.

Sarcocystosis was previously thought to be a common infection in Southeast Asia, but rarely resulting in symptomatic infection. Recent reports have identified *Sarcocystis nesbitti* to cause large outbreaks. It may be a grossly under-diagnosed cause of transient myositis. A unique syndrome of swelling of the jaw muscles was also described.[83]

As for congenital muscular dystrophy, Fukuyama congenital muscular dystrophy is more common in Japan while Ullrich congenital muscular dystrophy is more common elsewhere.[84]

Thyrotoxic periodic paralysis is a relatively common clinical complication of thyrotoxicosis among Southern Chinese, Japanese, Vietnamese, Filipino, and Korean patients. It mainly involves males with associated hypokalemia. Genetic mutations in the L-type calcium channel α1-subunit (Cav1.1) have been described in Southern Chinese patients.[85]

7.6.5 Toxic Neuropathies

Arsenic poisoning resulting in a predominantly sensory neuropathy has been associated with its use as herbicide in the plantation industry. This has reduced in recent years due to replacement by other herbicides. Instead there have been reports of chronic arsenic intoxication in South Asia from two sources: adulteration of opium or country liqueur, and contamination of ground water. In North India, arsenic is often added to opium to increase its potency. In some cases, this leads to distal sensory impairment of touch, pain and temperature, as well as painful paresthesia and muscle cramps.[86] West Bengal and Bangladesh are two areas with high arsenic contamination of ground water. In one study in West Bengal, this led to peripheral neuropathy in 29 out of 248 arsenic exposed patients.[87]

Muscle paralysis from tetrodotoxin toxicity is best known in Japan from consumption of inadequately processed puffer fish. It is also reported in Thailand from consumption of the eggs of horseshoe crabs.[88]

Among Southern Chinese, Thais and Malays, there is a high prevalence of NPC. Radiation to the cancer site, its adjacent tissues and the neck is an effective treatment for the cancer. Bulbar palsy is a common complication seen among the survivors, from delayed radiation neuropathy of the lower cranial nerves.[89]

7.7 MOVEMENT DISORDERS

7.7.1 Parkinson's Disease

The prevalence and incidence estimations of Parkinson's disease (PD) are affected by many factors including age, genetic susceptibility, environment, diagnostic criteria, as well as research methodology. Overall, the prevalence and incidence of PD reported in Asia are mostly at the lower end of those reported in the West, though there are exceptions.

A systematic review and meta-analysis on the prevalence of PD in Asia was published in 2009. Studies were divided into screening-based, or record-based. The age-standardized prevalence per 100,000 for all age groups in door-to-door surveys was 51.3/100,000 to 176.9/100,000. On the other hand, the prevalence estimated based on the record-based studies was lower at 35.8/100,000 to 68.3/100,000.[90] This discrepancy is thought to be due to under-diagnosis of PD. In the door-to-door surveys, between 14% and 78% of respondents found to have PD were previously undiagnosed.

A recent systematic review and meta-analysis to determine the trends of PD according to age, sex, and geographical location identified 21 studies in Asia. The prevalence of PD by age in Asia was 88/100,000 for people between 50 and 59 years of age, 376/100,000 (60−69 years), 646/100,000 (70−79 years), and 1418/100,000 (80 + years). The prevalence of PD by sex was 306/100,000 for females and 371/100,000 for males.[91] As the prevalence of PD increases with age, given the aging population across Asia, PD is likely to become an increasing clinical problem in the region.

With regard to the incidence of PD, a systematic review and meta-analysis found five studies conducted in Asia that met its inclusion criteria. A survey in Taiwan found the age-adjusted incidence rate to be 10.4/100,000,[92] similar to the mail survey in Wakayama, Japan, with an age-adjusted annual incidence rate of 10.5/100,000.[93] Overall, the age-adjusted annual incidence rate for Asia ranged from 5.71//100,000 to 32/100,000 person-years. Data that quote the incidence in older adults and the elderly usually report a higher estimated incidence; worldwide at 37.55/100,000 person-years in females above the age of 40, and 61.21/100,000 person-years in males above 40 years.[94] A recent study in Singapore found the age and sex-adjusted incidence rate for PD to be 32/100,000 person-years for individuals aged 50 years and above.[95] This study also identified a possible difference in the incidence rate for PD between ethnic groups, with higher incidences among the Indians and Chinese as compared to Malays, although the numbers were small.

In terms of the management of PD, one aspect that is unique among Asians is the use of *Tai Chi* or *taijiquan*. This was originally a Chinese martial art, and is now widely practiced as a balance-based exercise. A recent trial has demonstrated reduced balance impairments among the participants of *Tai Chi* as compared to the group who did conventional exercise of stretching or resistance training.[96] Another example is the use of the seed powder of the leguminous plant *Mucuna pruriens* as a traditional (Ayurvedic) treatment for PD.[97]

7.7.2 Atypical Parkinsonian Syndromes

Few epidemiological studies have looked at Parkinson-plus syndromes in Asia. One such study conducted in a rural Japanese district demonstrated an age-adjusted prevalence rate of 10/100,000, 13/100,000, and 6/100,000 for progressive supranuclear palsy (PSP), multiple system atrophy (MSA), and corticobasal degeneration (CBD), respectively.[98] In comparison, a study in the United Kingdom reported a lower age-adjusted prevalence of 6.4/100,000 for PSP and 4.4/100,000 for MSA.[99] However, it should be recognized that misdiagnosis (e.g., of MSA) is common.[100,101] In contrast to the situation in Europe and North America (where the parkinsonian variant of MSA or MSA-P predominates),[102,103] the cerebellar subtype of MSA (MSA-C) appears to be the predominant form of MSA in the Japanese population.[104]

One unique feature that can be found in Asia is the development of toxin-induced parkinsonism following exposure to joss paper.[105] Other toxic agents reported are manganese from welding fumes,[106] and carbon monoxide poisoning.[107,108]

7.7.3 Huntington's Disease

The prevalence of Huntington's disease (HD) in Asia is thought to be lower than in Western countries. This was demonstrated in a systematic review in 2012, in which the prevalence in Asia was found to be 0.40/100,000, whereas the prevalence in North America, Europe, and Australia was 5.70/100,000.[109] The incidence of HD in Asia was also lower compared to Western countries. The same systematic review found an incidence of 0.46−1.6/million per year in Asia, compared to 1.1−8/million per year in North America, Europe and Australia. As for ethnic variation of HD prevalence within Asia, in Malaysia, HD has been reported in all the major ethnic groups (Malay, Chinese, and Indian), although the numbers were too small to demonstrate any ethnic differences of prevalence.[110]

7.7.4 Other Movement Disorders

With regards to Wilson's disease in Asia, although there are no community-based prevalence and incidence data, a study in India showed that in a cohort of 282 Wilson's disease patients, more than two-thirds of them were male. The disease has also been reported to occur in the fifth and sixth decade of life (i.e. not necessarily during childhood or young adulthood). Interestingly, almost half of the patients in the cohort had a family history of consanguinity, which is known to increase the incidence of recessive disorders. Consanguinity of marriage is a common cultural practice particularly among some Muslim groups and South Indians. Studies in Korea and Taiwan looking at the carrier frequency of the more common mutations responsible for Wilson's disease suggest that the incidence of Wilson's disease in Asia may be higher than in other parts of the world.[111]

X-linked dystonia-parkinsonism (XDP) or "Lubag" affects males originating from the Panay Island in the Philippines. The disease usually presents in the third and fourth decades of life and is characterized by the development of dystonia (often with prominent involvement of the jaw and neck) and parkinsonism especially as the disease progresses. It is associated with two genes, DYT3 and TAF1.[112]

In terms of spinocerebellar ataxia (SCA) in Asia, there are differences in the predominant form(s) found in various countries. SCA 3 appears more commonly in Japan, China, Taiwan, and Singapore whereas SCA 2 is more prevalent in India and South Korea.[113] SCA 12, which frequently presents with action tremor, is more common in northern India.[114]

As for movement disorders associated with infections, a study in India found pure dystonia, dystonia with choreoathetosis, and parkinsonism to be the three most common clinical phenotypes in a cohort of patients with movement disorders secondary to infection. Viral encephalitis accounted for almost 90% of the infections responsible for the movement disorders.[115] Movement disorders have been reported in up to 70%−80% of JE patients.[116] Dengue, another very common mosquito-borne infection in this region (discussed further below) can rarely also present with movement disorders.[117,118]

Hyperglycemia-associated chorea/ballismus has been reported more commonly in Asian patients, especially women; the reason(s) for this are unclear.[119] It has been suggested that NMDA receptor encephalitis may be common in Asians.[120] A variety of movement disorders have been described with NMDA receptor encephalitis, most commonly orofacial dyskinesias.

Paroxysmal kinesigenic dyskinesia is a rare disorder in which brief episodes of dyskinesia are triggered by sudden movements. It is inherited in an autosomal dominant pattern with preponderance among males. On the other hand, a study in Malaysia demonstrated a lack of family history in most of the patients diagnosed, as compared to that reported in the literature.[121]

Hemifacial spasm is a fairly common movement disorder that is thought to have a high incidence in some Asian populations.[122] The mean age at the onset of symptoms is around 46 years in India and China.[123]

Latah is a culture-specific startle syndrome that can be found among Malays and Javanese in Indonesia and Malaysia. It is characterized by exaggerated startle responses and involuntary vocalizations, echolalia, and echopraxia, often to the amusement of bystanders.[124]

7.8 DEMENTIA

Dementia is characterized by a decline in memory or another cognitive ability, such as dysphasia or executive function, which causes significant impairment of function. According to the World Alzheimer Report 2015 on the Global Impact of Dementia, in 2015 there were 46.8 million people worldwide living with dementia. This number is expected to double every 20 years, to 74.7 million by 2030, and 131.5 million by 2050, due to population increase and aging. Of note, much of the increase in the prevalence of dementia is expected to take place in low- and middle-income countries, many of which are in Asia.

The estimated prevalence of dementia for people aged over 60 years in Asia ranges from 3.2% to 4.8% according to different regions. There is considerable variability in the reports from the individual countries. However, as a whole, there is no definite evidence that the prevalence in Asia is different from other continents. There are many possible reasons for the difference in prevalence between countries. Among these is the diagnostic criteria used. It has been shown that the estimation of dementia in similar populations would differ, whether the Diagnostic and Statistical Manual of Mental Disorders (third, fourth, or fifth edition) or the International Statistical Classification of Disease (9th or 10th edition) are being used.[125] One the other hand, a study in multi-ethnic Singapore has shown that the prevalence of dementia was lowest in the ethnic Chinese population, compared to the ethnic Malays and Indians. These differences were not accounted for by controlling for gender, age, education, and they remain when taking into account hypertension, diabetes, cardiovascular disease, stroke, and smoking. This suggests that genetic differences may also account for some of the variability in the prevalence of dementia found in Asia.[126]

7.8.1 Dementia Subtypes

Older studies conducted before 1990 in Asia seemed to suggest that there was a higher prevalence of vascular dementia as compared to Alzheimer's disease in the Chinese and Japanese population.[127] This is in contrast with studies done in Western populations whereby Alzheimer's disease was the most prevalent subtype of dementia. However, more recent studies have demonstrated that Alzheimer's disease is more common. This change was probably caused by a few factors, including differences in lifestyle, access to healthcare and management of vascular risk factors in the population.

The Hisayama study was a study to determine the incidence of dementia in a cohort of elderly Japanese. During the follow-up period, 275 (33.2%) out of 828 subjects developed dementia. The total incidence of dementia was 32.3 ($n = 275$). The incidence of the dementia subtypes were: Alzheimer's disease (14.6, $n = 124$), vascular dementia (9.5, $n = 81$), dementia with Lewy bodies (1.4, $n = 12$), combined dementia (3.8, $n = 33$), and others (3.1, $n = 16$) per 1000 person-years, respectively.[128]

7.9 CNS INFECTIONS

7.9.1 Bacteria

Until recently, the most common causes of bacterial meningitis worldwide are *Haemophilus influenzae*, *Streptococcus pneumonia*, and *Neisseria meningitides* in children; *S. pneumonia* and certain gram-negative bacteria in adults. With the recent introduction of the various vaccines for meningitis, there has been a dramatic reduction in bacterial meningitis in children, particularly those due to *H. influenzae*. However, largely due to differences in economic resources, access to and practices of vaccinations varies in different countries in Asia, leading to different impacts on the incidences of various bacterial meningitis cases. There have also been reports of specific bacteria being more important in certain regions, e.g., *Streptococcus agalactiae* in adult bacterial meningitis in Singapore and Hong Kong.[129,130]

TBM is a common meningitis particularly in the middle- and low-income countries in Asia, with India, Indonesia and China accounting for 23%, 10%, and 10% of global tuberculosis (TB) cases in 2014.[131] Out of the six main lineages of *Mycobacterium tuberculosis*, four lineages namely the Indo-Oceanic, East Asian (Beijing), East African-Indian, and Euro-American lineages predominate in Asia.[132] The Beijing lineage in particular is thought to be increasingly virulent and is less sensitive to BCG vaccination. Early diagnosis remains the most important factor determining the outcome of patients. As for clinical course, a study in Malaysia found that 56% of the human immunodeficiency virus (HIV)-negative TBM patients has paradoxical manifestation, defined as worsening of pre-existing tuberculous lesions or appearance of new lesions in patients whose condition initially improved with anti-tuberculous treatment, probably due to immune reaction.[133]

Leprosy remains prevalent in parts of Asia, with India accounting for half the global cases. Nepal and Myanmar are other countries with high prevalence. Melioidosis is an increasingly recognized infection in rural Southeast Asia and tropical Australia. Neuromelioidosis is seen in about 3% of patients with melioidosis, manifesting mainly as brain abscess. There may not be any sign of infection elsewhere. Tetanus, and neurological involvement in leptospirosis, typhus and typhoid are other important bacterial infections seen in parts of Asia.

7.9.2 Viruses

Of the virus infections that affect the CNS, JE is probably the most important. The annual incidence has been estimated to be close to 70,000 cases, with the disease largely confined to Asia.[134] It is fatal in 20%−30% of cases, with another 30%−50% of survivors having significant neurological complications. It is seasonal in areas with temperate climates such as in North Vietnam, but occurs throughout the year in warmer climates such as in South Vietnam. Herpes simplex encephalitis is another common cause of encephalitis widely distributed globally. In most Asian countries where the etiology is known, Herpes simplex often accounts for a fifth to a third of encephalitis cases. Rabies continues to cause 30,000 human deaths yearly in Asia, accounting for more than half the deaths globally.

Of the more recently discovered and emerging viral infections, the most important that affects the nervous system is the HIV, with the virus affecting the brain directly, or from secondary infection due to impaired immunity. An estimated 5.1 million in 2015 in Asia are living with HIV, with China, India, and Indonesia accounting for around three-quarters of the total. Enterovirus 71 (EV71) is another emerging viral encephalitis seen globally, but particularly common in Asia. It causes outbreaks of encephalitis among young children, the most severe being brainstem encephalitis resulting in high morbidity and mortality.

The WHO, in its report in 2009, declared dengue to be the most rapidly spreading mosquito-borne viral disease in the world. Asia accounts for more than 70% of the dengue globally. It is estimated that up to 5% of patients with dengue infection have neurological manifestations, attributed to the neurotropic effect of the virus, complications from the systemic illness, or post-infectious immune-mediated complications.[135] Nipah encephalitis is a newly discovered virus, forming a new genus henipah virus. It was first discovered following a fatal outbreak of encephalitis among pig farm workers in Malaysia and Singapore.[136] An unusual feature of the encephalitis is that up to 10% of survivors have relapsed or late-onset encephalitis which could occur years after the initial infection.[137] Since its discovery, outbreaks of Nipah encephalitis have been reported almost yearly in Bangladesh, and also in India and Southern Philippines. Pteropus bats are the reservoir of the virus, which is found widely in Asia, extending to Madagascar in Africa. The infection is thought to spread from bat to human through infected animals such as pigs and horses, and from human to human via respiratory secretions.

7.9.3 Parasites

For parasitic infection, the most important globally is malaria. According to the WHO, there were approximately 262 million cases of malaria globally leading to 839,000 deaths worldwide in 2000.[138] Most deaths are in children due to cerebral malaria. About a fifth of malaria cases are in Asia. Plasmodium knowlesi was previously thought to only affect primates, the long-tailed macaques monkey (Macaca fascicularis). In recent years, there have been increasing reports of human infection in Southeast Asia, including Malaysia, Thailand, Philippines, Vietnam, Myanmar, and Indonesia. It is now known to cause up to 70% of malaria cases in certain areas of Sarawak, Malaysia. Whether P. knowlesi is able to cause cerebral malaria is uncertain.

Neurocysticercosis is another important parasitic infection in Asia, particularly common in Nepal, India, Indo-China, parts of Philippine, China, Bali and Papua in Indonesia. The syndrome of seizures in association with single small enhancing computed tomography (CT) lesions has been reported as important manifestation of the disease in India, accounting for about 10% of epilepsy patients seen at major neurological centers in the country.

Sarcocystosis was previously thought to be a common infection in Southeast Asia, but rarely results in symptomatic infection. Recent reports have identified S. nesbitti as the cause of large outbreaks. It may be a grossly under-diagnosed cause of prolonged fever, and transient myositis.[83]

7.9.4 Fungus

As for fungal diseases, the most important clinically is Cryptococcus neoformans causing meningitis. Unlike in the West, most of the cryptococcal meningitis seen in Southeast Asia previously occurred among those who were healthy. With the advent of HIV, the infection is now seen more among those with HIV infection.

7.10 NEUROLOGY CARE

7.10.1 Number of Neurologists

According to WHO estimates, there are 100,000 neurologists worldwide with 20,000 located in Asia. Therefore, while Asia constitutes 60% of the world's population, it contains only 20% of the world's neurologists. This disparity is particularly evident in South and Southeast Asia. South Asia, for example, holds 20% (1.6 billion) of the world's population but less than 1.5% of the world's neurologists. Underserved countries, with one or fewer neurologists per million people, include Bangladesh, Cambodia, East Timor, India, Laos, Myanmar, Nepal, North Korea, Pakistan, and Papua New Guinea. As financing for neurological services in many parts of Asia is largely via out-of-pocket payments, neurologists tend to congregate in large capital cities and in private practice. The shortage of neurologists in the provincial towns and for public patients is therefore even more pronounced. A similar shortage of personnel is seen in many subspecialties including stroke, epilepsy and pediatric neurology.[139]

7.10.2 Neuroimaging, Clinical Neurophysiology Services, and Pharmaceuticals

Modern neuroimaging such as CT scans and magnetic resonance imaging (MRI) scans, as well as clinical neurophysiology services, are generally available in most of the capital cities, though their accessibility is highly variable depending on whether it is an urban or a rural setting, the level of economic development of the country and the system of healthcare financing. The most commonly used drugs for the treatment of neurological diseases are accessible in countries with good public health financing, but for the rest of Asia, which is the majority, patients have to pay out of pocket, limiting access to many drugs. As an example, there is very low use of interferon for treatment of MS in many countries in Asia, although its efficacy was demonstrated more than two decades ago.

7.10.3 Training Programs for Neurologists

To help remedy the shortage of neurologists in Asia, most countries with larger populations have established training programs in neurology. These programs are diverse. Countries previously under British rule tend to require internal medicine certification as a prerequisite for entry into training. The duration of training in most countries is 3 years.

7.10.4 Regional Networking and Publication

There are three regional organizations that act as vehicles for networking amongst neurologists in the region: the Asian & Oceanian Association of Neurology (AOAN), founded in 1961, the Asian & Oceanian Child Neurology Association (AOCNA), founded in 1983, and the ASEAN Neurological Association (ASNA), founded in 1994 for the Southeast Asian countries. Each organization holds biennial congresses. There are also subspecialty-based Asian organizations in stroke, epilepsy, clinical neurophysiology, movement disorders, muscle diseases and MS. They all organize regular regional congresses. To promote the development of Asian neurology, Neurology Asia (www.neurology-asia.org) was launched in 1996. It is the official journal of the AOAN, the AOCNA, and the ASNA.

References

1. Mehndiratta MM, Khan M, Mehndiratta P, Wasay M. Stroke in Asia: geographical variations and temporal trends. *J Neurol Neurosurg Psychiatry*. 2014;85(12):1308−1312.
2. Kulshreshtha A, Anderson LM, Goyal A, Keenan NL. Stroke in South Asia: a systematic review of epidemiologic literature from 1980 to 2010. *Neuroepidemiology*. 2012;38(3):123−129.
3. Feigin VL, Lawes CM, Bennett DA, Barker-Collo SL, Parag V. Worldwide stroke incidence and early case fatality reported in 56 population-based studies: a systematic review. *Lancet Neurol*. 2009;8(4):355−369.
4. Feigin V, Krishnamurthi R, Parmar P, et al. Update on the global burden of ischemic and hemorrhagic stroke in 1990-2013: the GBD 2013 study. *Neuroepidemiology*. 2015;45(3):161−176. Available from: http://dx.doi.org/10.1159/000441085.
5. Kubo M, Kiyohara Y, Kato I, et al. Trends in the incidence, mortality, and survival rate of cardiovascular disease in a Japanese community: the Hisayama study. *Stroke*. 2003;34(10):2349−2354.
6. Zhao D, Liu J, Wang W, et al. Epidemiological transition of stroke in China: twenty-one-year observational study from the Sino-MONICA-Beijing Project. *Stroke*. 2008;39(6):1668−1674.
7. Islam MN, Moniruzzaman M, Khalil MI, et al. Burden of stroke in Bangladesh. *Int J Stroke*. 2013;8(3):211−213.
8. Das SK, Banerjee TK, Biswas A, et al. A prospective community-based study of stroke in Kolkata, India. *Stroke*. 2007;38(3):906−910.
9. Kim JS, Kim YJ, Ahn SH, Kim BJ. Location of cerebral atherosclerosis: why is there a difference between East and West? *Int J Stroke* 2016;0(0):1−12.
10. Wasay M, Khatri IA, Kaul S. Stroke in South Asian countries. *Nat Rev Neurol*. 2014;10:135−143.
11. Tan KS, Tan CT, Churilov L, et al. Ischemic stroke in young adults: a comparative study between Malaysia and Australia. *Neurol Asia*. 2010;15(1):1−9.

12. Kim JS. Moyamoya disease: epidemiology, clinical features, and diagnosis. *J Stroke*. 2016;18(1):2−11.

13. Im SH, Yim SH, Cho CB, et al. Prevalence and epidemiological features of moyamoya disease in Korea. *J Cerebrovasc Endovasc Neurosurg*. 2012;14(2):75−78.

14. Wasay M, Farooq S, Khowaja ZA, et al. Cerebral infarction and tuberculoma in central nervous system tuberculosis: frequency and prognostic implications. *J Neurol Neurosurg Psychiatry*. 2014;85(11):1260−1264.

15. Li CS, Schminke U, Tan TY. Extracranial carotid artery disease in nasopharyngeal carcinoma patients with post-irradiation ischemic stroke. *Clin Neurol Neurosurg*. 2010;112(8):682−686.

16. Wang IK, Muo CH, Chang YC, et al. Risks, subtypes, and hospitalization costs of stroke among patients with systemic lupus erythematosus: a retrospective cohort study in Taiwan. *J Rheumatol*. 2012;39(8):1611−1618.

17. Wang D, Liu M, Lin S, et al. Stroke and rheumatic heart disease: a systematic review of observational studies. *Clin Neurol Neurosurg*. 2013;115(9):1575−1582.

18. Mac TL, Tran DS, Quet F, Odermatt P, Preux PM, Tan CT. Epidemiology, aetiology, and clinical management of epilepsy in Asia: a systematic review. *Lancet Neurol*. 2007;6(6):533−543.

19. Ding D, Wang W, Wu J, et al. Premature mortality in people with epilepsy in rural China: a prospective study. *Lancet Neurol*. 2006;5(10):823−827.

20. Banerjee TK, Ray BK, Das SK, et al. A longitudinal study of epilepsy in Kolkata, India. *Epilepsia*. 2010;51(12):2384−2391.

21. Loh NK, Lee WL, Yew WW, Tjia TL. Refractory seizures in a young army cohort. *Ann Acad Med Singapore*. 1997;26(4):471−474.

22. Lee WL, Low PS, Murugasu B, et al. Epidemiology of epilepsy in Singapore children. *Neurol J Southeast Asia*. 1997;2:31−35.

23. Berg AT, Scheffer IE. New concepts in classification of the epilepsies: entering the 21st century. *Epilepsia*. 2011;52(6):1058−1062.

24. Manonmani V, Tan CT. A study of newly diagnosed epilepsy in Malaysia. *Singapore Med J*. 1999;40(1):32−35.

25. Ngo AD, Rao C, Hoa NP, Hoy DG, Trang KT, Hill PS. Road traffic related mortality in Vietnam: evidence for policy from a national sample mortality surveillance system. *BMC Public Health*. 2012;12:561.

26. Misra UK, Tan CT, Kalita J. Seizures in encephalitis. *Neurol Asia*. 2008;13:1−13.

27. Ramasundrum V, Tan CT. Consanguinity and risk of epilepsy. *Neurol Asia*. 2004;9(Suppl 1):10−11.

28. Tran DS, Odermatt P, Le TO, et al. Prevalence of epilepsy in a rural district of central Lao PDR. *Neuroepidemiology*. 2006;26(4):199−206.

29. Lim KS, Li SC, Casanova-Gutierrez J, et al. Name of epilepsy, does it matter? *Neurol Asia*. 2012;17(2):87−91.

30. Lim KS, Lim CH, Tan CT. Attitudes toward epilepsy, a systematic review. *Neurol Asia*. 2011;16(4):269−280.

31. Lim KS, Wo SW, Wong MH, Tan CT. Impact of epilepsy on employment in Malaysia. *Epilepsy Behav*. 2013;27(1):130−134.

32. Khor AH, Lim KS, Tan CT, Wong SM, Ng CC. HLA-B*15:02 association with carbamazepine-induced Stevens-Johnson syndrome and toxic epidermal necrolysis in an Indian population: a pooled-data analysis and meta-analysis. *Epilepsia*. 2014;55(11):e120−e124.

33. Headache Classification Committee of the International Headache Society. Classification and diagnostic criteria for headache disorders, cranial neuralgias and facial pain. *Cephalalgia*. 1988;8(Suppl. 7):1−96.

34. Headache Classification Committee of the International Headache Society. The international classification of headache disorders: 2nd edition. *Cephalalgia*. 2004;24(Suppl. 1):1−160.

35. Stovner Lj, Hagen K, Jensen R, et al. The global burden of headache: a documentation of headache prevalence and disability worldwide. *Cephalalgia*. 2007;27(3):193−210.

36. Zhao F, Tsay JY, Cheng XM, et al. Epidemiology of migraine: a survey in 21 provinces of the People's Republic of China, 1985. *Headache*. 1988;28(8):558−565.

37. Cheng X. [Epidemiologic survey of migraine in six cities of China]. *Zhonghua Shen Jing Jing Shen KeZaZhi*. 1990;23(1):44−46:64.

38. Gourie-devi M, Gururaj G, Satishchandra P, Subbakrishna DK. Prevalence of neurological disorders in Bangalore, India: a community-based study with a comparison between urban and rural areas. *Neuroepidemiology*. 2004;23(6):261−268.

39. Peng KP, Wang SJ. Epidemiology of headache disorders in the Asia-Pacific region. *Headache*. 2014;54(4):610−618.

40. Alders EE, Hentzen A, Tan CT. A community-based prevalence study on headache in Malaysia. *Headache*. 1996;36(6):379−384.

41. Levy LM. An epidemiological study of headache in an urban population in Zimbabwe. *Headache*. 1983;23(1):2−9.

42. Stark RJ, Ravishankar K, Siow HC, Lee KS, Pepperle R, Wang SJ. Chronic migraine and chronic daily headache in the Asia-Pacific region: a systematic review. *Cephalalgia*. 2013;33(4):266−283.

43. Wang SJ, Liu HC, Fuh JL, et al. Prevalence of headaches in a Chinese elderly population in Kinmen: age and gender effect and cross-cultural comparisons. *Neurology*. 1997;49(1):195−200.

44. Tai ML, Jivanadham JS, Tan CT, Sharma VK. Primary headache in the elderly in South-East Asia. *J Headache Pain*. 2012;13(4):291−297.

45. Koh CW, Tan LP, Tan CT. A community based inter-cultural study on precipitating factors of headache. *Neurol J Southeast Asia*. 2002;7:19−24.

46. Multiple Sclerosis International Federation. *Atlas of MS 2013: Mapping Multiple Sclerosis Around the World*. London: Multiple Sclerosis International Federation; 2013:<https://www.msif.org/wp-content/uploads/2014/09/Atlas-of-MS.pdf>. Accessed 30.09.16.

47. Eskandarieh S, Heydarpour P, Minagar A, Pourmand S, Sahraian MA. Multiple sclerosis epidemiology in East Asia, South East Asia and South Asia: a systematic review. *Neuroepidemiology*. 2016;46(3):209−221.

48. Osoegawa M, Kira J, Fukazawa T, et al. Temporal changes and geographical differences in multiple sclerosis phenotypes in Japanese: nationwide survey results over 30 years. *Mult Scler*. 2009;15(2):159−173.

49. Houzen H, Niino M, Hata D, et al. Increasing prevalence and incidence of multiple sclerosis in northern Japan. *Mult Scler*. 2008;14(7):887−892.

50. Houzen H, Niino M, Hirotani M, et al. Increased prevalence, incidence, and female predominance of multiple sclerosis in northern Japan. *J Neurol Sci*. 2012;323(1−2):117−122.

51. Miyamoto K. Epidemiology of neuromyelitis optica [in Japanese]. *Nihon Rinsho*. 2015;73(Suppl 7):260−264.

52. Pandit L, Kundapur R. Prevalence and patterns of demyelinating central nervous system disorders in urban Mangalore, South India. *Mult Scler*. 2014;20(12):1651−1653.

53. Pandit L, Mustafa S, Kunder R, et al. Optimizing the management of neuromyelitisoptica and spectrum disorders in resource poor settings: experience from the Mangalore demyelinating disease registry. *Ann Indian Acad Neurol*. 2013;16(4):572−576.

54. Siritho S, Nakashima I, Takahashi T, Fujihara K, Prayoonwiwat N. AQP4 antibody-positive Thai cases: clinical features and diagnostic problems. *Neurology*. 2011;77(9):827−834.

55. Viswanathan S, Rose N, Masita A, et al. Multiple sclerosis in Malaysia: demographics, clinical features, and neuroimaging characteristics. *MultSclerInt*. 2013;2013:614716.

56. Lennon VA, Kryzer TJ, Pittock SJ, Verkman AS, Hinson SR. IgG marker of optic-spinal multiple sclerosis binds to the aquaporin-4 water channel. *J Exp Med*. 2005;202(4):473–477.

57. Shahrizaila N, Sobue G, Kuwabara S, et al. Amyotrophic lateral sclerosis and motor neuron syndromes in Asia. *J Neurol Neurosurg Psychiatry*. 2016;87(8):821–830. Available from: http://dx.doi.org/10.1136/jnnp-2015-312751.

58. Hirayama K. Non-progressive juvenile spinal muscular atrophy of the distal upper limb (Hirayama disease). In: de Jong JM, ed. *Handbook of Clinical Neurology*. Vol. 15. Amsterdam: Elsevier Science; 1991:107–120.

59. Gourie-devi M, Suresh TG, Shankar SK. Monomelic amyotrophy. *Arch Neurol*. 1984;41(4):388–394.

60. Foster E, Tsang BK, Kam A, Storey E, Day B, Hill A. Hirayama disease. *J Clin Neurosci*. 2015;22(6):951–954.

61. Gourie-Devi M, Nalini A. Madras motor neuron disease variant, clinical features of seven patients. *J Neurol Sci*. 2003;209 (1–2):13–17.

62. Kaji R, Izumi Y, Adachi Y, Kuzuhara S. ALS-parkinsonism-dementia complex of Kii and other related diseases in Japan. *Parkinsonism Relat Disord*. 2012;18(Suppl 1):S190–S191.

63. Bae JS, Yuki N, Kuwabara S, et al. Guillain-Barré syndrome in Asia. *J Neurol Neurosurg Psychiatry*. 2014;85(8):907–913.

64. McGrogan A, Madle GC, Seaman HE, Vries CS. The epidemiology of Guillain-Barré syndrome worldwide. *Neuroepidemiology*. 2008;32(2):150–163.

65. Ye Y, Zhu D, Wang K, et al. Clinical and electrophysiological features of the 2007 Guillain-Barré syndrome epidemic in northeast China. *Muscle Nerve*. 2010;42(3):311–314.

66. Mckhann GM, Cornblath DR, Ho T, et al. Clinical and electrophysiological aspects of acute paralytic disease of children and young adults in northern China. *Lancet*. 1991;338(8767):593–597.

67. Mckhann GM, Cornblath DR, Griffin JW, et al. Acute motor axonal neuropathy: a frequent cause of acute flaccid paralysis in China. *Ann Neurol*. 1993;33(4):333–342.

68. Ogawara K, Kuwabara S, Mori M, Hattori T, Koga M, Yuki N. Axonal Guillain-Barré syndrome: relation to anti-ganglioside antibodies and *Campylobacter jejuni* infection in Japan. *Ann Neurol*. 2000;48(4):624–631.

69. Shafqat S, Khealani BA, Awan F, Abedin SE. Guillain-Barré syndrome in Pakistan: similarity of demyelinating and axonal variants. *Eur J Neurol*. 2006;13(6):662–665.

70. Islam Z, Jacobs BC, Van belkum A, et al. Axonal variant of Guillain-Barre syndrome associated with Campylobacter infection in Bangladesh. *Neurology*. 2010;74(7):581–587.

71. Shahrizaila N, Yuki N. Guillain-Barreé syndrome, Fisher syndrome and Bickerstaff brainstem encephalitis: understanding the pathogenesis. *Neurol Asia*. 2010;15(3):203–209.

72. Yuki N, Susuki K, Koga M, et al. Carbohydrate mimicry between human ganglioside GM1 and *Campylobacter jejuni* lipooligosaccharide causes Guillain-Barre syndrome. *Proc Natl Acad Sci USA*. 2004;101(31):11404–11409.

73. Lyu RK, Tang LM, Cheng SY, Hsu WC, Chen ST. Guillain-Barré syndrome in Taiwan: a clinical study of 167 patients. *J Neurol Neurosurg Psychiatry*. 1997;63:494–500.

74. Mori M, Kuwabara S, Fukutake T, Yuki N, Hattori T. Clinical features and prognosis of Miller Fisher syndrome. *Neurology*. 2001;56:1104–1106.

75. Lai CH, Tseng HF. Nationwide population-based epidemiological study of myasthenia gravis in Taiwan. *Neuroepidemiology*. 2010;35(1):66–71.

76. Lee HS, Lee HS, Shin HY, Choi YC, Kim SM. The epidemiology of myasthenia gravis in Korea. *Yonsei Med J*. 2016;57(2):419–425.

77. Park SY, Lee JY, Lim NG, Hong YH. Incidence and prevalence of myasthenia gravis in Korea: a population-based study using the national health insurance claims database. *J Clin Neurol*. 2016;12(3):340–344.

78. Zhang X, Yang M, Xu J, et al. Clinical and serological study of myasthenia gravis in HuBei Province, China. *J Neurol Neurosurg Psychiatry*. 2007;78(4):386–390.

79. Au WL, Das A, Tl H. Myasthenia gravis in Singapore. *Neurol J Southeast Asia*. 2003;8:35–40.

80. Chiu HC, Vincent A, Newsom-davis J, Hsieh KH, Hung T. Myasthenia gravis: population differences in disease expression and acetylcholine receptor antibody titers between Chinese and Caucasians. *Neurology*. 1987;37(12):1854–1857.

81. Yu YL, Hawkins BR, Ip MS, Wong V, Woo E. Myasthenia gravis in Hong Kong Chinese. 1. Epidemiology and adult disease. *Acta Neurol Scand*. 1992;86(2):113–119.

82. Watanabe Y, Uruha A, Suzuki S, et al. Clinical features and prognosis in anti-SRP and anti-HMGCR necrotising myopathy. *J Neurol Neurosurg Psychiatry*. 2016;87(10):1038–1044.

83. Italiano CM, Wong KT, Abubakar S, et al. Sarcocystis nesbitti causes acute, relapsing febrile myositis with a high attack rate: description of a large outbreak of muscular sarcocystosis in Pangkor Island, Malaysia, 2012. *PloS Negl Trop Dis*. 2014;8(5): e2876.

84. Mercuri E, Muntoni F. Muscular dystrophies. *Lancet*. 2013;381 (9869):845–860.

85. Kung AW. Clinical review: Thyrotoxic periodic paralysis: a diagnostic challenge. *J Clin Endocrinol Metab*. 2006;91 (7):2490–2495.

86. Misra UK, Kalita J. Toxic neuropathies. *Neurol India*. 2009;57 (6):697–705.

87. Basu D, Dasgupta J, Mukherjee A, et al. Chronic neuropathy due to arsenic intoxication from geochemical sources: a 5 years follow up. *J Assoc Neurol Est Ind*. 1996;1:45–48.

88. Kanchanapongkul J. Tetrodotoxin poisoning following ingestion of the toxic eggs of the horseshoe crab Carcinoscorpius rotundicauda, a case series from 1994 through 2006. *Southeast Asian J Trop Med Public Health*. 2008;39(2):303–306.

89. Chew NK, Sim BF, Tan CT, Goh KJ, Ramli N, Umapathi P. Delayed post-irradiation bulbar palsy in nasopharyngeal carcinoma. *Neurology*. 2001;57(3):529–531.

90. Muangpaisan W, Hori H, Brayne C. Systematic review of the prevalence and incidence of Parkinson's disease in Asia. *J Epidemiol*. 2009;19(6):281–293.

91. Pringsheim T, Jette N, Frolkis A, Steeves TD. The prevalence of Parkinson's disease: a systematic review and meta-analysis. *Mov Disord*. 2014;29(13):1583–1590.

92. Chen RC, Chang SF, Su CL, et al. Prevalence, incidence, and mortality of PD: a door-to-door survey in Ilancounty, Taiwan. *Neurology*. 2001;57(9):1679–1686.

93. Morioka S, Sakata K, Yoshida S, et al. Incidence of Parkinson disease in Wakayama, Japan. *J Epidemiol*. 2002;12(6):403–407.

94. Hirsch L, Jette N, Frolkis A, Steeves T, Pringsheim T. The incidence of Parkinson's disease: a systematic review and meta-analysis. *Neuroepidemiology*. 2016;46(4):292–300.

95. Tan LC, Venketasubramanian N, Jamora RD, Heng D. Incidence of Parkinson's disease in Singapore. *Parkinsonism Relat Disord*. 2007;13(1):40–43.

96. Li F, Harmer P, Fitzgerald K, et al. Tai chi and postural stability in patients with Parkinson's disease. *N Engl J Med*. 2012;366 (6):511–519.

97. Katzenschlager R, Evans A, Manson A, et al. *Mucuna pruriens* in Parkinson's disease: a double blind clinical and pharmacological study. *J Neurol Neurosurg Psychiatry*. 2004;75(12):1672–1677.

98. Osaki Y, Morita Y, Kuwahara T, Miyano I, Doi Y. Prevalence of Parkinson's disease and atypical parkinsonian syndromes in a rural Japanese district. *Acta Neurol Scand*. 2011;124(3):182–187.

99. Schrag A, Ben-shlomo Y, Quinn NP. Prevalence of progressive supranuclear palsy and multiple system atrophy: a cross-sectional study. *Lancet*. 1999;354(9192):1771−1775.

100. Kim HJ, Jeon BS, Shin J, et al. Should genetic testing for SCAs be included in the diagnostic workup for MSA? *Neurology*. 2014;83(19):1733−1738.

101. Koga S, Aoki N, Uitti RJ, et al. When DLB, PD, and PSP masquerade as MSA: an autopsy study of 134 patients. *Neurology*. 2015;85(5):404−412.

102. Wenning GK, Geser F, Krismer F, et al. The natural history of multiple system atrophy: a prospective European cohort study. *Lancet Neurol*. 2013;12(3):264−274.

103. Low PA, Reich SG, Jankovic J, et al. Natural history of multiple system atrophy in the USA: a prospective cohort study. *Lancet Neurol*. 2015;14(7):710−719.

104. Stefanova N, Bücke P, Duerr S, Wenning GK. Multiple system atrophy: an update. *Lancet Neurol*. 2009;8(12):1172−1178.

105. Chew NK, Lee MK, Ali Mohd M, Tan CT. Parkinson's disease in occupational exposure to joss paper, a report of two cases. *Neurol J Southeast Asia*. 2003;8:117−120.

106. Huang CC. Parkinsonism induced by chronic manganese intoxication—an experience in Taiwan. *Chang Gung Med J*. 2007;30(5):385−395.

107. Choi IS. Parkinsonism after carbon monoxide poisoning. *Eur Neurol*. 2002;48(1):30−33.

108. Chai CH, Lim SY, Abdul Kadir KA, Goh KJ, Tan CT. Karaoke, power failure and carbon monoxide poisoning. *Neurol Asia*. 2011;16(3):255−257.

109. Pringsheim T, Wiltshire K, Day L, Dykeman J, Steeves T, Jette N. The incidence and prevalence of Huntington's disease: a systematic review and meta-analysis. *Mov Disord*. 2012;27(9):1083−1091.

110. Ng WK, Teh BT, Malmberg I, et al. Huntington's disease in Malaya: a clinical and genetic study. *Neurol J Southeast Asia*. 1997;2:57−63.

111. Zhang Y, Wu ZY. Wilson's disease in Asia. *Neurol Asia*. 2011;16(2):103−109.

112. Rosales RL. X-linked dystonia parkinsonism: clinical phenotype, genetics and therapeutics. *J Mov Disord*. 2010;3(2):32−38.

113. Tan EK. Autosomal dominant spinocerebellar ataxias: an Asian perspective. *Neurol J Southeast Asia*. 2002;7:1−8.

114. Kalia LV, Rockman-Greenberg C, Borys A, Lang AE. Tremor in spinocerebellar ataxia type 12. *Mov Disord Clin Pract*. 2014;1(1):76−78.

115. Jhunjhunwala K, Netravathi M, Pal PK. Movement disorders of probable infectious origin. *Ann Indian Acad Neurol*. 2014;17(3):292−297.

116. Misra UK, Kalita J. Spectrum of movement disorders in encephalitis. *J Neurol*. 2010;257(12):2052−2058.

117. Carod-artal FJ, Wichmann O, Farrar J, Gascón J. Neurological complications of dengue virus infection. *Lancet Neurol*. 2013;12(9):906−919.

118. Tan AH, Linn K, Sam IC, Tan CT, Lim SY. Opsoclonus-myoclonus-ataxia syndrome associated with dengue virus infection. *Parkinsonism Relat Disord*. 2015;21(2):160−161.

119. Lim SY, Ramli N, Tai SM, Nair SR. Involuntary movements in an elderly woman with recently diagnosed diabetes. *J Clin Neurosci*. 2011;18(4):539:590.

120. Abdullah S, Lim SY, Goh KJ, Lum LCS, Tan CT. Anti-N-methyl-D-aspartate receptor (NMDAR) encephalitis: a series of ten cases from a university hospital in Malaysia. *Neurol Asia*. 2011;16(3):241−246.

121. Tai ML, Lim SY, Tan CT. Idiopathic paroxysmal kinesigenic dyskinesia in Malaysia, a multi-racial Southeast Asian country. *J Clin Neurosci*. 2010;17(8):1089−1090.

122. Wu Y, Davidson AL, Pan T, Jankovic J. Asian over-representation among patients with hemifacial spasm compared to patients with cranial-cervical dystonia. *J Neurol Sci*. 2010;298(1-2):61−63.

123. Wang L, Hu X, Dong H, et al. Clinical features and treatment status of hemifacial spasm in China. *Chin Med J*. 2014;127(5):845−849.

124. Bakker MJ, Van dijk JG, Pramono A, Sutarni S, Tijssen MA. Latah: an Indonesian startle syndrome. *Mov Disord*. 2013;28(3):370−379.

125. Sahadevan S, Saw SM, Gao W, et al. Ethnic differences in Singapore's dementia prevalence: the stroke, Parkinson's disease, epilepsy, and dementia in Singapore study. *J Am Geriatr Soc*. 2008;56(11):2061−2068.

126. Venketasubramanian N, Sahadevan S, Kua EH, Chen CP, Ng TP. Interethnic differences in dementia epidemiology: global and Asia-Pacific perspectives. *Dement Geriatr Cogn Disord*. 2010;30(6):492−498.

127. Narasimhalu K, Lee J, Auchus AP, Chen CP. Improving detection of dementia in Asian patients with low education: combining the Mini-Mental State Examination and the Informant Questionnaire on Cognitive Decline in the Elderly. *Dement Geriatr Cogn Disord*. 2008;25(1):17−22.

128. Matsui Y, Tanizaki Y, Arima H, et al. Incidence and survival of dementia in a general population of Japanese elderly: the Hisayama study. *J Neurol Neurosurg Psychiatry*. 2009;80(4):366−370.

129. Chong HT, Tan CT. Epidemiology of central nervous system infections in Asia, recent trends. *Neurol Asia*. 2005;10:7−11.

130. Wilder-Smith E, Chow KM, Kay R, Ip M, Tee N. Group B streptococcal meningitis in adults: recent increase in Southeast Asia. *Aust N Z J Med*. 2000;30(4):462−465.

131. World Health Organization. *Global Tuberculosis Report 2015*. Geneva, Switzerland: WHO; 2015.

132. Gagneux S, Small PM. Global phylogeography of *Mycobacterium tuberculosis* and implications for tuberculosis product development. *Lancet Infect Dis*. 2007;7:328−337.

133. Tai ML, Nor HM, Kadir KA, et al. Paradoxical manifestation is common in HIV-negative tuberculous meningitis. *Medicine (Baltimore)*. 2016;95(1):e1997.

134. Campbell GL, Hills SL, Fischer M, et al. Estimated global incidence of Japanese encephalitis: a systematic review. *Bull World Health Organ*. 2011;89(10):766−774:774A.

135. Murthy JM. Neurological complication of dengue infection. *Neurol India*. 2010;58(4):581−584.

136. Chua KB, Goh KJ, Wong KT, et al. Fatal encephalitis due to Nipah virus among pig-farmers in Malaysia. *Lancet*. 1999;354(9186):1257−1259.

137. Tan CT, Goh KJ, Wong KT, et al. Relapsed and late-onset Nipah encephalitis. *Ann Neurol*. 2002;51(6):703−708.

138. World Health Organization. *World Malaria Report 2015*. Geneva, Switzerland: WHO; 2015.

139. Tan CT. Neurology care in Asia. *Neurology*. 2015;84:623−625.

III. TROPICAL NEUROEPIDEMIOLOGY BY LARGE AREAS OF THE WORLD

Neurologic Diseases in Tropical Oceania

Jacques Joubert and Craig Barnett

University of Melbourne, Melbourne, VIC, Australia

8.1 PREAMBLE

Tropical Oceania, for the purposes of the following discussion, includes Tropical Australia, as well as the Micronesian, Melanesian, and Polynesian Islands. Although variably described as either part of either Australasia or Polynesia, New Zealand is not considered here as it lies outside the Tropics. The following chapter is an attempt to outline the particular characteristics of neurological illnesses that manifest within the Tropical Oceania region, with an emphasis on features that may distinguish it from Neurological practice in other parts of the world. The reader is reminded that common illnesses occur commonly, in all parts of the world, and that this clinical adage is equally true for the region under consideration. As such, the following text should be considered a description of the peculiarities of the region rather than a balanced examination of all causes of neurological illness.

8.2 TROPICAL AUSTRALIA

8.2.1 Overview

The practice of neurology in the Tropical North of Australia is confounded by many challenges. The epidemiological and clinical features of neurological disease in this region, differ from other parts of the world in a number of ways. The vast distances, and the harsh and changeable climatic conditions, along with isolation from peer support and lack of facilities, make neurological practice difficult and similar to third-world

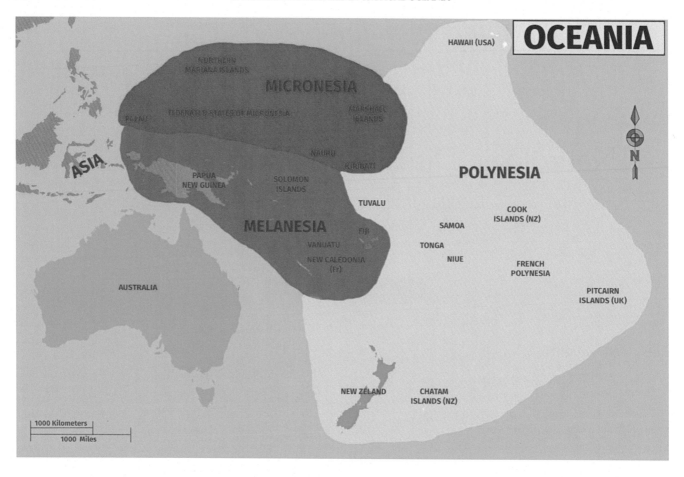

conditions. However, these challenges also provide opportunities for investigation and innovative research that can be translated into effective and rewarding practice.

8.2.2 Geography and Health Resources

8.2.2.1 *The Northern Territory*

The Northern Territory (NT) comprises a large part of the center of the continent as well as the central area of the northern coastline. The Gulf of Carpentaria lies between the NT coastline and the Cape York Peninsula of Queensland to the East. Further west, the coastline is washed by the Arafura and Timor Seas. What is termed colloquially as the "top end" of the NT lies 11 degrees from the equator, and is therefore in a tropical latitude.

The capital city of the NT is Darwin, which lies on the coast, with the smaller town of Nhulunbuy also on the northern coastline, but some 30 hours' drive to the East. Tennant Creek, Katherine and Alice Springs are large towns found progressively further inland, with the latter around 17 hours' drive south of Darwin.

Outside these more urban areas, the population is distributed in a mixture of homelands, outstations and remote rural communities. The NT is the most sparsely populated area in Australia with a population of only around 220,000 despite its massive size (1,349,129 km^2). The greatest concentration is along the Stuart Highway which runs from Darwin in the north down through Alice Springs in Central Australia, to Port Augusta on the South Australian coast. According to 2012 figures the population of Darwin was 116,215, Alice Springs 25,186, and Katherine around 10,000. The population is made up of 68% persons of European descent and 31% Aboriginal people, the Indigenous population that has occupied the region for at least the last 40,000 years.

The "Top End" has marked seasonal variation, with a "dry" season that is almost rain-free, and a "wet" season that lasts about 4 months. "The Wet" is preceded by violent electrical storms and is associated with cyclones and torrential monsoonal rains. These factors create barriers to accessing remote communities for aircraft and road vehicles, especially at the end of

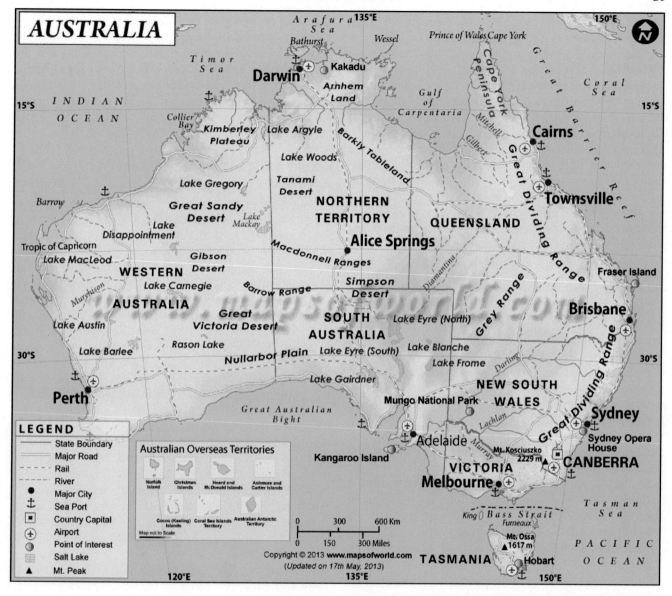

the year when "the Wet" has set in. The climate in this area is tropical with a high humidity. This contrasts with the central region, around Alice Springs, where desert conditions prevail.

The Royal Darwin Hospital (RDH) serves the Darwin Health Region (DHR); an area comparable in size to England (120,000 km²); as well as acting as a tertiary referral hospital for the Northern parts of both the Northern Territory and Western Australia (a land mass of 2 million km²).

It is a teaching hospital which has a wide range of specialist services, including neurology and neurosurgery, as well as a specialist infectious diseases unit. There is one resident neurologist. The RDH is affiliated

with Flinders University of South Australia. On site is the Centre for Evidence Based Nursing as well as the Menzies School of Health Research, the latter being a multidisciplinary research center for the study of chronic and infectious diseases, particularly those related to Aboriginal health. Indigenous Australians usually comprise around 40% of inpatients at RDH.

There are two resident neurosurgeons in Darwin, but neurosurgical cases that need more than routine care, time permitting, are usually transported to Adelaide in the south. In central Australia, Alice Springs has a hospital with intensive care facilities but no neurological or neurosurgical services. Katherine Hospital lies between these two, and is a 60-bed general hospital servicing the local community and surrounding remote aboriginal communities often via a flying medical service.

Healthcare in outlying areas of the NT is provided by government community health clinics. Most clinics are serviced by doctors on an infrequent basis, ranging from weekly to monthly, who often attend by airplane. Clinic staff often include Aboriginal health workers who tend to be residents in the community, as well as nursing staff who may accompany visiting doctors. Communications with the hospital are by telephone or radio.

Often facilities are very limited. For instance, in isolated communities there may be only a single solar-powered telephone and some may have no public telephone communication at all.

8.2.2.2 Queensland and Far North Queensland

Queensland is, after Western Australia, the largest state in Australia. It is bordered by the Northern Territory to the West, and South Australia and New South Wales to the South. The Pacific Ocean lies off the eastern coast of Queensland. The most northern part of the state is the Cape York Peninsula which is separated from New Guinea by the Torres Strait. The peninsula has the Gulf of Carpentaria and Arafura Sea to the West, and the Coral Sea to the East. The population of Queensland is the third largest of all the Australian states, being 4.691 million in 2013. The climate varies, with a dry low-rainfall western region, a temperate east coast, and hot and humid conditions in Far North Queensland and the Torres Strait Islands which have monsoonal rains during summer.

The capital city of Queensland is Brisbane, with Townsville and Cairns lying, respectively, more than 1300 and 1600 km to the north. Far North Queensland includes the Cape York Peninsula, and stretches from the region between Townsville (population 172,000) and Cairns (population 142,000), to the Torres Strait, which separates Papua New Guinea from Australia by 150 km at its narrowest part. The Torres Strait Islands are a group of at least 274 small islands which lie within the Torres Strait. Torres Strait Islanders are of Melanesian ancestry, culturally linked to Papua New Guinea, and distinct from Australian Aboriginal people. Most of the 274 islands are governed by the state of Queensland.

Cairns Base Hospital (CBH) serves an area the size of the United Kingdom. The population of Cairns is about 150,000, of whom 13% are Aboriginal people. The catchment area for CBH extends to the Torres Strait, a distance of 1000 km. CBH has magnetic resonance imaging (MRI), computed tomography (CT), and electroencephalogram (EEG) facilities. These are free services to all populations, however, there is no resident neurologist or neurosurgeon, the nearest being in Townsville, which has several neurosurgeons and one neurologist. Complex neurosurgical cases are transported to Brisbane, which is a 3-hour flight. When this is not possible within a suitable time frame, patients may be transported to Townsville which is 4 hours by road or 1.5 hours by helicopter. The Royal Australian Flying Doctor service is legendary.

8.2.2.3 The Kimberleys and North Western Australia

The Kimberley region lies in the northern part of Western Australia and is bordered by the Northern Territory to the east, the Indian Ocean to the west and the Timor sea to the north. To the south lie the Great Sandy and Tanami Deserts. The Kimberley area is large, covering 423,517 km^2, a land mass three-times the size of England. It is the hottest region in Australia, and has a tropical monsoonal climate with the wet weather occurring from November to April. Broome, the largest town, is known for periodic cyclones. The landscape and geography consist mainly of ancient mountain ranges and deep valleys, as well as savannas dotted with baobab and eucalypt trees.

The population of the Kimberleys is around 41,000. The three main towns are Broome (population 15,000), Kununurra (population 5000), and Derby (population 3600). Tourism is the main industry. Neurological services are provided by the Northern Territory (Darwin Base Hospital) and the capital city of Western Australia, Perth (more than 2100 km by road south of Broome). In this area 33% of the population are of Aboriginal descent. The local health clinic system is similar to those in the Northern Territory, namely clinics run by Aboriginal Health Workers and periodically and variably visited by outside doctors and nursing staff.

8.2.3 Indigenous Health

The Australian Aboriginal population could be considered to be in early epidemiological transition. Notably, the low birth-weight rate in Aboriginal children is twice that of non-Aboriginal children. Infant and Maternal mortality remains significantly higher in Aboriginal than non-Aboriginal Australians despite reductions in the gap in recent decades. Infant malnutrition exists beside high rates of adult obesity, diabetes, cardiovascular and chronic kidney disease; and alcoholism is widespread.

Premature adult mortality is endemic with life expectancies of 10–15 years less than in non-indigenous Australians. Scabies and pyoderma is rife in children (up to 70% affected) and the highest rate of acute rheumatic fever in the international literature comes from the "Top End" of Australia. Mental health disorders are highly prevalent and suicide rates are as high as five-times the rates observed in similar non-indigenous age groups. Dementia rates have also been found to be significantly higher than in non-Aboriginal people. The standardized stroke mortality rate in Aboriginals is twice that of non-Aboriginal persons. Although the prevalence of epilepsy is similar in both populations, serious presentations such as status epilepticus are more common in Aboriginal populations.

Medical management is beset by multiple difficulties such as lack of facilities, widespread illiteracy, a culture of "learned helplessness", an entrenched history of mistrust for the government and non-Aboriginal Australians and geographic isolation. Cultural issues play a significant role at all levels. Avoidance of eye contact is seen as a gesture of respect, and for some (not all), to make direct eye contact can be viewed as being rude, disrespectful or even aggressive. In many, a high degree of intra- and inter-regional mobility is prevalent. Moreover, in the Northern Territory alone, at least 30 different languages and dialects are spoken. Children will change their names for long periods and often have dual house status in a single community, as well as between different communities. School attendance is often intermittent.

8.2.4 Neurology in Tropical Australia

8.2.4.1 Infectious Diseases: Protozoal and Nematode Infections

8.2.4.1.1 MALARIA

Although endemic malaria is considered to have been eradicated in Australia (last indigenous case reported in 1962), travelers returning from nearby endemic areas are frequent, and it therefore needs to be included as a diagnostic differential in Tropical Australia. Vectors for the malaria parasite certainly exist in this region (Anopheles farauti Laveran and Anopheles annulipes Walker) and the potential for cerebral malaria remains ever present.

Most of the 274 Torres Strait Islands are geographically part of Tropical Australia, however a few islands close to mainland New Guinea belong to the Western Province of Papua New Guinea. The potential for re-introduction of malaria into these islands remains very real. Malarial fatalities (from imported Plasmodium falciparum infections) have been reported in the Torres Strait islands in 1990 and 1992. Active detection of carriers in groups such as refugees is therefore imperative. Since East Timor gained independence in the last decade of the 20th century, significantly more malaria

cases have been imported into Darwin. At least one malarial fatality has occurred there.

It has been calculated that by 2030, with a projected 1.5°C increase in overall temperature and a 10% increase in summer rainfall, the main malarial vector in Australia, A. farauti, could reach as far south as Gladstone, 800 km further south of the current boundary of its existence.

8.2.4.1.2 NEMATODE-INDUCED MYOSITIS

Three patients have been described in far north Queensland suffering from myositis caused by direct infection of nematode Haycocknema perplexum. The clinical features were chronically progressive muscle wasting and weakness, with all three also having significant dysphagia.

8.2.4.2 Infectious Diseases: Vector-Borne Viral Infections

8.2.4.2.1 FLAVIVIRAL INFECTIONS

The flavivirus genus of RNA viruses are important worldwide and the impact of these is particularly marked in neighboring Asia. There, as many as 50,000 cases of encephalitis and 10,000 deaths per year have been reported resulting from Japanese Encephalitis.

The flavivirus infections encountered in tropical Australia are encephalitis caused by the Murray Valley encephalitis virus, Kunjin virus (KUNV; subtype of West Nile virus (WNV)) and hemorrhagic fever caused by Dengue Fever virus.

HOST IMMUNITY Host immunity plays an important role in flavivirus infections. The effect of cross-reacting antibodies from one flavivirus infection may result in either protection or increased severity of a secondary flaviviral infections.

TARGET POPULATIONS Murray Valley encephalitis tends to affect more persons with no immunity, as opposed to those residing in endemic areas. It particularly affects children and travelers from non-endemic areas. KUNV tends to affect adults, particularly those suffering from chronic diseases and immunosuppression, and the elderly.

NEUROLOGICAL FEATURES OF FLAVIVIRAL INFECTIONS The incubation period of flaviviral infections leading to neurological sequelae may be as short as 2–3 days, but is more commonly 5–15 days. There is usually a short febrile illness followed by one or more of the following:

- **Encephalitis and meningitis**: these manifest as reduced levels of consciousness with, if there is meningeal involvement, pleocytosis in the cerebrospinal fluid (CSF).

- **Seizures**: these are common in Murray Valley encephalitis as well as in cases of Kunjin fever. Serial seizures or true status epilepticus carrying a poor prognosis. Focal seizures, generalized seizures and a subtle form of ongoing convulsive activity presenting as recurrent twitching of, for example, a finger or the corner of the mouth may occur. EEG demonstrates periodic lateralized discharges. Status epilepticus may result from raised intracranial pressure, and brainstem signs from brainstem compression. Brainstem encephalitis may also, on occasion, result in a "locked-in" syndrome.
- **Movement disorders**: these reflect the presence of basal ganglia involvement, sometimes demonstrated on MRI, as part of flaviviral encephalitis. Clinically the presentation may be of typical parkinsonian features such as a rhythmic tremor of the limbs, akinesis and rigidity. Other movement disorders may occur such as myoclonic jerks, the Opsoclonus Myoclonus syndrome, jaw dystonia, choreiform movements, orofacial dyskinesias and general rigidity of the limbs and body.
- **Cerebellar dysfunction**: this may, rarely, complicate the clinical picture.
- **Myelitis**: this is characterized by anterior horn cell damage. In fact, early outbreaks of Murray Valley virus infection were thought to be due to an aberrant poliomyelitis outbreak. Typically, flaccid motor paralysis with muscle wasting develops resulting in a lower motor neuron weakness of the limbs. Bulbar and respiratory muscle involvement develops in 60%–70% of cases presenting with anterior horn cell involvement.
- **Peripheral nerve involvement**: very occasionally there will be neurophysiological evidence of peripheral nerve involvement beyond the anterior horn cell in the form of demyelination with slowed conduction. Nerve conduction tests generally demonstrate normal motor and sensory conduction, but evidence of muscle denervation (fasciculation and spontaneous fibrillation) is common, because of the anterior horn cell involvement. Similarly, there may be bladder involvement with acute urinary retention. Sensory involvement and radiculopathy may also be apparent clinically.
- **Long-term neurological and neuropsychological sequelae**: these are common, occurring in about 50% of cases who survive to leave hospital. Advanced age, presence of co-morbidities, alcoholism and evidence of encephalitis are risk factors for persistent sequelae and/or death.

DIAGNOSIS OF FLAVIVIRAL INFECTION The diagnosis of a flaviviral infection should be suspected in any person visiting or resident of an area endemic for flaviviruses, presenting with one or a combination of fever, rash, arthralgia and neurological symptoms as described above, unless there is an obvious alternative diagnosis. In the elderly, alcoholic or chronically ill, KUNV infection should be suspected, whereas in children, Murray Valley encephalitis viral (MVEV) infection should be considered earlier. Serological tests may confirm the clinical suspicion. The immunoglobulin M (IgM) capture enzyme-linked immunosorbent assay (ELISA) is the standard test. This may be done on CSF or serum, and serial testing is advised in cases where the diagnosis is strongly suspected but the initial tests are negative. Flavivirus vaccination or exposure to other flaviviruses in an area may give false-positive results. In these cases, parallel testing against other flaviviruses may be helpful. Flavivirus reverse transcriptase polymerase chain reaction (PCR) or real-time PCR have been useful particularly on CSF samples. Neutralizing antibodies to flaviviruses are more specific, but are not performed routinely in all centers.

TREATMENT Currently there is no specific antiviral agent used against flaviviruses. A potential treatment is interferon-α, but results are variable. Other possible treatments are ribavirin and intravenous immune globulin, however these are empiric, and unproven in humans. Treatment at present is general support such as hydration and treating seizures if they occur. Currently there are no vaccines in routine use against Murray Valley Encephalitis or Kunjin fever, whereas for Japanese encephalitis there is a three-dose regime of inoculation recommended for travelers who spend long periods in endemic Asian areas.

EPIDEMIOLOGY OF ARBOVIRUSES In 1974, another large outbreak of viral encephalitis occurred in the Murray Valley region. In this outbreak, most cases of encephalitis were due to MVEV, but a small proportion was caused by KUNV, now known to be a subtype of WNV. This mosquito-borne flavivirus is endemic in the Kimberley Region of Western Australia as well as the adjoining Northern Territory.

During this epidemic there were 58 cases among with 13 fatalities, indicating the potentially devastating effect of this infection.

CLIMATE AND ANIMAL VECTORS AND OTHER MODES OF TRANSMISSION A natural enzootic cycle exists between birds (the night heron) and bird-biting mosquitos (e.g., *Culex annulirostris*, *Culex quinquefasciatus*, and *Aedes normanensis*). In the NT there are 20 million feral pigs, and these herds are suspected of acting as vertebrate hosts like the birds. Humans are infected when they enter this cycle. The lack of prolonged viremia in humans prevents them acting as transmitters of the virus. Recently, however, transplacental transmission, infected blood products and infected transplanted organs have also been implicated in viral transmission.

The virus has been detected in mosquitoes in Tropical Northern Queensland as well. There, during the wet season (February to July) MVEV presence is most likely. Spread outside these areas occurs when infected birds and wind-blown mosquitoes carry the virus outside enzootic areas during times of unusually heavy rainfall and flooding. Spread can occur as far as southern areas of Western Australia and also to the dry areas of Central Australia, as occurred in 1974 and 2001.

Weather conditions play a significant role as the vector population (*C. annulirostris*) increases after flooding. Rare in Central Australia, MVEV infections were recorded after extreme rainfall in 2000. The main months for epidemics are between January and May, often coinciding with vector population migration or bird movement to non-endemic areas. Both ecological and climatic changes have been implicated in MVEV activity in the northern area of Western Australia. Global warming may result in expansion of the present MVEV endemic area to central and southern Australia in the foreseeable future.

SEROPREVALENCE Between 1975 and 2002, epidemiological data indicated that the majority of cases of MVEV occurred in the Kimberley region of Western Australia, with the second most commonly affected area being the "Top End" and a few cases in southern Western Australia, as well as northern and middle Queensland. There was a cluster in Central Australia in 2001/2 on the border between the NT and South Australia and the NT and Queensland. In both the NT and Western Australia, the rate of seropositivity is higher in the Indigenous population, with indications that most children are infected in the first decade of life. During the epidemic seasons, high seroconversion rates take place. In areas of infrequent epidemics, seropositivity is low. For instance in the Murray Valley, only 5% of the population were seropositive 7 years after the last epidemic. Seropositivity for MVEV is very high in Aboriginal communities in the endemic areas of northern WA and adjoining NT.

There is evidence that most infections occur before the age of 10 years. In areas that are not endemic, seropositivity is rare. However, there are indications that there is an increase in both MVEV and KUNV in

Australia. Between March and July 2000, there were nine cases of encephalitis identified in Western Australia. Moreover, for the first time since serological testing began two decades ago, seroconversion has occurred in sentinel chickens in western New South Wales.

8.2.4.2.2 MURRAY VALLEY ENCEPHALITIS VIRUS

The incubation period is between days and weeks, the majority being from 1 to 4 weeks. Different from Japanese encephalitis virus (JEV) and WNV, where a mild febrile illness is common, MVEV infection is probably asymptomatic in most cases, only presenting as a clinically recognizable syndrome with the development of signs of encephalitis.

There is no age difference in susceptibility to the development of encephalitis. Most commonly encephalitis occurs in either adults recently entering, or young children residing in endemic areas.

Encephalitis usually presents with headache, cognitive impairment, confusion and tremor. As opposed to the clinical presentation in outbreaks of West Nile fever, where the triad of arthralgia, rash and fever in humans is common, this is not so for MVEV. In fact, in Murray Valley viral infections, the ratio of symptomatic compared to asymptomatic infections is 1 in 700 to 1 in 1200. However, in symptomatic patients with Murray Valley viral infection, 50% present with either encephalitis or meningitis. The case fatality in these is between 15% and 30% and long-term neurological sequelae occur in 50% of cases.

Between 1987 and 1996, 18 cases of encephalitis were identified, the majority being children. Fever was a constant finding and some patients had diarrhea, cough or a rash in the prodromal phase. Seizures were frequent. There may be rapid resolution or the condition may evolve into a more severe condition with clinical features indicating cerebral hemisphere, brainstem and spinal cord involvement. The mortality in children is about 25% with the risk being particularly high in those under 2 years of age. The risk of permanent sequelae is also high, again particularly in patients under the age of 2 years.

8.2.4.2.3 KUNJIN VIRUS

KUNV is a subtype of WNV, with 99.2% homology. WNV has invaded North, Central and South America as well as Europe and Africa. It is likely that it will invade the eastern seaboard of Australia where there is little endemic immunity among vertebrates. Cross-immunity exists between KUNV and WNV and attenuated KUNV will act protectively against WNV. KUNV is both endemic and epizootic in Western Australia, especially in the Kimberleys and the Gascoyne region, and in the NT and Queensland.

Wading birds such as the rufous night heron are the natural reservoirs for the virus, and the freshwater mosquito *C. annulirostris* (Skuse) is the principal epidemic vector. Every year there are isolated cases reported in endemic areas. KUNV is milder than Murray Valley Encephalitis and encephalitis is probably rare.

KUNV is similar to MVEV, but is more widespread and results in less severe clinical manifestations. KUNV is endemic in tropical north Australia and is carried from a bird reservoir by mosquitoes, particularly *C. annulirostris*. There is no evidence of person-to-person transmission, and infection confers a lifelong immunity. KUNV infects humans and domestic animals. It was first isolated from *C. annulirostris* in northern Queensland in 1960, and named after an Aboriginal clan living in the area of the Mitchell River. Serological surveys have shown that KUNV is widespread, but it is particularly prevalent in Tropical Australia. It has been found in south eastern Australia on occasion; however very few epidemiological studies have been carried out looking at the life cycle, nature and frequency of KUNV infection in Australia.

The incubation period is similar to that of MVEV and most infections are subclinical. Two forms of clinical disease have been documented: a condition characterized by fever, rash and fatigue and sometimes lymphadenopathy, and secondly, a meningoencephalitis which occurs infrequently, and is rarely fatal.

Diagnosis can be made in the following ways; isolation of KUNV from clinical material; detection of KUNV RNA in clinical material; immunoglobulin G (IgG) seroconversion or a significant increase in antibody level or a fourfold rise in titer of KUNV-specific IgG proven by neutralization or another specific test; KUNV-specific IgM detected in the CSF; KUNV-specific IgM detected in serum in the absence of IgM to Murray Valley encephalitis, Japanese encephalitis or Dengue viruses.

Confirmation of laboratory results by a second arbovirus reference laboratory is required to differentiate KUNV from WNV and MVEV if the case occurs in areas of Australia not known to have established enzootic/endemic activity or regular epidemic activity.

8.2.4.2.4 DENGUE FEVER

Worldwide, dengue fever is the second most common mosquito-borne infection afflicting humans after malaria. Dengue fever has occurred, and still occurs, in the NT and far north Queensland and is found more commonly in urban areas than in rural ones. The virus causes a hemorrhagic fever, and epidemics have been recorded in Australia since the 1880s which spread down as far as northern New South Wales. The virus is carried by a mosquito vector, *Aedes aegypti*. This mosquito has been eradicated from Western Australia and the NT, but between 1990 and 1998 seven epidemics of Dengue have occurred in north Queensland. Boat-borne refugees from East Timor have carried Aedes larvae and mosquitoes to Tropical Australia in recent years, and many of the recent dengue cases in Darwin are from this source.

Dengue fever has also been reported in New South Wales and in 2004 the dengue mosquito was detected in Tennant Creek in the NT, but for all practical purposes northern Queensland is the only area at risk as being endemic to the virus. Recent dengue epidemics in islands of the South Pacific appear to be associated with La Nin driven increases in rainfall and ambient temperature. The factors promoting the resurgence of dengue in Tropical Australia are climatic change, migration of infected populations, unsanitary urbanization with poor public health control, and lack of closed water systems. Open rainwater tanks provide excellent breeding grounds for *A. aegypti* and in potentially endemic regions may become a major public health hazard.

Dengue affects mainly older children and adults. It presents with fever, intense headache and joint pains. Rarely evidence of a true hemorrhagic fever with shock and bleeding supervenes, and even more rarely an encephalopathic picture develops.

CLINICAL FEATURES Dengue fever must be considered in the differential diagnosis of a patient presenting with an encephalopathy of unknown etiology, in an area where the *A. aegypti* mosquitos are known to exist—especially if there are indications of a bleeding tendency and a lowered platelet count. Neurological manifestations of Dengue fever, however, are not common, and the World Health Organization (WHO)-endorsed guidelines only added neurological features in 2009.

The most common neurological presentation attributed to Dengue fever is an encephalopathy that may be secondary to renal and hepatic failure or a true encephalitis due to the virus. In endemic areas between 4% and 47% of cases of encephalitis have been attributed to Dengue fever. This may be characterized by confusion with decreased level of consciousness and seizures. Pure motor weakness is the second most common presentation and is due to a Guillain—Barré syndrome. Myositis, myelopathy and myoclonus may also occur. Encephalitis has the worst prognosis while pure motor weakness may recover completely.

8.2.4.2.5 JAPANESE ENCEPHALITIS VIRUS

In Asia, Japanese encephalitis is the foremost cause of viral encephalitis with a mortality rate approaching

60%. The JEV is endemic in the whole of India and China, as well as South East Asian countries as far south as Indonesia.

The 274 Torres Strait islands lie to the north of Australia between Cape York and Papua New Guinea, and cover an area of 48,000 km². The JEV, a mosquito-borne flavivirus, was first identified in the Torres Strait Islands in 1995. In the Torres Strait islands, sero-epidemiological studies in sentinel feral pigs have confirmed that Japanese encephalitis is ubiquitous. On at least nine of the Torres Strait Islands sero-epidemiological studies have shown widespread infection of both the pig population and humans.

Fortunately, mainland Australia has been relatively spared. Two fatal human cases were identified in 1998 in Northern mainland Australia, but no cases have been identified since then. It is thought that the JEV was brought from Papua New Guinea, by infected mosquitoes (Culex tritaeniorhynchus) carried by the monsoon winds. East Timor is another potential source. The most concerning and least manageable potential reservoir, is the vast number (20 million) of feral pigs roaming the NT and Cape York. These pigs are often located close to isolated Aboriginal communities. Sentinel pigs on the Australian mainland (Cape York) have remained free from the virus since 1998; however, the perennial danger remains passage of Torres Strait islanders to mainland Australia.

Past infection with Japanese encephalitis confers lifelong protection. Currently several vaccines are available: SA14-14-2, IC51 (marketed in Australia and New Zealand as JESPECT and elsewhere as IXIARO) and ChimeriVax-JE (marketed as IMOJEV). A formalin-inactivated vaccine is recommended for travelers spending 30 days or more in endemic areas, as well as residents of endemic areas. Boosters should be given every 3 years for those at risk. Confirmation of disease by detection of JE antibodies in serum or CSF.

After the initial nonspecific features such as fever and headache, neurological dysfunction may be experienced with photophobia, lethargy, irritability, drowsiness, neck stiffness, confusion, ataxia, aphasia, intention tremor, convulsions, coma and death. Although only a small proportion of infections result in clinical disease, a quarter of symptomatic cases of JEV are fatal and a further quarter result in permanent complications. Mortality from JEV is much higher in the young, and CNS infection may produce lasting neurological deficit in the form of cognitive impairment and deafness. No transmission occurs between humans, so isolation of patients is not necessary. Treatment is supportive, with attention to development of seizures and raised intracranial pressure.

8.2.4.2.6 NEWER VIRAL INFECTIONS ASSOCIATED WITH ENCEPHALITIS

Hendra virus infection was first identified in Australia in 1994. Three people were infected and two died. All had contact with either patients with (symptomatic) pneumonia or asymptomatic horses. Serological evidence of horse and fruit bat infection have been found from Melbourne to Madang, Papua New Guinea.

Australian Bat Lyssavirus (ABL), is also a bat-borne virus, which is closely related to rabies virus, but unique to Australia. Once clinically apparent, ABL results in an acute, progressive and fatal neurological condition. In 1996, a 39-year-old woman developed a progressive neurological condition characterized by bulbar involvement, dying on the 21st day after becoming ill. She had been exposed to bats. In 1998, a 37-year-old woman succumbed 27 months (an atypically long incubation period) after being bitten by a fruit bat, from a rabies-like condition. Serological evidence has shown that bats are infected from as far south as Melbourne, to as far north as Darwin. Both the rabies vaccine and pooled immunoglobulin provide significant protection against ABL and should be administered after a patient has been scratched or bitten by a bat in Australia. No treatment is available once the condition manifests. PCR in the spinal fluid and serology is diagnostic.

8.2.4.3.1 MELIOIDOSIS

Melioidosis is a significant public health concern in both Tropical Australia and Southeast Asia. This infection from Burkholderia pseudomallei (a saprophytic water and soil bacterium) is the most common cause of community-acquired septicemic pneumonia in the "Top End" of the NT. It is acquired not by ingestion, but by inoculation through abraded skin or nasal inhalation in the form of dust (as occurred in helicopter pilots in Vietnam) or water droplets (during monsoonal rains). The organism typically produces a marked bacteremia and exudes exotoxins. It has a predilection for direct invasion of tissues and organs ranging from eye, prostate, bone, skin, central nervous system tissue and meninges.

EPIDEMIOLOGY Melioidosis is the most frequent cause of fatal community acquired pneumonia at the RDH, in the NT, and, in fact, the overall mortality rate for melioidosis in Australia is about 20%. It also occurs in northern Queensland, the Kimberleys and the Torres Strait Islands. Flooding has resulted in extension of the infection to areas as far south as the Brisbane River Valley in Ipswich, Queensland, and

Tennant Creek in the Northern Territory (at a latitude of 19.5 degrees south). Prevalence increases with increased rainfall and 85% of cases occur during the wet season (November through April). In the endemic region for melioidosis, outbreaks, often with fatalities, have also occurred after disturbed weather conditions (Cyclone Thelma, Tiwi islands; Katherine floods in the NT). Changed climatic conditions associated with global warming, particularly if there is increased rainfall, may expand the currently endemic areas for melioidosis, particularly if soil ecology shift is a part of these disturbances.

In Australia, *B. pseudomallei* has been isolated in the NT, Queensland and northern Western Australia, but is hyperendemic in the "Top End" of the NT, where the incidence is one of the highest in the world. Compared to northeast Thailand where the estimated annual incidence between 1987 and 1991 was 4.4/100,000, in northern Australia and the Torres Straits Islands the incidence is 16.5/100,000 per year, with this figure rising steeply during increased wet conditions to 40.1. In Thailand, paddy fields are common sources of the organism, but outbreaks in northern Western Australia have been associated with the contamination of portable water. The organism can survive for years in distilled water and is difficult to destroy; being resistant to desiccation, antiseptics and detergents. It has an intrinsic resistance to many antibiotics.

Susceptible groups are persons suffering from diabetes mellitus, alcoholism, renal disease, liver cirrhosis, chronic lung disease and thalassemia.

CLINICAL PICTURE Melioidosis can occur at any age, but the peak incidence is in the fourth and fifth decades. Significantly more men are affected than women. Worldwide as many as 50% of patients with melioidosis have diabetes mellitus, usually of the adult-onset type, and often poorly controlled.

PATHOLOGY Melioidosis is a septicemic condition that leads to death through a gram-negative bacteremia, and may disseminate to a variety of organs that include prostate, skeletal muscle, spleen and liver as well as the brain. In the brain, seeding may result in mycotic aneurysms, cerebral abscesses or meningitis. The bacterium may infect the cornea, causing corneal ulceration and, in children, especially in Thailand, may infect the parotid gland leading to a suppurative parotitis. In a review of 252 cases of melioidosis in the NT, Currie found that 88% were acute and 12% were chronic. Acute pulmonary infections represent the most frequent manifestation of melioidosis resulting in either a pneumonic picture or multiple lung abscesses. Patients are often febrile, tachypnoeic and have a productive cough. The purulent sputum is, unlike the case of tuberculosis, seldom bloody. Lung abscesses may remain localized or rupture into the pleural cavity causing an empyema.

In an analysis of 232 cases that were seen over a 9-year period (1989–98) at the RDH, Currie described 12 cases that presented with neurological features.

8.2.4.3.2 CRYPTOCOCCUS

Cryptococcosis is an important infection of the CNS encountered worldwide. Cryptococci are saprophytic, encapsulated fungi that are passed to humans from birds or the environment. The neurological importance of this fungal infection in Tropical Australia is that it is a cause of chronic meningitis.

Worldwide, the most common variety of cryptococcus infecting humans is *Cryptococcus neoformans* which in itself is the commonest cause of fungal meningitis.

However, in Tropical Australia, *Cryptococcus bacillisporus* (formerly known as *C. neoformans* var. *gattii* seotypes B and C) is more common. In Australia, there is a disproportionate infection rate for both *C. neoformans* and *C. bacillisporus* varieties in Aboriginal people. In fact in Arnhem land, in the far North of Australia, a rural area which is traditionally Aboriginal, the relative risk for cryptococcal disease both CNS and pulmonary is 20.6 for Aboriginal versus non-Indigenous people.

In the NT there have been three case series of cryptococcal disease reported. The total combined number of cases detected in the NT over a period of 43 years being 79. Most cases were due to *C. bacillisporus* and the majority of infected people were Aboriginal, mostly from rural areas.

Cryptococcus bacillosporus has been isolated from the river red gum trees *Eucalyptus camaldulensis* and *Eucalyptus tereticornis*. However, the highest incidence of infection by *C. bacillosporus* occurs in the NT of Australia and Papua New Guinea, where these eucalypt varieties are not endemic. There may therefore be other environmental foci. In fact, *C. bacillosporus* has been detected in other eucalypt species in southern, temperate Australia. Whereas *C. neoformans* spores are typically inhaled in an aerosol form from pigeon-droppings, the mechanism of human infection by *C. bacillosporus* is uncertain.

The Australiasian Cryptococcal Study Group described 312 cases of cryptococcal disease detected in Australia and New Zealand between 1994 and 1997. The overall majority (85%) of infections were due to *C. neoformans* infections, and the most common clinical presentation was meningoencephalitis. Of those patients with CNS disease, the majority (79%) were immunocompromised. Only 47 of 312 infections were due to *C. bacillisporus*. There are significant differences between cryptococcal disease as it presents in Tropical Australia and elsewhere in Australia and New Zealand. The clinical and other features of 18 consecutive patients treated for cryptococcal infections in the RDH in NT between 1993 and 2000, was reported in 2004. Among 18 patients, there was only one immuno-compromised patient, and all of the 15 patients tested for the immunodeficiency virus (HIV) antibody were negative. Moreover, in this series only 27% of patients were infected with *C. neoformans*. Only four of the 18 patients had CNS involvement alone. The majority of those with neurological infections had pulmonary lesions as well.

8.2.4.3.3 NAEGLERIA

Naegleria fowleri, a free-living amoeba occurring naturally in the NT and is found in stagnant water such as ponds and poorly chlorinated swimming pools. Optimal temperatures for this organism lie between 25 and 40°C. It enters the body through the nose or perforated eardrums and causes primary amoebic meningoencephalitis (PAM) which is often fatal. Only 19 cases have been reported in Australia.

8.2.4.3.4 ECHINOCOCCUS

Although not common in Tropical Australia compared to South Eastern Australia, *Echinococcus granulosus* is part of the predator/prey cycle existing between wild dogs (dingoes) and kangaroos. The possibility exists of wild dogs defecating in habited rural areas, and thus infecting cattle. Other reservoirs are feral pigs and wombats.

8.2.4.4 Non-infectious Neurological Diseases
8.2.4.4.1 KAVA

For centuries, Kava, a recreational beverage made from an extract from *Piper methysticum* Frost has been used in the Pacific island populations as a sedative. In recent decades, the use of Kava has spread to tropical north Australia, particularly among the indigenous peoples. Long-term Kava users can suffer from recognized side effects such as a Kava dermopathy (a scaly, yellow skin rash), general emaciation and elevation of liver enzymes. as a result of liver toxicity. Neurological sequelae of acute and chronic Kava use have been documented in case reports and include:

- severe choreiform or athetoid movements;
- seizures (possibly as a direct effect of the Kava or a consequence of withdrawal); and
- hallucinations and psychotic symptoms.

Interestingly, studies have shown that both chronic or excessive acute Kava use has not been shown to result in either cognitive or saccadic abnormalities in humans. This contrasts with the effects of chronic intake of petrol, alcohol or cannabis.

8.2.4.4.2 INHALANTS

The abuse of petrol and other volatile substances as an inhalant is a worldwide phenomenon which is often found in urban environments but is unusually prevalent in rural Australia, even in remote isolated communities in the tropical North. Leaded petrol is a readily available source of inhalant, and several studies from tropical Australia have documented the neurological effects of petrol sniffing.

Aromatic hydrocarbons such as xylene, toluene and benzene found in petrol (gasoline), if inhaled, produce effects such as a sensation of relaxation and euphoria. Neurological symptoms and signs such as ataxia, slurred speech and diplopia may also occur. These are usually reversible with cessation of inhalation. However, if the petrol contains tetraethyl lead, more dramatic sequelae of inhalation such as visual hallucinations, distortion of vision and psychosis have been reported. Chronic recreational users of petrol may have impairment of memory and demonstrate attention deficits. In severe cases, death may occur from aspiration of vomitus while unconscious, cardiac arrest or profound hypoxia due to respiratory depression.

Significantly, continuous and repeated inhalation of leaded petrol can produce the features of lead encephalopathy, the duration of which extends beyond the acute effects of petrol sniffing. In lead encephalopathy, combinations of neurological abnormalities such as hyperreflexia, nystagmus, seizures, and altered levels of consciousness may be seen. The condition may also be fatal. Hospitalization and treatment with chelating agents may therefore be necessary. At autopsy, lead encephalopathy has been shown to result in cortical, cerebellar, basal ganglionic and brainstem damage.

8.2.5 Cerebrovascular Disease

Although there is an acknowledged paucity of reliable data, epidemiological work indicates that Indigenous Australians have a significantly higher rate of stroke compared to non-Indigenous Australians. In fact, compared with non-Indigenous Australians, the Indigenous rate of stroke is increased 2.6-fold in men and 3.0-fold in women. The mean age for the greatest disparity is in the range of 35—54 years with a six to sevenfold increase in incidence. Similarly the disability-adjusted life years (DALYs) between the ages of 35 and 44 years in Indigenous Australians was eightfold that of non-Indigenous Australians. High risk is apportioned to hypertension, obesity, diabetes, smoking and alcoholism, all highly prevalent in Indigenous Australian populations. Theories abound as to the causes of these, but clearly, the Indigenous community as a whole has rapidly transitioned from nomadic hunter-gatherer lifestyles to a sedentary lifestyle replete with a western diet, alcohol consumption and smoking.

8.2.6 Rheumatic Fever and Rhematic Heart Disease

The rate of both these conditions in the Aboriginal communities of the tropical north of Australia is among the highest in the world. Chronic and recurrent staphylococcal pyoderma, resulting from chronic recurrent scabies infections, are the major putative causes for the high rate of rheumatic fever. This contrasts with the situation in other countries where pharyngeal infection (streptococcal) is most frequently implicated.

A study from the "Top End" of Australia published in 1999, shows that between 1935 and 1996, a total of 555 documented cases of acute rheumatic fever met the revised Jones criteria, and revealed that Sydenham's chorea is very frequent. About 30% of all Aboriginal persons with acute rheumatic fever at the "Top End" developed Sydenhams chorea. Of these 98% were Aboriginal, and 73% were female. The mean time to development of chorea after an attack of rheumatic fever was 2.1 years. One-quarter of patients with chorea experienced recurrent episodes. Three-quarters of all episodes presented clinically with isolated chorea. In patients with chorea, acute rheumatic carditis was present in 25% and arthritis was present in 8%. Subsequently 58% developed rheumatic valvular heart disease. This was particularly likely if rheumatic carditis had been diagnosed in the patient.

8.2.7 Otitis Media

Otitis media is very common in Aboriginal children. Extension of the infection in the temporal bone can lead to mastoiditis, petrositis, labyrinthitis, and paralysis of the facial nerve. Extension of the infection intracranially can cause extradural abscess, brain abscess, subdural abscess, sigmoid sinus thrombophlebitis, otic hydrocephalus, and meningitis. Although the use of antibiotics in otitis media makes intracranial complications of otitis media less common, one should be constantly aware of the possibility of such a complication.

8.3 MELANESIA

This region lies to the North of Australia, extending from the Arafura Sea and Western Papua in the West, to the Pacific Ocean in the East, as far as the Fiji

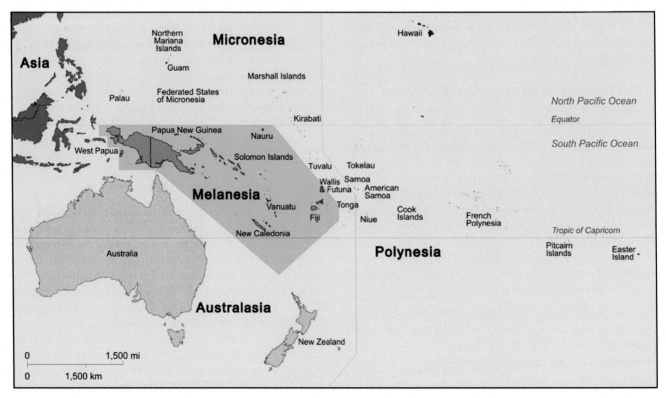

Islands. In addition to the four independent countries of Papua New Guinea, Vanuatu, Solomon Islands and Fiji, there is also Western Papua and the Maluku Islands governed by Indonesia, and New Caledonia, a Territory of France.

8.3.1 Papua New Guinea

8.3.1.1 Trauma

In Port Moresby craniocerebral trauma is common due to endemic criminal violence.

8.3.1.2 Vector-Borne Viral Disease

8.3.1.2.1 MALARIA

Malaria is endemic in this region. At least 90% of the population are at risk. In 2009 of a total population of 7 million there were 143,395 suspected cases with 22,896 hospitalizations. Malaria is transmitted by mosquitoes, the major plasmodium species being *P. falciparum* (56%) and *Plasmodium vivax* (41%), while anophelese mosquitos species are *Anopheles punctulatus, A. farauti,* and *Anopheles kollensis.*

Severe cerebral malaria (associated with *P. falciparum*) has a mortality ranging from 8% to 50% (depending on associated organ involvement) with most deaths occurring in the first 48 hours. The condition is characterized by severe headache, obtundation, seizures and progressive coma. See Table 8.1 for

falciparum-affected areas of Oceania including Papua New Guinea.

8.3.1.2.2 ZIKA VIRUS

Zika is a mosquito-borne flavivirus, which in 2013–14 was responsible for a large outbreak in French Polynesia. The importance of this infection lies in the associated Guillain–Barré syndrome which may result, as well as the association with microcephaly in the developing fetus of an infected mother.

Zika from French Polynesia is considered to have spread to Easter Island and further.

8.3.1.2.3 KURU

Kuru is a form of spongiform encephalitis, related to Creutzfeldt–Jakob disease and bovine spongiform encephalopathy. It has strong historical roots in the endocannabalistic culture prevalent in New Guinea as funeral rites. This is a prion disease transmissible not only to humans through ingestion but also to chimpanzees. The practice of cannibalism was widespread among the Fore people of New Guinea and between 1960 and 1970 reached epidemic proportions which necessitated construction of a "Kuru Hospital". The condition was characterized by tremors and mainly cerebellar symptoms, but typically there was no dementia.

TABLE 8.1 Populations at Risk of *Plasmodium falciparum* Malaria in 2010

Country	Age group	Unstable Pf risk[a]	Stable transmission			Total stable[c]	Total PAR[d]
			PfPR2-10 ≤ 5%[b]	5% > PfPR2-10 ≤ 40%[c]	PfPR2-10 > 40%[b]		
Papua New Guinea	All	1.61	36.95	11.12	0.00	48.07	49.68
Papua New Guinea	15 +	0.98	22.52	6.78	0.00	29.30	30.27
Papua New Guinea	5–14	0.40	9.24	2.78	0.00	12.03	12.43
Papua New Guinea	0–4	0.23	5.18	1.56	0.00	6.75	6.97
Solomon Islands	All	0.00	3.25	2.09	0.00	5.34	5.34
Solomon Islands	15 +	0.00	1.96	1.26	0.00	3.22	3.22
Solomon Islands	5–14	0.00	0.81	0.52	0.00	1.33	1.33
Solomon Islands	0–4	0.00	0.48	0.31	0.00	0.79	0.79
Vanuatu	All	0.01	2.41	0.00	0.00	2.41	2.42
Vanuatu	15 +	0.01	1.49	0.00	0.00	1.49	1.49
Vanuatu	5–14	0.00	0.59	0.00	0.00	0.59	0.59
Vanuatu	0–4	0.00	0.33	0.00	0.00	0.33	0.33

[a]*Population in 100,000s living in areas of unstable* P. falciparum *risk (PfAPI <0.1/1000 people p.a.).*
[b]*PfPR2-10 is the* P. falciparum *parasite rate in 2- to 10-year-olds.*
[c]*Population in 100,000s living in areas of stable* P. falciparum *risk (PfAPI ≥0.1/1000 people p.a.).*
[d]*The total population at risk (PAR) in 100,000s living in areas of* P. falciparum *malaria transmission.*
Courtesy of the Malaria Atlas Project. https://en.wikipedia.org/wiki/Malaria_Atlas_Project.

Because the culture demanded that women were the family members to practice the endocannabalism, in some villages there were almost no women left. It has now been largely confined to history, but has provided significant scientific data in the study of Creutzfeldt–Jakob Disease and spongiform encephalopathy of cattle.

8.3.1.2.4 CYSTICERCOSIS

Recent (2015) studies have revealed that Papua is still highly endemic, with most of the neurocystercicosis (NCC) cases also presenting with subcutaneous cysticercosis. *Taenia solium* completes its life cycle using pig intermediate hosts and human definitive hosts. However, dogs can also become infected with cysticerci through the ingestion of parasite eggs. Although it is conceived that humans were only obligatory intermediate hosts when this parasite emerged as a human parasite without involvement of pigs; humans are now paratenic hosts (with the exception of cannibalism). In humans, NCC resulting in epileptic seizures can be a life-threatening disease. Taeniasis occurs through eating uncooked or undercooked pork contaminated with cysticerci, the metacestode stage of *T. solium*. Adult tapeworms then mature in the intestines of the infected person. Eggs from mature adult worms can infect pigs, dogs and humans resulting in cysticercosis.

8.3.2 New Caledonia

Zika virus has been identified.

8.3.3 Fiji

Zika virus has been identified.

Chikungunya infections have been documented. Neurological sequelae occur in about 16% and range from encephalitis and myelitis, to peripheral nerve palsy and myopathy.

8.4 MICRONESIA

This Northernmost group of islands within the Oceania region comprises literally thousands of small islands in the western Pacific Ocean. These are governed by five independent sovereign nations: the Federated States of Micronesia, Palau, Kiribati, Marshall Islands, and Nauru—as well as three U.S. territories in the northern part: Northern Mariana Islands, Guam, and Wake Island.

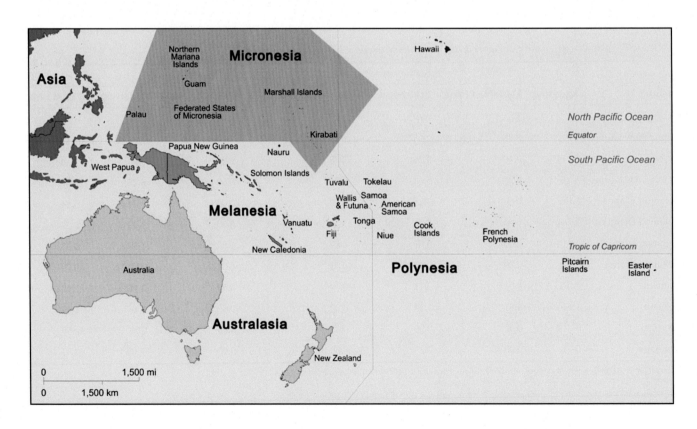

8.4.1 Guam

8.4.1.1 ALS–Parkinsonism–Dementia Complex

In Guam, one of the Mariana Islands, several decades ago it was found that both the clinical and pathological features of Amyotrophic Lateral Sclerosis (ALS) and Parkinson's disease (PD) occurred with a startlingly high frequency. One in five Chamorros from this island, over the age of 25 years, died from this condition. This was 50- to 100-times the incidence in either Continental Europe or the United States. With the passage of time, the incidence of these conditions has fallen almost to the level of the United States and Europe. The opinion is that in this population there is a metabolic defect that disturbs calcium metabolism, the chronic deficiency of which allows transintestinal absorption of neurotoxins in the form of heavy metals. Although there are clinical differences between presentations of Guanian ALS and PD, neuronal degeneration occurs in both, both have development of neurofibrillary tangles and in PD, there is depigmentation of the substantia nigra. Interestingly, in two other foci of abnormally high incidences of ALS and PD, namely the Kii Peninsula of Japan and a Western part of Papua New Guinea, during the same decades of observed decline in Guam, the incidence has also dramatically declined. The reason for this is unknown, but altered food habits are implicated.

8.4.2 Marshall Islands

Zika virus has been identified.

8.4.3 Kiribati and Nauru

Chikungunya has been identified.

8.5 POLYNESIA

Made up of over a thousand Islands, Polynesia is the western most region of Oceania, lying in the central and southern Pacific ocean. The tropical regions of Polynesia include but are not limited to Samoa, American Samoa, Tonga, French Polynesia, the Cook Islands, Hawaii in the far north and Easter Island in the far east.

8.5.1 Samoa and Tonga

Zika virus has been identified.

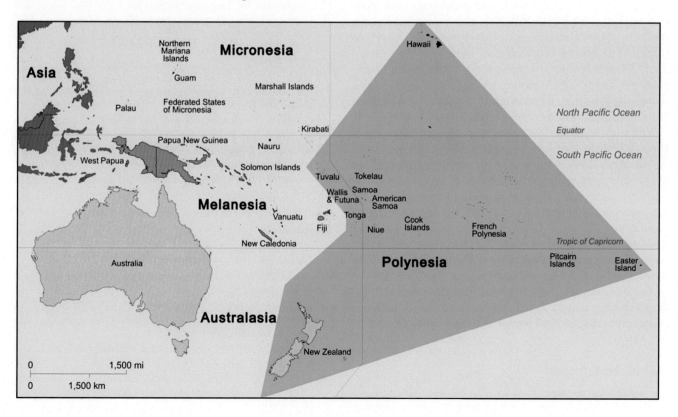

8.6 CEREBROVASCULAR DISEASE IN OCEANIA OUTSIDE AUSTRALIA

In the Asia-Pacific region, and particularly in Oceania, there has been a substantial increase in cerebrovascular disease which has been associated with the sociocultural epidemiological transition. For instance, in terms of deaths per 100,000 people, in Australian males this is 136, while in Samoa it is 477, in Fiji it is 580 and in Nauru 922. The breakdown of risk into obesity, diabetes, hypertension, smoking, and alcohol abuse indicates that much of the cerebrovascular disease which results from these risk factors can be averted.

8.7 THE FUTURE

Mosquito-borne disease is highly prevalent both in Tropical Australia and Oceania. Malaria is transmitted by a number of mosquito vectors while the *A. aegypti* species is the carrier of Dengue virus, West Nile Fever virus, Chikungunya as well as Zika virus. Traditionally insecticides have been used to eradicate the mosquitoes, but their efficacy is hampered by physical conditions such as the breeding of mosquitoes in deep clefts of trees. Insecticides must come into contact with the vector to kill it. Now, a novel development in molecular biology offers promising ways to inhibit the carriage of pathogens, both viral and protozoal. This advance allows precise deletion and rearrangement of DNA sequences in all living cells. The technology is that of clustered regularly interspaced short palindromic repeats (CRISPR). CRISPR-CAS9 consists of a RNA guide that unerringly seeks out the nucleotide that has been targeted combined with an enzyme, Cas9, which with exquisite deftness slices away the genomic sequence. The goal is to use CRISPR-CAS9 to delete a natural gene sequence in the mosquito, replacing it with one which confers sterility.

Mosquito generations are short—often only 2−3 weeks—and within several months, the species would carry the "sterility gene" resulting in total eradication of the line. Using CRISPR to edit the genomes in mosquito species would at one blow change the landscape of infective disease in Tropical Australia and Oceania.

Acknowledgments

We would like to thank Professor Vernon Marshall, Dr Ralph Poppenbeek and Dr Rodrigo Azubel for valuable advice in the preparation of the manuscript.

References and Selected Reading

Newer Viral Encephalitides

1. McCormack G, Allworth AM. Emerging viral infections in Australia. *Med J Aust*. 2002;177(1):45−49.
   A succinct overview of emerging and well-known viral illnesses that can cause neurological symptoms in Australia.
2. O'Sullivan JD, Allworth AM, Paterson DL, et al. Fatal encephalitis due to novel Paramyxovirus transmitted from horses. *Lancet*. 1997;349:93−95.
   Important description of Hendra virus encephalitis.
3. Allworth A, Murray K, Morgan J. A human case of encephalitis due to a Lyssavirus recently identified in fruit bats. *Commun Dis Intell*. 1996;20:504.
   A description of Lyssavirus encephalitis.

Murray Valley Encephalitis

4. Cordova SP, Smith DW, Broom AK, et al. Murray Valley encephalitis in Western Australia in 2000, with evidence of southerly spread. *Commun Dis Intell*. 2000;24:368−372.
   Indications of spread of areas potentially affected by MVEV, possibly due to climatic change.
5. Burrow JN, Whelan PI, Kilburn CJ, et al. Australian encephalitis in the Northern Territory: clinical and epidemiological features, 1987−1996. *Aust N Z J Med*. 1998;28:590−596.
   An important documentation of features of MVEV in Australia.
6. Douglas MW, Stephens DP, Burrow JNC, Anstey NM, Talbot K, Currie BJ. Murray Valley encephalitis in an adult traveler complicated by long-term flaccid paralysis: case report and review of the literature. *Trans R Soc Trop Med Hyg*. 2007;101:284−288.

Dengue Fever

7. Misra U, Kalita J, Syam U, Dhole T. Neurological manifestation of dengue virus infection. *J Neurol Sci*. 2006;244 (1−2):117−122.
   A report of neurological manifestation of dengue fever in Indian patients.
8. Solomon T, Dung N, Vaughn D, et al. Neurological manifestations of dengue infection. *Lancet*. 2000;355(9209):1053−1059.
   A report of neurological manifestations of dengue fever in Vietnamese patients.

Melioidosis

9. Currie B, Fisher D, Howard DM, Burrow JNC. Neurological melioidosis. *Acta Trop*. 2000;74:145−151.
   An intensive review of neurological experience of melioidosis in the NT.
10. Koszyca B, Currie BJ, Blumbergs PC. The neuropathology of melioidosis: two cases and a review of the literature. *Clin Neuropathol*. 2004;23(5):195−203.
    A description of two cases and extensive discussion of the pathogenesis of CNS melioidosis.
11. Cheng AC, Currie BJ. Melioidosis: epidemiology, pathophysiology, and management. *Clin Microbiol Rev*. 2005;18 (2):383−416.
    An extensive review of melioidosis.

12. White NJ. Melioidosis. *Lancet*. 2003;361:1715–1722.
A concise review of melioidosis.

Neurocysticercosis

13. Del Brutto OH, García HH. *Taenia solium* Cysticercosis: the lessons of history. *J Neurol Sci*. 2015;359(1–2):392–395. Available from: http://dx.doi.org/10.1016/j.jns.2015.08.011.

Malaria

14. Gething PW, Patil AP, Smith DL, et al. A new world malaria map: *Plasmodium falciparum* endemicity in 2010. *Malar J*. 2011;10:378. Available from: http://dx.doi.org/10.1186/1475-2875-10-378.

9

Neuroepidemiology in Latin America

Carlos N. Ketzoian, Abayubá Perna, and Heber J. Hackembruch

Neuroepidemiology Section, Institute of Neurology, School of Medicine, University of the
Republic, Montevideo, Uruguay

9.1 INTRODUCTION

The development of neuroepidemiology in Latin America has a history of more than 30 years. The minutes of incorporation of the Panmerican Society of Neuroepidemiology (*Sociedad Panamericana de Neuroepidemiología*, SPNE) were subscribed on July 6, 1984, in the city of Caracas, Venezuela. Together with Professor Bruce Schoenberg, colleagues from different Latin American countries created the Panamerican Society of Neuroepidemiology, thus opening a path that we have since then followed and still follow—with times of growth and times of lesser activity.

Some of the founders of the SPNE that must be mentioned are Prof. Dr. Pedro Ponce Ducharne (Venezuela), Dr. Marcelo Cruz (Ecuador), Dr. Felipe García Pedroza (Mexico), Dr. Gustavo Pradilla (Colombia), Dr. Nelly Chiófalo (Chile), and Dr. Beatriz González (Venezuela). They subscribed the minutes of incorporation of the SPNE with Professor Bruce Shoenberg, who served as its first President and who, at that time, was Head of Neuroepidemiology at the US National Institutes of Health.

The goals of this Society created in 1984 are as follows: to sponsor annual international conferences to establish scientific relationships with the Pan American societies acting in different countries; to promote the international transfer of scientific knowledge through the creation of a journal of the SPNE; and to build strong bonds with the International World Organization in order to receive information and to provide advice to local organizations of each of the member countries.

DOI: http://dx.doi.org/10.1016/B978-0-12-804607-4.00009-5

First Executive Board of the Panmerican Society of Neuroepidemiology (*Sociedad Panamericana de Neuroepidemiología*, SPNE)

President: Dr. Bruce Schoenberg, USA
Vice-President: Dr. Marcelo E. Cruz, Ecuador
Secretary: Dr. Pedro Ponce Ducharne, Venezuela
Treasurer: Dr. Beatriz González, Venezuela

Board Members:

First: Dr. Felipe García Pedroza, Mexico
Second: Dr. Juan Cabrera, Peru
Third: Dr. Gustavo Pradilla Ardila, Colombia
Fourth: Dr. Max Sánchez, Bolivia
Fifth: Dr. Patricia Barberis, Ecuador

Substitute Board Members:

First: Dr. Nelly Chiofalo, Chile
Second: Dr. Jaime Gómez, Colombia
Third: Dr. Arsencio Zúñiga, Colombia
Fourth: Dr. Barry Gordan, USA
Fifth: Dr. N. Zúñiga, Costa Rica
Sixth: Dr. Álvaro Vellegas, Venezuela

Honorary Members of the Panamerican Society of Neurology:

Dr. Liana Bolis, WHO
Dr. Shu Chino Lu, People's Republic of China
Dr. Hansten Held, Federal Republic of Germany
Dr. Rene González, PAHO

SPNE Congress Steering Committee
—October 1986

Steering and Research Committee, with one delegate per country:

Ecuador: Dr. Marcelo Cruz
Panama: Dr. Fernando Gracia
Argentina: Dr. Ernesto Herskovitz—Dr. Lucía Bonomi
Chile: Dr. Nelly Chiófalo
Peru: Dr. Eduardo Beteta
Bolivia: Dr. Touchy Marinkovic
Colombia: Dr. Ivan Jiménez
Brazil: Dr. Ivao Baptista Dos Reis Filho
USA: Dr. Andrés Salazar
Canada: Dr. Vladimir Hachinski
Venezuela: Dr. Beatriz González—Dr. Pedro L. Ponce
Mexico: Dr. Rubio Donadieu
Uruguay: Dr. Carlos Ketzoian
Costa Rica: Dr. Nelly Zúñiga
Puerto Rico: Dr. Luis Sánchez-Longo

Representatives from outside the Pan American Region:

Spain: Dr. Luis Oller Daurella
Germany: Dr. Harsten Held
France: Dr. Michael Dumas
Africa: Dr. Benjamín Osomtukum
United Kingdom: Dr. Marta Eleam
China: Dr. Shi-Chuo-Lu
Japan: Dr. Hisa Susuki

Since the creation of the SPNE, different neuro-epidemiological surveys have been conducted in different Latin American countries—most of the initial ones were pilot studies aimed at determining the prevalence of neurological diseases. At the same time, conferences and scientific meetings began to be organized.

The First Neuroepidemiology Congress was held in Bogotá, Colombia, in April 1985, with 120 participants. The minutes of this first conference highlight the relevance of this type of meeting, which included workshops and presentations of descriptive papers by researchers from Venezuela, Colombia, Ecuador, Peru, Chile, and Mexico. Most of these studies followed the World Health Organization (WHO) Research Protocol for Measuring the Prevalence of Neurological Disorders to identify the statistics for neurological disorders in the different regions of Latin America. The second conference was held in Santiago, Chile, in October 1986.

For this conference, a Steering Committee was set up with representatives of the different countries, among whom the participation of Prof. Michel Dumas (France) is noteworthy, along with representatives from Africa, China, and of different Latin American and European countries.

The early passing of Prof. Bruce Shoenberg at the age 44 on July 14, 1987, in Bethesda, Maryland, led to a downsizing that ended with the closing of the Neuroepidemiological Branch of the National Institute of Neurological and Communicable Disorders and Stroke, National Institutes of Health, where Professor Shoenberg had worked since 1975.

Past Presidents

Dr. Bruce Schoenberg (USA) 1984—87
Dr. Marcelo Cruz (Ecuador) 1987—91
Dr. Gustavo Román (Colombia) 1991—95
Dr. Manuel Somoza (Argentina) 1995—99, 1999—2003
Dr. Carlos N. Ketzoian (Uruguay) 2003—07, 2007—12

SPNE Present Executive Board
(Appointed in La Paz, Bolivia, in March 2012)

President: Dr. Violeta Díaz (Chile)
Vice-President: Dr. Beatriz González (Venezuela)
Secretary: Dr. Miguel Ernesto Córdova Ruiz (Peru)
Treasurer: Dr. Briceida Feliciano (Puerto Rico)

Board members:

Dr. Carlos Laforcada Rios (Bolivia)
Dr. Ana Robles (Dominican Republic),
Dr. Martha Galeano (Paraguay),
Dr. Abayuba Perna (Uruguay),
Dr. Ernesto Triana (Panama)

List of Country Representatives to the Panamerican Society
of Neuroepidemiology

- Juan Carlos Fernández (Bolivia)
- Carlos Mario Melcon (Argentina)
- Santiago Fontiveros (Venezuela)
- Fernando Gracia (Panama)
- Leonardo Bartoloni (Argentina)
- Princesa Rodriguez (Honduras)
- Ángel Chinea (Puerto Rico)
- Mario Tolentino (Dominican Republic)
- Norberto Cabral (Brazil)
- Mario Muñoz Collazos (Colombia)
- Federico Silva (Colombia)
- Marco Tulio Medina (Honduras)
- Hernan Bayona (Colombia)
- Salvador Gonzales Pavo (Cuba)
- Claudia Valencia (El Salvador)
- Felipe García Pedroza (Mexico)
- Lilian Núñez Orozco (Mexico)
- Sheila Ourquies (Brazil)
- Heber Jochen Hackembruch (Uruguay)

Nevertheless, neuroepidemiology meetings continued to be held throughout Latin America: San Juan, Puerto Rico (1987); Montevideo, Uruguay (October 1991); Guatemala (October 1995); Cartagena de Indias, Colombia (October 1999); and Santiago, Chile (October 2003). These events were scheduled as pre-conference meetings of the Pan American Neurology Conferences. In October 2007, the pre-conference meeting was held at the Dominican Republic as part of the Pan American Neurology Conference of Santo Domingo, and lastly, the third and last Pan American Neuroepidemiology Conference took place in November–December 2010, in Punta del Este, Uruguay. The latest SPNE meeting was held in La Paz, Bolivia, in March 2012, as a pre-conference meeting of the Pan American Neurology Conference that took place in that country. The current Board of the Panamerican Society of Neuroepidemiology was elected therein.

In the years that followed 1999, the SPNE established close links with the *Institut d'Épidemiologie Neurologique et de Neurologie Tropicale*, at Limoges, France, presided over by Prof. Michel Dumas. This was the starting point for a prolific exchange with the French Neurology community through close bonds with different Latin American countries.

Thus, during the sixth Meeting of the Panamerican Society of Neuroepidemiology, which took place in Chile on October 8, 2003, the First Sessions of French-Pan American Neurological Friendship were held and presided over by Profs Manuel Somoza, Gustavo Román, Maurice Collard, and Michel Dumas.

The Second Sessions of French-Pan American Neurological Friendship took place on October 7, 2007, in Santo Domingo, Dominican Republic.

On November 30, 2010, during the 3rd Pan American Neuroepidemiology Conference of Punta del Este, Uruguay, the Third Sessions of French–Pan American Neurological Friendship took place under the Honorary Presidency of Professors Michel Dumas and Mario Tolentino-Dipp (Dominican Republic).

For several years, Latin American neurologists have had the opportunity to serve internships, and to follow Masters and PhD programs co-sponsored by different Latin American universities and the University of Limoges, France. Furthermore, several neuroepidemiology surveys have been conducted with the participation of colleagues from the University of Limoges and from different Latin American countries, as will be specified on the next pages.

9.2 NEUROEPIDEMIOLOGY IN LATIN AMERICA

At the time of writing, there are research groups and researchers working in Neuroepidemiology in virtually all Latin American countries. We present data on the research studies conducted in recent years and published in international journals, regional journals, meetings, scientific meetings, and conferences in the area. We sent surveys to renowned Latin American researchers who work in the field of neuroepidemiology. Our findings are presented in this chapter.

This list is non-exhaustive; it features 61 researchers or research teams working in neuroepidemiology, distributed across 17 countries throughout Latin America (Fig. 9.1).

The number of diseases and conditions studied by these teams is variable.

As far as we know, there are three different types of researchers and/or teams:

1. Some teams research into a single disorder, such as multiple sclerosis (MS) or epilepsy, and the performance of epidemiological surveys is motivated by the researchers' interest in such disorders, which is their research line. We have identified 26 researchers or research teams (43%) in this category.
2. Other researchers are interested in more than one disorder, and their research lines are associated with various neurological diseases. These researchers use the epidemiological method to increase their knowledge on the distribution and the determinants of certain disorders at their respective countries.
3. Lastly, some researchers in neuroepidemiology have dedicated their scientific activities based on the epidemiological method to a great number of neurological disorders in their respective countries. While this is the smallest category in terms of number of members, the fact is that the interaction among these researchers and other researchers focused on a single disorder has enhanced the

FIGURE 9.1   Neuroepidemiology research groups. Non-exhaustive list of researchers in Neuroepidemiology in Latin America (last 5 years).

development of this field of knowledge in the region. This model has been followed in several of our countries. Furthermore, the participation of these researchers at scientific meetings held throughout the region has enabled the conduction of multicenter neuroepidemiological studies involving different Latin American countries, as we will see below.

We have identified 35 neuroepidemiology researchers or research teams who work with two or more disorders (Categories 2 and 3, 57%).

What are the main neurological diseases these teams are studying? There is a wide variability in the diseases studied by the different teams working in Latin American countries (Fig. 9.2).

The list is long and varies widely across the different countries. Some diseases are studied in neuroepidemiology surveys conducted in most Latin American countries, such as epilepsy, MS, stroke or dementia syndromes. We also identified teams working in rare neurological disorders. Some examples of the latter include the teams studying meningoencephalitis due to dengue fever in Paraguay; sleep disorders in Ecuador; neurocysticercosis in Peru and Ecuador, or

Pompe disease in Uruguay and Brazil. Their interest is generally based on the high prevalence of these conditions at the relevant countries (for instance, dengue fever or neurocysticercosis), or else, on the interest of the researchers in a specific issue, which leads them to generate specific knowledge in such disease in their home countries (for instance, sleep disorders in Ecuador or Pompe disease in Uruguay or Brazil, where the disease has a low prevalence) (Fig. 9.3).

Other interesting examples concern specific diseases that may only be studied in some of our countries, for example, neurological disorders associated with altitude. Some research teams based in Bolivia and Peru are studying this issue, which is highly clinically relevant, not only for their respective populations, but also for the international community as a whole.

Fig. 9.3 presents the diseases most frequently studied by neuroepidemiological surveys in Latin America.

Epilepsy is the most widely studied disease in the Latin American region. Also, across several Latin American countries, there are research teams working on neurological diseases such as MS, headache, stroke, and dementia syndromes. The research into extrapyramidal   diseases   (including   Parkinson's   disease),

FIGURE 9.2   Neuroepidemiology in Latin America: main neurological diseases under study by country.

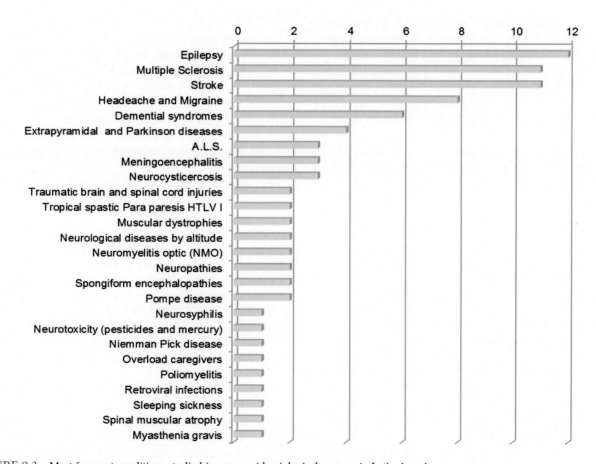

FIGURE 9.3   Most frequent conditions studied in neuroepidemiological surveys in Latin America.

amyotrophic lateral sclerosis (ALS) and meningoen-cephalitis is less frequent. Neurocysticercosis continues to be prevalent in several of our countries.

The studies into human T-lymphotropic virus-1 (HTLV-1) conducted in countries of the region, such as Colombia and Panama, were among the first to describe this disease worldwide.

9.3 EPILEPSY

Most Latin American epidemiological studies on epi-lepsy have been population-based prevalence studies.

In a door-to-door survey in two phases conducted in Migues, a semi-rural area of Canelones, Uruguay, in 1990, the prevalence of epilepsy was determined at 9.1 cases per 1000 persons.[1] Three years later, a door-to-door survey conducted among 20,965 inhabitants of Villa del Cerro-Casabó, Montevideo, Uruguay, with prevalence point set at September 1, 1993, we found 245 cases of epilepsy (prevalence: 1.6 cases of epilepsy per 1000 persons), with a slightly higher number of cases among men (12.5 cases per 1000 persons) than among women (10.9 cases per 1000 persons). The prev-alence of active epilepsy in this population was 6.4 cases per 1000 persons. When adjusted per age group, the prevalence was higher among participants aged 20–29 and among those aged 70–79, respectively. With respect to treatment, 46.9% of epilepsy cases were under treatment; of them, 40.4% were active epi-lepsy cases under treatment, and 6.5% were cases on non-active epilepsy under treatment. 36.7% of epilepsy patients presented remitting epilepsy and were not under treatment. 14.3% of patients with active epilepsy were left untreated (treatment gap). There were no data available for 2.1% of cases. In the Villa del Cerro-Casabó study, only 7% of patients with epilepsy had experienced seizures in their lifetime.

In terms of the types of epileptic seizures, 65.3% had experienced generalized tonic-clonic seizures; 9.44% had experienced simple partial seizures; 9.44% had experienced secondarily generalized partial crisis; a lower percentage (7.87%) had experienced complex partial seizures; and only 2.4% had experienced petit mal absence crises. Data were not available in 5.55% of cases. Patients that had presented with febrile seizures during childhood (<5 years of age) were at a higher risk of developing epilepsy at the adult age (odds ratio (OR) = 6.73, 95% confidence interval (CI): 4.73–10.40).

In other studies, such as the one conducted by Agnes Fleury et al.[1] in a Mexican rural population, a lower epilepsy prevalence than the one found among the urban population of Montevideo was found. Their prevalence was 3.7 cases per 1000 persons (95% CI: 2.3–6.5), when adjusted to the world's population. This was also a population-based study conducted in

two phases to determine the prevalence of the main neurological diseases based on the WHO protocol.

The report submitted to the International League Against Epilepsy (ILAE)[2] by a work group led by Dr. Patricio Abad in Ecuador is extremely interesting. This study presents epidemiological surveys on epilepsy throughout Latin America (Fig. 9.4; Table 9.1). Prevalence varies across the different countries and also across different studies performed in a single country.

The studies present data on lifetime prevalence, active prevalence, prevalence by gender and, in some cases, prevalence in children, although not all data are available for all countries.

Lifetime prevalence ranges from 3.2 to 44.3, with a median of 16.2 cases per 1000 persons. In turn, for active epilepsy, prevalence ranges between 3.8 (Junín, Argentina, M. Melcon) and 57 (Panama, F. Gracia).

Such variability is due to the following:

1. socioeconomic and cultural factors prevailing in these populations, with different ethnic backgrounds (higher or lower presence of native communities and of immigrant populations from Europe or Africa);
2. the different methodologies used by the different researchers;
3. the high prevalence of neurocysticercosis in some of Latin American countries.

No differences were found in terms of the preva-lence of epilepsy among rural and urban populations, or among genders, based on the ILAE Report.

There are fewer epilepsy incidence studies in Latin America (Table 9.2). Again, the variability in incidence across the different studies is significant, even though most studies report incidence rates ranging between 113 and 190 new cases of epilepsy per 100,000 person-years.

The treatment gap of patients with active epilepsy in Latin America also shows a great variability (7%–90%) according to the different authors, with the low-est rates found in two studies conducted in Argentina (7% M. Somoza; 22% S. Kochen), and the highest reported in Bolivia (90%, A. Nicoletti) (Table 9.3).

Regarding the burden of epilepsy in our countries, Colombian research published in 2015[3] showed that epilepsy accounted for 0.88% of the total mortality in that country, with a total of 5.35 Disability-Adjusted Life Years (DALYs), 75% of which (3.91 DALYs) were due to early death, with a greatest burden among men than among women.

It is necessary to increase our knowledge of the epi-demiology of epilepsy in our countries. There are few studies aimed at identifying the risk factors for this disease, especially in children, even though most of these factors are preventable, and their identification

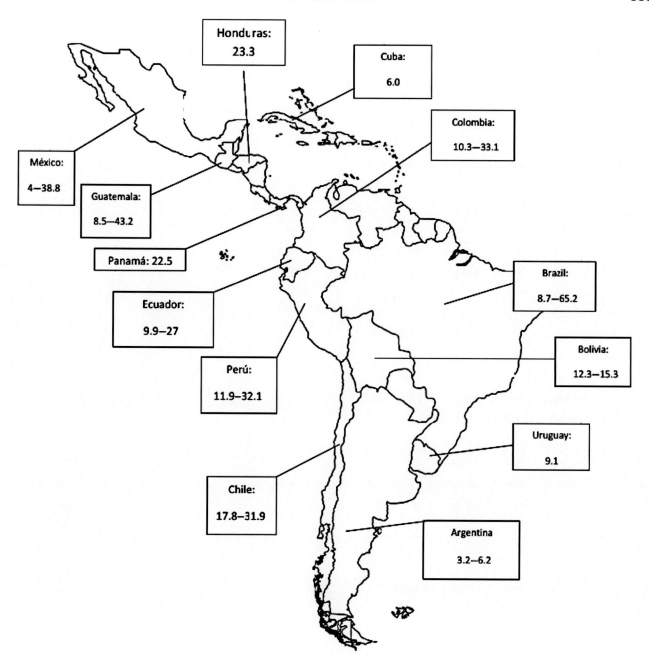

FIGURE 9.4 Lifetime prevalence rates of epilepsy in different community-based studies in Latin America (range shown when more than one study per country is available).

and treatment would mean a significant improvement to the health of the Latin American populations.

9.4 MULTIPLE SCLEROSIS

As for epilepsy, many Latin American researchers have focused their interests in the epidemiology of MS in the region.

A prevalence study conducted among the population of Uruguay with majority Caucasian ethnics[4] with prevalence point set on September 1, 1997 and using the capture—recapture method, found a prevalence of 20.9 cases of MS per 100,000 persons (95% CI: 20.9—21.5). According to the 2006 population survey, Uruguay has a Caucasian population of 96.9%, with 9.1% of Afro-American descent; 3.8% of indigenous descent; and 0.3% of Asian descent (the numbers do not add up to 100% because participants were allowed

TABLE 9.1 Epilepsy—Prevalence Surveys in Latin America (Cases per 1000)

Studies	Country	Year	Population	N	Lifetime prevalence	Prevalence active epilepsy	Children	Male	Female	Method of ascertainment
Rural										
Nicoletti	Bolivia	1994	All ages	9955	12.3	11.1		11.4	13.1	Q + E
Borges	Brasil	2000	All ages	481	18.6	12.4				Q + E + T
Li	Brasil	2007[a][b]	All ages	n/a	9.2	5.4				Q
Pradilla	Colombia	1996	All ages	544	33.1			43.1	25.6	Q + E
Pradilla	Colombia	2002[b]	All ages	544	33.1					Q + E
Zuñiga	Colombia	1984	All ages	1053	19.9					Q
Basch	Ecuador	1995	All ages	221	24					Q + E
Carpio	Ecuador	1983	All ages	935	18.2			27.8	6.9	Q + E
Cruz	Ecuador	1992	All ages	2723		11.4		10.3[c]	12.3[c]	Q + E
Cruz	Ecuador	1982	All ages	1113	27	17.1		16.5	17.6	Q + E
Del Brutto	Ecuador	2003	All ages	2415	9.9					Q
Placencia	Ecuador	1987	All ages	72,121	14.3	8		12.1	16.3	Q + E
García-Noval	Guatemala	1992	All ages	2292	43.2	17.9		46	34	Q + E + T
Mendizabal	Guatemala	1991	All ages	1882	8.5	5.8		7.4	9.5	Q + E
Medina	Honduras	1997	All ages	6473	23.3	15.4		13.7	17.1	Q + E + T
Gutierrez	Mexico	1980[b]	Children	360			25			Q + E + T
Quet	Mexico	2007	All ages	5195	2.3–4[d]		3.9			Q + E + T
Gracia	Panamá	1988	All ages	337		57				Q + E
Gracia	Perú	2006/07	All ages	17,450	17.3	10.8				Q + E + T
Montano	Perú	2000	All ages	1004	32.1	16.6				Q + E + T
Ketzoian	Uruguay	1990	All ages	1975	9.1					Q + E
Urban										
Melcon	Argentina	1993	All ages	17,049	6.2	3.8		3.5	4	Q + E
Somoza	Argentina	1991	Children	26,270			3.2	3.4	3	Q + E
Marinkovic	Bolivia	1988[b]	All ages	915	15.3					Q + E
Borges	Brasil	2001	All ages	17,293	18.6	8.2				Q + E
Gomes	Brasil	2000	All ages	982	16.3	5.1		17.9	15	Q + E
Marino	Brasil	1984	All ages	7603	11.9					Q + E
Noronha	Brasil	2002	All ages	54,102	9.2	5.4				Q + E
Sampaio	Brasil	2005–06	Children[e]	10,405			8.7/9.7[e]			Q + E + T
Chiófalo	Chile	1975	Children	2085			31.9			Q + E + T
Chiófalo	Chile	1992[b]	All ages	7195		17				Q + E
Lavados	Chile	1988	All ages	17,694		17.8	17[c]	17.7[c]	17.8[c]	Health records
Gomez	Colombia	1974	All ages	8658	19.5			15.5	22.9	Q + E + T
Pradilla	Colombia	2002[b]	All ages	622	16.1					Q + E
Pradilla	Colombia	2002[b]	All ages	288	17.4					Q + E

(*Continued*)

TABLE 9.1 (Continued)

Studies	Country	Year	Population	N	Lifetime prevalence	Prevalence active epilepsy	Children	Male	Female	Method of ascertainment
Pradilla	Colombia	1996	All ages	8910	10.3			10.1	13.8	Q + E
Zuloaga	Colombia	1983	All ages	4549	21.4			22.9	20.2	Q + E
Pascual	Cuba	1980[b]	All ages	45,537	6		7.5	8.4	6.5	Q + E
Caraveo-Anduaga	Mexico	1988	Adults	1984	38.8			34	48	Q
Garcia-Pedroza	Mexico	1975	Children	2027			44.3			Q + E + T
Gutierrez-Avila	Mexico	1978	Children	1042			16			Q + E + T
Cruz-Alcalá	Mexico	2000	All ages	9082	6.8			8.3	5.5	Q + E
Gracia	Panamá	1986	All ages	955	22.5			14	32.2	Q + E
Reyes	Perú	1992	Children	2016			11.9			Q
COMBINED: RURAL/URBAN										
Velez	Colombia	2006[b]	All ages	8910	11.3	10.1				Q

[a]Lifetime/active prevalence.
[b]Year of publication.
[c]Prevalence of active epilepsy.
[d]Household/individual Q.
[e]Persons younger than 17 years.
E, neurological evaluation; N, sampled population; Q, questionnaire; T, tool = EEG or CT of the head.

TABLE 9.2 Epilepsy—Incidence Surveys in Latin America (Cases per 100,000 People-Years)

Studies	Country	Year	Population	N	Incidence	Rural vs urban	Method of ascertainment
Nunes	Brasil	2003	Children	2285	7	Urban	Q + E
Lavados	Chile	1988	All ages	17,694	113	Urban	Health records
Placencia	Ecuador	1987	All ages	72,121	122–190	Rural	Q + E
Villaran	Perú	1999–2004	All ages	817	162.4	Rural	Q + E + T

TABLE 9.3 Epilepsy Treatment Gap

				Treatment gap of epilepsy in Latin American countries				
Author, year	Methods[a]	Sources	Country	Target population	Population size	Prevalence of active epilepsy (per 1000)	Treatment gap (%)	Comments
Nicoletti A, 1999	Direct method	Screening of residents with epilepsy in the selected population (door-to-door survey)	Bolivia	People with epilepsy in Cordillera province of Bolivia	10,124	12.3	90	This study was designed to search prevalence of epilepsy in a rural population
Placencia M, 1994	Direct method	Screening of residents with epilepsy in the selected population (door-to-door survey)	Ecuador	People with epilepsy in Carchi Province of Ecuador	72,121	12.2	63	This is a pioneer study designed to analyze the characteristics of epilepsy in a largely untreated population of a developing country

(Continued)

TABLE 9.3   (Continued)

Treatment gap of epilepsy in Latin American countries

Author, year	Methods[a]	Sources	Country	Target population	Population size	Prevalence of active epilepsy (per 1000)	Treatment gap (%)	Comments
Meidna M, 2005	Direct method	Screening of residents with epilepsy in the selected population (door-to-door survey)	Honduras	People with epilepsy in Salama Country of Honduras	7834	15.4	52	The selected population was previously advised regarding how to treat epilepsyly since a similar study was carried out some years before the present study
Del Brutto O, 2005	Direct method	Screening of residents with epilepsy in the selected population (door-to-door survey)	Ecuador	People with epilepsy in Atahualpa, a rural village of Ecuador	2415	9.9	43	At the time of this study only 13 patients (57%) were taking antiepileptic medication, which were taken irregularly in all cases
Noronha AL, 2004	Indirect method	Data based on antiepileptic drug available at the Municipal Health Secretaries in Campinas and São José do Rio Preto (public health system covers only 70% of the population in these cities)	Brazil	Population of Campinas and São José do Rio Preto, Brazil	4,400,000	18.6[b]	50	Campinas and São José do Rio Preto are two cities that have reasonable living conditions, infrastructure and healthcare systems
Somoza MJ, 2005	Direct method	Screening of children with epilepsy	Argentina	Primary school population in Buenos aires, Argentina	302,032	2.6	7	This study was performed in a highly selected urban population
Mendizabal JE, 1996	Direct method	Screening of residents with epilepsy in the selected population (door-to-door survey)	Guatemala	People with epilepsy in a rural village of Guatemala	2111	8.5	69	This study was designed to search prevalence of epilepsy in a rural population
Carpio A, 2012	Direct method	Screening of residents with epilepsy in the selected population (door-to-door survey)	Ecuador	People with epilepsy in Azuay Province of Ecuador	42,294	13.5	73	This study was designed to evaluate a model on primary health care in epilepsy
Kochen S, 2005	Direct method	Screening of residents with epilepsy in a population based study	Argentina	People with epilepsy in Junin, a urban area of Argentina	17,049	3.8	22	This study was carried out in a selected urban population

[a]See definitions of direct and indirect method to calculate treatment gap in the text.
[b]Number assumed, based on a previous study addressed to analize prevalence.

to report more than one racial background). Using the diagnostic criteria of Poser et al.,[5] 92.4% of the cases were clinically definite MS, 4.9% were laboratory-supported definite MS, and 2.7% were clinically probable MS. We found no cases with laboratory-supported probable MS. The prevalence of MS in our population was higher within the 35–39 years and 40–44 years age groups, respectively, and among female patients (prevalence among women: 26.5 cases per 100,000 persons; prevalence among men: 13.5 cases per 100,000 persons; women/men gender ratio: 2:1). 68.5% of the cases were remitting-relapsing MS; 12.8% were secondarily progressive MS; 11.4% were progressive remitting-relapsing MS; and 6.2% cases were progressive primary MS. There were no data available for 1.2% of cases. The syndromes most frequently found in our population were the following: pyramidal syndrome (68.6%); cerebrellar syndrome (52.1%); retrobulbar optic neuritis (41.2%); medular syndromes (35.5%); brainstem syndromes (33.1%); sphincteral-genital syndromes (23.1%); sensorial syndromes (20.8%); cognitive-behavioral syndromes (9.2%); and other syndromes (8.7%). In 1.5% of cases the neurological examination was normal. The most frequent age at onset ranged between 25 and 45 years. Regarding the time to progression, over 35% of cases had progression at 5–9 years; almost 20% showed progression at 1–4 years; and the remaining cases had progression at over 10 years. Regarding MS-related disability, at the time of the research, 64.3% of participants had preserved ambulation; 22% walked with aids; and 7.1% of patients were confined to bed or armchair. There were no data available for 6.6% of cases.

In May–June 2006, Patricio Abad et al.[6] developed an observational, cross-sectional research study in the three major Ecuadorian cities, a country located over the Equator line. This research was conducted in Quito, at 2816 m above sea level and with 2,036,260 inhabitants, Cuenca, an Andean city at 2800 m above sea level and with 666,085 inhabitants, and in Guayaquil, a city located above the sea level as well and with 2,206,213 inhabitants. In the first two cities, the population is mainly of *Mestizo* ethnics, with a few European descendants and a minority of Native and Afro-Ecuadorian populations. In Guayaquil, on the contrary, the *Mestizo* and European-descent populations prevail, with lower Native and Afro-Ecuadorian populations than those in Quito and Cuenca. Patricio Abad et al. found a 5.05 prevalence of MS (95% CI: 4.08–6.03) cases for 100,000 persons in Quito; 1.62 (95% CI: 1.62–2.91) cases per 100,000 persons in Guayaquil; and 0.75 (95% CI: 0.024–0.175) cases per 100,000 persons in Cuenca. These prevalence rates were lower than the ones found in Uruguay with the same research protocol. This may be due to differences

in the ethnic composition of our populations and to environmental factors: while Uruguay is a flat country, Ecuador has varied climates and three clearly defined regions: the hill and Andes mountain region; the Pacific Ocean coastal area at sea level; and the Eastern or Amazon area. In terms of age groups, the highest prevalence rates were found in the 30–49 years age group, as in Uruguay, with a predominance of female patients (women/men gender ratio: 2.3 in Quito; 4.4 in Guayaquil; and 4.0 in Cuenca). The distribution of MS clinical forms was similar to the one noted in Uruguay: 71.32% were cases of remitting-relapsing MS; 15.38% were secondarily progressive MS; 8.32% were secondarily progressive remitting-relapsing MS; and 4.89% were cases of primary progressive MS.

In a multicenter study conducted by more than 50% of neurologists practicing in Puerto Rico, in the Caribbean area (San Juan Bautista School of Medicine, San Juan MS Center)[7] that used multiple data sources, the incidence was found at 5 cases per 100,000 persons, with a higher rate among female patients (women: 6.9 cases per 100,000 persons; men: 2.8 cases per 100,000 persons). The women/men gender ratio was 3:1. The median age at diagnosis was 40.8 years, while the onset of disease was reported at 35.5 years for both genders.

In several papers, Regina Papais Alvarenga et al. reported on epidemiological studies on MS and other demylienating disorders.

In a survey conducted in 2012 in the city of Volta Redonda, Rio de Janeiro, Brazil, using the capture–recapture method,[8] the raw prevalence of MS was determined to be at 15.3 cases per 100,000 persons. When adjusted by the capture–recapture method, the prevalence was of 30.7 cases per 100,000 persons (95% CI: 20.4–61.1). These rates are among the highest reported both in Brazilian and in other Latin American studies.

In a systematic review of the prevalence of MS in Brazil,[9] R. Alvarenga et al. selected 19 MS prevalence surveys that met their inclusion criteria. They included surveys from the Southeast, Mid-West, Southern, and Northeastern regions. No study was found in the Northern region. These authors found a high variability in the prevalence rates reported by the different studies. A positive correlation was found between latitude and the prevalence of MS ($r^2 = 0.54$; $p = 0.02$): the higher the Southern latitude, the higher the prevalence of MS. Using the random-effects model, the authors found a prevalence of 8.69 cases per 100,000 persons (95% CI: 6.0–12.6). In this review, a bivariate analysis found a significant association with latitude (OR = 1.09, 95% CI: 1.04–1.14); with ethnic background (percentage of Caucasian population, OR = 1.03, 95% CI: 1.01–1.05), and with climate, with the tropical equatorial climate being a protective factor (OR = 0.55, 95% CI: 0.05−0.46). Using multivariate analysis models, only the tropical equatorial climate was statistically significant as a protective factor.

A multicenter study conducted by Regina Alvarenga Papais et al.[10] reported cases of neuromyelitis optica and MS in white, *Mestizo*, and African populations living in areas of low prevalence like South America. They performed a cross-sectional multicenter study. Only the individuals followed in 2011 with a confirmed diagnosis of inflammatory demyelinating disorder were considered eligible. A total of 1917 patients from 22 MS centers from across Latin America were included. The main disease categories were MS 76.9%, neuromyelitis optica 11.8%, other neuromyelitis optica syndromes 6.5%, clinical isolated syndrome 3.5%, acute disseminated encephalomyelitis 1.0%, and acute encephalopathy 0.4%. Females predominated in all main categories. White ethnicity also predominated except in patients with neuromyelitis optica. The relative frequency of neuromyelitis optica among patients

with relapsing-remitting MS + neuromyelitis optica cases in South America was 14.0%. MS affects three-quarters of all patients with idiopathic inflammatory demyelinating disorders in South America. A higher frequency of neuromyelitis optica was found among non-white populations.

At the 2010 LACTRIMS conference held in the city of Santiago, Chile, a research group on the epidemiology of MS in Latin America was created: the GEEMAL or Multi-Center Collaborative Study Group of Multiple Sclerosis in Latin America.

In 2012, Dr. Mario Melcon, one of the members of the Group, published a review in *Multiple Sclerosis* entitled "Towards establishing MS prevalence in Latin America and the Caribbean".[11]

As for epilepsy surveys, in the case of MS there are wide variations in the prevalence rates reported across the different Latin American countries (Fig. 9.5). The lowest rates were found in surveys conducted in Cochabamba, Bolivia (1.5 cases per 100,000 persons) and the highest rates in research conducted in Mexico

Location of survey	Latitude	Prev./ 100,000
Mexico, San Pedro García Garza	25°N	30
Mexico, Chihuahua, DF, Jalisco, Potosí	22°N 28°N	7
Mexico, Monterrey, Sinaola,	25°N 17°N	7.5
Cuba, Cienfuegos	21°N	10–25.5
French West Indies	14°N	14.8
Costa rica, Central America	10°N	6
Venezuela, Isla Margarita	10°N	5.26
Panama, Central America	8°N	5.24
Colombia, Bogota	4°N	4.41
Ecuador, Quito	0°N –S	5.05
Perú, Lima	12°S	7.6
Brazil, cuiaba, M. Grosso	15°S	4.41
Bolivia, Cochabamba	17°S	1.5
Paraguay, Asuncion	23°S	5.7
Brazil, Santos, SP	23°S	15.5
Chile, All Country	17°S 56 °S	5.69
Uruguay, Montevideo	34°S	20.5
Argentina, Buenos Aires	34°S	25.6
Argentina, Junin	34°S	12
Chile, Punta Arenas	53°S	13–14
Argentina, Patagonia	36°–55°S	17.2
Antartic	60°S	

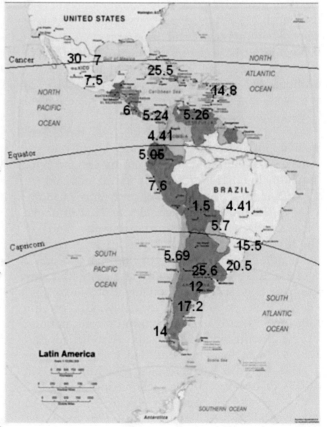

FIGURE 9.5   The multiple sclerosis prevalence and distribution study area in Latin American and Caribbean countries is divided into two regions: the intertropical area and the Southern Cone.

(30 cases per 100,000 persons at San Pedro García Garza), Cuba (25.5 cases per 100,000 persons at Cienfuegos), Argentina (25.6 cases per 100,000 persons at Buenos Aires), and Uruguay (20.5 cases per 100,000 persons, countrywide).

As in the case of epilepsy surveys, this is due to variations in the methodology used in the research into this disease, as well as to the variability in terms of population characteristics and of the local conditions prevailing in each country, which could account for the variations noted in the frequency of the disease.

In the paper published by the GEEMAL Group, two different MS prevalence regions were identified in Latin America: the inter-tropical area (an area extending between the Tropic of Cancer and the Tropic of Capricorn, to both sides of the Equator), and the area below the Tropic of Capricorn.

The GEEMAL Group conducted a multivariate analysis considering different factors, which they presented at the 20th World Congress of Neurology, Marrakesh, Morocco, November 12–17, 2011.

The Group performed bivariate and a multivariate analyses of the prevalence of MS and different factors (latitude, ultraviolet (UV) radiation, temperature, the degree of sun exposure, altitude, ethnics, and the level of education as a socioeconomic indicator). The prevalence of MS was included as dependent variable and the potential predictors were included as independent factors.

The bivariate analysis (linear regression) of the association existing between the prevalence of MS expressed in cases per 100,000 persons (dependent variable) and different factors (independent variables) found a statistically significant association with the following factors:

- latitude ($r^2 = 0.473$; $p = 0.001$): the higher the Northern or Southern latitudes, the higher the prevalence, with the lowest prevalence rates located close to the Equator;
- ethnics ($r^2 = 0.381$; $p = 0.013$): the higher the proportion of European descent population, the higher the prevalence;
- altitude ($r^2 = 0.257$; $p = 0.029$): the higher the altitude, the lower the prevalence.

Given that the Native Latin American population usually lives in higher areas (European immigrants usually establish themselves at sea-level areas), this is likely to act as a confounding factor.

A multivariate analysis was subsequently conducted. The MS prevalence in Latin America was modeled with generalized linear models (GLMs), and the following were factored as potential predictors of prevalence: latitude, UV radiation, temperature, the degree of sun exposure, altitude, and ethnics. Said predictors were modeled in bivariate and multivariate analyses, and adjusted in a linear and non-linear model. In order to compare the different models and to select the best of these models, the criteria used were the significance of predictors and the percentage of explained deviance. The authors opted for the model with the lowest p-value and the highest explained deviance.

In the multivariate analysis, the best GLM to explain the prevalence of MS was the one using the log-normal and Gaussian error distribution. In this model, the best predictors of MS were latitude (polynomial grade 2), temperature (polynomial grade 0), and ethnics. The model explained 85.30% of deviance (Fig. 9.6).

The analysis of the graphics evidences the following:

- the lowest prevalence of MS in Latin America is located at 15–18 degrees of latitude South and increases as we move North or South;
- the higher the proportion of Caucasian ethnics, the higher the prevalence of MS;
- there is an "ideal" temperature for the development of MS, of about 18°C. Below or above this temperature, the prevalence of MS is lower.

The other predictors used in the survey were not included in the model.

9.5 DEMENTIA SYNDROMES

There are multiple neuroepidemiological studies on dementia syndromes conducted in Latin American countries, among which we highlight the ones developed in Argentina, Brazil, Chile, Colombia, Cuba, Ecuador, Peru, Puerto Rico, Uruguay, and Venezuela.

In a door-to-door observational, population-based survey among adults +60 years old with cognitive impairment and dementia in the social vulnerable area of the Matanza Riachuelo Basin, in Buenos Aires, Argentina, L. Bartoloni et al.[12] found a prevalence of cognitive impairment of 26.4% (18.1% with non-dementia cognitive impairment, and 8.3% with dementia). They found a higher prevalence of dementia in younger individuals than rates reported in other surveys, probably due to the lower control of vascular risk factors in this socially vulnerable population of Buenos Aires, Argentina.

Briseida Feliciano-Astacio et al. developed The Puerto Rico Alzheimer's Disease Register Study, and identified 3915 cases between May and October 2015. The number of cases evidenced an exponential growth with age +64 years, with a clear predominance of female patients. The authors concluded that direct resources were needed for the early detection of the disease and to improve the quality of life of patients

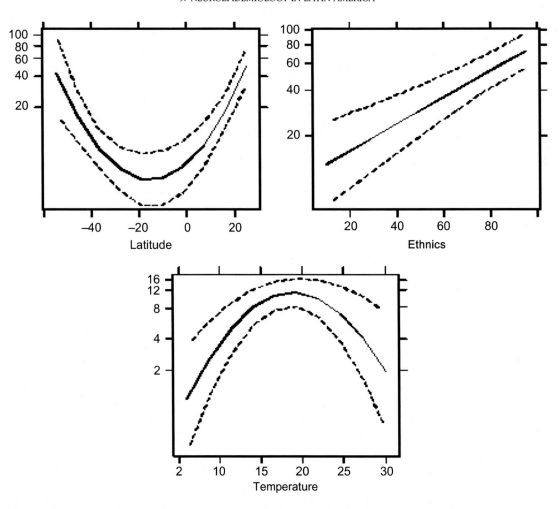

FIGURE 9.6   Multivariate analysis with generalized linear models to explain the prevalence of multiple sclerosis (dependent variable) as a function of latitude, ethnics, and temperature (independent variables). The y-axis shows the prevalence of multiple sclerosis in the log scale; the x-axis shows independent variables. GEEMAL (*Grupo Colaborativo Multicéntrico para el Estudio de la Esclerosis Múltiple en América Latina*), 2012.

with Alzheimer's disease, their caregivers and their families.

In Colombia, Gustavo Pradilla et al.[13] (EPINEURO) found a prevalence of dementia of 13.1 cases per 1000 persons (95% CI: 8.5–19.3) in a cross-sectional population-based study of two phases conducted between September 1995 and August 1996. These rates are significantly higher than those reported in other studies of the region.

In the survey conducted at Villa del Cerro-Casabó of Montevideo, Uruguay, which was a door-to-door, two-phase survey carried out in a population of 20,965 individuals, 85 cases of dementia syndrome were reported. The prevalence was of 4.0 cases per 1000 persons and was higher among women (6.2 cases per 1000 persons) than among men (1.8 cases per 1000 persons). When only the population aged +65 years was included, the raw prevalence was 3.11 cases per 100 persons. Using the Hachinski Score, the probable etiologies were: 56% degenerative, 16% vascular, 12%

mixed, 4% other etiologies, and 8% undetermined. The increase in the prevalence of dementia in this population was exponential with age—as reported in other studies—with a prevalence of 97 cases per 1000 persons among +80 year-old patients. This survey found a statistically significant association between the education level and dementia syndrome (the lower the education level, the higher the prevalence of dementia, $p < 0.0001$).

Oscar Del Brutto et al.[14] aimed to assess the effects of oily fish consumption on cognitive performance in a population of frequent fish consumers living in rural coastal Ecuador, within the Atahualpa Project that included 2478 residents of Native/*Mestizo* ethnic groups. They performed a door-to-door survey and evaluated the cognitive performance among the residents of +60 years old using the Montreal Cognitive Assessment. Using generalized multivariate linear models adjusted for demographics, cardiovascular risk factors, edentulism and symptoms of depression, they found a positive correlation between the number of fish servings per week and the Montreal Cognitive Assessment Score ($p = 0.038$).

Ricardo Nitrini and Paulo Caramelli[15] published a report on the prevalence of dementia in Latin America based on population studies.

The raw prevalence of dementia among the +65-year-old populations varied in these countries as follows: 3.11 (Uruguay) and 13.09 (Venezuela) cases per 1000 persons, and the standardized prevalence ranged from 2.66 to 12.16, respectively (Table 9.4).

A comparison of the data available for Latin American countries with systematic reviews or with the data of European surveys concluded that there are no significant differences between the prevalence of dementia in our countries and in other countries (Tables 9.5 and 9.6), that the prevalence is higher among women as found in other international research studies (Table 9.5), and that in Latin America, as in other countries, the prevalence is higher among illiterate and uneducated populations (Table 9.7).

There are few Latin American studies on the incidence of dementia.

Mexican authors[16] reported dementia incidence rates of 27.3 cases of dementia per 1000 person-years, and of 223 cases of cognitive impairment per 1000 person-years. These authors also found a higher prevalence of dementia in people with lower education levels, while gender has a differential effect according to the age group. They also reported that high blood pressure, diabetes and depression were risk factors for dementia but not for cognitive impairment.

There are few Latin American studies aimed at estimating the frequency of frontotemporal dementia. Nilton Custodio et al.[17] concluded that the prevalence of frontotemporal dementia in Latin America ranged between 0.13 and 0.18 per 100 persons aged +65 years. These rates evidence that, in our region, the disease has a mid prevalence when compared with studies performed among Western and Eastern populations.

TABLE 9.4 Prevalence of Dementia (%) and 95% Confidence Interval in Eight Latin American Studies, by Age Groups (Nitrini et al.[15])

Country	Age groups (years)						≥ 65 (Crude prevalence)	≥ 65 (Standardized prevalence)
	65–69	70–74	75–79	80–84	85–89	90 +		
Uruguay	0.88	0.67	2.94	5.88	11.41	24.68	3.11	2.66
	(0.38–1.72)	(0.22–1.57)	(1.61–4.88)	(3.72–8.78)	(6.79–17.67)	(15.57–35.86)	(2.50–3.85)	(2.61–2.71)
Chile	1.25	2.39	5.48	11.93	16.67[a]	NA	4.38	4.12
	(0.60–2.28)	(1.35–3.92)	(3.51–8.10)	(8.15–16.66)	(10.48–24.57)		(3.57–5.33)	(4.06–4.18)
Brazil[1]	1.63	3.19	7.89	15.15	34.67	48.48	7.13	7.07
	0.78–2.97	(1.79–5.22)	(4.96–11.79)	(10.46–20.92)	(24.02–46.57)	(30.81–66.45)	(5.94–8.49)	(6.99–7.15)
Venezuela	4.53	5.46	19.14	24.7	39.51	54.55	13.09	12.16
	(2.75–6.99)	(3.50–8.08)	(14.52–24.45)	(17.98–32.41)	(28.80–50-96)	(36.32–71.89)	(11.35–15.03)	(12.06–12.26)
Cuba	3.16	4.39	7.01	12.26	20.30	30.47	8.17	6,47
	(2.72–3.67)	(3.78–5.05)	(6.22–7.87)	(11.01–13.58)	(18.50–22.26)	(26.98–34.11)	(7.75–8.55)	(6.40–6.55)
Brazil[2]	0.12	1.23	2.59	3.13	12.05[a]	NA	2.03	1.76
	(0.00–0.66)	(0.53–2.42)	(1.19–4.86)	(1.27–6.33)	(7.51–18.01)		(1.48–2.71)	(1.72–1.80)
Peru	1.03	2.1	8.33	14.53	38.24	49.12	6.72	6.75
	(0.38–2.23)	(0.96–3.95)	(5.24–12.47)	(9.73–20.54)	(22.17–56.41)	(35.67–62.74)	(5.53–8.08)	(6.67–6.82)
Brazil[3]	4.06	7.1	9.52	13.28	15.28	42.31	8.84	8.12
	(2.18–6.85)	(4.44–10.49)	6.18–13.84	(7.93–20.42)	(7.88–25.65)	(23.37–63.09)	(7.25–10.69)	(8.04–8.20)
All studies	2.40	3.57	7.04	11.88	20.20	33.07	7.13	5.97
	(2.11–2.72)	(3.18–4.00)	(6.41–7.69)	(10.87–12.91)	(18.62–21.78)	(29.98–36.20)	(6.86–7.42)	(5.91–6.06)

[a]*Prevalence in individuals aged 85 or over.*
NA, not available.
Brazilian studies: [1]Herrera et al.[18]; [2]Ramos-Cerqueira et al.[19]; [3]Bottino et al.[20].

TABLE 9.5　Prevalence of Dementia According to Age (Pooled Data of Eight Latin American Studies From Six Countries) and Comparison With a Systematic Review of Dementia Prevalence Studies by Lopes and Bottino[21] and Lopes et al.[22] (Nitrini et al.[15])

Age	LA studies				Systematic review	
	N (Studies)	Dementia (N)	Participants (N)	Prevalence (%) (95% CI)	N (Studies)	Prevalence (%) (95% CI)
65–69	8	238	9902	2.40 (2.11–2.72)	17	1.2 (0.8–1.5)
70–74	8	276	7725	3.56 (3.18–4.00)	19	3.7 (2.6–4.7)
75–79	8	428	6110	7.04 (6.41–7.69)	21	7.9 (6.2–9.5)
80–84	8	482	4058	11.88 (10.87–12.91)	20	16.4 (13.8–18.9)
85–89	6	463	2204	20.20 (18.62–21.78)[a]	16	24.6 (20.5–28.6)
90–94	6	294	890	33.07 (29.98–36.20)[a,b]	6	39.9 (34.4–45.3)
>95	–			–	6	54.8 (45.6–63.9)

[a]Fort he Chilean study (Albala et al., 1997)[23] and one Brazilian study (Herrera et al., 2002)[18], only data for subjectsup to 84 years old were included.
[b]Prevalence in the 90 years oro ver age group.
CI, confidence interval.

TABLE 9.6　Comparison of the Prevalence of Dementia by Gender Between Pooled Data of Seven Latin American Studies and Pooled Data From European Studies as Reported by Lobo et al.[24] (Nitrini et al.[15])

Age	LA studies						European studies	
	Women			Men			Women	Men
	Dem. (N)	Partic. (N)	Prevalence (%) (95% CI)	Dem. (N)	Partic. (N)	Prevalence (%) (95% CI)	Prevalence (%) (95% CI)	Prevalence (%) (95% CI)
65–69	149	5620	2.65 (2.25–3.10)	79	3479	2.27 (1.80–281)	1.0 (0.7–1.4)	1.6 (1.2–2.0)
70–74	196	4781	4.10 (3.55–4.69)	65	2317	2.81 (2.17–3.57)	3.1 (2.5–3.6)	2.9 (2.3–3.5)
75–79	293	3802	7.71 (6.89–8.59)	112	1888	5.93 (4.90–7.09)	6.0 (5.3–6.7)	5.6 (4.8–6.4)
80–84[b]	2326		12.51 (11.17–13.94)	162	1489	10.88 (9.34–2.55)	12.6 (11.5–13.8)	11.0 (9.7–12.3)
85–89	281	1244	22.59 (20.30–24.97)	182	960	18.96 (16.49–21.55)	20.2 (18.4–21.9)	12.8 (10.9–14.7)
90 + [a]	189	500	37.80 (33.56–42.28)	105	390	26.92 (22.54–31.67)	30.8 (28.1–33.4)	22.1 (18.1–26.1)

[a]Prevalence in the 90 years or over age group.
[b]For one Brazilian study (Herrera et al.[18]), only data for subjects up to 84 years old were included.
CI, confidence interval; Dem., dementia; Partic., participants.

TABLE 9.7　Prevalence of Dementia Among Illiterate and Literate Subjects in Latin American Studies (Nitrini et al.[15])

Country	Illiterate			Literate			
	Dem. (N)	Total	Prevalence (%) (95% CI)	Dem. (N)	Total	Prevalence (%) (95% CI)	P
Cupa	128	355	36.06 (31.06–41.30)	1371	17,996	7.62 (7.23–8.03)	<0.0001
Chile	39	775	5.03 (3.60–6.82)	58	1438	4.03 (3.07–5.18)	0.2735
Brazil[1]	68	567	11.99 (9.41–14.93)	49	1089	4.50 (3.36–5.92)	<0.0001
Venezuela	67	286	23.43 (18.61–28.77)	105	1054	9.96 (8.24–11.95)	<0.0001
Brazil[2]	40	192	20.83 (15.35–27.22)	56	915	6.12 (4.66–7.86)	<0.0001
Peru	41	269	15.24 (11.16–20.07)	62	1263	4.91 (3.79–6.24)	<0.0001
Pooled data	383	2444	15.67 (14.21–17.18)	1701	23,755	7.16 (6.84–7.50)	<0.0001

CI, confidence interval; Dem., dementia.
Brazilian studies: [1]Herrera et al.[18]; [2]Bottino et al.[20].
Data from the Uruguay study (Ketzoian et al., 1997)[25] and one Brazilian study (Ramos-Cerqueira et al.[19]) are not included.

9.6 STROKE

Most Latin American epidemiological studies on stroke have been population-based prevalence and mortality studies.

In Uruguay, in the door-to-door survey conducted at Villa del Cerro-Casabó, Montevideo, to which we referred above, in a population of 20,965 inhabitants, the prevalence of stroke was determined at 8.68 cases per 1000 persons (prevalence point, September 1993).

In Colombia, Gustavo Pradilla et al.[13] (EPINEURO) found a prevalence of stroke of 19.9 cases per 1000 persons (95% CI: 14.3–27.4) in a cross-sectional population-based study of two phases conducted between September 1995 and August 1996.

Other studies have explored the access to thrombolytic treatment among patients with ischemic stroke. The access of the Latin American population to this type of treatment—which is widely available in the so-called first world—is far from being generalized in our countries. Some publications report on the difficulties of implementing acute stroke treatment centers in Latin America, and on the importance of their implementation to provide equal access to treatment for the Latin American populations.

In a hospital registry of cerebrovascular diseases at two referral hospitals located in the city of Panama, Fernando Gracia et al.[26] conducted a prospective assessment of the use of intravenous (i.v.) thrombolytic therapy as a public health policy between June 2005 and June 2006. These authors found 63.3% cases of ischemic stroke, 24.2% cases of hemorrhagic stroke, 6.4% of transient ischemic stroke, and 4.0% cases of subarachnoid hemorrhages. High blood pressure was the most frequent risk factor in this population. Only 3% of ischemic stroke cases were eligible for i.v. thrombolytic therapy, and nosocomial mortality was high (28.4%). The authors concluded that stroke is a disease associated with high mortality and morbidity in Panama, and that the risk factors and disease subtypes were similar to those reported in other Latin American studies. The findings of this study led to the implementation in 2013 of i.v. thrombolytic therapy in Panama as public health policy.

In Puerto Rico, in a survey conducted at three stroke treatment centers, Briselda Feliciano et al. found that out of 821 patients with stroke, 82 (10%) could benefit from i.v. thrombolytic therapy. Stroke mortality in Puerto Rico due cerebrovascular diseases for 2004 was 39.3%, 39.8% for men and 38.9% for women. The authors concluded that thrombolytic therapy is an underused modality for the treatment of acute ischemic stroke in Puerto Rico. Given the social burden of stroke and the demonstrated efficacy of thrombolytic therapy, effective methods to improve the provision of acute stroke care should be found.

In Brazil, a stroke prevalence survey was conducted in the rain forest, at the Ribeirinha Community and at an urban population in the Brazilian Amazon.[27] The aim of this survey was to determine the prevalence of cerebrovascular disease in a town in the Brazilian Amazon basin, and to compare the prevalence between a riparian population and an urban population. This door-to-door survey was carried out in 2011 among a population of 6216 residents aged +35 years, in Coari, Amazonas, Brazil. They found a prevalence of stroke of 6.3% in the rural population, and 3.7% in the urban population, with differences after gender and age adjustment. The riparian population had less access to medical care in comparison to the urban population.

Also in Brazil, at the Municipio de Vassouras, Rio de Janeiro, Ana Beatriz Calmon et al.[28] estimated the prevalence of stroke among the elderly population based on data from the Family Health Program. They found a prevalence of 2.9% (3.2% in men, and 2.7% in women). The prevalence of stroke increased with the age. There were no differences between rural and urban populations.

Pablo Lavados et al.[29] carried out a Stroke Research Project in Chile, entitled Iquique Stroke Study or PISCIS. As part of the PISCIS Project, the incidence of spontaneous intracerebral hemorrhages as first event was evaluated in a Hispanic-*Mestizo* population of Iquique (Chile) over a 2-year period (2000–02). The authors found an age-adjusted incidence of 13.8 cases per 100,000 person-years for non-lobar hemorrhages, and 4.9 cases per 100,000 person-years for lobar hemorrhages. Non-lobar hemorrhages were more frequent as first events among young men, while lobar hemorrhages were more frequent as first events among adult women. The authors concluded that the incidence of spontaneous brain hemorrhage as a first event was high in this population when compared with white populations, but lower than the one noted among non-Caucasian populations.

Pablo Lavados, Violeta Diaz, Liliana Jadue et al.[30] assessed the variations in mortality adjusted by stroke at the different regions of Chile to find the potential socioeconomic and cardiovascular variables that would explain these variations, in an ecological survey published in 2011. The highest risk of stroke mortality was associated with age, female sex and residing in certain regions. Sixty-two percent of the mortality rate found in this study was explained by a higher prevalence of poverty, diabetes, sedentary lifestyle, and overweight.

In Uruguay, Heber J. Hackembruch, Abayubá Perna and Carlos Ketzoian studied stroke mortality trends in

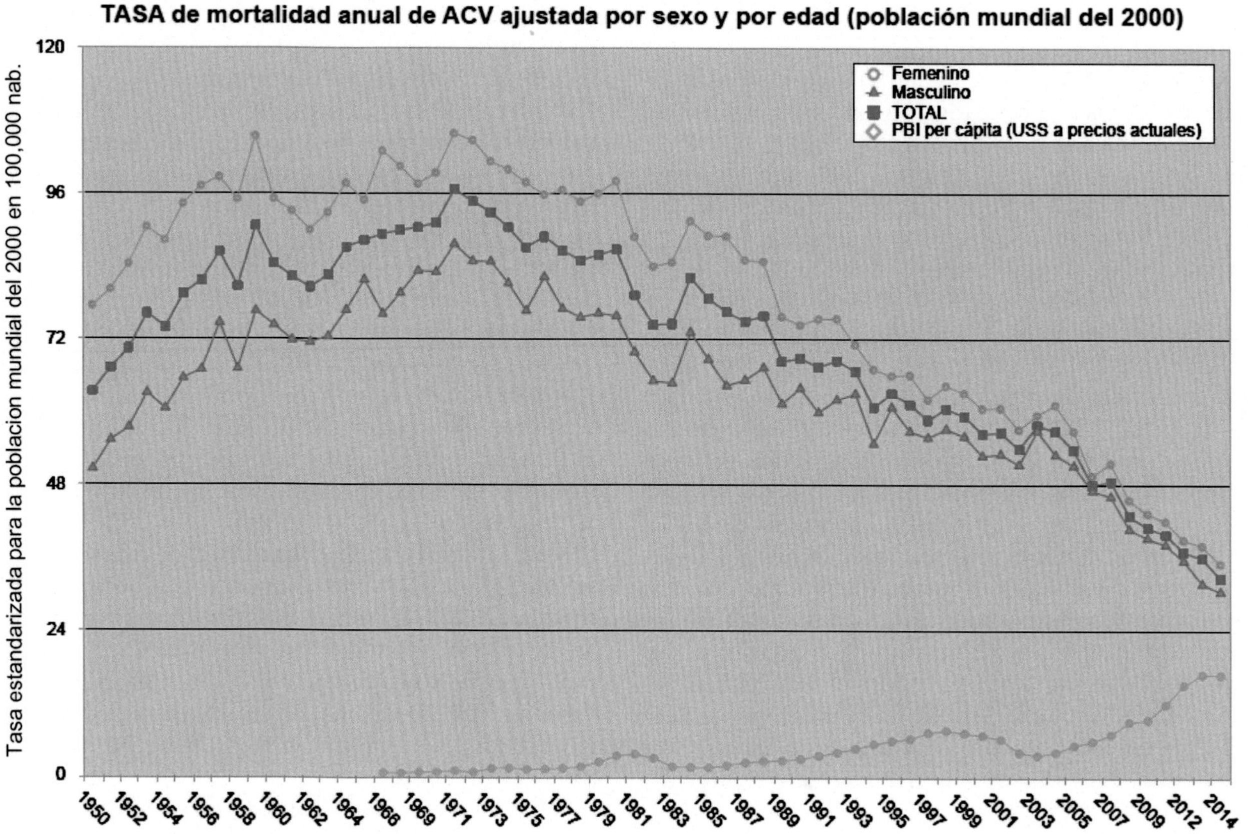

FIGURE 9.7   Annual mortality rate due to stroke in Uruguay (1950–2014) adjusted by age and gender to the 2000 world population.

the Uruguayan population between 1950 and 2014. The purposes of this study were to describe and analyze the stroke mortality trends in Uruguay, adjusted by age and gender, in the 1950–2014 period (64 years), to compare the findings with the mortality rates noted in other countries, and to assess the association between stroke mortality trends and socio-economic indicators. To this extent, the authors carried out an ecological, descriptive, and time-adjusted stroke mortality trends survey based on different data sources. Stroke mortality was adjusted by age and gender based on the 2000 estimated world population. The Uruguayan stroke mortality rates as at the year 2000 was compared to the one noted in other countries (2000–02). A time-adjusted trend analysis was performed (linear regression) with the data obtained for the 1950–2014 period. The association between stroke mortality and the gross domestic product (GDP) in 1965–2014 was also analyzed.

Mortality evidenced a decreasing trend throughout the period studied in the survey ($r^2 = -0.600$; $p < 0.0001$). Two different periods were noted: one of growth (years 1950–71, when the peak of 96 deaths per 100,000 persons was recorded), and another one of decline (1971–2014, with 32 deaths per 100,000 persons

in 2014; Fig. 9.7). This turning point coincides with the incorporation of Intensive Care Units in 1970 and with the advent of the computerized tomography scan of the head, which has been available in Uruguay since 1980. Stroke mortality by gender throughout these 64 years was consistently higher among female patients. The ratio of stroke mortality to the total causes of death in Uruguay ranged from 11% to 13% in the last 20 years—with stroke being the third cause of death in the country. The decreasing stroke mortality trend shows a negative correlation with GDP ($r = -0.949$; $p < 0.0001$), which could mean that the higher the socioeconomic level, the lower the mortality rate (Fig. 9.7). These data have also been found in other international research studies.

When comparing standardized stroke mortality rates of several Latin American countries with the one found in international studies, we may note that Uruguay and Brazil have the highest rates (Fig. 9.8). Mortality tends to decrease in countries like Venezuela, Cuba, Chile, and Argentina. The countries with the lowest stroke mortality in the region are Mexico, Ecuador, Costa Rica and Puerto Rico, with rates similar to those found in developed countries.

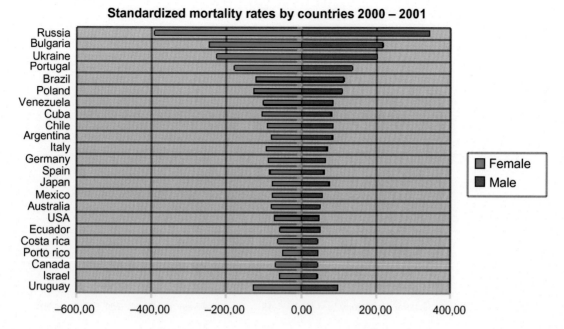

Standardized mortality rates by countries 2000 – 2001

FIGURE 9.8 Mortality rates by gender. Comparison between Latin American countries and other regions. Rates adjusted to the 2000 world population.

As part of the Atahualpa Project, Oscar Del Brutto et al. have published several papers on stroke. The Atahualpa Project is a study conducted in a rural population of Ecuador with a high proportion of Native/*Mestizo* inhabitants.

In a door-to-door survey, the authors found that the rate of cerebral microbleeds in stroke-free older adults was the same as that reported from other ethnic groups.[31] They further studied the association between ankle-brachial index and silent cerebral small vessel disease, and did not find an association between that index and cerebral microbleeds, although there was an association between this index and silent lacunar infarction.[32]

As part of this project, the authors studied intracranial arterial stenosis and found a similar prevalence as the one recorded in Asian populations.[33] The risk of stroke in patients with intracranial arterial stenosis was eight-times higher when adjusted for several confounding factors.

9.7 HEADACHE AND MIGRAINE

In a survey conducted in an urban area of Salvador de Bahia, Brazil, the global prevalence of headaches was 14.8%—being significantly higher among women, older and less-educated individuals, migrants, the unemployed, people of low socioeconomic status, and people married or living with a domestic partner.[34] The survey was based on a study of 1511 people. The presence of a "psychiatric-disorder diagnosis" increased the risk of having headaches by 4.2-times.

The findings of this survey were similar to those found in other studies. In a rural population of Mexico, Agnes Fleury et al.[35] found headache prevalence numbers in line with those of the world's population, at 22.4 cases per 1000 persons (95% CI: 17.7−28.2). These numbers are significantly lower than the ones noted in other studies.

In Colombia, Gustavo Pradilla et al.[13] (EPINEURO) found a prevalence of migraine of 71.2 cases per 1000 persons (95% CI: 65.5−76.8) in a cross-sectional population-based study of two phases conducted between September 1995 and August 1996.

Several Latin American research teams have focused their interest on headaches. The Latin American Headache Society has the purpose of advancing knowledge on this disease in our countries.

Table 9.8 presents the results of prevalence surveys conducted in several Latin American countries. The variability in the methodology employed, as well as the diversity of the age groups included in the different studies, may partly explain the differences found across the different countries.

Annual headache prevalence rates ranged between 58.3% (Uruguay) and 93.3% (Brazil), with a higher prevalence among women than among men in all studies.

In turn, the prevalence of migraine (Table 9.9) ranged from 5% (Uruguay) to 16.8% (Venezuela, Mérida) and was also higher among women than among men.

TABLE 9.8   Prevalence of Headache in Different Latin American Studies

Author, year, country	Population	Age (years)	Period	Male (%)	Female (%)	Total (%)
Barea et al. (1993)[36] Brazil	School children (n = 538)	10–18	Lifetime	92.30	94.40	93.30
			1 year	77.90	87.90	82.90
Miranda et al. (2000)[37] Puerto Rico	General community (n = 1610)	All	1 year	27	40	35.90
Rondon et al.[38] Venezuela	Student (n = 1714)	10–21	Undetermined			84.40
C. Perez, A. Perna, et al.[39] Uruguay	General community (n = 335)	+15	1 year	49	66	58.30

TABLE 9.9   Prevalence of Migraine in Different Latin American Studies

Author, year, country	Population	Age (years)	Period	Male (%)	Female (%)	Total (%)
Pablo M. Lavados (1993)[40] Chile	General community (n = 1385)	+15	1 year	2.00	11.90	7.30
Barea et al. (1993)[36] Brazil	Students (n = 538)	10–18	1 year	9.60	10.30	9.90
Miranda et al. (2000)[37] Puerto Rico	General community (n = 1610)	All	1 year	6.00	16.70	13.50
Rondon et al.[38] Venezuela	Students (n = 1714)	10–21	Undetermined	9.33	22.20	16.80
C. Perez, A. Perna, et al.[39] Uruguay	General community (n = 335)	+15	1 year	1	8	5

9.8 EXTRAPYRAMIDAL DISEASE AND PARKINSON'S DISEASES

In Colombia, Gustavo Pradilla et al.[13] (EPINEURO) found a prevalence of Parkinson's disease of 4.7 cases per 1000 persons (95% CI: 2.2–8.9) in a cross-sectional population-based study of two phases conducted between September 1995 and August 1996. This number is relatively high when compared to the one found in other surveys conducted in the region.

Mario Melcon et al.[41] conducted a survey on the prevalence of the main neurological diseases in Junín, Province of Buenos Aires, Argentina. They found raw prevalence rates of Parkinson's disease of 399 per 100,000 persons for all ages, and of 876 per 100,000 persons among participants +40 years old. 34% of cases were diagnosed *de novo* in the survey, i.e., one-third of cases had gone undiagnosed.

The rates for Parkinson's disease and parkinsonian syndromes in Junín are significantly higher than those recorded in other countries. It should be noted that Junín is a city in the Province of Buenos Aires with extensive agricultural areas and widespread use of sprayed agrotoxics. On the other hand, drug-induced Parkinsonian syndromes evidenced numbers that were lower than those found in other studies (0.47 cases per 1000 persons). In the Villa del Cerro-Casabó study, we found a prevalence of 3.00 cases per 1000 persons for parkinsonian syndromes; 1.42 for Parkinson's disease; and 0.76 for drug-induced parkinsonian syndromes.

9.9 ALS

The first prospective, population-based study on ALS was conducted in Uruguay, with the purpose of determining the incidence and prevalence of this condition in our population, using the capture–recapture method.[42]

The incidence of ALS in Uruguay for 2002–03 was found to be 1.42 cases per 100,000 person-years. For the 45–74 age group, the annual incidence was 3.74 cases per 100,000 person-years. The raw prevalence found for Uruguay in this survey was 1.9 cases per 100,000 persons as at December 21, 2002, and 2.01 cases per 100,000 persons when estimated with the capture–recapture method. This first study led to a compared epidemiology survey between Uruguay and Limousin, France, whose findings were published in 2009.[43] The authors concluded that the survival of ALS patients in Uruguay was 9 months shorter than for patients in Limousin, probably due to the heterogeneity of medical care and the absence of an ALS referral center in Uruguay.

9.10 NEUROINFECTIOUS DISEASES

The research into neuroinfectious diseases in Latin America has a long history, with Prof. Antonio Spina França Netto (1927–2010), from Brazil, as its main authority. Prof. Spina França Netto promoted the neurosciences by supplementing his clinical training with a strong basic education. His contribution to cerebrospinal fluid testing to assess for neurological diseases, especially central nervous system infectious and inflammatory conditions, has been a turning point in neurology—both in Latin America and worldwide. Prof. Spina França Netto served as representative of the Brazilian Academy of Neurology at the World Federation of Neurology, where he acted as Vice President.

In Latin America, further to the interest in acute neurological infections, there is a concern over the management of neurological diseases such as Creutzfeldt–Jakob Disease. This condition represents a special concern due to the identification of a new variant in Europe some years ago, whose spread to our continent would be catastrophic, not only for public health but also for the livestock-dependent economies of many of our countries. In Argentina, the Ministry of Public Health implemented a Creutzfeldt–Jakob Disease Surveillance Program that included the creation of a Neuropathology Referral Center between 1983 and 1996, and a countrywide Referral Center for this disease since 1997.[44] Each potential case is tested, diagnosed, and classified according to internationally accepted diagnostic criteria. The assessment is based on clinical data, electroencephalograph findings, imaging tests (magnetic resonance imaging), cerebrospinal fluid testing for 14-3-3 protein, and necropsy or specimen neuropathology biopsy. Testing includes immunohistochemistry and western blot for prion protein, plus blood DNA tests. Out of the 517 cases assessed until 2008, no cases of the Creutzfeldt–Jakob Disease variant or iatrogenic cases had been reported. The authors concluded that the incidence of the disease is higher in the city of Buenos Aires, probably due to under-recording practices in the rest of the country. To our knowledge, there are no other surveillance programs for spongiform encephalopathies like the one available in Argentina in other countries of the region.

9.11 NEUROEPIDEMIOLOGY TRAINING AND EDUCATION

The training of neurologists in Latin America does not usually include the study of neuroepidemiology. In Argentina, neurologists may follow a Neuroepidemiology Certification Program at the Argentine Society of Neurology. In Venezuela, Dr. Beatriz González leads a professors' training program at Universidad Central de Venezuela. In Uruguay, neurologists' training includes an 8-hour course taught by the Neuroepidemiology Section of the Institute of Neurology at the School of Medicine.

The experience of the University of Rio de Janeiro (UNIRIO) is most interesting. At said school, Professor Regina Alvarenga Papais leads a Neuroepidemiology Masters Program (2 years) and a PhD program (4 years). This program has benefited many researchers and is an outstanding opportunity for academic education in this field of knowledge, with eight openings for the Masters and for the PhD program each year.

Several countries have opted to sponsor neuroepidemiology training courses by renowned professors in this field, with courses organized in Uruguay, Brazil, Ecuador, Mexico, and Peru, among others, with the participation of professors from the Neurological Epidemiology and Tropical Neurology Institute of Limoges, France; Prof. Walter Rocca from the Mayo Clinic, Rochester, Minnesota; and Prof. Giancarlo Logroscino, from University of Bari, Italy, to name but a few initiatives of this kind.

9.12 CONCLUSIONS

Latin America has a large number of research studies into the epidemiology of neurological diseases. Different neurological diseases are being studied, most of them with particular interest in the region. In the last few years, multicenter studies developed by different researchers have yielded interesting results.

The challenges for Latin American neuroepidemiology are as follows:

- to improve the methodology of neuroepidemiological studies in Latin America;
- to implement research methodology courses in the region with the participation of experienced researchers;
- to promote the implementation of more multicenter studies among different researchers in Latin America;
- to integrate Latin American surveys with those conducted by more experienced teams (i.e., Compared Epidemiology);
- to improve the training of neurologists in neuroepidemiological research methodology; and
- to promote Masters and PhD programs among young neurologists in Latin America.

References

1. Ketzoian C, et al. The prevalence of main neurological diseases in a population of Uruguay: a pilot study. *Rev Neurol Argent.* 1994;19:21–31.
2. Epidemiology of epilepsy in Latin America: current status. Report for the I.L.A.E. Chair: Abad P (Ecuador); Task Force Members: Barragan E (Mexico), Beghi E (Italy), Burneo JG

(Canada), Carpio A (Ecuador), Sander J (UK & Netherlands), Tellez Zenteno J (Canada); August 2014.

3. Méndez-Ayala A, Nariño D, Rosselli D. Burden of epilepsy in Colombia. *Neuroepidemiology*. 2015;44:144–148.

4. Ketzoian C, Oehninger C, Rega I, et al. Neuroepidemiology section & demyelinating diseases section, Institute of Neurology, School of Medicine, University of the Republic, Uruguay. Institut d'Épidemiologie et Neurologie Tropical, Université de Limoges, France. Presented at the World Federation of Neurology Meeting, Neuroepidemiology Branch, Boston, MA; May 4, 2007.

5. Poser CM, Paty DW, Scheinberg L, et al. New diagnostic criteria for multiple sclerosis: guidelines for research protocols. *Ann Neurol*. 1983;13:227–331.

6. Abad P, Pérez M, Castro E, Alarcón T, Santibáñez R, Díaz F. Prevalencia de esclerosis múltiple en Ecuador. *Neurología*. 2010;25(5):309–313.

7. Feliciano B. Esclerosis múltiple en Puerto Rico. San Juan Bautista School of Medicine, San Juan MS Center. Personal communication. 2014.

8. Calmon AB, Pereira F, Marin B, Preux PM, Papais Alvarenga R. Prevalence of multiple sclerosis in the city of Volta Redonda-Rio de Janeiro, Brazil using the capture-recapture method. *Neuroepidemiology*. 2016;46:88–95.

9. da Gama Pereira AB, Sampaio Lacativa MC, da Costa Pereira FF, Papais Alvarenga RM. Prevalence of multiple sclerosis in Brazil: a systematic review. *Mult Scler Relat Disord*. 2015;4: 572–579.

10. Papais-Alvarenga RM, Vasconcelos CCF, Carra A, de Castillo IS, Florentin S, Díaz de Bedoya FH, et al. Central nervous system idiopathic inflammatory demyelinating disorders in South Americans: a descriptive, multi-center, cross-sectional study. *PLoS One*. 2015;10(7):e0127757:doi 10.1371/journal.pone.0127757

11. Melcon MO, Melcon CM, Bartoloni L, et al. Towards establishing MS prevalence in Latin America and the Caribbean. *Mult Scler J*. 2013;19(2): The online version of this article can be found at: http://msj.sagepub.com/content/early/2012/04/03/1352458512441985>

12. Bartoloni L, Blatt G, Insua I, et al. A population-based study of cognitive impairment in socially vulnerable adults in Argentina. The Matanza Riachuelo Study. Preliminary results. *Dement Neuropsychol*. 2014;8(4):339–344.

13. Pradilla AG, Vesga BE, León-Sarmiento FE, Grupo GENECO. National neuroepidemiological study in Colombia (EPINEURO). *Rev Panam Salud Pública*. 2003;14(2):104–111:<http://dx.doi.org/10.1590/S1020-49892003000700005>

14. Del Brutto OH, Mera RM, Gillman J, Zambrano M, Ha J-E. Oily fish intake and cognitive performance in community-dwelling older adults: the Atahualpa Project. *J Community Health*. 2016;41 (1):82–86. Available from: http://dx.doi.org/10.1007/s10900-015-0070-9.

15. Nitrini R, Bottino C, Albala C, et al. Prevalence of dementia in Latin America: a collaborative study of population-based cohorts. *Int Psychogeriatr*. 2009;21(4):622–630.

16. Mejia-Arango S, Gutierrez LM. Prevalence and incidence rates of dementia and cognitive impairment no dementia in the Mexican population: data from the Mexican Health and Aging Study. *J. Aging Health*. 2011;23(7):1050–1074.

17. Custodio N, Herrera-Perez E, Lira D, Montesinos R, Bendezu L. Prevalence of frontotemporal dementia in community-based studies in Latin America: a systematic review. *Dement Neuropsychol*. 2013;7(1):27–32.

18. Herrera Jr. E, Caramelli P, Silveira ASB, Nitrini R. Epidemiologic survey of dementia in a community-dwelling Brazilian population. *Alzheimer Dis AssocDisord*.. 2002;16:103–108.

19. Ramos-Cerqueira AT, et al. Identification of dementia cases in the community: a Brazilian experience. *J Am Geriatr Soc*. 2005;53:1738–1742.

20. Bottino C.M., et al. Estimate of dementia prevalence in a community sample from S˜ao Paulo, Brazil. Dement Geriatr Cogn Disord. 26, 291–299.

21. Lopes MA, Bottino CMC. Prevalencia de demencia em diversas regioes do mundo: analise dos estudos epidemiologicos de 1994 a 2000. *Arquivos de Neuro-Psiquiatria*. 2002;60:61–69.

22. Lopes MA, Hototian SR, Reis GC, Elkis H, Bottino CMC. Systematic review of dementia prevalence 1994 to 2000. *Dement Neuropsychol*. 2007;1:230–240.

23. Albala C, Quiroga P, Klaassen G, Rioseco P, Pérez H, Calvo C. Prevalence of dementia and cognitive impairment in Chile. (Abstr). *World Congress of Gerontology*, Adelaide, Australia, 1997:483 in Nitrini R, Bottino C, Albala C, et al. Prevalence of dementia in Latin America: a collaborative study of population-based cohorts. Int Psychogeriatr. 2009;21 (4):622–630.

24. Lobo A, et al. Prevalence of dementia and major subtypes in Europe: a collaborative study of population-based cohorts. Neurologic Diseases in the Elderly Research Group. *Neurology*. 2000;54(Suppl. 5):S4–S9.

25. Ketzoian C. et al. Estudio de la prevalencia de las principales enfermedades neurológicas en una población del Uruguay. La Prensa Medica Uruguaya, 1997:17, 9–26. In Nitrini R, Bottino C, Albala C, et al. Prevalence of dementia in Latin America: a collaborative study of population-based cohorts. Int Psychogeriatr. 2009;21(4):622-630.

26. Gracia F, Benzadon A, Gonzalez-Castellon M, et al. Registro hospitalario de la enfermedad cerebrovascular en Panamá: Estudio prospectivo en dos hospitales de referencia en la Ciudad de Panamá y programa de terapia trombolítica endovenosa como política de salud pública. Academia Panameña de Medicina y Cirugía. Vol. 33 N⍛ 2. 2013.

27. Gomes Fernandes T, Martins Benseñor I, Carvalho Goulart A, et al. Stroke in the rain forest: prevalence in a Ribeirinha community and an urban population in the Brazilian amazon. *Neuroepidemiology*. 2014;42:235–242.

28. da Gama Pereira AB, Alvarenga H, Silva Pereira Jr R, Serrano Barbosa MT. Stroke prevalence among the elderly in Vassouras, Rio de Janeiro State, Brazil, according to data from the Family Health Program. *Cad Saúde Pública*. 2009;25 (9):1929–1936.

29. Lavados PM, Sacks C, Prina L, et al. Incidence of lobar and nonlobar spontaneous intracerebral haemorrhage in a predominantly Hispanic-Mestizo Population—the PISCIS stroke project: a community-based prospective study in Iquique, Chile. *Neuroepidemiology*. 2010;34:214–221.

30. Lavados PM, Díaz V, Jadue L, Olavarría VV, Cárcamo DA, Delgado I. Socioeconomic and cardiovascular variables explaining regional variations in stroke mortality in Chile: an ecological study. *Neuroepidemiology*. 2011;37:45–51.

31. Del Brutto VJ, Zambrano M, Mera R, et al. Population-based study of cerebral microbleeds in stroke-free older adults living in rural Ecuador: The Atahualpa Project. *Stroke*. 2015;46:1984–1986.

32. Del Brutto OH, Sedler MJ, Mera RM, et al. The association of ankle-brachial index with silent cerebral small vessel disease: results of the Atahualpa Project. *Int J Stroke*. 2015;10:589–593.

33. Del Brutto OH, Mera RM, Lama J, Zambrano M, Del Brutto VJ. Intracranial arterial stenosis in Ecuadorian Natives/Mestizos. A

population-based study in older adults. The Atahualpa Project. *Arch Gerontol Geriatr*. 2015;61:480–483.

34. Bastos BS, de Almeida-Filho N, Sousa Santana V. Prevalence of headache as a symptom in the urban area of Salvador, Bahia, Brazil. *Arq Neuropsiquiatr*. 1993;51(3):307–312.

35. Quet F, Preux PM, Huerta M, et al. Determining the burden of neurological disorders in populations living in tropical areas: who would be questioned? Lessons from a Mexican rural community. *Neuroepidemiology*. 2011;36:194–203.

36. Barea LM, Tannhauser M, Rotta NT. An epidemiologic study of headache among children and adolescents of southern Brazil. *Cephalalgia*. 1996;16(8):545–549.PubMed PMID: 8980856.

37. Miranda H, Ortiz G, Figueroa S, Peña D, Guzmán J. Prevalence of headache in Puerto Rico. *Headache*. 2003;43(7):774–778. PubMed PMID: 12890132.

38. Rondon J, Padròn-Freytez A, Rada R. Prevalencia de la migraña en estudiantes de educación bàsica y media de Mérida, Venezuela. *Rev Panam Salud Pùblica*. 2001;9(2):73–76.

39. Perez C, Perna A, Saldaña M, Riera F, Pereira E, Calvo J, Cohelo J, Rovira S, Aguirrezábal X, Rega I. Grupo de Cefalea.

Sección Neuroepidemiología. Instituto de Neurología. Hospital de Clínicas. Montevideo. Uruguay. <http://www.sociedadpanamericananeuroepidemiologia.org/cefalea_uruguay.pdf>

40. Lavados PM, Tenhamm E. Epidemiology of migraine headache in Santiago, Chile: a prevalence study. *Cephalalgia*. 1997;17 (7):770–777.PubMed PMID: 9399008.

41. Melcon M, Vergara R, Anderson D. Prevalence of Parkinsonism in Junín (B.A.) Argentina. *Neuroepidemiology*. 1995;14:36.

42. Vázquez MC, Ketzoian C, Legnani C, et al. Incidence and prevalence of amyotrophic lateral sclerosis in Uruguay: a population-based study. *Neuroepidemiology*. 2008;30:105–111.

43. Gil J, Vázquez MC, Ketzoian C, et al. Prognosis of ALS: comparing data from the Limousin referral centre, France, and a Uruguayan population. *Amyotroph Lateral Scler*. 2009;10:355–360: doi:10.1080/17482960902748686. <http://dx.doi.org/10.1080/17482960902748686>

44. Begué C, Martinetto H, Schultz M, et al. Creutzfeldt–Jakob disease surveillance in Argentina, 1997-2008. *Neuroepidemiology*. 2011;37:193–202.

10

Epidemiology of Neurological Disorders in Sub-Saharan Africa

Redda Tekle-Haimanot[1], Edgard B. Ngoungou[2,3], and Dawit Kibru[4]

[1]Addis Ababa University, Addis Ababa, Ethiopia [2]University of Health Sciences, Libreville, Gabon
[3]University of Limoges, Limoges, France [4]Bahir Dar University, Bahir Dar, Ethiopia

OUTLINE

10.1 INTRODUCTION

The term "sub-Saharan Africa" (SSA) carries out an artificial geographical delimitation and covers very different socioeconomic and environmental realities in different countries. However, there are many common denominators that justify carrying out the analysis of neurological epidemiological studies at this scale.

Many studies in SSA were formerly published in regional or national newspapers, with little circulation and not indexed in international databases, making

them difficult for the scientific community to access. In addition, factors such as the reduced number of neurologists and neuroepidemiologists, the lack of attention paid to the impact of infectious diseases (human immunodeficiency virus (HIV), malaria, tuberculosis (TB), etc.), lack of exploration tools and the difficulties in obtaining baseline data on health indicators all limit the knowledge of the distribution of neurological diseases in this part of the world. Yet, over the past 40 years, a mutation has gradually emerged and continues at an accelerated rate. The conditions of practice have

Neuroepidemiology in Tropical Health
DOI: http://dx.doi.org/10.1016/B978-0-12-804607-4.00010-1

profoundly changed with, on the one hand, the development of epidemiology, which, in better analyzing the facts, orientates preventive measures and, on the other hand, the multiplication of neuroradiological tools. In addition, the development of genetic, immunological, therapeutic, technological and computational knowledge, which conditions complementary explorations for the diagnosis and treatment of patients, is accessible in some countries of this sub-continent.

Finally, SSA is currently in epidemiological transition. The development of chronic diseases is linked to changes in lifestyle, behavior and environment, and aging populations. Thus, the frequency of chronic diseases, in particular neurological diseases, will dramatically increase in the future.

10.2 INFECTIONS OF THE NERVOUS SYSTEM

Infections of the central nervous system (CNS) are the most common neurological disorders in SSA. With the high prevalence of HIV infection, there has been an exponential rise in neurological infectious conditions.

10.2.1 HIV-Related Neurological Diseases

HIV infection is a major cause of morbidity and mortality in SSA. The infection has a predilection for the nervous system. Therefore, neurological complications are common and carry a high case mortality. As shown in a study in Cameroon, the most common clinical presentations were headache (83.0%), focal signs (40.6%), and fever (37.7%). *Toxoplasma* encephalitis and cryptococcal meningitis were leading etiologies of HIV-associated CNS disease in 32.2% and 25.5% of patients, respectively. Overall mortality was 49.0%. Primary CNS lymphoma and bacterial meningitis had the highest mortality rates of 100.0% followed by tuberculous meningitis (79.8%). Low CD4 count was an independent predictor of fatality.[1]

HIV infection can also affect a spectrum of neuropsychiatric performances of the patients. In a Malawian study, the overall prevalence of suspected HIV-associated dementia (HAD) was 14.0%. Male gender and low education level were independent risk factors of suspected HAD.[2] Another study from Zambia reported that 20.0% HIV positive individuals who were antiretroviral (ARV)-naïve met the criteria for International HIV Dementia Scale (IHDS)-defined neuropsychological impairment.[3]

10.2.2 Neurosyphilis

It was relatively rare in Africa until the HIV epidemic. However, investigations among Ethiopian Jewish immigrants (Falasha) in Israel revealed a seropositivity of 12% for syphilis among the adults, while all the children examined were seronegative. A search for neurosyphilis did not reveal neuropsychiatric complications, even in the three persons whose cerebrospinal fluid tested positive. Therefore, it can be concluded that despite the high luetic seropositivity reported, syphilis in Africa used to run a relatively benign course and that neurosyphilis was a very rare complication until the outbreak of the acquired immune deficiency syndrome (AIDS) pandemic.[4,5] A retrospective study from Kenya reported that only one or two neurosyphilis cases were annually seen in a hospital setting.[6] Other studies have shown that the incidence of early and late syphilis in SSA has dropped considerably over the past few decades. The number of reported cases of early syphilis in certain urban areas, however, appears to be high. It is suggested that the downward trend in the incidence of syphilis in Africa is related to the increased and often indiscriminate use of penicillin.[7]

10.2.3 Poliomyelitis

In Africa, poliomyelitis used to be a major cause of motor disability. In 1988, the World Health Organization (WHO) and its partners established the Polio Eradication Initiative. During the subsequent long campaign, the Africa Region has had its share of a continuous struggle to intensify the eradication efforts, In East African countries (Eritrea, Ethiopia, Djibouti, Kenya, Somalia, South Sudan, and Uganda) the challenges in polio eradication were mainly due to low routine immunization coverage, across the region, porous borders and large population movements. As a result, wild poliovirus transmission has been reported up until 2014. The region remains at risk and a sustained outbreak response is being given priority in the eradication effort.[8]

In 2011, one Wild Polio Virus type-1 (WPV1) case was detected in Nyanza Province in western Kenya; the isolate was most closely related to the WPV1 circulating during 2010 in eastern Uganda and was distantly related to the WPV1 circulating in northern Kenya during 2009 that was imported from Sudan (with origin in Nigeria). Genetic sequencing of WPV1 isolates indicated that undetected transmission occurred during two periods of at least 8 months each during 2009−11 in the Kenya−Uganda border area.[9]

In 2017, poliomyelitis remains endemic in three countries in the world, one in Africa: Nigeria. The country is also affected by circulating vaccine-derived poliovirus type 2 outbreaks. The Global Polio Eradication Initiative is focusing on refining surveillance and closing immunity gaps, and maintaining political commitment, financial resources, and technical support at all levels. Nigeria continues to implement emergency outbreak response, in response to the detected WPV1 strain and circulating vaccine-derived poliovirus type 2 strains affecting the country. The response is part of a broader regional outbreak response, coordinated with neighboring countries, in particular the Lake Chad sub-region, including northern Cameroon, parts of Central African Republic, Chad, and southern Niger. Detection of polio cases underscores the risk posed by low-level undetected transmission and the urgent need to strengthen subnational surveillance.

10.2.4 Leprosy

Leprosy is the most common cause of peripheral neuropathy in Ethiopia. The infectious disease is curable with multidrug therapy (dapsone and rifampicin for paucibacillary leprosy and rifampicin, dapsone and clofazimine for multibacillary leprosy). Globally in 2012, the number of chronic cases of leprosy was 189,000, down from some 5.2 million in the 1980s.[10] In a review of records of the Ministry of Health in Ethiopia from 2000 to 2011, over 5034 leprosy cases were recorded nationally every year. Although the Leprosy Control Program has been successful, there is still evidence of active transmission and the existence of new infections within the country.[11] The situation in other East African countries is similar.

In a community camp-based cross-sectional study from Malawi between 2006 and 2011, trends of leprosy prevalence and detection increased, prevalence/detection ratios were over 1 and cure rates by cohort analysis of 2009 multibacillary and 2010 paucibacillary cases were 33% and 63%, respectively; far below the expected 80%, although the national prevalence remained at less than 1 case per 10,000 population. Leprosy was still an important public health problem in Malawi. Improving knowledge and skills of health workers, registration and recording of data, contact tracing, decentralization and integration of treatment to health centers and introduction of leprosy awareness days and community-based surveillance could help to improve early detection, treatment, case holding and prevention of disabilities.[12]

The two neighboring West African countries of Nigeria and Benin are still getting hundreds and thousands of cases of leprosy. In Nigeria, in 2015, a total of 2892 new cases were notified among which 9% were children and 15% had disabilities because of late presentation at hospitals. In Benin, between 150 and 200 new cases of leprosy are recorded every day, of which 10% are children. That number has not changed much in a decade.

A recent publication highlights an investigation identifying a cluster of Guinean patients with drug-resistant leprosy infections. Some of the patients were identified as having dapsone-resistant *Mycobacterium leprae*, as well as a single case demonstrating rifampicin resistance.[13] This finding is significant because while the prevalence of leprosy has declined greatly, there are still approximately 200,000 new cases of leprosy reported globally each year. The emerging threat of drug-resistant leprosy infection in West Africa could pose a serious public health risk if left unaddressed. More efforts may have to be put into educating people about this disease. Prompt diagnosis may help to avoid further spread to other people who might be in close contact with the infected person.

10.2.5 Tuberculous Spondylitis

Several causes of myelopathies and myeloneuropathies have been identified in SSA; however, tuberculous spondylitis is the primary cause of paraparesis. Approximately 10% of patients with extra-pulmonary TB have skeletal involvement, the spine being the most common site. Spinal spondylosis accounts for 50% cases of skeletal TB and it is more common is children and young adults.[14] In a retrospective study performed in Ethiopia over 2 years, TB was diagnosed in 11.2% of medical admissions of which 12.1% were diagnosed with TB spondylitis.[15]

TB is the most common HIV-related opportunistic infection. In a Nigerian study, 10% of 1320 HIV-infected patients were co-infected with TB. Of the 35% patients with extra-pulmonary TB 14% had spinal TB.[16]

In spinal spondylosis, the level of spinal involvement determines the extent of the neurological manifestations. Neurological deficits are common with involvement of thoracic and cervical regions. Formation of a cold abscess around the vertebral lesion is a characteristic feature of spinal TB. Diagnosis of spinal TB depends on the presence of the characteristic clinical and neuroimaging findings.

10.3 TROPICAL SPASTIC PARAPARESIS

The tropical myeloneuropathies are classified into Tropical Ataxic Neuropathy (TAN) and Tropical Spastic Paraparesis (TSP). TSP is also known as human T-cell lymphotrophic virus type-1 (HTLV-1)-associated myelopathy (HAM) or HTLV-1 associated spastic paraparesis (HAM/TSP). TSP manifests with predominant spastic motor weakness of the legs and minimal sensory signs and bladder dysfunction. A strong association was discovered between this type of spastic paraparesis in Africa, South America, and Japan and HTLV-1.[17−19] In most of the studies, seropositivity for HTLV-1 in those with TSP tends to correspond to the level of seropositivity in the general population. In Ethiopia, HTLV-1 appears not to be strongly associated with TSP, but HTLV-1 seropositivity in the general population is very low.[20,21]

10.4 NEUROFLUOROSIS

Excessive exposure to high fluoride leads to dental fluorosis and skeletal fluorosis. When water with very high fluoride levels is consumed over many years, there is deposition of fluoride in bones causing generalized sclerosis, osteophytosis and narrowing of vertebral foramina and intervertebral disk spaces. Clinical manifestations are due to mechanical compression of spinal cord and nerve roots by osteophytes and vertebral column compression. These changes results in radiculopathy and myelopathy.

The East African Rift System which is part of the Afro-Arabian Rift System which extends from the Dead Sea in the North, along the Red Sea and then through Ethiopia, Kenya, Djibouti, Uganda−Zaire border region, Tanzania, northern Zambia, Malawi, and Mozambique. The East African Rift System covers a large part of Ethiopia and a significant portion of Djibouti, Kenya, Uganda, and Tanzania. Waters in the rift valley portion of these countries are high in fluoride because of the volcanic rocks of East Africa.

In a study in Ethiopia, skeletal and crippling fluorosis were found predominantly among male workers of the sugar estates within the Rift Valley. Skeletal fluorosis invariably occurred in those who were consuming water with fluoride levels of more than 4 mg-F/L for over 10 years. The most common incapacitating neurological complication of crippling skeletal fluorosis was cervical radiculomyelopathy.[22,23]

Endemic fluorosis has also been associated with hypothyroidism due to the inhibition of thyroid-stimulating hormone output from the pituitary gland by fluoride. High levels of fluoride have also been incriminated in the production of secondary hyperparathyroidism.[24,25]

10.5 KONZO

It is a paralytic disease that occurs in outbreaks in rural areas of low income countries that include Democratic Republic of Congo, Mozambique, Tanzania, and Central African Republic. The disorder is manifested with a sudden onset of spastic paraparesis which is permanent. The clinical picture ranges from isolated hyperreflexia of the lower limbs to that of a bedridden state with spastic paraparesis and associated weakness of the trunk and arms, impairment of eye movements, speech and vision. Once the disease has set in, there is no treatment for Konzo other than physical rehabilitation. The disease onset is associated with the consumption of a monotonous diet of inadequately processed bitter cassava containing high levels of cyanide and low sulfur amino acids. It has been observed that the cyanide content of cassava increases with drought. The best method for processing cassava consists of soaking (retting period) and sun drying the wet flour for about 2 hours.[26−28] Epidemics occur when these conditions coincide at times of severe food shortage. Up until 1993, outbreaks in poor rural areas in Africa contributed to more than 3700 cases of konzo. The number of affected people is underestimated. From unofficial reports, the number of cases was estimated to be at least 100,000 in 2000, in contrast to the 6,788 cases reported up to 2009 from published papers.[29]

10.6 STROKE

Data on the epidemiology of stroke in SSA are scarce and fragmented due to lack of registries, neuroimagery, poor access to care, and lack of specialists.[30] Most studies are hospital-based except for a few community surveys conducted in some countries that reveal high frequency of stroke affecting young people. Overall, prevalence is estimated at about 300 cases per 100,000 population in SSA.[31] In the absence of preventive measures, SSA will face an epidemic of noncommunicable diseases over the years, among them stroke. In Nigeria, the prevalence of 114/100,000 inhabitants was reported in a door-to-door study in an urban community.[32] In other door-to-door studies in rural and semi-rural populations the prevalences were 58/100,000,[33] 131/100,000,[34] 163/100,000,[35] and as high as 1331/100,000 inhabitants.[36] In the latter study, the age-adjusted prevalence of stroke survivors was 1460/100,000, about seven-times higher than previous estimates. In Benin, the prevalence was estimated to be 460/100,000 inhabitants in urban areas, much higher than in Nigeria.[37]

Incidence rates in Nigeria were generally high, ranging from 25.2/100,000 person-years in Lagos to 26.0/100,000 person-years in Ibadan.[38,39] For the definition of stroke cases, these studies are consistent with the WHO case definition or modified definition and only one used cranial computed tomography (CT) to confirm the diagnosis.[37] The diagnosis, treatment and overall management of stroke is very much hampered by the very limited availability of cranial CT imaging in many African countries.[40] This situation has improved in some countries such as Tanzania.[41] In hospital specialized services in some countries, stroke may be the leading cause of hospitalization.

Most studies describe an average age higher than 55 years. However, younger subjects (younger than 45 years) were also observed. All of the studies generally report that ischemic strokes have a higher frequency than hemorrhagic strokes.[42,43] The initial mortality ranges from 20.3% in Mauritania[44] to more than 40.0% in other countries.[43,45–48] At 1 year, mortality was 62% in Nigeria[46] and 50.6% in Senegal.[49] In comatose stroke, a more severe prognosis was observed, close to 82.9%.[49]

Since rheumatic heart diseases are common, emboli from cardiac sources are other important causes of cerebrovascular accidents. It is believed that these cases are underdiagnosed. Additionally, studies from Uganda have demonstrated that the embolic lesions were discovered only at autopsy.[50]

African studies have indicated that atherosclerosis is rare in the African black.[51] Hypertension is the major risk factor in all studies.[52–54] The hypertension factor is more prominent among young Africans who present with stroke, unaware of their hypertensive status.[55] The INTERSTROKE study has shown that hypertension is the main risk factor of all stroke subtypes with an odds ratio of 2.67.[56] Next to thrombotic accidents, hypertensive strokes are the neurological problem most commonly encountered.[57,58] In Sierra Leone, Lisk showed that hypertension was present in 68% of patients, but only in those over 30 years.[53] Other factors, including HIV, may explain the development of stroke in young people.[59] Diabetes, obesity,[60] dyslipidemia,[61] sickle cell anemia,[62] and periodontitis in Senegal[63] were significantly associated with stroke. Some studies highlight the impact of seasonality and climate change on the occurrence of stroke in SSA's western and central regions.[64–66]

HIV infection also emerged as a new risk factor of cerebrovascular accidents (CVAs). This is reported in a study performed in Malawi. HIV infection and hypertension were strongly linked to stroke. HIV was the predominant risk factor for young stroke (≤45 years), with a prevalence of 67% and an adjusted odds ratio of 5.6 (2.4–12.8). There was an increased risk of a stroke in patients with untreated HIV infection, but the highest risk was in the first 6 months after starting antiretroviral therapy. In older participants (HIV prevalence 17%), HIV was associated with stroke, but with a lower population attributable fraction than hypertension (5% vs 68%). There was no interaction between HIV and hypertension on stroke risk.[67]

The majority of studies were in favor of male dominance. A study in Congo showed that in the Congolese context, gender did not significantly influence the different risk factors.[68] In terms of knowledge and perception studies, in the West and Central areas of SSA, most people have a lack of knowledge about stroke.[69] Finally, post-stroke life involves suffering in survivors, explained by disturbing physical and emotional symptoms. Several studies have shown a depressive state after stroke. The cost of stroke in SSA varies from country to country, and even from region to region, but remains very high.[70]

10.7 EPILEPSY

Next to infection, epilepsy is most prevalent non-traumatic neurological disorder in SSA. Although prevalence and incidence data of the disease are scarce, Preux and Druet-Cabanac in their review of the epidemiology and etiology of the disorder found the disease was two- to three-times more common than in the industrialized countries.[71] In SSA, high frequency of epilepsy has been found in isolated tribes of Liberia,[72] Tanzania,[73] and Ethiopia.[74] There are suggestions that genetic factors may play a role in these high prevalence rates. In the general population of SSA, the prevalence from door-to-door surveys is on average 13‰.[75] The prevalence of epilepsy in east Africa varies widely from 5.0‰ to 18.2‰.[76]

10.7.1 Etiologies of Epilepsy

10.7.1.1 CNS Infections

Epilepsy can result from CNS infection by viruses such as measles, secondary to bacterial meningitis or encephalitis, fungi and parasites, most commonly neurocysticercosis (NCC). There are few published data on the association of viral infection and epilepsy. An association was not found between arboviruses (toga virus and bunya viruses) and epilepsy. Measles commonly presents with neurological complications and epilepsy. Seizures in patients with HIV can occur during opportunistic infections including cryptococcosis, herpes simplex virus, toxoplasmosis, and TB. HIV can also produce seizures through direct invasion of the CNS.[71] Bacterial meningitis (meningococcal) and encephalitis commonly cause epilepsy. In Sudan, epilepsy still occurs in 11% of infants, 3 years after meningococcal meningitis.[77] TB meningitis causes long-term epilepsy in 8%–14% of patients.[71]

A significant association has been recognized between cysticercosis, parasitic infection with *Taenia solium* tapeworm larvae, and epilepsy. The prevalence of cysticercosis is higher in developing countries where sanitation facilities are less advanced and vectors include pigs, food, and fecal contamination of the water supply. Results of studies on the impact of NCC on epilepsy in Africa are inconsistent. A study carried out in Burundi highlighted the importance of cysticercosis in the area as 31.5% of the control subjects screened positive for this parasite. The attributable risk for cysticercosis was 50% (95% confidence interval, 42–57) in this population.[78] A similar association of NCC and epilepsy is reported from Tanzania with a large-scale neuro-imaging study.[79] In a very recent worldwide meta-analysis, the common odds ratio was 2.7 (95% CI 2.1–3.6, $p < 0.001$). Etiologic fraction was estimated to be 63% in the exposed group among the population. Cysticercosis is a significant contributor to late-onset epilepsy in tropical regions around the world, in particular SSA and its impact may vary depending on transmission intensity.

Malaria is highly prevalent in many areas of east Africa, and contracting *Plasmodium falciparum* malaria in childhood is associated with nearly a 50% chance of developing cerebral malaria and a 37.5% chance of developing epilepsy.[80,81]

Filarial infections such as loasis and Bancroft filariasis may cause seizures but their effect on epilepsy is not clear. Epilepsy has been shown to be more frequent in areas highly endemic for onchocerciasis (*Onchocerca volvulus*) although the importance of onchocerciasis in epilepsy is a subject of debate, and epidemiological studies have produced discordant results.[82]

Trypanosomiasis due to *Trypanosoma brucei gambiense* can cause epilepsy in the CNS phase of the disease. Cerebral toxoplasmosis has been reported to cause seizures in about 25% of infected cases and the prevalence is much higher in patients with HIV. Paragonimiasis, schistosomiasis, echinococcosis, angiostrongylosis, amoebiasis, and trichinosis may also cause epilepsy. A relation between toxocariasis and epilepsy has been mentioned recently.[83,84]

10.7.1.2 Birth Complications

In SSA, two-thirds of women requiring emergency obstetric care are unable to receive it.[77] Untreated adverse perinatal complications are associated with an increased risk of suffering a period of hypoxic–ischemic encephalopathy and subsequently developing epilepsy.[85,86] In a study from Tanzanian children, children with epilepsy predominantly had focal features that support the suggestion that most epilepsy in this region has a symptomatic etiology. Adverse perinatal events were strongly associated with epilepsy. Genetic and social factors may also be important. Epilepsy may be preventable in a significant proportion of children with better antenatal and perinatal care.[85]

10.7.1.3 Febrile Convulsions

Children under the age of 5 years may also experience febrile convulsions in the event of high fever and, whilst this is not considered epilepsy, a small number of these children go on to develop epilepsy.[71] There is a recognized association between febrile seizures and epilepsy in SSA. Epilepsy and febrile seizures are responsible for a significant burden of disease in rural Zambia.[87]

10.7.1.4 Head Injuries

A common cause of childhood epilepsy is perinatal head injury. Later in life, a traumatic head injury or surgery may also result in epilepsy. It has been noted that even mild head trauma leads to an increased risk of epilepsy and this risk is present for more than 20 years following the injury. Indeed, the study found the most significant risk of developing epilepsy to be associated with war injuries. In SSA, head injuries have been identified as risk factors for epilepsy, and in this region, road traffic accidents and violent conflict are important risk factors for head injury.

Many other risk factors exist including tumors, vascular disease and potentially brain stressors: malnutrition and low economic conditions.

10.7.2 Genetic Factors

A proportion of people with epilepsy also have no identifiable cause. There are also several known genetic causes of epilepsy and, although a genetic mutation cannot be identified in many individuals with epilepsy, a family history of epilepsy is often present.[88] It is noted in 6%–60% of patients studied in SSA, whereas only 5% have such a history in the US. In Tanzania, studies found a complex transmission of epilepsy among the Wapagoro people. In certain communities, consanguinity is common.[71] Very high prevalence of epilepsy has been found in some isolated tribes in this area. In the Zay Society of Ethiopia a prevalence of 29.5/1000 was found. Due to its historical isolation, epilepsy genes may have become widely disseminated throughout the Zay Society, accounting for the elevated prevalence.[74] The marginalization of patients with epilepsy by society potentially forces people with epilepsy to intermarry. Valid data are difficult to obtain and require careful drawing of family trees, laboratory investigations and technical support. Genetic interaction with environmental factors is highly probable.[71]

10.7.3 Burden

Epilepsy is associated with huge number of admissions to hospital, considerable duration of admission, and mortality. The overall incidence of admissions was 45.6/100,000 person-years of observation (95% confidence interval (95% CI: 43.0–48.7)) in a Kenyan study. Improved supply of anti-epileptic drugs (AEDs) in the community, early initiation of treatment, and adherence would reduce hospitalization of people with epilepsy and thus the burden of epilepsy on the health system.[80]

Stigma was found to be a common problem among patients suffering from epilepsy, and their relatives. The results reinforce the need for creating awareness among patients, relatives and the community at large about epilepsy and addressing misconceptions attached to it.[89] Consequences of epilepsy such as burns, lack of education, poor marriage prospects, and unemployment need to be addressed.[90]

10.7.4 Prognosis

The prognosis of epilepsy depends on the underlying cause and co-morbidities. Epilepsy-related deaths were proportionately higher after drug supply was stopped and among patients who were receiving drugs irregularly or who had only partial seizure control. Patients with epilepsy showed an increased mortality rate, which was twice that of the general rural Tanzanian population of similar age. Management of epilepsy in rural Africans should also emphasize methods to prevent epilepsy-related causes of death among patients with epilepsy.[89]

10.7.5 Treatment Options

Serious medical complications often result from seizures, especially if untreated for more than 2 years. Social stigma decreases educational opportunities and misperceptions regarding seizures may result in delayed care. Some evidence suggests that epilepsy is under-reported, under-recognized, and under-treated in this population.[87]

Phenobarbitone is most commonly used antiepileptic drug in Africa and is widely affordable. Despite the availability of this cheap and effective treatment and health education, more than 90% of patients living in rural areas remain untreated.[91] Potential reasons for this include cultural factors, lack of awareness of medical treatment and inaccessibility of medical services.[92]

In addition, this situation, where a large number of people with epilepsy do not have access to any treatment at all, is likely to result in feelings of helplessness and frustration. Given that 70% of people with epilepsy could lead full, seizure-free lives if treated, it is important to consider the role that AEDs could play in decreasing this association and therefore decreasing the likelihood that people with epilepsy move into lower socioeconomic groups.[76]

There were significant differences in the comorbidities across sites in SSA. Focal features are common in acute convulsive epilepsy, suggesting identifiable and preventable causes. Malnutrition and cognitive and neurologic deficits are common in people with active convulsive epilepsy and should be integrated into the management of epilepsy in this region.[90]

10.8 NODDING SYNDROME

The epidemic disease referred to by the names "nodding disease", "head nodding disease", and "nodding Syndrome" has been reported in southern Tanzania, northern Uganda and South Sudan. It was first described by Jilek and Jilek-Aall among children in the Mahenge village of Southern Tanzania.[93]

There have been numerous publications on the occurrence of nodding syndrome is East Africa. One could broadly define the disorder as an epilepsy disease that occurs in children in the age range of 5–15 years manifested by repeated/head nodding episodes triggered by eating and cold temperatures. The progressive condition is accompanied by cognitive impairment. The seizure episodes are mainly of the atonic focal nature although there are evidences of progression to generalization with loss of consciousness. Some studies have demonstrated family history of seizures with more than one child being affected in the same family.

Nodding syndrome has been recognized in areas of high onchocerciasis infestation.[94,95] However, relationship with onchocerciasis could only be speculated on. Tumwine et al. observed that not all studies have shown positive results on the relationship of epilepsy and onchocerciasis.[96]

The study of Winkler et al. which aimed at classifying nodding Syndrome clinically and at the same time determining its possible cause made several important observations.[97] The seizures were of "head nodding only" in some and in others there was another manifestation of the "head nodding plus" type which included other types of seizures with loss of consciousness in 17.7%. Precipitating factors were found in 17.7%. What was most significant was the presence of family history in 90.3% and hippocampal pathologies (sclerosis and gliotic changes) on magnetic resonance imaging (MRI) in some of the patients. Electroencephalograms (EEG) showed interictal epileptic activities in very few and specific changes.[97] On the other hand, Dowell et al. reported marked abnormality on EEG and nonspecific changes on MRI.[98] What was intriguing in the Winker et al. study was that skin polymerase chain reaction (PCR) was positive for *O. volvulus* in those patients whose MRI demonstrated pathological changes. However, PCR in cerebrospinal fluid was negative in all patients.[97]

Kaiser made the interesting observation that the absence of evidence for the association of onchocerciasis infection and nodding syndrome among the investigated patients reported by Winkler et al. may be due to mass treatment of the study group with ivermectin before their investigation.[99] Furthermore, the role of mass distribution of ivermectin against onchocerciasis the pathogenesis of nodding syndrome is further supported by Colebunders et al., who found that in northern Uganda nodding syndrome was on the decline since 2008 and no case was reported in 2013. The decline coincided with the mass distribution of ivermectin and the use of larvacides against black-fly.[100] Very recently, Johnson et al. discovered that people with nodding syndrome have higher levels of antibodies against a protein called leiomodin-1 than do healthy people from the same village. These antibodies were found to be neurotoxic. This protein is structurally like those made by *O. volvulus*, and antibodies that react against *O. volvulus* do the same against leiomodin-1, suggesting that nodding syndrome is an autoimmune disease triggered by *O. volvulus*.[101]

Landis et al. found an association in the peaks in the occurrence of the nodding syndrome among children with wartime conflict and temporary internal displacement conditions where insecurity, household internment, infectious diseases, and food insecurity were rampant.[102]

10.9 LATHYRISM

Lathyrism is a neurotoxic disorder caused by excessive, prolonged consumption of grass pea (*Lathyrus sativus*). Much of our knowledge about this pulse and the disabling disease it causes comes from studies in India.[103,104] The neurotoxic culprit, beta-*N*-oxalylamino-L-alanine acid (BOAA) has been identified and characterized.[105] Although in the past the disease occurred in Europe, North Africa, the Middle East and Asia, it is at present endemic in India,[106] Bangladesh,[107] and Ethiopia.[108] It is notable that the disorder has not been reported in other parts of East Africa, and for that matter, any other SSA country. The disorder remains endemic in the highland regions of the northern part of Ethiopia where outbreaks have occurred in association with both flooding and drought. Grass pea

can be cultivated in extreme climatic conditions. In Ethiopia, it has been known for many years. Its leg-paralyzing features have gained descriptive local names such as "sebre", "gayu", and "gwaya". Although known to be associated with *L. sativus*, the paralysis has given rise to false local beliefs concerning the method by which the disease is acquired.[108,109]

The crippling disease develops after heavy consumption of grass pea for over 2 months. It is uniformly manifested by a predominantly motor spastic paraparesis with varying degrees of disability. The male to female ratio of affected people has been found to be 2.5:1. It is also evident that females tend to exhibit the milder form of the disease. The disease affects predominantly the productive young member of the society.

In a prevalence study performed in Northern Ethiopia, neurolathyrism patients were detected in 9.5% households (prevalence of 2.38%). The mean number of affected family members per household was 1.27 (standard deviation 0.65, range 1−3). Most (77.5%) patients developed the disability during the epidemic (1995−99). Younger people were more affected during the epidemic than during the nonepidemic period ($p = 0.01$). The presence of a neurolathyrism patient in the family was associated with illiteracy (adjusted odds ratio (aOR) of 2.23 (1.07−5.10)) of the head of household, with owning a grass pea farm (aOR 2.01 (1.04−3.88)) and with the exclusive cooking of grass pea foods using handmade traditional clay pots (aOR 2.06 (1.08−3.90)).[110]

Those that have succumbed to the disease remain permanently disabled and without opportunities for social and physical rehabilitation. Since productive young men are commonly affected, the result is a major loss of critical manpower for peasant families. The common precipitating factors identified through different studies were heavy physical labor, febrile illness, and diarrhea.

10.10 DEMENTIA

Information on dementia is very limited in Africa. In a review of dementia in Africa, Inechen made the following very revealing and important observations that the reported low frequencies of the disorder may be due to differential survival rates, hiding of cases by relatives due to stigma, reluctance to seek treatment, poor access to medical care, the feeling that the old person has come to the end of his useful life and also defective case-finding techniques.[111] A systematic analysis in Africa estimated the overall prevalence of dementia in people older than 50 years in about 2.4%

(2.76 million people), the majority of them living in SSA (76%).

In more recent surveys in Central Africa, the prevalence was much higher. Researches focused in particular on two countries: Central African Republic and Republic of Congo.[112,113] In total, 2002 community-dwelling participants aged over 65 years old were included. The prevalence of all-type dementia was estimated at 7.2% in rural Central African Republic, 5.5% in urban Central African Republic, 4.5% in rural Congo, and 4.8% in urban Congo.[112] After 2 years of follow-up, researchers recorded 101 (9.8%) deaths. Compared to participants with normal cognition at baseline, mortality risk was more than 2.5-times higher among those with dementia (hazard ratio (HR) = 2.53, 95% CI: 1.42–4.49, $p = 0.001$). Among those with dementia, only clinical severity of dementia was associated with an increased mortality risk (HR = 1.91; CI 95%, 1.23–2.96; $p = 0.004$). Age (per 5-year increase), male sex, and living in an urban area were independently associated with increased mortality risk across the full cohort.[114] Among the dementia-free cohort, the crude incidence of dementia was estimated at 15.79 (95% CI 10.25–23.32) per 1000 person-years. The estimated standardized incidence (on the 2015 SSA population) was 13.53 (95% CI 9.98–15.66) per 1000 person-years. Regarding baseline characteristics, old age ($p = 0.003$) and poor social engagement (assessed by community activity) ($p = 0.028$) at baseline were associated with increased dementia incidence among Congolese older adults.

Prevalence in general was the highest among females aged 80 years or over (19.7%) and there was little variation between regions.[115] Similar observations have also been made by Georges-Carey et al.[116] Their meta-analysis showed that the majority (57.2%) of dementia cases belong to Alzheimer's disease and 26.9% were vascular dementia.

Rare causes of dementia were also reported from East Africa. A study performed in Kenya reported that there were four definite, seven probable and two possible cases of Creutzfeldt–Jakob disease (CJD) in Kenya between January 1990 and May 2004. The electroencephalographic and histological features were typical of sporadic CJD. Sporadic CJD occurs in Kenya and the clinical, encephalographic, and histological features were not different to those described elsewhere. Although they did not see variant, hereditary and iatrogenic forms of CJD, neurologists should not exclude these in making diagnoses.[117]

Regarding social representations, data showed in Congo that because of changes in behavioral and psychological symptoms related to dementia, people with dementia (especially women) were severely punished or abused. Traditional rules of kinship and economic hardship experienced by families were partly responsible for the weakening in the status of these older people. Thus, the accusation of witchcraft was the major form of stigmatization.[118]

10.11 PERIPHERAL NEUROPATHIES

A hospital-based study in Nigeria pointed out that the primary etiological factors in peripheral neuropathies were Guillain–Barré syndrome, diabetes mellitus, nutrition deficiencies, malignancy, and leprosy. This concurs with the experience in Ethiopia. However, when the community at large in tropical countries in considered, leprosy is by far the most common cause of peripheral neuropathy in areas where it is endemic. The afflicted persons often face rejection and ostracism because of myths and prejudices.[119]

Peripheral neuropathic complications of diabetes in Africa have prevalences ranging from 22% to 58%. In Ethiopia, one study reported a prevalence of 54%. The high prevalence of neuropathy in Africans may be due to poor biochemical control of diabetes because of late diagnosis and inadequate treatment.[120]

Peripheral neuropathies also can be caused by HIV infection and its associated illnesses. A hospital-based study from Kenya reported that as high as 60% of admitted HIV patients had peripheral neuropathies. Sixty percent of the patients had signs of peripheral neuropathy, especially impaired sense of vibration; 54% of these patients had both signs and symptoms. Electromyographic and nerve conduction velocity revealed peripheral neuropathy in 40% of AIDS patients. The types of peripheral neuropathy included distal symmetrical peripheral neuropathy (37.5%), polyneuropathy, and mononeuritis multiplex. When the symptoms, signs, and electroneurophysiological test findings were considered, all AIDS patients had evidence of peripheral neuropathy.[121]

10.12 PARKINSON'S DISEASE

Parkinson's disease is understudied and underdiagnosed in the sub-Saharan population. Several studies have shown that the prevalence is lower in Africans than in populations of European origin. The few existing studies point to a prevalence of 7/100,000 to 20/100,000, much lower than in Caucasians.[122] According to Richards and Chaudhuri, the low prevalence in Africans may be due to the confounding effects of high selective mortality and low case ascertainment. According to their hypothesis, the high mortality may be due to Africans being more vulnerable to vascular

PD, which is associated with high mortality. Conversely or in addition, Africans may be protected against PD.[123] Dotchin et al. found a prevalence of 20/100,000 among the rural population of Tanzania, higher than in previous studies reported in SSA.[124]

In one retrospective study in a university-based neurology clinic in Ethiopia, over a 1-year period, a total of 15.1% of the neurological patients were seen for movement disorders. Of these, most were for parkinsonism (47.7%).[125] In a door-to-door survey done by Tekle-Haimanot in rural Ethiopia, which included a population of 60,820, only four cases of PD were found (all male), giving a crude prevalence of 7/100,000. The authors state that the reason for the difference in the prevalence of PD among white populations is due to the difference in population age structure.[126] However, age-standardized prevalences were not calculated.

With the world population aging, PD is predicated to become an increasing problem.

10.13 CONCLUSION

The burden of both transmissible and non-transmissible diseases impacts greatly on the economic development of the region. Neuroepidemiological studies are still rare in many domains in SSA. We need more studies to improve knowledge on the frequency and risk factors, to understand also sociocultural issues and to propose adapted and adequate prevention and management strategies.

References

1. Luma HN, Tchaleu BCN, Temfack E, et al. HIV-associated central nervous system diseases in patients admitted at the Douala General Hospital between 2004 and 2009: a retrospective study. *AIDS Res Treat.* 2013;2013:709810.
2. Patel VN, Mungwira RG, Tarumbiswa TF, et al. High prevalence of suspected HIV-associated dementia in adult Malawian HIV patients. *Int J STD AIDS.* 2010;21:356–358.
3. Holguin A, Banda M, Elizabeth J, et al. HIV-1 effects on neuropsychological performance in a resource-limited country, Zambia. *AIDS Behav.* 2010;15:1895–1901.
4. Verner E, Shteinfeld M, Raz R, et al. Diagnostic and therapeutic approach to Ethiopian immigrants seropositive for syphilis. *Isr J Med Sci.* 1988;24:151–155.
5. Buck AA, Spruyt DJ. Seroreactivity in the Venereal Disease Research Laboratory test and the fluorescent treponemal antibody test. *Am J Hyg.* 1964;80:91.
6. Kwasa TO. The pattern of neurological disease at Kenyatta National Hospital. *East Afr Med J.* 1992;69:236–239.
7. Rampen F. Venereal syphilis in tropical Africa. *Br J Vener Dis.* 1978;54:364–368.
8. UNICEF. Eradicating polio, updated January 13, 2017. https://www.unicef.org/immunization/polio/. Accessed 07.03.17.
9. Centers for Disease Control and prevention (CDC). Progress toward global polio eradication-Africa, 2011. *MMWR Morb Mortal Wkly Rep.* 2012;61:190–194.
10. Anonymous. Global leprosy situation. *Wkly Epidemiol Rec.* 2012;87:317–328.
11. Baye S. Leprosy in Ethiopia: epidemiological trends from 2000 to 2011. *Adv Life Sci Health.* 2015;2:1–44.
12. Msyamboza KP, Mawaya LR, Kubwalo HW, et al. Burden of leprosy in Malawi: community camp-based cross-sectional study. *BMC Int Health Hum Rights.* 2012;6:12.
13. Avanzi C, Busso P, Benjak A, et al. Transmission of drug-resistant leprosy in Guinea-Conakry detected using molecular epidemiological approaches. *Clin Infect Dis.* 2016;63:1482–1484.
14. Garg RK, Somvanshi DS. Spinal tuberculosis: a review. *J Spinal Cord Med.* 2011;34:440–454.
15. Hodes RM, Seyoum B. The pattern of tuberculosis in Addis Ababa, Ethiopia. *East Afr Med J.* 1989;66:812–818.
16. Iliyasu Z, Babashani M. Prevalence of tuberculosis co-infection among HIV-seropositive patients attending the Aminu Kano Teaching Hospital, northern Nigeria. *J Epidemiol.* 2009;19:81–87.
17. de The G, Gessain A, Gazzole L, et al. Comparative seroepidemiology of HTLV-I and HTLV-III in the French West Indies and some African countries. *Cancer Res.* 1985;45:4633–4636.
18. Roman GC, Schoenberg BS, Madden DL, et al. Human T-lymphotropic virus type 1 antibobies in the serum of patients with tropical spastic paraparesis in the Seychelles. *Arch Neurol.* 1987;44:605–607.
19. Johnson RD, Griffin A, Arregui A, et al. Spastic paraparesis and HLV-1 infection in Peru. *Ann Neurol.* 1988;23(Suppl):S151–S155.
20. Karpas A, Maayayan S, Raz R. Lack of antibodies to adult T-cell leukemia virus and to AIDS virus in Israeli Felashas. *Nature.* 1986;19:794.
21. Abebe M, Tekle-Haimanot R, Gustafsson A, Forsgren L, Denis F. Low HTLV-1 seroprevalence in endemic tropical spastic paraparesis in Ethiopia. *Trans R Soc Trop Med Hyg.* 1990;85:109–112.
22. Tekle-Haimanot R, Fekadu A, Bushera B, Mekonnen Y. Fluoride levels in water and endemic fluorosis in Ethiopian Rift Valley. In: *International Workshop on Fluorosis Prevention and Defluoridation of Water.* de-fluoride.net/1stproceedings/12-16.pdf. Accessed 04.03.17.
23. Tekle-Haimanot R. Neurological complication of endemic skeletal fluorosis with special emphasis on radiculomyelopathy. *Paraplegia.* 1990;28:244–251.
24. Susheela AK, Bhatnagar M, Vig K, Mondal NK. Excess fluoride ingestion and thyroid hormone derangements in children living in Delhi, India. *Fluoride.* 2005;38:98–108.
25. Teotia SPS, Teotia M. Hyperactivity of the parathyroid glands in endemic osteofluorosis. *Fluoride.* 1972;5:115–126.
26. Howlett WP, Brubaker GR, Mingi N, Rosting H. Konzo, an epidemic upper motor neuron disease studied in Tanzania. *Brain.* 2005;113:223–235.
27. Tylleskar T, Banea M, Bikangi N, et al. Cassava cyanogens and Konzo, an upper motor neuron disease found in Africa. *Lancet.* 1992;339:208–211.
28. World Health Orhganization. Konzo: a distinct type of upper motor neuron disease. *Wkly Epidemiol Rec.* 1996;71:225–232.
29. Nzwalo H, Cliff J. Cassava and cyanogen intake to toxico-nutritional neurological disease. *PLoS Negl Trop Dis.* 2010;5:e1051.
30. Sagui E, M'Baye PS, Dubecq C, et al. Ischemic and hemorrhagic strokes in Dakar, Senegal: a hospital-based study. *Stroke.* 2005;36:1844–1877.
31. Connor MD, Walker R, Modi G, Warlow CP. Burden of stroke in black populations in sub-Saharan Africa. *Lancet Neurol.* 2007;6:269–278.

32. Danesi M, Okubadejo N, Ojini F. Prevalence of stroke in an urban, mixed-income community in Lagos, Nigeria. *Neuroepidemiology.* 2007;28:216–223.

33. Osuntokun BO, Adeuja AOG, Schoenberg BS. Neurological disorders in Nigerian Africans: a community-based study. *Acta Neurol Scand.* 1987;75:13–21.

34. Sanya EO, Desalu OO, Adepoju F, Aderibigbe SA, Shittu A, Olaosebikan O. Prevalence of stroke in three semi-urban communities in middle-belt region of Nigeria: a door to door survey. *Pan Afr Med J.* 2015;20:33.

35. Enwereji KO, Nwosu MC, Ogunniyi A, Nwani PO, Asomugha AL, Enwereji EE. Epidemiology of stroke in a rural community in Southeastern Nigeria. *Vasc Health Risk Manag.* 2014;10:375–388.

36. Ezejimofor MC, Uthman OA, Maduka O, et al. Stroke survivors in Nigeria: a door-to-door prevalence survey from the Niger Delta region. *J Neurol Sci.* 2017;372:262–269.

37. Cossi MJ, Gobron C, Preux PM, et al. Stroke: prevalence and disability in Cotonou, Benin. *Cerebrovasc Dis.* 2012;33:166–172.

38. Danesi MA, Okubadejo NU, Ojini FI, Ojo OO. Incidence and 30-day case fatality rate of first-ever stroke in urban Nigeria: the prospective community based Epidemiology of Stroke in Lagos (EPISIL) phase II results. *J Neurol Sci.* 2013;331:43–47.

39. Osuntokun BO, Bademosi O, Akinkugbe OO, Oyediran AB, Carlisle R. Incidence of stroke in an African City: results from the Stroke Registry at Ibadan, Nigeria, 1973–1975. *Stroke.* 1979;10:205–207.

40. Fabiyi OO, Garuba OE. Geo-spatial analysis of cardiovascular disease and biomedical risk factors in Ibadan, South-Western Nigeria. *J Settlements Spat Plann.* 2015;6:61–69.

41. Matuja W, Janabi M, Kazema R, et al. Stroke subtypes in Black Tanzanians: a retrospective study of computerized tomography scan diagnosis at Muhimbili National Hospital, Dar Es Salaam. *Trop Doct.* 2004;34:144–146.

42. Touré K, Thiam A, Sène Diouf F, et al. Epidemiology of stroke at the Clinic of Neurology, Fann University Teaching Hospital, Dakar-Senegal. *Dakar Med.* 2008;53:105–110.

43. Keita AD, Toure M, Diawara A, et al. Epidemiological aspects of stroke in CT-scan department of the Point-G Hospital in Bamako, Mali. *Med Trop.* 2005;65:453–457.

44. Diagana M, Traore H, Bassima A, Druet-Cabanac M, Preux PM, Dumas M. Contribution of computerized tomography in the diagnosis of cerebrovascular accidents in Nouakchott, Mauritania. *Med Trop.* 2002;62:145–149.

45. Thiam A, Sene-Diouf F, Diallo AK, et al. Aetiological aspects of neurological diseases in Dakar: follow-up after 10 years (1986–1995). *Dakar Med.* 2000;45:167–172.

46. Garbusinski JM, Van Der Sande MA, et al. Stroke presentation and outcome in developing countries: a prospective study in the Gambia. *Stroke.* 2005;36:1388–1393.

47. Walker RW, Rolfe M, Kelly PJ, George MO, James OF. Mortality and recovery after stroke in the Gambia. *Stroke.* 2003;34:1604–1609.

48. Zabsonre P, Yameogo A, Millogo A, Dyemkouma FX, Durand G. Risk and severity factors in cerebrovascular accidents in west african Blacks of Burkina Faso. *Med Trop.* 1997;57:147–152.

49. Sene Diouf F, Basse AM, Toure K, et al. Prognosis of stroke in department of neurology of Dakar. *Dakar Med.* 2006;51:17–21.

50. James PD. Cerebrovascular disease in Uganda. *Trop Geogr Med.* 1975;27:125–131.

51. Williams AO, Resch JA, Loewensen RB. Cerebral atherosclerosis: a comparative autopsy study between Nigerian Negroes and American Negroes and Caucasians. *Neurology.* 1969;19:205–210.

52. Assengone-Zeh Y, Ramarojaona R, Ngaka D, Kombila P, Loembe PM. Diagnostic problems with cerebral vascular accidents in Gabon. *Med Trop.* 1991;51:435–440.

53. Lisk DR. Hypertension in Sierra Leone stroke population. *East Afr Med J.* 1993;70:284–287.

54. Lester FT. Blood pressure levels in Ethiopian outpatients. *Ethiop Med J.* 1973;11:145–154.

55. Walker RW. Hypertension and stroke in sub-Saharan Africa. *Trans R Soc Trop Med Hyg.* 1994;88:609–611.

56. O'Donnell MJ, Xavier D, Liu L, et al. Risk factors for ischaemic and intracerebral haemorrhagic stroke in 22 countries (the INTERSTROKE study): a case control study. *Lancet.* 2010;376: 112–123.

57. Lester FT. Neurological diseases in Addis Ababa, Ethiopia. *Afr J Neurol Sci.* 1979;8:7–11.

58. Abebe M, Tekle-Haimanot R. Cerebrovascular accidents in Ethiopia. *Ethiop Med J.* 1990;28:53–61.

59. Longo-Mbenza B, Longokolo Mashi M, Lelo Tshikwela M, et al. Relationship between younger age, autoimmunity, cardiometabolic risk, oxidative stress, HAART, and ischemic stroke in Africans with HIV/AIDS. *ISRN Cardiol.* 2011;2011: 897908.

60. Ugoya OS, Ugoya AT, Agaba IE, Puepet HF. Stroke in persons with diabetes mellitus in Jos, Nigeria. *Niger J Med.* 2006;15: 215–218.

61. Karaye KM, Nashabaru I, Fika GM, et al. Prevalence of traditional cardiovascular risk factors among Nigerians with stroke. *Cardiovasc J Afr.* 2007;18:290–294.

62. Lagunju IA, Brown BJ, Sodeinde OO. Stroke recurrence in Nigerian children with sickle cell disease treated with hydroxyurea. *Niger Postgrad Med J.* 2013;20:181–187.

63. Diouf M, Basse A, Ndiaye M, Cisse D, Lo CM, Faye D. Stroke and periodontal disease in Senegal: case-control study. *Public Health.* 2015;129:1669–1673.

64. Ansa VO, Ekott JU, Essien IO, Bassey EO. Seasonal variation in admission for heart failure, hypertension and stroke in Uyo, South-Eastern Nigeria. *Ann Afr Med.* 2008;7:62–66.

65. Tshikwela ML, Londa FB, Tongo SY. Stroke subtypes and factors associated with ischemic stroke in Kinshasa, Central Africa. *Afr Health Sci.* 2015;15:68–73.

66. Kintoki Mbala F, Longo-Mbenza B, Mbungu Fuele S, et al. Impact of seasons, years El Nino/La Nina and rainfalls on stroke-related morbidity and mortality in Kinshasa. *J Mal Vasc.* 2016;41:4–11.

67. Benjamin LA, Corbett EL, Connor MD, et al. HIV, antiretroviral treatment, hypertension, and stroke in Malawian adults: a case-control study. *Neurology.* 2016;86:324–333.

68. Ossou-Nguiet PM, Gombet TR, Ossil Ampion M, Otiobanda GF, Obondzo-Aloba K, Bandzouzi-Ndamba B. Genre et accidents vasculaires cérébraux à Brazzaville. *Rev Epidemiol Sante Publ.* 2014;62:78–82.

69. Cossi MJ, Preux PM, Chabriat H, Gobron C, Houinato D. Knowledge of stroke among an urban population in Cotonou (Benin). *Neuroepidemiology.* 2012;38:172–178.

70. Adoukonou T, Kouna-Ndouongo P, Codjia JM, et al. Direct hospital cost of stroke in Parakou in northern Benin. *Pan Afr Med J.* 2013;16:121.

71. Preux P-M, Druet-Cabanac M. Epidemiology and aetiology of epilepsy in sub-Saharan Africa. *Lancet Neurol.* 2005;4:21–31.

72. Van der Waals F, Goudsmir J, Gajdusck DC. See-ee: clinical characteristics of highly prevalent seizure disorders in the Gbawein and Wroughbarh Clan region of Grand Bassa Country, Liberia. *Neuroepidemiology.* 1983;2:35–44.

73. Aall-Jilek L. Epilepsy in the Wapogoro tribe in Tanganyika. *Acta Psychiatr Scand.* 1965;41:57–86.

74. Alemu S, Tadesse Z, Cooper P, et al. The prevalence of epilepsy in the Zay Society, Ethiopia. An area of high prevalence. *Seizure.* 2006;15:211–213.

75. Ba-Diop A, Marin B, Druet-Cabanac M, Ngoungou EB, Newton CR, Preux PM. Epidemiology, causes, and treatment of epilepsy in sub-Saharan Africa. *Lancet Neurol.* 2014;13:1029–1044.

76. Abigail P, Adeloye D, Carey RG, et al. Estimate of the prevalence of epilepsy in Sub–Saharan Africa: a systematic analysis. *J Glob Health.* 2012;2:020405.

77. Pearson L, Larsson M, Fauveau V, Standley J. *Childbirth care. World Health Organization: Opportunities for Africa's Newborns: Practical Data, Policy and Programmatic Support for Newborn Care in Africa.* Geneva, Switzerland: WHO; 2007.

78. Nsengiyumva G, Druet-Cabanac M, Ramanankandrasana B, et al. Cysticercosis as a major risk factor for epilepsy in Burundi, East Africa. *Epilepsia.* 2003;44:950–955.

79. Winkler AS, Blocher J, Auer H, et al. Epilepsy and neurocysticercosis in rural Tanzania—an imaging study. *Epilepsia.* 2009;50:987–993.

80. Idro R, Ndiritu M, Ogutu B, et al. Burden, features, and outcome of neurological involvement in acute falciparum malaria in Kenyan children. *JAMA.* 2007;297:2232–2240.

81. Ngoungou EB, Preux PM. Cerebral malaria and epilepsy. *Epilepsia.* 2008;49(Suppl. 6):19–24.

82. Druet-Cabanac M, Boussinesq M, Dongmo L, et al. Review of epidemiological studies searching for a relationship between Onchocerciasis and epilepsy. *Neuroepidemiology.* 2004;23:144–149.

83. Nicoletti A, Bartoloni A, Sofia V, et al. Epilepsy and toxocariasis: a case-control study in Burundi. *Epilepsia.* 2007;48:894–899.

84. Quattrocchi G, Nicoletti A, Marin B, et al. Toxocariasis and epilepsy: systematic review and meta-analysis. *PLoS Negl Trop Dis.* 2012;6:e1775.

85. Burton KJ, Rogathe J, Whittaker R, et al. Epilepsy in Tanzanian children: association with perinatal events and other risk factors. *Epilepsia.* 2012;53:752–760.

86. Mung'ala-Odera V, White S, Meehan R, et al. Prevalence, incidence and risk factors of epilepsy in older children in rural Kenya. *Seizure.* 2008;17:396–404.

87. Birbeck GL. Seizures in rural Zambia. *Epilepsia.* 2000;41:277–281.

88. Yemadje LP, Houinato D, Quet F, Druet-Cabanac M, Preux PM. Understanding the differences in prevalence of epilepsy in tropical regions. *Epilepsia.* 2011;52:1376–1381.

89. Shibru T, Alemu A, Tekle-Haimanot R, et al. Perception of stigma in people with epilepsy and their relatives in Butajira, Ethiopia. *Ethiop J Health Dev.* 2006;20:170–176.

90. Kariuki SM, Matuja M, Akpalu A, et al. Clinical features, proximate causes, and consequences of active convulsive epilepsy in Africa. *Epilepsia.* 2014;55:76–85.

91. Tekle-Haimanot R, Forsgren L, Abebe M, et al. Clinical and electroencephalographic characteristics of epilepsy in rural Ethiopia: a community based study. *Epilepsy Res.* 1990;7:230–239.

92. Tekle-Haimanot R, Forsgren L, Ekstedt J. Incidence of epilepsy in rural central Ethiopia. *Epilepsia.* 1997;38:541–546.

93. Jilek WG, Jilek-Aall LM. The problem of epilepsy in a rural Tanzanian tribe. *Afr J Med Sci.* 1970;1:305–307.

94. Ovuga E, Kipp W, Mungherera M, et al. Epilepsy and retarded growth in a hyperendemic focus of onchocerciasis in rural western Uganda. *East Afr Med J.* 1992;69:554–556.

95. Kaiser C, Benninger C, Asaba G, et al. Clinical and electroclinical classification of epileptic seizure in west Uganda. *Bull Soc Pathol Exot.* 2000;93:255–259.

96. Tumwine J, Vandemaele K, Chungong S, et al. Clinical and epidemiologic characterisctics of Nodding Syndrome in Mundri Country, South Sudan. *Afr Health Sci.* 2012;12:242–248.

97. Winkler AS, Friedrich K, Konig R, et al. The Head Nodding Syndrome: clinical manifestations and possible causes. *Epilepsia.* 2008;49:2008–2015.

98. Dowell SF, Sejvar JJ, Riek L, et al. Nodding syndrome. *Emerg Infect Dis.* 2013;19:1374–1384.

99. Kaiser C. Head nodding syndrome and river blindness: a parasitologic perspective. *Epilepsia.* 2009;50:2325.

100. Colebunders R, Hendy A, Mokili JL, Wamala JF, Kaducu J, Kur L. Nodding syndrome and epilepsy in onchocerciasis endemic regions: comparing preliminary observations from South Sudan and the Democratic Republic of the Congo with data from Uganda. *BMC Res Notes.* 2016;9:182.

101. Johnson TP, Tyagi R, Lee PR, et al. Nodding syndrome may be an autoimmune reaction to the parasitic worm *Onchocerca volvulus. Sci Transl Med.* 2017;9:377.

102. Landis JL, Palmer VS, Spencer PS. Nodding syndrome in Kitgum District, Uganda: association with conflict and internal displacement. *BMJ Open.* 2014;4:e006195.

103. Acton HW. An investigation into the causation of lathyrism in man. *Indian Med Gaz.* 1922;57:241–247.

104. Dwivedi MP, Prasad BG. An epidemiological study of lathyrism in the district of Rewa, Madya Pradesh. *Indian J Med Res.* 1964;52:81–116.

105. Rao SLN. A sensitive and specific calorimetric method for the determination of L.B. diaminoproprionic acid and the *Lathyrus sativus* neurotoxin. *Anal Biochem.* 1978;864:386–395.

106. Dwivedi MP. Epidemiological aspects of lathyrism in India—a changing scenario. In: Spencer PS, ed. *Grass-Pea: Threat and Promise.* New York: Third World Medical Research Foundation; 1989:1–26.

107. Hague A, Mannan MA. The problems of lathyrism in Bangaldesh. In: Spencer PS, ed. *Grass-Pea: Threat and Promise.* New York: Third World Medical Research Foundation; 1989:27–35.

108. Gebre-Ab T, Wolde-Gabriel Z, Maffi M, et al. Neurolathyrism—a review and a report of an epidemic. *Ethiop Med J.* 1978;16:1–11.

109. Tekle-Haimanot R, Kidane Y, Wuhib E, et al. Lathyrism in rural northwestern Ethiopia: a highly prevalent neurotoxic disorder. *Int J Epidemiol.* 1990;19:664–672.

110. Getahun H, Lambein F, Vanhoorne M, et al. Pattern and associated factors of the neurolathyrism epidemic in Ethiopia. *Trop Med Int Health.* 2002;7:118–124.

111. Inechen B. The epidemiology of dementia in Africa: a review. *Soc Sci Med.* 2000;50:1673–1677.

112. Guerchet M, Bandzouzi B, Mbelesso P, Clément JP, Dartigues JF, Preux PM. Prevalence of dementia in two countries of Central Africa: comparison of rural and urban areas in the EPIDEMCA study. *Neuroepidemiology.* 2013;41:253.

113. Guerchet M, Mbelesso P, Ndamba-Bandzouzi B, et al. Epidemiology of dementia in Central Africa (EPIDEMCA): protocol for a multicentre population-based study in rural and urban areas of the Central African Republic and the Republic of Congo. *SpringerPlus.* 2014;3:338.

114. Samba H, Guerchet M, Ndamba-Bandzouzi B, et al. Dementia-associated mortality and its predictors among older adults in sub-Saharan Africa: results from a 2-year follow-up in Congo (the EPIDEMCA-FU study). *Age Ageing.* 2016;45:681–687.

115. Rizzi L, Rosset I, Roriz-Cruz M. Global epidemiology of dementia: Alzheimer's and vascular types. *Biomed Res Int.* 2014;2014:908915.

116. Georges-Carey R, Adeloye D, Chan KY, et al. An estimate of the prevalence of dementia in Africa: a systematic analysis. *J Glob Health.* 2012;2:020401.

117. Adam AM, Akuku O. Creutzfeldt–Jakob disease in Kenya. *Trop Med Int Health*. 2005;10:710–712.

118. Kehoua G, Dubreuil CM, Ndamba-Bandzouzi B, et al. From the social representation of the people with dementia by the family carers in Republic of Congo towards their conviction by a customary jurisdiction, preliminary report from the EPIDEMCA-FU study. *Int J Geriatr Psychiatr*. 2016;31:1254–1255.

119. Osuntokun BO, Adeuja AOG. Some epidemiological aspects of peripheral neuropathies in Africa. *Afr J Neurol Sci*. 1983;2:49–53.

120. Tekle-Haimanot R, Jemal A, Doyle D. Light and electron microscopic changes in sural nerves in Ethiopian diabetes. *Ethiop Med J*. 1989;27:1–8.

121. Mbuya SO, Kwasa TO, Amayo EO, et al. Peripheral neuropathy in AIDS patients at Kenyatta National Hospital. *East Afr Med J*. 1996;73:538–540.

122. Leboubou A, Echouffo-Tcheugu JB, Kengne A. Epidemiology of neurodegenerative diseases in Sub-Saharan Africa: a systemic review. *BMC Public Health*. 2014;14:653.

123. Richards M, Chaudhuri KR. Parkinson's disease in populations of African origin: a review. *Neuroepidemiology*. 1996;15:214–221.

124. Dotchin C, Masuya O, Kissima J, et al. The prevalence of Parkinson's disease in rural Tanzania. *Mov Dis*. 2008;23:1567–1672.

125. Ayele W, Nokes DJ, Messele T, et al. Higher prevalence of Anti-HCV antibodies among HIV-positive compared to HIV-negative inhabitants of Addis Ababa, Ethiopia. *J Med Virol*. 2002;68:12–17.

126. Tekle-Haimanot R, Abebe M, Gebre-Mariam A, et al. Community-based study of neurological disorders in rural central Ethiopia. *Neuroepidemiology*. 1990;9:263–277.

PART IV

FOCUS ON SPECIFIC NEUROLOGICAL SYNDROMES OR DISEASES IN TROPICAL AREAS

11

Epilepsy

Charles R. Newton[1,2] and Pierre-Marie Preux[3]

[1]KEMRI/Wellcome Trust Research Programme, Kilifi, Kenya [2]University of Oxford, Oxford, United Kingdom
[3]Institute of Epidemiology and Tropical Neurology, University of Limoges, Limoges, France

11.1 INTRODUCTION

Tropical countries carry a significant proportion of the global burden of epilepsy, since the incidence and mortality of epilepsy is highest in these regions. In addition, epilepsy has profound social consequences, engendering stigma, and frequent seizures limit opportunities for education, employment and marriage. People with epilepsy (PWE) are often not able to access biomedical diagnosis or treatment. Although inexpensive anti-epileptic drugs (AED) exist and control seizures in about 70% of patients, most PWE in tropical regions do not receive appropriate AED—the so-called epilepsy treatment gap.

The epidemiology of epilepsy in tropical regions is complicated by cultural beliefs of epilepsy limiting use of biomedical facilities, lack of reliable medical records and expertise in the diagnosis of epilepsy, different definitions used and limited investigative facilities to determine risk factors and causes.

11.2 DEFINITIONS

11.2.1 Epileptic Seizures

Epileptic seizures are the clinical manifestation of an abnormal and excessive discharge of a group of neurons in the brain. The manifestations of a seizure are determined by the anatomical localization and spread of the abnormal neuronal discharges, and may involve transient alteration of consciousness, motor, sensory, autonomic, or psychic symptoms. Epileptic seizures may occur spontaneously or be provoked by acute cerebral or systemic insults such as central nervous system (CNS) infections of the brain, traumatic brain injury (TBI), cerebrovascular disease, metabolic perturbations, and withdrawal from alcohol or drugs. They may be an isolated event, or recurrent if the underlying acute disorder recurs.

Seizures may be provoked or unprovoked:

1. **Provoked seizures** occur during, or are closely associated with acute cerebral or systemic insults as outlined above. They are usually termed acute symptomatic seizures. Febrile seizures are a form of provoked seizure occurring in childhood after the age of 1 month and usually before the age of 6 years. They are associated with a febrile illness not caused by CNS infection and do not meet criteria for other provoked seizures.
2. **Unprovoked seizures** may be a consequence of an acute brain insult, such as CNS infection, stroke or head injury, or there may be no clear antecedent etiology.

11.2.2 Epilepsy

The International League Against Epilepsy (ILAE) has traditionally defined epilepsy as recurrent, i.e., two or more, unprovoked seizures which are at least 24 hours apart. More recently, ILAE defined also epilepsy by the presence of an epileptic syndrome or if the risk of having another seizure in the next 10 years after a single unprovoked seizure is greater than 60%. The ILAE defines an epileptic seizure as "a transient occurrence of signs and/or symptoms due to abnormal excessive or synchronous neuronal activity in the brain".[1] The definition of epilepsy is a clinical one based on the history; clinical examinations are not needed, but are helpful for classification (see below).

11.3 CLASSIFICATION

11.3.1 Classification of Epilepsies and Epileptic Syndromes

Epilepsy is a symptom of cerebral conditions, not a disease. It has many causes which the ILAE have categorized into three broad etiological groups: genetic, structural/metabolic and unknown cause. The revised ILAE classification of the epilepsies[2] is complex, includes many rare syndromes and it relies on the availability of electroencephalography (EEG) and neuroimaging, which are difficult to apply outside of well-resourced specialist clinics.

11.3.2 Classification by Seizure Type

The classification of seizures and epilepsy relies on clinical criteria, and if available EEG. The ILAE has provided a useful system for classifying the types of seizures and the etiology.[3] The classification of seizures is divided into two main categories: generalized and focal. Generalized seizures are classified predominately into convulsive and absences. Focal seizures are categorized as focal motor, dyscognitive (formerly known as complex partial seizures), other focal sensory and secondarily generalized seizures. This simple classification can be used easily in tropical countries, and diagnosis can be made on history alone, although EEG data does help identify those with focal onset.

11.3.2.1 Focal Seizures

A focal seizure has onset in one cerebral hemisphere, detected by the history or by EEG. Focal seizures may occur with or without impairment of consciousness, and they may evolve to a bilateral convulsive seizure.

The clinical manifestations of focal seizures depend on the anatomical localization of the neuronal discharge. In focal seizures without impairment of consciousness, there may be only subjective sensory or psychic phenomena (an aura) or conspicuous motor features (e.g., focal jerking or posturing). Sensory phenomena may be somatosensory (e.g., focal paresthesia), special sensory (e.g., visual, auditory, olfactory or gustatory hallucination), autonomic (e.g., rising epigastric sensation) or psychic phenomena which include déjà-vu.

Focal seizures that are associated with impairment of consciousness may begin with an aura, typically lasting seconds, before consciousness is affected, or they may start abruptly. The impairment of consciousness is characterized by staring, which may be accompanied by automatisms, such as: ambulation (e.g., wandering or circling), manual (e.g., fiddling, finger tapping or more complex such as removal of items of clothing), orofacial (e.g., lip smacking or chewing), or verbal (e.g., humming, whistling or meaningless words). The seizures typically last 1 or 2 minutes and are followed by a period of confusion and drowsiness lasting up to 15 minutes.

Focal seizures that evolve into a bilateral convulsive seizure usually take the form of a tonic-clonic seizure. Tonic and clonic seizures occur less frequently. If the seizure spread is rapid there may be no obvious focal onset. Focal onset can, however, be suggested if the patient has focal seizures occurring independently or if there are focal epileptiform discharges on the EEG.

11.3.2.2 Generalized Seizures

Generalized seizures are characterized by widespread involvement of both cerebral hemispheres from the onset. Typically, the seizure starts with abrupt alteration of consciousness without warning. Electroencephalographic discharges are bilateral, grossly synchronous and symmetrical over both cerebral hemispheres.

11.3.2.2.1 GENERALIZED CONVULSIVE SEIZURES

This type of seizure is the most recognized and well known in tropical areas of the world. Generalized convulsive are the most conspicuous and include seizure types formerly known as tonic-clonic or "Grand mal". These seizures usually begin with sudden loss of consciousness, sometimes accompanied by a scream. The patient falls to the ground. Occasionally generalized tonic-clonic seizures are preceded by increasing myoclonic jerks. If there is an aura, the seizure should be classified as focal becoming generalized. Loss of consciousness is followed by the tonic phase with extension of the spine, clenching of the jaw and stiffness of the limbs. Respiration ceases and cyanosis may develop.

This is succeeded by the clonic phase which is characterized by jerking of the limbs and facial muscles. Breathing is stertorous and saliva may froth at the mouth. If the tongue is bitten, the saliva may be bloodstained. At the end of the clonic phase there may be incontinence. Although the duration of the tonic and clonic phases varies, in most seizures the tonic-clonic phase lasts less than 5 minutes. During the final phase the patient is unconscious with flaccid muscles. Consciousness is slowly regained, but invariably the patient is confused and drowsy and will often need to sleep. On waking, there may be headache and muscle soreness.

11.3.2.2.2 ABSENCE SEIZURES

Absence seizures (previously known as "petit mal") occur without warning and are characterized by arrest of ongoing activity and a blank stare. Muscle tone is preserved and the patient does not fall. The eyes may deviate upwards with flickering of the eyelids. More than 80% of attacks last less than 10 seconds. Minor automatisms are seen in more prolonged attacks. Recovery is rapid without drowsiness or confusion, and the patient will typically resume normal activity as if nothing has happened. Absence seizures can often be precipitated by hyperventilation.

The most useful clinical features in distinguishing absence from focal seizures with impairment of consciousness are: duration of <10 seconds; lack of aura; and a rapid return to full consciousness.

11.3.2.2.3 OTHER MOTOR SEIZURES

Myoclonic seizures are characterized by a brief twitch or jerk of the limbs which may occur singly or repetitively and are often asymmetrical. Recovery is immediate and alteration of consciousness may not be obvious. Other motor seizures include tonic seizures (generalized tonic contraction with loss of consciousness and no clonic phase), clonic seizures (loss of consciousness and generalized clonic jerking, not preceded by a tonic phase), atonic seizures (loss of consciousness with loss of muscle tone). These seizures are usually seen in diffuse cerebral disorders often associated with intellectual disability, but may also be seen as a result of a partial response to treatment.

11.3.2.2.4 UNCLASSIFIABLE SEIZURES

Unclassifiable describes those seizures which do not conform to the descriptions of focal and generalized seizures outlined above.

11.4 EPIDEMIOLOGY

11.4.1 Methodological Considerations

The incidence (number of new cases per unit time) and prevalence (number of cases at a particular point in time) may be difficult to determine,[4] particularly in tropical countries because of difficulties: (1) in identifying those with epilepsy in the community; (2) with differing definitions of epilepsy used by different researchers; and (3) with the expertise required in making a diagnosis of epilepsy.

11.4.1.1 Difficulties With Identification

It is often difficult to identify PWE in the community in tropical countries, particularly during epidemiological studies, since the cultural beliefs and stigma often impede the identification. Surveys are difficult to conduct because of the lack of infrastructure. The accuracy of results obtained depends on the sensitivity and specificity of the method used. These will vary for different seizure types: surveys designed to identify tonic-clonic seizures, by asking about specific symptoms, are likely to be poor at detecting focal seizures. Studies from a hospital or clinic may provide more accurate diagnosis, particular to etiology and seizure types; but will be biased towards people with severe epilepsy. As many patients, may seek traditional healers, hospital-based studies grossly underestimate the prevalence of epilepsy. In studies based on the recall of seizures, the prevalence is often underestimated, since history of seizures may be "forgotten", particularly in areas with stigma or another social disadvantage. Some studies have used other methods, such as that of notification by a "key informant" within the village,[5] but it seems likely that this too underestimates the true prevalence of epilepsy. Other methods such as capture—recapture lead to a better estimation.[6]

11.4.1.2 Definition

Epilepsy is defined as "the tendency to recurrent, unprovoked, seizures", but differing criteria have been used by researchers in assessing incidence and prevalence in tropical countries. Some studies have deemed single seizures to be "epilepsy", while others have included seizures provoked by fever or alcohol, and those occurring in the context of acute illness. The question of definition is important in determining prevalence. The term "people with epilepsy" could include all those who have ever had a seizure (or at least two unprovoked seizures). However, it may be limited to those with active epilepsy. The ILAE defines active as those who have had either a seizure, or have taken AEDs to prevent seizures, in the last 5 years, but many researchers and clinicians use a time period of 1 year, since this is the time period used to initiate treatment in many tropical countries.[7]

11.4.1.3 The Problem of Diagnosis

The diagnosis of epilepsy is often problematic. It relies on a clinical description, which may not be available if no witnesses are present and the patient becomes unconscious during the seizure. Minor seizures may be difficult to recognize and may be misdiagnosed as sensory symptoms, panic attacks, or mental illness. The diagnosis is often suspected but the diagnosis is not made for a considerable length of time. In tropical countries, people with non-convulsive seizures are unlikely to seek help, particularly where biomedical facilities are not accessible either because of cost or distance. The epidemiological measures are, therefore, skewed towards those with more convulsive types of epilepsy or those with frequent seizures.

11.4.2 The Prevalence of Epilepsy

In non-tropical countries, the prevalence of epilepsy has been estimated to be 5.8/1000, from 2.7‰ to 12.4‰.[4] In tropical countries, the prevalence is likely to be higher, since the populations are younger, and the incidence of the risk and causes are higher. In systematic reviews of epilepsy, the median prevalence was 14.2/1000 (range 2.0—74.4) in Africa,[8] 6.0/1000 (range 1.5—14.2) in Asia[9] and 15.8 (range 4.6—23.3) in Latin America.[10] This discrepancy between Asia and other parts of tropical world is not explained but hypotheses have been raised.[11]

Door to door surveys are the best screening method in areas with limited medical records, but estimated prevalence rates still vary widely. Some of this variation may be explained by differences in sample size. In a population of 150,000 in a rural district of Kenya, Edwards et al. noted substantial heterogeneity of prevalence across small geographical areas.[7] Studies with sample sizes of less than 1000, which may be influenced by local genetic and environmental factors, have found prevalence rates of up to 74/1000, whereas surveys of greater than 15,000 have found prevalence rates of 5.3—12.5/1000.[12]

11.4.3 The Incidence of Epilepsy

The incidence of epilepsy lies between 20/100,000 and 70/100,000 per annum in non-tropical countries. Few incidence studies have been reported from tropical countries. In sub-Saharan Africa the incidence ranged from 64/100,000 to 187/100,000 per year based upon six studies.[8] In Asia incidence is only available from China and India, where the rates ranged from 29/100,000 to 60/100,000 per year based upon five publications.[9] In Latin America, the incidence was measured only in five studies and the median incidence was estimated to be 138.2/100,000 per year.[10] The incidence studies are complicated by mobile populations, the lack of medical records, the difficulties of a precise diagnosis, and the fact that many PWE will not seek medical help or will be treated by a traditional healer. Again, it is surprising that the incidence in Asia is much lower than in other tropical areas, but this comparison is based on relatively few studies.

11.4.4 Seizure Type

Accurate classification of seizure type depends on a detailed description of the seizures and to some extent will be influenced by the level of experience of the person taking the history. Cultural factors and language may make it difficult to obtain a clear description of the seizures, and an aura or focal onset is often not reported by the patient or their family, so that focal seizures are incorrectly classified as generalized seizures. If an EEG is not available, this will also lead to underdiagnosis of focal seizures.

The proportion of patients with focal and generalized seizures varies considerably between studies, depending on the case ascertainment and the availability of EEG. In Ethiopia only 20% of patients had seizures with evidence of focal onset,[13] whereas in Nigeria the figure is 52%.[14] In a more recent study of active convulsive epilepsy in five sites across Africa, 53% had abnormal EEGs (adjusted prevalence of 2.7 (95% confidence interval (CI): 2.5–2.9) per 1000), and 52% of the abnormal EEGs had focal features (75% with temporal lobe involvement).[15] The frequency and pattern of changes differed with site. Abnormal EEGs were associated with adverse perinatal events (risk ratio (RR) = 1.19 (95% CI: 1.07–1.33)), cognitive impairments (RR = 1.50 (95% CI: 1.30–1.73)), use of AED (RR = 1.25 (95% CI: 1.05–1.49)), focal seizures (RR = 1.09 (95% CI: 1.00–1.19)), and seizure frequency (RR = 1.18 (95% CI: 1.10–1.26) for daily seizures; RR = 1.22 (95% CI: 1.10–1.35) for weekly seizures; and RR = 1.15 (95% CI: 1.03–1.28) for monthly seizures).

11.5 THE CAUSES OF EPILEPSY

11.5.1 Identifying the Etiology

The increase in neuroimaging facilities (computerized tomography (CT) and magnetic resonance imaging (MRI)) in tropical areas has increased the identification of the underlying causes of epilepsy in this region. However, history of preceding events can help, although cultural factors often make it difficult to obtain an adequate history. Lack of medical records reduces the proportion of patients in whom a clear etiology is identified. Many studies attribute the cause of epilepsy based on history or physical examination, and this gives rise to a wide range of presumptive causes of epilepsy in tropical regions.

The most common method for identifying possible causes is the use of case–control studies to study established risk factors. In tropical countries, the main risk factors for the development of epilepsy are family history of seizures, history of febrile seizures or acute symptomatic seizures, perinatal trauma, head injury and infections of the CNS. The studies from sub-Saharan Africa have found CNS infections, head injury and perinatal complications to be more important than stroke and cerebral tumors,[8] compared to a population-based study in the U.K.[16]

11.5.2 Infection of CNS

The infections of the CNS that cause epilepsy are very different between the regions.

11.5.2.1 Acute Bacterial Meningitis

Bacterial meningitis carries a 5.4% risk of subsequent epilepsy, and the proportion developing sequelae is higher in Africa and Asia, than in Europe.[17] Risk factors for developing epilepsy after bacterial meningitis include etiology and the presence of seizures during the acute illness. People who survive pneumococcal meningitis have the highest risk, followed by *haemophilus* meningitis, and least in meningococcal meningitis.[17] About 5.4% of the survivors of neonatal meningitis develop epilepsy.[18] Other risk factors for the development of epilepsy include persistent neurological deficits, EEG abnormalities, low cerebrospinal fluid glucose levels on admission to hospital.

11.5.2.2 Other Bacterial Suppurative Infections

Up to 70% of survivors of acute bacterial cerebral abscess develop focal epilepsy. The incidence of epilepsy following subdural empyema is less well established.

11.5.2.3 Tuberculosis

Epilepsy occurs in 8%–14% of patients after tuberculosis meningitis, and tuberculoma within the brain often presents with focal seizures.[19]

11.5.2.4 Parasitic

Taenia solium, the cause of neurocysticercosis, is found throughout tropical regions, but the endemicity is very focal. It is the most frequent cause of acquired epilepsy in certain regions of the tropics, in particular Latin America, parts of Asia and Africa.[20] In a very recent systematic review and meta-analysis, in which 37 studies from 23 countries were included (n = 24,646 subjects, 14,934 with epilepsy and 9,712 without epilepsy). the common odds ratio between cysticercosis and epilepsy was 2.7 (95% CI: 2.1–3.6).[21] The differences in the epilepsy associated with *T. solium* may be caused by differences in endemicity, although there is some suggestion that the parasite may also differ between continents. Other factors such as availability of latrines, proximity of humans to pigs, and consumption of undercooked pork may determine prevalence of epilepsy.

Malaria is a major parasitic disease, which occurs in many tropical areas. The five species that infect humans are transmitted by the *Anopheles* species, but *Plasmodium falciparum* is responsible for most of the CNS disease. Most cases of *falciparum* malaria occur in sub-Saharan Africa, where it infects mainly young children. A number of studies have shown that children who survive cerebral malaria than have an increased risk of developing epilepsy: with an odds ratio of 4.4; (95% CI 1.4–13.7) in Kenyan children,[22] 3.9 (95% CI: 1.7–8.9) in Gabonese[23] and a significant incalculable OR in Malawian children.[24] The RR of developing epilepsy was 9.4 (95% CI: 1.3–80.3) in a Malian cohort study.[25] There is also evidence that the risk of epilepsy is increased after less severe forms of *falciparum* malaria.[22] There is no data from Asia, where *Plasmodium* vivax is more common and it is mainly adults who are infected with *P. falciparum*.

Onchocerciasis is caused by *Onchocerca volvulus* transmitted by the black fly. It is endemic in some parts of Africa and to a lesser extent in Latin America. The association between onchocerciasis and epilepsy has been difficult to establish, partly because the putative mechanisms of epileptogenesis are unclear. A number of systematic reviews examined this association.[26,27] Pion et al. performed a meta-regression analysis, which showed that prevalence of epilepsy increased on by 0.4% for each 10% increase in onchocerciasis prevalence.[27] A recent and very large study conducted in several countries of sub-Saharan Africa found an odds ratio of 2.2 (CI 95%: 1.6−3.2).[28]

Toxocara canis and *Toxocara cati* are ubiquitous, transmitted by dogs and cats, respectively, and cause toxocariasis. The association between toxocariasis and epilepsy was documented in Burundi (OR 2.1, 95% CI: 1.2−3.8)[29] and in Bolivia (OR 2.7, 95% CI: 1.4−5.2)[30] and more recently in five sites across Africa.[28] There are no studies reported from Asia.

Toxoplasma gondii is another ubiquitous organism, but is more prevalent in humid regions of South America and Africa. Exposure to toxoplasmosis (as measured in seroprevalence studies) appears to be lower in Asia compared to Africa. A recent review using a random effects model on 2888 subjects, of whom 1280 had epilepsy (477 positive for toxoplasmosis) and 1608 did not (503 positive for toxoplasmosis), calculated the OR to be 2.25 (95% CI: 1.27−3.90).[31]

11.5.2.5 Viral Infections

Viral infections are common in tropical areas. Besides the ubiquitous viruses, e.g., herpes virus, many tropical countries are endemic for arboviruses, i.e., viruses transmitted by mosquitos. Furthermore, the greatest burden of human immunodeficiency virus (HIV) is in Africa. Viruses infect the CNS, most commonly causing an encephalitis. The cause of encephalitis does vary considerably from region to region. So in Asia the most common encephalitis is caused by Japanese B virus, whilst other viruses such as West Nile, chikungunya, dengue, Murray Valley, and Nipah are found in many regions of South East Asia. In Africa, HIV, West Nile, Rift Valley, chikungunya, and Epstein Barr are common causes of CNS infections.

Encephalitis is an important cause of epilepsy, but the prevalence of epilepsy following viral encephalitis varies according to the etiology. Thus, unprovoked seizures and epilepsy develops in 40%−65% of people who survive herpes simplex encephalitis. However, there is very little data published on the development of epilepsy following the other viruses common in tropical areas. Japanese encephalitis is probably an important cause of epilepsy in Asia, and we do not know the role of emerging viral diseases, e.g., chikunkunya, zika.

11.5.2.5.1 HUMAN IMMUNODEFICIENCY VIRUS

Most seizures that occur in HIV-positive patients are acute symptomatic seizures secondary to opportunistic infections, brain tumors, metabolic derangements or immune reconstitution syndrome. Seizures occur in about 4%−17% of patients with HIV infection, but prevalence of epilepsy varies according to age, stage of HIV disease and treatment. HIV infection may cause epilepsy through direct invasion of the brain and the seizures tend to be generalized convulsive seizures.

11.5.3 Traumatic Brain Injury

The data on TBI in tropical areas is difficult to interpret. Most studies have reported it as an etiological factor in 1%−19% of cases of epilepsy, but criteria for what constitutes a significant head injury vary, so studies are not comparable. TBI was a significant risk factor in case−control studies in Africa[8] and Asia.[9]

11.5.4 Perinatal Factors

It is often assumed that poor perinatal care will result in a high frequency of epilepsy, but this is not definitely proven. It is possible that in Africa those with severe cerebral damage, and the greatest risk of epilepsy, do not survive. Perinatal complications increased the risk of developing epilepsy by 1.9- to 7.3-times in case−control studies performed in Burundi, Tanzania, and Kenya. A study of active convulsive epilepsy in five African countries suggests that perinatal insults based upon the history of the state of the baby in the first 6 hours may be responsible for about a third of the epilepsy in these areas.[28] Prenatal infection and developmental malformations of the brain are likely to be relevant as well as birth trauma and hypoxia. Detailed knowledge of birth history and obstetric records is often not available, and cerebral palsy and intellectual disability have been used as markers of perinatal insults. In a community study performed in a rural area of central Ethiopia, intellectual disability was found in 7.9% of patients with epilepsy.[32]

11.5.5 Cerebrovascular Disease

As life expectancy increases, it is likely that stroke will become a more important cause of epilepsy in tropical countries, but to date most studies have found a history of stroke in only 1%−2% of cases of epilepsy.

11.5.6 Tumors

Seizures are the presenting feature in about 40% of patients with cerebral tumors and, where neuroimaging is possible, about 6% of patients with newly diagnosed epilepsy are found to have a tumor.[16] With low-grade tumors, seizures may be present for many years before other features develop. In sub-Saharan Africa most studies have reported cerebral tumors in 1%–2% of patients with epilepsy. A study performed in Nigeria with CT of the brain found cerebral tumors in 15% of adolescents and adults with epilepsy,[33] but this study was performed in a selected group of patients and there are no population-based data.

11.5.7 Family History of Epilepsy

A family history of epilepsy is reported by 5%–60% of patients in studies from sub-Saharan Africa.[8] In a community-based study in rural Ethiopia, 32% of patients with epilepsy had a relative who had one or more seizures and 12% had affected first-degree relatives. Family history was a significant risk factor for epilepsy in case–control studies in many tropical countries with odds ratios of 2.6–3.5. Family history may represent underlying genetic susceptibility, but could be explained by common exposure to environmental factors.

In isolated communities, specific diseases may be an important cause of epilepsy. Examples include Grand Bassa County in Liberia[34] and the Wapogoro people in a rural area of Tanzania.[35] In both cases, possible inherited neurodegenerative disorders may explain some of the highest local prevalences of epilepsy reported in Africa.

In addition to these rare genetic causes of epilepsy, there is a strong genetic component to many age-related epilepsy syndromes. There is even a genetic contribution to some of the more common epilepsies with structural causes. For example, the risk of epilepsy following a head injury is greater in those with a family history of epilepsy than those without. With advances in molecular biology, increasingly cases previously classified as "of unknown cause" are found to have genetic causes.

11.6 COMORBIDITY

Epilepsy is associated with cognitive impairment or psychiatric conditions, but it is not clear whether the prevalence of these conditions differs in tropical areas compared to non-tropical areas, since there is little data. In Africa about a quarter of PWE have cognitive impairment,[36] and this may be aggravated by the AED

that they are taking. Short-term memory, psychomotor speed and sustained attention are particularly impaired. Cognitive impairment will also reduce adherence, unless the patient is supported by the family. Psychiatric conditions occur in about a third,[37,38] but may reach up to 80% in specialized clinics. The most common disorders are anxiety and depression, with psychosis being reported less frequently. They are associated with seizure frequency, polypharmacy and stigmatization.

11.7 MORTALITY

PWE are at higher risk of death. The situation in low- and middle-income countries, which comprise mostly tropical countries, appears to be worse than in high-income countries. In a recent review,[39] the annual mortality rate (AMR) was estimated at 19.8 (range 9.7–45.1) deaths per 1000 PWE with a weighted median standardized mortality ratio (SMR) of 2.6 (range 1.3–7.2) among higher-quality population-based studies. In the clinical cohort studies the AMR was 7.1 (range 1.6–25.1) deaths per 1000 people. Higher SMRs occurred in studies among children and adolescents, those with symptomatic epilepsies, and those reporting less adherence to treatment. The main causes of death in PWE living in LMICs include those directly attributable to epilepsy, which yield a mean proportional mortality ratio (PMR) of 27.3% (range 5%–75.5%) derived from population-based studies. These direct causes comprise status epilepticus (PMRs ranging from 5% to 56.6%), and sudden unexpected death in epilepsy (PMRs ranging from 1% to 18.9%). Important causes of mortality indirectly related to epilepsy include drowning, head injury, and burns.

11.8 SOCIOCULTURAL ASPECTS

For PWE, the impact of the disease is not only clinical. The occurrence of public seizures, the existence of associated neurological and/or psychiatric disorders, have a direct impact on their quality of life and their integration into society. To understand the impact of this stigma, it is essential to take into account the beliefs of individuals and of the society in which they live.

Nearly 20 years after the launch of the ILAE initiative, "Out of the Shadows", and despite the many efforts made to talk about epilepsy in the public arena, this disease remains a taboo subject. Everywhere epilepsy arouses fear. Numerous works concerning the sociocultural experience of epilepsy have been carried out in tropical areas. The approaches used were most often ethnological, anthropological, and sociological.

In these studies, despite the differences between countries and cultures, epilepsy is most often seen as a punishment either for acts committed in the past or for failure to respect local rules imposed by society (consumption of forbidden foods, non-respect of rituals, etc.). In other studies, epilepsy is a sign of a sorcerer's revenge. The perceived supernatural nature of the disease frightens individuals who avoid talking about it. In many societies, the notion of contagiousness or impurity generates guilt and shame for parents and those affected by the disease and fear of being rejected by others in the social group. Epilepsy is a complex disease with many manifestations, the most spectacular of which is the generalized tonic-clonic seizure. The name given to epilepsy in some societies refers only to this type of seizure because it is the only one recognized, that maintains a terrifying image of the disease and the stigma of the people affected. Like all frightening or disturbing subjects, individuals avoid talking about it. The demand of PWE will be more than having a treatment to no longer present epileptic seizures (forget that they have been epileptic) than to have information and understand their disease.

Epilepsy is often considered a disorder in the relationships between the living and the dead, leading to possession by a spirit. It is through this disorder that the patient encounters the world of spirits. This supernatural explanation is found in almost all studies, to varying degrees, and often implies the responsible cause of the disease. The origin of the disease is rarely mentioned, testifying to its profound ignorance in the populations. If human-to-human infection is not found in the supposed causes of the disease, the majority of the population fears it. Epilepsy is often considered contagious by drool or urine emitted during seizures. A fly on a patient's mouth during a seizure may be contaminated. The preparation of food in the same dishes or the fact of drinking from the same glass are also dreaded. In about 40% of cases, epilepsy is considered an incurable disease. The stigma of the disease, such as burn scars (following falls into an open fire), identifies the patient and is often considered as a criterion of incurability; having a seizure in public may be another.

One of the social realities encountered in all societies is the stigmatization of patients with epilepsy.[40,41] Stigma is a social process that discredits and devalues the individual. The stigmatized person will be perceived by the social group as deviant (because she is not in the social norm). Despite the heterogeneity of cultures, epilepsy is seen everywhere as a source of social disorder due to illness. The peculiarity of epilepsy, in relation to other diseases, is that it has for centuries been regarded as a curse of the gods, demonic possession or a form of madness, arousing fear and rejection of those who were affected. The dramatic manifestations of certain types of seizures (such as generalized tonic-clonic seizures) are fertile ground for the development of fears and myths surrounding the disease. All these factors have contributed to disrupting the social interactions of patients with epilepsy. The unpredictable and uncontrollable aspect of seizures are also important factors that modify these social interactions. Physicians and traditional healers have an essential social role in restoring order by arresting the most visible and disturbing manifestations of the disease.

11.9 TREATMENT

Medical facilities to manage epilepsy in countries in tropical areas are very limited. Access to care is often difficult. The intricacy of sociocultural aspects, poverty, lack of medical resources and education are all factors that will contribute to the increase in the stigmatization of PWE and create difficulties for their social integration.

All the sociocultural factors mentioned previously and the supposed causes of epilepsy can explain the use of traditional therapists and healers in tropical areas. Traditional practitioners form a heterogeneous group; their therapeutic methods are numerous and derive directly from their cultural representations of the disease. In general, this is the first recourse, and it is only after several therapeutic failures that a consultation in the hospital is carried out. A long period may occur between the onset of the seizures and the medical consultation, which can lead to complications.

But the patient and his entourage can also benefit from being taken care of by a traditional therapist. Epileptic seizures are sources of deep anxiety which can be an essential element in the repetition of fits and then their behavioral consequences. By breaking this morbid cycle, the traditional healer can lessen the distress of the patient and his family. The plants used by traditional healers have not been studied so far and some may possess interesting anti-convulsant properties. It is essential not to oppose traditional medicine to modern medicine. They are complementary, with traditional healers providing psychosocial support. A better understanding of their interactions would help patients.

Currently, anti-epileptic treatments available could eliminate seizures in two-thirds of patients with epilepsy and ensure an almost normal life. As with any chronic disease, long-term treatment is difficult to accept culturally as patients expect modern drug therapy to cure the disease rapidly. In case of failure, the patient may be again led to consult the healers with often a feeling of discouragement and resignation.

The World Health Organization has drawn up a list of essential medicines corresponding to the minimum needs of a health system. This list specifies the molecules that have the best efficacy, tolerance and cost-effectiveness for the diseases considered to be priorities according to the region of the world in which the country is located. Epilepsy is one of these priority diseases.

For some authors, the major problem in the treatment of epilepsy is the availability of antiepileptic drugs. However, the management of patients will primarily depend on the competence, the experience of the doctor and the quality of his diagnosis. There is no ideal antiepileptic drug for developing countries, a drug that would have a broad spectrum of efficacy, without drug interaction, with a pharmacokinetic profile suitable for malnourished and multi-infected individuals and with few side effects.

The treatment gap is defined as the percentage difference between the number of patients with active epilepsy and the number of patients whose seizures are treated appropriately in a given population at a given time. This definition therefore includes both the diagnostic and therapeutic deficit. The treatment gap is estimated to be over 80% in many countries in the tropics. The reasons for the lack of treatment in epilepsy are multiple (sociocultural aspects, level of development of the country, economic level, remoteness of medical structures, cost of AED).[42,43] These factors must be analyzed at several levels (country, community, patients). Among the AED arsenal, phenobarbital remains the drug most often available and most prescribed in tropical areas because it is the cheapest and on the list of essential drugs. It has many advantages (its simplicity by taking one tablet per day, it is inexpensive, it is effective on many types of seizures, it has side effects that are not dangerous) but also presents some disadvantages such as the possible appearance of cognitive disorders, a potential status epilepticus when abruptly discontinued or its possible use by drug addicts. There are other AEDs available at reasonable prices in tropical areas (phenytoin, diazepam and ethosuximide). Since 1993, carbamazepine and sodium valproate were included among the essential drugs. Despite the availability of AEDs, patients are not always compliant and the effectiveness of treatments is diminished. Knowledge of this non-compliance should be considered in integrated primary healthcare programs. These programs must be accompanied by training of health personnel to enable them to better identify patients and to set up better management in involving the community to modify behaviors, attitudes and sociocultural representations of the disease in tropical areas.

11.10 CONCLUSION

Epilepsy in tropical areas has many specificities due to its impact on the health of these patients (morbidity and mortality), the existence of specific (especially parasitic) risk factors and negative sociocultural representations. Many initiatives have been put in place to get people with this condition "out of the shadows", since it represents a heavy burden for patients, their families, and society. While many studies have improved the knowledge of this disease in tropical areas (epidemiology, sociology, anthropology), the challenge now is to implement integrated actions involving the health policies of these countries, with patients and populations, education of health professionals, primary prevention of specific risk factors, and improvements of the health systems.

References

1. Fisher RS, Van Emde BW, Blume W, et al. Epileptic seizures and epilepsy: definitions proposed by the International League Against Epilepsy (ILAE) and the International Bureau for Epilepsy (IBE). *Epilepsia*. 2005;46:470−472.
2. Berg AT, Berkovic SF, Brodie MJ, et al. Revised terminology and concepts for organization of seizures and epilepsies: report of the ILAE Commission on Classification and Terminology, 2005−2009. *Epilepsia*. 2010;51:676−685.
3. Thurman DJ, Beghi E, Begley CE, et al. Standards for epidemiologic studies and surveillance of epilepsy. *Epilepsia*. 2011;52 (Suppl 7):2−26.
4. Sander JW, Shorvon SD. Incidence and prevalence studies in epilepsy and their methodological problems: a review. *J Neurol Neurosurg Psychiatry*. 1987;50:829−839.
5. Pal DK, Das T, Sengupta S. Comparison of key informant and survey methods for ascertainment of childhood epilepsy in West Bengal, India. *Int J Epidemiol*. 1998;27:672−676.
6. Debrock C, Preux PM, Houinato D, et al. Estimation of the prevalence of epilepsy in the Benin region of Zinvie using the capture-recapture method. *Int J Epidemiol*. 2000;29:330−335.
7. Edwards T, Scott AG, Munyoki G, et al. Active convulsive epilepsy in a rural district of Kenya: a study of prevalence and possible risk factors. *Lancet Neurol*. 2008;7:50−56.
8. Ba-Diop A, Marin B, Druet-Cabanac M, Ngoungou EB, Newton CR, Preux PM. Epidemiology, causes, and treatment of epilepsy in sub-Saharan Africa. *Lancet Neurol*. 2014;13:1029−1044.
9. Mac TL, Tran DS, Quet F, Odermatt P, Preux PM, Tan CT. Epidemiology, aetiology, and clinical management of epilepsy in Asia: a systematic review. *Lancet Neurol*. 2007;6:533−543.
10. Bruno E, Bartoloni A, Zammarchi L, et al. Epilepsy and neurocysticercosis in Latin America: a systematic review and meta-analysis. *PLoS Negl Trop Dis*. 2013;31:e2480.
11. Yemadje LP, Houinato D, Quet F, Druet-Cabanac M, Preux PM. Understanding the differences in prevalence of epilepsy in tropical regions. *Epilepsia*. 2011;52:1376−1381.
12. Preux PM, Druet-Cabanac M. Epidemiology and aetiology of epilepsy in sub-Saharan Africa. *Lancet Neurol*. 2005;4:21−31.

13. Tekle-Haimanot R, Forsgren L, Ekstedt J. Incidence of epilepsy in rural central Ethiopia. *Epilepsia*. 1997;38:541−546.

14. Osuntokun BO, Adeuja AO, Nottidge VA, et al. Prevalence of the epilepsies in Nigerian Africans: a community-based study. *Epilepsia*. 1987;28:272−279.

15. Kariuki SM, White S, Chengo E, et al. Electroencephalographic features of convulsive epilepsy in Africa: a multicentre study of prevalence, pattern and associated factors. *Clin Neurophysiol*. 2015;127:1099−1107.

16. Sander JW, Hart YM, Johnson AL, Shorvon SD. National General Practice Study of Epilepsy: newly diagnosed epileptic seizures in a general population. *Lancet*. 1990;336:1267−1271.

17. Edmond K, Clark A, Korczak VS, Sanderson C, Griffiths UK, Rudan I. Global and regional risk of disabling sequelae from bacterial meningitis: a systematic review and meta-analysis. *Lancet Infect Dis*. 2010;10:317−328.

18. Stevens JP, Eames M, Kent A, Halket S, Holt D, Harvey D. Long term outcome of neonatal meningitis. *Arch Dis Child Fetal Neonatal Ed*. 2003;88:F179−F184.

19. Senanayake N, Roman GC. Aetiological factors of epilepsy in the tropics. *J Trop Geogr Neurol*. 1991;1:69−79.

20. Ndimubanzi PC, Carabin H, Budke CM, et al. A systematic review of the frequency of neurocyticercosis with a focus on people with epilepsy. *PLoS Negl Trop Dis*. 2010;4:e870.

21. Debacq G, Moyano LM, Garcia HH, et al. Systematic review and meta-analysis estimating association of cysticercosis and neurocysticercosis with epilepsy. *PLoS Negl Trop Dis*. 2017;11: e0005153.

22. Carter JA, Neville BG, White S, et al. Increased prevalence of epilepsy associated with severe falciparum malaria in children. *Epilepsia*. 2004;45:978−981.

23. Ngoungou EB, Koko J, Druet-Cabanac M, et al. Cerebral malaria and sequelar epilepsy: first matched case-control study in Gabon. *Epilepsia*. 2006;47:2147−2153.

24. Birbeck GL, Molyneux ME, Kaplan PW, et al. Blantyre Malaria Project Epilepsy Study (BMPES) of neurological outcomes in retinopathy-positive paediatric cerebral malaria survivors: a prospective cohort study. *Lancet Neurol*. 2010;9:1173−1181.

25. Ngoungou EB, Dulac O, Poudiougou B, et al. Epilepsy as a consequence of cerebral malaria in area in which malaria is endemic in Mali, West Africa. *Epilepsia*. 2006;47:873−879.

26. Druet-Cabanac M, Boussinesq M, Dongmo L, Farnarier G, Bouteille B, Preux PM. Review of epidemiological studies searching for a relationship between onchocerciasis and epilepsy. *Neuroepidemiology*. 2004;23:144−149.

27. Pion SD, Kaiser C, Boutros-toni F, et al. Epilepsy in onchocerciasis endemic areas: systematic review and meta-analysis of population-based surveys. *PLoS Negl Trop Dis*. 2009;3(6):e461.

28. Ngugi AK, Bottomley C, Kleinschmidt I, et al. Prevalence of active convulsive epilepsy in sub-Saharan Africa and associated risk factors: cross-sectional and case-control studies. *Lancet Neurol*. 2013;12:253−263.

29. Nicoletti A, Bartoloni A, Sofia V, et al. Epilepsy and toxocariasis: a case-control study in Burundi. *Epilepsia*. 2007;48:894−899.

30. Nicoletti A, Bartoloni A, Reggio A, et al. Epilepsy, cysticercosis, and toxocariasis: a population-based case-control study in rural Bolivia. *Neurology*. 2002;58:1256−1261.

31. Ngoungou EB, Bhalla D, Nzoghe A, Darde ML, Preux PM. Toxoplasmosis and epilepsy--systematic review and metaanalysis. *PLoS Negl Trop Dis*. 2015;9:e0003525.

32. Tekle-Haimanot R, Abebe M, Gebre-Mariam A, et al. Community-based study of neurological disorders in rural central Ethiopia. *Neuroepidemiology*. 1990;9(5):263−277.

33. Ogunniyi A, Adeyinka A, Fagbemi SO, Orere R, Falope ZF, Oyawole SO. Computerized tomographic findings in adolescent and adult Nigerian epileptics. *West Afr J Med*. 1994;13:128−131.

34. Goudsmit J, Van, der Waals FW. Endemic epilepsy in an isolated region of Liberia. *Lancet*. 1983;1:528−529.

35. Jilek-Aall L, Jilek W, Miller JR. Clinical and genetic aspects of seizure disorders prevalent in an isolated African population. *Epilepsia*. 1979;20:613−622.

36. Munyoki G, Edwards T, White S, et al. Clinical and neurophysiologic features of active convulsive epilepsy in rural Kenya: a population-based study. *Epilepsia*. 2010;51:2370−2376.

37. Adewuya AO, Ola BA. Prevalence of and risk factors for anxiety and depressive disorders in Nigerian adolescents with epilepsy. *Epilepsy Behav*. 2005;6:342−347.

38. Akinsulore A, Adewuya A. Psychosocial aspects of epilepsy in Nigeria: a review. *Afr J Psychiatr*. 2010;13:351−356.

39. Levira F, Thurman DJ, Sander JW, et al. Premature mortality of epilepsy in low-and middle-income countries: a systematic review from the Mortality Task Force of the International League Against Epilepsy. *Epilepsia*. 2017;58:6−16.

40. Rafael F, Houinato D, Nubukpo P, et al. Sociocultural and psychological features of perceived stigma reported by people with epilepsy in Benin. *Epilepsia*. 2010;51:1061−1068.

41. Birbeck G, Chomba E, Atadzhanov M, Mbewe E, Haworth A. The social and economic impact of epilepsy in Zambia: a cross-sectional study. *Lancet Neurol*. 2007;6:39−44.

42. Mbuba CK, Ngugi AK, Newton CR, Carter JA. The epilepsy treatment gap in developing countries: a systematic review of the magnitude, causes, and intervention strategies. *Epilepsia*. 2008;49:1491−1503.

43. Meyer AC, Dua T, Ma J, Saxena S, Birbeck G. Global disparities in the epilepsy treatment gap: a systematic review. *Bull World Health Organ*. 2010;88:260−266.

12

Dementia

Maëlenn Guerchet and Martin Prince

King's College London, London, United Kingdom

12.1 BACKGROUND

Dementia is a syndrome, usually chronic or progressive by nature, due to disease of the brain that affects memory, learning, orientation, language, comprehension and judgment. It is characterized by a progressive, global deterioration in intellect leading to difficulties in the ability to perform everyday activities. Older people are mainly affected, but 2%−10% of all cases are estimated to start before the age of 65 years. Despite predominantly affecting older people, it is not a normal part of aging.

Dementia syndrome is linked to a large number of underlying brain pathologies, of which Alzheimer's disease (AD), vascular dementia (VaD), dementia with Lewy bodies, and frontotemporal dementia are the most common. Subtypes can be difficult to define, and mixed forms exist. The link between the pathological lesions found in the brain and the severity or the symptoms of dementia is unclear. The co-existence of other conditions such as cerebrovascular disease can be important.

12.1.1 Brain Pathology

The two core pathological hallmarks of AD are senile plaques and neurofibrillary tangles. Plaques, located outside the neurons, are composed of amyloid beta (Aβ) and called amyloid plaques, and the tangles within the brain cells are made of hyperphosphorylated tau protein. They are both associated with the progressive loss of neurons and synapses, brain atrophy and dilatation of the lateral ventricles due to loss of brain tissue. Increased levels of inflammation, oxidative stress, and nerve cell death are associated with these changes.[1] Another important cause of brain damage in AD is ischemia, which can be due to cerebral amyloid angiopathy, cerebral atherosclerosis and small-vessel disease. Brain changes underlying AD probably start to develop at least 20−30 years before the onset of symptoms, with first signs around the base of the brain from 50 years old, plaques and tangles later spreading up to the cortical regions.[2]

VaD, the second most common form of dementia after AD, shows different neuropathological signs.

Neural networks are affected by multifocal and or diffuse lesions (lacunes to microinfarcts) caused by systemic, cardiac, and local large- and small-vessel disease.[3] VaD may be the result of infarcts or insults detectable with brain imaging; however, the presence of those on imaging is neither sufficient nor necessary for the presence of VaD. The development of cerebrovascular pathology is likely to follow more variable patterns than in AD. Symptoms and signs develop gradually with the brain pathology, ultimately leading to the diagnosis of dementia.

According to clinico-pathological studies, mixed pathologies are much more common than "pure" forms of dementia (particularly for AD and VaD, and AD and dementia with Lewy bodies).[4] The presence and severity of cognitive, behavioral and psychological symptoms of dementia (BPSD) in the population is only explained to a limited extent by post-mortem measurement of the classic dementia pathological hallmarks.[5]

12.1.2 Burden

Among chronic diseases, dementia is one of the most important contributors to dependence and disability at older ages and, in high-income countries (HICs), transition into residential and nursing home care.[6] BPSD, which usually occur later in the course of the disease, impact on the quality of life of the older person and are an important cause of strain for the relatives or caregivers. Around half of all people with dementia need personal care while the others will probably develop similar needs over time.[6] Early in the course of the condition, older people will develop needs for care and supervision, which will intensify over time until more advanced stages of dementia leave them relying fully on caregivers for their basic needs.[6] The intensity of needs for care of people with dementia usually exceeds the demands associated with other conditions. A study in the USA showed that caregivers of people with dementia were more likely than caregivers of people with other conditions to provide help with getting in and out of bed (54% vs 42%), dressing (40% vs 31%), toileting (32% vs 26%), bathing (31% vs 23%), managing incontinence (31% vs 16%), and feeding (31% vs 14%).[7] Reports from the 10/66 Dementia Research Group in Dominican Republic and China confirmed those findings with people with dementia needing more care and also dementia caregivers experiencing more strain than caregivers of those with other health conditions.[8,9]

The Global Burden of Disease (GBD), using Institute of Health Metrics and Evaluation (IHME) disability weights, estimates that dementia is the ninth most burdensome condition for people aged 60 years and over. In essence, the GBD estimated that there were 3.9 million years of life lost (YLL) among older people aged 60 years and over, and 6.2 million years lived with disability (YLD), resulting in 10.0 million disability-adjusted life years (DALY), attributed to dementia.[10] In contrast with other conditions, the impact of dementia comes mainly from years lived with disability, rather than YLL from premature mortality. The health loss attributed to dementia in this last GBD (using IHME disability weights) is much smaller compared to the previous estimates,[11] and can mostly be explained by the change in methodology (i.e., new disability weights). Consequently, those last estimates have been critiqued for failing to capture the full impact of chronic diseases on disability, needs for care and attendant societal costs. This limitation reveals itself to be significant for older people and conditions like dementia, where most of the impact is due to disability rather than premature mortality.

The estimates of global societal economic costs of dementia were recently estimated,[12] using costs estimated at the country level and then aggregated. The global cost of dementia was estimated to be US$818 billion in 2015, which represents an increase of 35.4% since the previous estimates.[12,13] In essence, global direct medical costs of dementia represented US$159.2 billion (19.5% of the total cost), while direct social costs were estimated at US$327.9 billion (40.1% of the total) and informal care costs at US$330.8 billion (40.4% of the total). As reported in 2010, the proportional contribution of direct medical care costs was modest, particularly in HICs. However, there was an increasing relative contribution of direct social care sector costs and a decreasing relative contribution of informal care costs with increasing country income level. Cost estimates increased for all world regions, with the greatest relative increase occurring in the African and in East Asia regions (largely driven by the upwards revision of the prevalence estimates for those regions). The distribution of total costs in sub-categories (direct medical care costs, social care costs, and informal care) varied by region, consistently with country income level variation. The relative contribution of informal care was greatest in the African regions and lowest in North America, Western Europe and some South American regions, while the reverse was true for social sector costs.[12]

12.1.3 Prevention

Although brain damage accumulates and cognitive function declines progressively with age, most older adults will never develop dementia up to the time of death. Nevertheless, marked differences in cognitive health between individuals in late life are observed at a population-level. These differences may in part be a function of the level of exposure to a number of factors across the entire life course. They are usually associated with an increased or reduced future likelihood of cognitive impairment and dementia in populations, and therefore referred to as risk or protective factors. The rate of development of the underlying brain pathologies maybe influenced by certain modifiable risk factors, or the detrimental effects of brain pathology on cognitive function may be neutralized by those factors. Hence, the clinical onset of dementia may be advanced or delayed by the presence or absence of these factors.

For now, as no established diagnostic biomarkers of dementia-related brain damage exist,[14] and as the mechanisms linking this damage to the expression of dementia symptoms are not fully understood,[15] prevention of dementia is typically conceived as the delay of the clinical onset of the disease rather than a slowing or avoidance of the development of the underlying neuropathology. As for other chronic diseases, primary prevention of dementia corresponds to ideally "delay until death" the symptomatic onset, or, a delaying or deferring of onset to older ages than that at which it would otherwise have occurred. The population prevalence could be reduced by 50% with an average 5-year delay in the age of onset, which would greatly reduce its impact in the general population.

Potential risk and protective factors for dementia have been investigated in epidemiological studies and some of these have also been tested in experimental studies.[16] The 2014 World Alzheimer Report,[17] published by Alzheimer's Disease International (ADI), focused on modifiable risk factors, because of their potential to be targeted for prevention, in four key domains (developmental, psychological and psychosocial, lifestyle and cardiovascular risk factors). After critically examining the evidence, it was concluded that there was convincing evidence that the dementia risk for populations could be modified through reduction in tobacco use and better control and detection of hypertension and diabetes as well as cardiovascular risk factors. Evidence for possible causal associations with dementia was the strongest for low education in early life, hypertension in midlife and smoking and diabetes across the life course. Nevertheless, nonmodifiable risk factors (eminently age, sex, and genetic factors) are also very important. The apolipoprotein E (APOE) gene is the only common genetic risk factor for non-familial, late-onset AD identified to date as having a major impact on dementia risk,[18] with its ε4 allele increasing and its ε2 allele decreasing the risk.[19] However, it is a marker of susceptibility not an autosomal deterministic gene, which means that it is neither necessary nor sufficient to cause disease.

12.1.4 An Aging and Changing World

In 2015, 901 million people aged 60 or over were currently living in the world, representing 12% of the global population. Most older people are living in less-developed regions (602 million).[20] Some regions like Europe already comprise a great percentage of its population aged 60 or over (24%), but aging will also occur rapidly in other regions of the world over the next decades. By 2050, all major areas of the world except Africa will have nearly a quarter or more of their populations aged 60 or over. The global number of older persons is expected to reach 1.4 billion by 2030 and to more than double by 2050, reaching 2.1 billion. Between 2015 and 2050, 66% of this increase will occur in Asia, 13% in Africa, 11% in Latin America and the Caribbean, and the remaining 10% in other areas. Although Africa has the youngest age distribution of any major area, it is also projected to age rapidly over the next 35 years, with the percentage of its older population rising from 5% in 2015 to 9% by 2050.[20]

Ongoing population aging (or "demographic transition") is due to falling mortality rates among older people, and increasing life expectancy from age 60, with no upper limit in sight. The demographic transition is characterized by a shift from high fertility/high mortality states to low fertility/low mortality states. Child mortality is falling thanks to economic and social development, as well as improvements in public health. The population grows rapidly for a period of time, followed by a fall in fertility rates (people have fewer children) until it reaches replacement rates (two births per woman), so population growth slows and stops. This demographic transition is occurring much more rapidly in some developing middle-income countries than was the case in the developed countries. Thus, the transition from 7% of the total population aged 65 years and over, to 14% which took 115 years in France (1865–1980), 69 years in the USA (1944–2013), and 45 years in the United Kingdom (1930–75), will be completed in just 21 years in Brazil (2011–32), 23 years in Sri Lanka (2004–27) and 26 years in China (2000–26).

Partly because of the demographic transition and population aging, changes in the profile and patterning of disease happen. Chronic diseases, which tend to be strongly age-associated, become more prevalent and have a bigger impact. The prevalence and incidence of dementia, for example, doubles with every 5-year increase in age. This trend is exacerbated by changes in lifestyles and behaviors that predispose towards them.

The "epidemiologic transition" is linked to changes in behaviors and lifestyles, towards more "westernized" lifestyle patterns (sedentary, high dietary consumption of salt, fat and sugars, smoking and alcohol use). These changes are often driven by several factors including globalization, industrialization and urbanization. These changes lead to an increase in the incidence of some chronic diseases including particularly cardiovascular diseases and cancers. There are likely also adverse effects on brain aging in populations undergoing this transition as the same risk factors may also increase risk for dementia. Simultaneously, economic and social developments, and improvements in the health sector, contribute to bringing infectious diseases under control, and improving childhood and maternal health. Infectious, infant and maternal diseases which account for mortality at young ages, but also result in much childhood/lifelong disability, are mostly preventable. Consequently, in every world region, chronic diseases have now taken over as the leading cause of death.

Given the global aging of the population, the demographic and epidemiological transitions, as well as the impact of dementia on the older people affected, their relatives and societies, with no known cure on the horizon, health and care systems will be increasingly challenged. Estimating the scale of the dementia epidemic and understanding its characteristics and trends is therefore of great importance.

In this chapter, we will therefore provide an overview of the main epidemiological parameters of dementia in tropical areas: prevalence, incidence, mortality, and their trends over time. Most of the tropical regions comprise countries that are classified as low- and middle-income countries (LMICs) by the World Bank, we will then focus on the different regions as defined by the GBD: Asia (mainly Asia South, Asia Southeast and Oceania), Latin America and the Caribbean, and sub-Saharan Africa (SSA).

12.2 PREVALENCE

ADI, the international federation of Alzheimer associations around the world, has been gathering evidence on the global prevalence of dementia, its projections over time and its impact since 2004.

For its first review of the global evidence on the prevalence, numbers affected and its impact, ADI convened a panel of experts and conducted a consensus exercise in 2004.[21] Those estimates were then revised on two occasions, for the 2009 and 2015 World Alzheimer Reports,[22,23] when a systematic review of the world literature on the prevalence of dementia was conducted (and updated) and a quantitative meta-analysis was performed to synthesize the evidence for some regions. Over the years, and since the first estimates from the consensus were published, the evidence base has greatly expanded. According to the 2015 World Alzheimer Report,[23] while three GBD regions (East Asia, Western Europe and Asia Pacific-High Income) accounted for the majority of the world's studies, coverage showed the greatest improvements in Central SSA and Western SSA. The proportion of studies conducted in LMICs through to 2015 was 52%, a clear improvement compared to the 39% estimated in 2009. The regions that manifestly lack evidence in research relative to population size are currently Central Europe, Eastern and Southern SSA, and South and Southeast Asia as well as Central Asia (with no studies at all). Overall quality of studies was judged to be especially high in Latin America and SSA while it was particularly poor in East Asia, Southeast Asia, Central Europe and the Asia-Pacific—high-income regions. However, overall study quality tended to improve over time.[23]

Based on the age-specific, or age- and sex-specific estimates of the prevalence applied to the United Nations (UN) population projections, the latest estimate of the global number of people with dementia was 46.8 million in 2015. This number is expected to almost double every 20 years, reaching 74.7 million in 2030 and 131.5 million in 2050. Most of the people with dementia are currently living in LMICs, with 58% of all people in 2015. A figure that will rise to 63% of all people with dementia living in LMICs in 2030, and 68% in 2050.[23] Details of the total population of people over 60, crude estimated prevalence of dementia (2015), estimated number of people with dementia (2015, 2030, and 2050) and proportionate increases (2015—30) in tropical regions are given in Table 12.1.

12.2.1 Asia

A total of 17 studies matching inclusion criteria (population-based studies among people aged 60 years and over, following established diagnostic criteria, for which the field work started on or after January 1, 1980) were identified, of which 11 were carried out in Asia South (India[24–32] and Bangladesh[33]) and six in Asia Southeast (Sri Lanka,[34] Thailand,[35–37] and Malaysia[38,39]); only one study was found in Oceania,[40] which included only indigenous people and was therefore ineligible for the review. Most of the studies in those two regions (64% for Asia South and 60% for Asia Southeast) were conducted between 2000 and 2009. A two-phase design was mostly used (73% of the studies in Asia South and 83% in Asia Southeast) while they generally included between 500 and 2999 participants. Response rates were good in Asia South (64% of

TABLE 12.1 Total Population of People Over 60, Crude Estimated Prevalence of Dementia (2015), Estimated Number of People With Dementia (2015, 2030, and 2050), and Proportionate Increases (2015—30) in Tropical Regions

GBD region	Over 60 population (millions, 2015)	Crude estimated prevalence (%, 2015)	Number of people with dementia			Proportionate increases (%)	
			2015	2030	2050	2015–30	2015–50
Asia (total)	485.83	4.7	22.85	38.534	67.18	69	194
Oceania	0.64	3.5	0.02	0.04	0.09	83	289
Asia, South	139.85	3.7	5.13	8.61	16.65	68	225
Asia, Southeast	61.72	5.8	3.60	6.55	12.09	82	236
Americas (total)	147.51	6.4	9.44	15.75	29.86	67	216
Caribbean	5.78	6.5	0.38	0.60	1.07	60	183
LA, Andean	5.51	6.1	0.34	0.64	1.43	88	322
LA, Central	26.64	5.8	1.54	2.97	6.88	93	348
LA, Southern	9.88	7.6	0.75	1.15	2.05	52	172
LA, Tropical	24.82	6.7	1.66	3.11	6.70	88	305
Africa (total)	87.19	4.6	4.03	6.99	15.76	74	291
SSA, Central	4.78	3.3	0.16	0.26	0.54	60	238
SSA, East	19.86	3.5	0.69	1.19	2.77	72	300
SSA, Southern	6.06	3.9	0.24	0.35	0.58	46	145
SSA, West	17.56	3.1	0.54	0.85	1.84	58	241
World	897.14	5.2	46.78	74.69	131.45	60	181

GBD, Global Burden of Disease; LA, Latin America; SSA, sub-Saharan Africa.
From Prince M, Wimo A, Guerchet M, Ali G-C, Wu Y-T, Prina M. World Alzheimer Report 2015: The Global Impact of Dementia: An Analysis of Prevalence, Incidence, Costs and Trends. *London: Alzheimer's Disease International; 2015.*

the studies had 80%—100%) but half of the studies in Asia Southeast did not specify this information in their publications. The diagnostic criteria predominantly used in these regions were the Diagnostic and Statistical Manual of Mental Disorders (DSM)-IV/III-R. After meta-analyzing the results from 10 of those studies, the standardized prevalence of dementia was estimated to 5.63% in Asia South and 7.15% in Asia Southeast, which represents a total number of 8.75 million people living with dementia in those regions in 2015. Those numbers are projected to go through a proportionate increase of 225%—289% between 2015 and 2050.[23]

12.2.2 Latin America and Caribbean

The coverage of Latin America and the Caribbean is considered good. Eighteen population-based studies from eight different countries were identified (Cuba,[29] Dominican Republic,[29] Jamaica,[41] Peru,[29,42] Venezuela,[29,43] Mexico,[29,44] Chile,[45] and Brazil[46—49]). They were mainly conducted between 2000 and 2009 (72%) and 10 of them followed a one-phase design

(most being from the 10/66 Dementia Research Group) including between 1500 and 2499 participants (47% of the studies) or 500—1499 participants (35%). Response rate was 80%—100% for 12 of the studies (66%). The DSM-IV/III-R criteria were still the most commonly used in these regions, with 61% of the studies (n = 11). The crude prevalence of dementia ranged from 5.8% to 7.6% in those GBD regions, while the standardized prevalence for the whole Latin American region (inc. Caribbean) was estimated to be 8.41%. The number of people with dementia living there was therefore estimated to be 4.67 million, and is expected to undergo important proportionate increases from 2015 to 2050 (172%—322%).[23]

12.2.3 Sub-Saharan Africa

It is the tropical region that presented the worst coverage, but it has shown great improvements over the last decade. While only West SSA had population-based studies in the 2009 World Alzheimer Report, the 2015 update of the systematic review found studies for two additional regions: Central SSA and East SSA while Southern SSA remains with a lack of evidence. Nine studies, of high quality, were identified (of which seven included in the quantitative meta-analysis) from Nigeria,[50] Benin,[51,52] Central African Republic,[53,54] the Republic of Congo,[53,54] and Tanzania[55] (Table 12.2). The majority were conducted after 2000 (45% between 2000 and 2009 and 33% from 2010 onwards) and had a two-stage design (78%). The number of participants included seemed lower than that observed in studies in Latin America, with again 78% interviewing between 500 and 1499 older people, however studies had good response rates (89% between 80 and 100%). The DSM-IV/III-R remained the preferred diagnostic criteria in this region (78% of the studies). After collating all the evidence, the standardized prevalence for SSA was estimated to be 5.47%, leading to an estimated 1.63 million people over 60 living with dementia. The expected proportionate increases in numbers were quite similar to the ones that will affect Latin America, with 145%−300% between 2015 and 2050.[23]

Since the publication of those estimates, one study carried out in Lalupon (Nigeria) was published[56] and reported an age-adjusted prevalence of dementia of 2.9% (95% CI 1.6%−4.4%).

12.3 INCIDENCE

The first systematic review of the global incidence of dementia was published in the 2012 World Health Organization report *Dementia: A Public Health Priority*.[57] While the evidence base had clearly improved, no further systematic reviews of the incidence of dementia had been published until the update conducted and reported in the 2015 World Alzheimer Report.[23] A meta-analysis was also performed to estimate the global incidence and numbers of new cases of dementia, aged 60 years and over in 21 GBD regions. Including again only population-based studies for which dementia was defined according to DSM-IV (or International Classification of Diseases (ICD)-10 or similar clinical criteria), 61 studies in total were eligible of which 46 could be included in the global meta-analysis. Twenty-two (of 61) studies were conducted in LMICs, and the proportion of new studies conducted in LMICs increased since the initial report (40% from just 32% of studies in the original review). Unfortunately, no studies at all were found from the following tropical regions: Oceania, Southeast Asia, Southern, Central and Eastern SSA and Latin America Southern. Two studies, both from India, have now been published for South Asia.[58,59]

The global incidence of dementia increased exponentially with increasing age with incidence doubling with every 6.3-year increase in age, from 3.9/1000 person-years (pyr) at age 60−64 years to 104.8/1000 pyr at age 90+ years. The incidence of dementia seemed to be lower in LMICs (doubling every 8.6 years from 5.2/1000 pyr to 58.0/1000 pyr) than in HICs (doubling every 5.8 years from 3.5/1000 pyr to 124.9/1000 pyr). However, the incidence of dementia in LMICs was only 10% lower than in HICs and it was not statistically significant.[23]

A total of over 9.9 million new cases of dementia each year worldwide were anticipated, implying one new case every 3.2 seconds. In these updated estimates, the proportion of new cases arising in Asia, the Americas and Africa have increased while it has decreased in Europe.

12.3.1 Asia

Two studies on the incidence of AD were conducted in Kerala, India,[58,59] 10 years apart. Over a follow-up period of 8.1 years, 104 developed dementia (98 with AD) among the 1066 eligible participants who were cognitively normal at baseline. The incidence rates for AD was 15.5 (95% CI: 14.6−16.5) per 1000 pyr for those aged ≥65 years.[59] The incidence of AD increased significantly and proportionately with increasing age.

According to the latest estimates,[23] the annual number of incident cases of dementia was 1,875,793 for South and Southeast Asia.

12.3.2 Latin America and the Caribbean

The incidence of dementia in Latin America and the Caribbean is mainly informed by three studies (Table 12.3): one in Brazil,[64] the nationally representative Mexican Health and Aging Study[44] and the second wave of the 10/66 Dementia Research Group's studies in Mexico, Peru, Venezuela, Cuba, and the Dominican Republic.[61] Collating the data from those studies, the age- and sex-standardized incidence of dementia was estimated to be 15.1/1000 pyr for people aged 60 and over. Once applied to the number of people at risk of dementia living in those regions (total numbers in each age group, minus numbers with prevalent dementia), the annual number of incident cases of dementia in Latin America and the Caribbean was 750,383, of which 254,059 and 261,361, respectively, were living in the GBD Latin America Central and Latin America Tropical regions.[23]

12.3.3 Sub-Saharan Africa

The 2015 World Alzheimer Report included the evidence from a single country available at that time: Nigeria[62,63] (Table 12.3). In the Indianapolis-Ibadan Dementia Project,[62] the age-standardized annual incidence of dementia was significantly lower among

TABLE 12.2 Characteristics of Population-Based Studies on the Prevalence of Dementia in Sub-Saharan Africa, 2016

Reference	Country, region/city	Rural/ urban area	Design	Sampling	Residents of institutions	Lower and upper age limits	Dates of fieldwork	Sample size	Numbers interviewed (proportion responding)	Screening instruments	Diagnostic criteria	Multidomain cognitive assessment? Disability? Neuroimaging?
SSA WEST												
Hendrie et al. [50]	Nigeria, Ibadan	Urban	2 stage	Door-to-door	No	≤65	01/05/92–31/03/94	2535	2494 (98.4%)	CSI-D	DSM III-R/ICD10	Clinical assessment/ADL/CT scans
Yusuf et al. (2011)	Nigeria, Zaria	Urban	1 stage	Systematic random sampling	No	≤65	03/07–10/07	322	322 (100%)	CSI-D, CERAD, BDS	DSM-IV/ICD10	CERAD/Stick Design Test
Ogunniyi et al. [56]	Nigeria, Lalupon	Rural	2 stage	Catchment area[a]	No	≤65	05/13–02/14	642	642 (100%)	IDEA	DSM-IV	Clinical assessment/CHIF
Guerchet et al. [51]	Benin, Djidja	Rural	2 stage	Catchment area[a]	No	≤65	03/07–04/07	514	502 (97.6%)	CSI-D	DSM-IV	Clinical assessment/Psychometric tests/ADL
Paraiso et al. [52]	Benin, Cotonou	Urban	2 stage	Random sampling (Proportional)	No	≤65	05/08–09/08	1162	1139 (98.0%)	CSI-D	DSM-IV	Clinical assessment/Psychometric tests/ADL
SSA CENTRAL												
Guerchet et al. [53]	CAR, Bangui	Urban	2 stage	Catchment area[a]	No	≤65	10/09–12/09	509	496 (97.4%)	CSI-D/Five-word test	DSM-IV	Clinical assessment/Psychometric tests/NPI
	Congo, Brazzaville	Urban	2 stage	Catchment area[a]	No	≤65	01/10–03/10	546	520 (95.2%)	CSI-D/Five-word test	DSM-IV	Clinical assessment/Psychometric tests/ADL/NPI
Guerchet et al. [54]	CAR, Nola	Rural	2 stage	Catchment area[a]	No	≤65	11/11–03/12	501	473 (94.4%)	CSI-D	DSM-IV	Clinical assessment/Psychometric tests/ADL/NPI
	CAR, Bangui	Urban	2 stage	Random sampling (Proportional)	No	≤65	01/12–03/12	514	500 (97.3%)	CSI-D	DSM-IV	Clinical assessment/Psychometric tests/ADL/NPI
Guerchet et al. [54]	Congo, Gamboma	Rural	2 stage	Catchment area[a]	No	≤65	08/12–12/12	529	520 (94.3%)	CSI-D	DSM-IV	Clinical assessment/Psychometric tests/ADL/NPI
	Congo, Brazzaville	Urban	2 stage	Random sampling (proportional)	No	≤65	09/12–12/12	537	500 (93.1%)	CSI-D	DSM-IV	Clinical assessment/Psychometric tests/ADL/NPI
SSA EAST												
Longdon et al. [55]	Tanzania, Hai	Rural	2 stage	Catchment area[a]	No	≤70	?	1260	1198 (95.1%)	CSI-D	DSM-IV	CERAD

[a]Door-to-door.

ADL, activities of daily living; BDS, Blessed Dementia Scale; CERAD, Consortium to Establish a Registry for Alzheimer's Disease; CHIF, Clinician home-based assessment of impairment in functioning; CSI-D, Community Screening Interview for Dementia; CT, computed tomography; DSM, Diagnostic and Statistical Manual of Mental Disorders; ICD, International Classification of Diseases; IDEA, Identification and Intervention for Elderly Africans; NPI, Neuropsychiatric Inventory; SSA, sub-Saharan Africa. Psychometric tests = Free and Cued Selective Reminding Test, Zazzo's cancellation Task, Isaac's Set Test of verbal fluency.

TABLE 12.3 Characteristics of the Incidence Studies Identified in Tropical Regions Included in the WAR2015 MA

Study, reference	Setting	Country	Mid-year of baseline survey	Follow-up time (mean or *median)	Lower age limit	Criterion	Institutions included in sampling?	Cohort at risk	Person years
LATIN AMERICA									
Brazil, Nitrini et al. [60]	Catanduva, Sao Paulo	Brazil	1997	3.3	65	DSM-IV	Not stated	1538	3649
10/66 Cuba [61]	Havana, Matanzas	Cuba	2004	4.5*	65	10/66 Dementia/ DSM-IV	No	2517	8679
10/66 Dominican Republic [61]	Santo Domingo	DR	2005	5.1*	65	10/66 Dementia/ DSM-IV	No	1769	5561
10/66 Peru [61]	Lima, Canete	Peru	2005	2.9*/3.7*	65	10/66 Dementia/ DSM-IV	No	1767	3913
10/66 Venezuela [61]	Caracas	Venezuela	2005	4.3*	65	10/66 Dementia/ DSM-IV	No	1820	5269
10/66 Mexico [61]	Mexico City, Morelos	Mexico	2006	3.0*	65	10/66 Dementia/ DSM-IV	No	1823	4164
MHAS [44]	Nationally representative	Mexico	2001	2	60	DSM-IV	No	6847	12,980
SOUTH ASIA									
India, Mathuranath et al. [59]	Trivandrum City, Kerala	India	2001	8.1	55	DSM-IV	No	1066	8528
SUB-SAHARAN AFRICA									
Indianapolis-Ibadan Dementia Project [62]	Ibadan	Nigeria	1994	5.1	65	DSM-III/ ICD10	No	2459	Not stated
Ibadan Study of Aging [63]	Ibadan	Nigeria	2004	3.3	65	Clinical/ NS	No	1225	3890

DSM, Diagnostic and Statistical Manual of Mental Disorders; ICD, International Classification of Diseases; MHAS, Mexican Health and Aging Study; NS, not stated.

Yoruba than among African Americans (Yoruba, 1.3% (95% CI, 1.1%−1.6%); African Americans, 3.24% (95% CI, 2.1%−4.4%)). The same pattern was shown for AD (Yoruba, 1.1% (95% CI, 0.9%−1.3%); African Americans, 2.5% (95% CI, 1.4%−3.6%)). While more recently, after a 3-year follow-up, the estimated incidence of dementia was 21.8/1000 pyr (95% CI, 17.7−27.0) in the Ibadan Study of Aging.[63]

This evidence from Nigeria led to the estimation of a total annual number of 446,569 new cases of dementia in SSA.[23] The latest dementia incidence estimated in the Ibadan Study of Aging (Nigeria) after 5 years of follow-up confirmed the previous estimates, with 20.9/1000 pyr (95% CI, 17.7−24.9).[65]

More recently, the EPIDEMCA-FU (Epidemiology of Dementia in Central Africa—Follow-up) study has reported preliminary results from a 2-year follow-up in Central Africa.[66] Among older people living in both urban and rural areas of the Republic of Congo, the standardized incidence of dementia was estimated at 13.5/1000 pyr (95% CI, 9.9−15.7).

The lack of longitudinal evidence on dementia is striking in LMICs and/or tropical regions with no evidence at all from Southeast Asia and a clear underrepresentation of the African continent. More research is needed to provide information on regions with no evidence or more up to date information.

12.4 MORTALITY

It is well known that people with dementia have shorter lives. A systematic review on survival in dementia showed that the median survival time from age of onset of dementia was 3.3–11.7 years, with most of studies in the 7- to 10-year period, while median survival times from age of disease diagnosis ranged from 3.2 to 6.6 years for dementia or AD cohorts.[60] Among all the studies included in this review, only one had been conducted in a low- and middle-income setting (Brazil).[67]

Obviously, much variability can occur among individuals living with dementia. Assessing the independent contribution of dementia to mortality is particularly challenging as death certificates are often not reliable and other health comorbidities may be related to dementia and death among older people. Death of people with dementia can therefore not be systematically attributed to dementia.[68]

While a meta-analysis of studies mostly from HICs estimated a 2.5-fold increased risk for people with dementia[69] in people over 65 living in the community, estimates from LMICs point to a slightly higher mortality risk. In Latin America and the Caribbean, evidence is coming from Brazil[67] and the 10/66 Dementia Research Group population-based studies[61] once again. Three to five years after inception in urban sites in Cuba, the Dominican Republic, and Venezuela, and rural and urban sites in Peru, Mexico, and China, mortality hazards were 1.6- to 5.7-times higher in individuals with dementia at baseline than in those who were dementia-free.[61] The estimates from Brazil were on the upper band of this range, with a hazard ratio (HR) of 5.2 (95% CI, 3.7–7.1).[67] The only evidence from Asia relies on the study of predictors of mortality among older people living in a South Indian urban community (Chennai), which showed a hazard of 2.8 (95% CI, 1.8–4.5) for dementia, slightly lower HR = 2.3 (95% CI, 1.5–2.7) when adjusted on age and sex.[70] In the sub-Saharan region, Nigeria was the first country to provide estimates on mortality risk among people with dementia. In the Indianapolis Ibadan Dementia Project, dementia was significantly associated with increased mortality at both sites, with a relative risk of 2.8 (95% CI, 1.1–7.3) in Ibadan,[71] which is higher than the hazard reported later in the Ibadan Study of Aging (adjusted HR = 1.5 (95% CI, 1.2–1.8)).[65] The risk reported among older people living in the Republic of Congo was similar, more than 2.5-times higher among those with dementia (adjusted HR = 2.5 (95% CI, 1.4–4.4)),[72] while a larger relative risk has been recorded in Tanzania after a 4-year follow-up (HR = 6.3 (95% CI, 3.2–12.6) after adjustments).[73]

12.5 TRENDS

All the projections in the number of people living with dementia presented earlier assume that the age- and sex-specific prevalence of dementia will remain constant (i.e., not vary over time) and that the increases are only driven by population aging. This scenario is actually improbable and secular trends in dementia prevalence are very plausible. The prevalence of a condition, being the product of its incidence and its average duration, could be affected by a change in either or both indicators.

Over time, societal changes occur and will potentially have an effect on population health. During successive generations, risks (e.g., vascular diseases) and protective factors (e.g., education) have changed considerably.[74] As a result of changes in life expectancy and risk profiles, secular trends may vary among world regions, and among different population subgroups within one country.[75]

There has been an increased interest recently in a possible decline of age-specific dementia prevalence or incidence in HICs. A review of the epidemiological evidence in Western Europe[76] showed that four of the five studies identified reported non-significant changes in overall dementia occurrence. Additionally, a global review of secular trends in dementia prevalence, incidence and mortality was conducted,[75] in which studies applying constant methods to defined populations were included. Among the nine studies tracking dementia prevalence, seven tracking dementia incidence and four tracking mortality among people with dementia, all but one were conducted in HICs. The evidence was considered too inconsistent to reach sound and generalizable conclusions regarding the existence of trends in dementia prevalence, incidence, and mortality. The unique study assessing dementia incidence over time in Nigeria[77] found no significant difference in dementia or AD incidence between the 1992 and 2001 Yoruba cohorts.

While public health strategies (improving the control of cardiovascular outcomes for example) alongside improvements in education could progressively lead to a decline in age-specific incidence of dementia in HICs, this decline will also depend on simultaneous changes in survival/mortality patterns for people living with dementia. On the other hand, the current evidence on the rise of non-communicable diseases (increase in cardiovascular risk factors and morbidity) and changing lifestyles in LMICs are more consistent for now with an increase in dementia age-specific incidence and prevalence over the next decades in those regions.[75]

Assessing the secular trends in dementia epidemiology is the next step research needs to undertake. More studies using a stable and fixed methodology—in defined and representative populations, with good response rate—to estimate variations in dementia prevalence, incidence, and mortality over time need to be developed.

References

1. Ballard C, Gauthier S, Corbett A, Brayne C, Aarsland D, Jones E. Alzheimer's disease. *Lancet*. 2011;377:1019–1031.

2. Braak H, Braak E, Bohl J, Bratzke H. Evolution of Alzheimer's disease related cortical lesions. *J Neural Transm, Suppl*. 1998;54:97–106.

3. Jellinger KA. The pathology of "vascular dementia": a critical update. *J Alzheimer's Dis*. 2008;14:107–123.

4. Neuropathology Group. Medical Research Council Cognitive Function and, Aging Study. Pathological correlates of late-onset dementia in a multicentre, community-based population in England and Wales. Neuropathology Group of the Medical Research Council Cognitive Function and Ageing Study (MRC CFAS). *Lancet*. 2001;357:169–175.

5. Matthews FE, Brayne C, Lowe J, McKeith I, Wharton SB, Ince P. Epidemiological pathology of dementia: attributable-risks at death in the Medical Research Council Cognitive Function and Ageing Study. *PLoS Med*. 2009;6:e1000180.

6. Alzheimer Disease International. *Journey of Caring: An Analysis of Long-Term Care for Dementia*. London: Alzheimer's Disease International; 2013:92 p.

7. Alzheimer's Association. *2013 Alzheimer's Facts and Figures*. Chicago, IL: Alzheimer's Association; 2013.

8. Liu Z, Albanese E, Li S, et al. Chronic disease prevalence and care among the elderly in urban and rural Beijing, China: a 10/66 Dementia Research Group cross-sectional survey. *BMC Public Health*. 2009;9:394.

9. Acosta D, Rottbeck R, Rodriguez G, Ferri CP, Prince MJ. The epidemiology of dependency among urban-dwelling older people in the Dominican Republic; a cross-sectional survey. *BMC Public Health*. 2008;8:285.

10. Murray CJ, Vos T, Lozano R, et al. Disability-adjusted life years (DALYs) for 291 diseases and injuries in 21 regions, 1990–2010: a systematic analysis for the Global Burden of Disease Study 2010. *Lancet*. 2012;380:2197–2223.

11. World Health Organization. *The Global Burden of Disease, 2004 update*. Geneva, Switzerland: WHO; 2008.

12. Wimo A, Guerchet M, Ali GC, et al. The worldwide costs of dementia 2015 and comparisons with 2010. *Alzheimer's Dement*. 2017;13:1–7.

13. Wimo A, Winblad B, Jonsson L. The worldwide societal costs of dementia: estimates for 2009. *Alzheimer's Dement*. 2010;6:98–103.

14. Frisoni GB, Bocchetta M, Chételat G, et al. Imaging markers for Alzheimer disease Which vs how. *Neurology*. 2013;81:487–500.

15. Savva GM, Wharton SB, Ince PG, et al. Age, neuropathology, and dementia. *N Engl J Med*. 2009;360:2302–2309.

16. Hughes TF, Ganguli M. Modifiable midlife risk factors for late-life cognitive impairment and dementia. *Curr Psychiatry Rev*. 2009;5:73–92.

17. Prince M, Albanese E, Guerchet M, Prina AM. *World Alzheimer Report 2014: Dementia Risk Reduction: An Analysis of Protective and Modifiable Risk Factors*. London: Alzheimer's Disease International; 2014.

18. Corder E, Saunders A, Strittmatter W, et al. Gene dose of apolipoprotein E type 4 allele and the risk of Alzheimer's disease in late onset families. *Science*. 1993;261:921–923.

19. Raber J, Huang Y, Ashford JW. ApoE genotype accounts for the vast majority of AD risk and AD pathology. *Neurobiol Aging*. 2004;25:641–650.

20. United Nations. Department of Economic and Social Affairs, Population Division. World Population Prospects: The 2015 Revision, Key Findings and Advance Tables. Vol Working Paper No. SA/P/WP. 241; 2015.

21. Ferri CP, Prince M, Brayne C, et al. Global prevalence of dementia: a Delphi consensus study. *Lancet*. 2005;366:2112–2117.

22. Alzheimer's Disease International. *World Alzheimer Report 2009*. London: Alzheimer's Disease International; 2009.

23. Prince M, Wimo A, Guerchet M, Ali G-C, Wu Y-T, Prina AM. *World Alzheimer Report 2015: The Global Impact of Dementia: An Aanalysis of Prevalence, Incidence, Costs and Trends*. London: Alzheimer's Disease International; 2015.

24. Shaji S, Promodu K, Abraham T, Roy KJ, Verghese A. An epidemiological study of dementia in a rural community in Kerala, India. *Br J Psychiatry*. 1996;168:745–749.

25. Rajkumar S, Kumar S, Thara R. Prevalence of dementia in a rural setting: a report from India. *Int J Geriatr Psychiatry*. 1997;12:702–707.

26. Chandra V, Ganguli M, Pandav R, Johnston J, Belle S, DeKosky ST. Prevalence of Alzheimer's disease and other dementias in rural India: the Indo-US study. *Neurology*. 1998;51:1000–1008.

27. Vas CJ, Pinto C, Panikker D, et al. Prevalence of dementia in an urban Indian population. *Int Psychogeriatr*. 2001;13:439–450.

28. Shaji S, Bose S, Verghese A. Prevalence of dementia in an urban population in Kerala, India. *Br J Psychiatry*. 2005;186:136–140.

29. Llibre Rodriguez JJ, Ferri CP, Acosta D, et al. Prevalence of dementia in Latin America, India, and China: a population-based cross-sectional survey. *Lancet*. 2008;372:464–474.

30. Mathuranath PS, Cherian PJ, Mathew R, et al. Dementia in Kerala, South India: prevalence and influence of age, education and gender. *Int J Geriatr Psychiatry*. 2010;25:290–297.

31. Raina S, Razdan S, Pandita K. Prevalence of dementia in ethnic Dogra population of Jammu district, North India: a comparison survey. *Neurol Asia*. 2010;15:65–69.

32. Seby K, Chaudhury S, Chakraborty R. Prevalence of psychiatric and physical morbidity in an urban geriatric population. *Indian J Psychiatry*. 2011;53:121–127.

33. Palmer K, Kabir ZN, Ahmed T, et al. Prevalence of dementia and factors associated with dementia in rural Bangladesh: data from a cross-sectional, population-based study. *Int Psychogeriatr*. 2014;26:1905–1915.

34. de Silva HA, Gunatilake SB, Smith AD. Prevalence of dementia in a semi-urban population in Sri Lanka: report from a regional survey. *Int J Geriatr Psychiatry*. 2003;18:711–715.

35. Phanthumchinda K, Jitapunkul S, Sitthi-amron C, Bunnag SC, Ebrahim S. Prevalence of dementia in an urban slum population in Thailand: validity of screening methods. *Int J Geriatr Psychiatry*. 1991;6:1099–1166.

36. Senanarong V, Harnphadungkit K, Poungvarin N, Thongtang O, Sukhatunga K, Vannasaeng S. Prevalence of dementia, including vascular dementia, in 1,070 Thai elderly in Bangkok. *J Stroke Cerebrovasc Dis*. 2000;9:121–122.

37. Wangtongkum S, Sucharitkul P, Silprasert N, Inthrachak R. Prevalence of dementia among population age over 45 years in Chiang Mai, Thailand. *J Med Assoc Thailand*. 2008;91:1685–1690.

38. Krishnaswamy S, Kadir K, Ali R, Sidi H, Mathews S. Prevalence of dementia among elderly Malays in an urban settlement in Malaysia. *Neurol J Southeast Asia*. 1997;2:159–162.

39. Hamid TA, Krishnaswamy S, Abdullah SS, Momtaz YA. Sociodemographic risk factors and correlates of dementia in older Malaysians. *Dementia Geriatr Cognit Disord*. 2010;30:533–539.

40. Galasko D, Salmon D, Gamst A, et al. Prevalence of dementia in Chamorros on Guam: relationship to age, gender, education, and APOE. *Neurology*. 2007;68:1772–1781.

41. Neita SM, Abel WD, Eldemire-Shearer D, James K, Gibson RC. The prevalence and associated demographic factors of dementia from a cross-sectional community survey in Kingston, Jamaica. *Int J Geriatr Psychiatry*. 2014;29:103–105.

42. Custodio N, García A, Montesinos R, Escobar J, Bendezú L. Prevalencia de demencia en una población urbana de Lima-Perú: estudio puerta a puerta. *An Fac Med*. 2008;69:233–238.

43. Molero AE, Pino-Ramirez G, Maestre GE. High prevalence of dementia in a Caribbean population. *Neuroepidemiology*. 2007;29:107–112.

44. Mejia-Arango S, Gutierrez LM. Prevalence and incidence rates of dementia and cognitive impairment no dementia in the Mexican population: data from the Mexican Health and Aging Study. *J Aging Health*. 2011;23:1050–1074.

45. Albala C, Quiroga P, Klaassen G, Rioseco P, Perez H, Calvo C. Prevalence of dementia and cognitive impairment in Chile. In: *World Congress of Gerontology*. Adelaide, Australia; 1997.

46. Herrera Jr. E, Caramelli P, Silveira AS, Nitrini R. Epidemiologic survey of dementia in a community-dwelling Brazilian population. *Alzheimer Dis Assoc Disord*. 2002;16:103–108.

47. Scazufca M, Menezes PR, Vallada HP, et al. High prevalence of dementia among older adults from poor socioeconomic backgrounds in Sao Paulo, Brazil. *Int Psychogeriatr*. 2008;20:394–405.

48. Bottino CM, Azevedo Jr. D, Tatsch M, et al. Estimate of dementia prevalence in a community sample from Sao Paulo, Brazil. *Dementia Geriatr Cognit Disord*. 2008;26:291–299.

49. Lopes MA, Ferrioli E, Nakano EY, Litvoc J, Bottino CM. High prevalence of dementia in a community-based survey of older people from Brazil: association with intellectual activity rather than education. *J Alzheimer's Dis*. 2012;32:307–316.

50. Hendrie HC, Osuntokun BO, Hall KS, et al. Prevalence of Alzheimer's disease and dementia in two communities: Nigerian Africans and African Americans. *Am J Psychiatry*. 1995;152:1485–1492.

51. Guerchet M, Houinato D, Paraiso MN, et al. Cognitive impairment and dementia in elderly people living in rural Benin, west Africa. *Dementia Geriatr Cognit Disord*. 2009;27:34–41.

52. Paraiso MN, Guerchet M, Saizonou J, et al. Prevalence of dementia among elderly people living in Cotonou, an urban area of Benin (West Africa). *Neuroepidemiology*. 2011;36:245–251.

53. Guerchet M, M'Belesso P, Mouanga AM, et al. Prevalence of dementia in elderly living in two cities of Central Africa: the EDAC survey. *Dementia Geriatr Cognit Disord*. 2010;30:261–268.

54. Guerchet M, Ndamba-Bandzouzi B, Mbelesso P, et al. Comparison of rural and urban dementia prevalences in two countries of central africa: the EPIDEMCA study. *Alzheimer's Dementia*. 2013;1:P688.

55. Longdon AR, Paddick SM, Kisoli A, et al. The prevalence of dementia in rural Tanzania: a cross-sectional community-based study. *Int J Geriatr Psychiatry*. 2013;28:728–737.

56. Ogunniyi A, Adebiyi AO, Adediran AB, Olakehinde OO, Siwoku AA. Prevalence estimates of major neurocognitive disorders in a rural Nigerian community. *Brain Behav*. 2016;6:e00481.

57. World Health Organization. *Alzheimer's Disease International. Dementia: A Public Health Priority*. Geneva, Switzerland: WHO; 2012.

58. Chandra V, Pandav R, Dodge HH, et al. Incidence of Alzheimer's disease in a rural community in India: the Indo-US study. *Neurology*. 2001;57:985–989.

59. Mathuranath PS, George A, Ranjith N, et al. Incidence of Alzheimer's disease in India: a 10 years follow-up study. *Neurol India*. 2012;60:625–630.

60. Todd S, Barr S, Roberts M, Passmore AP. Survival in dementia and predictors of mortality: a review. *Int J Geriatr Psychiatry*. 2013;28:1109–1124.

61. Prince M, Acosta D, Ferri CP, et al. Dementia incidence and mortality in middle-income countries, and associations with indicators of cognitive reserve: a 10/66 Dementia Research Group population-based cohort study. *Lancet*. 2012;380:50–58.

62. Hendrie HC, Ogunniyi A, Hall KS, et al. Incidence of dementia and Alzheimer disease in 2 communities: Yoruba residing in Ibadan, Nigeria, and African Americans residing in Indianapolis, Indiana. *JAMA*. 2001;285:739–747.

63. Gureje O, Ogunniyi A, Kola L, Abiona T. Incidence of and risk factors for dementia in the Ibadan study of aging. *J Am Geriatr Soc*. 2011;59:869–874.

64. Nitrini R, Caramelli P, Herrera Jr. E, et al. Incidence of dementia in a community-dwelling Brazilian population. *Alzheimer Dis Assoc Disord*. 2004;18:241–246.

65. Ojagbemi A, Bello T, Gureje O. Cognitive reserve, incident dementia, and associated mortality in the Ibadan Study of Ageing. *J Am Geriatr Soc*. 2016;64:590–595.

66. Samba H, Guerchet M, Bandzouzi-Ndamba B, et al. Incidence of dementia among older adults in Central Africa: first results from the republic of congo in the EPIDEMCA-FU study. *Alzheimer's Dementia*. 2015;11:P221–P222.

67. Nitrini R, Caramelli P, Herrera Jr. E, et al. Mortality from dementia in a community-dwelling Brazilian population. *Int J Geriatr Psychiatry*. 2005;20:247–253.

68. World Health Organization. *Dementia: A Public Health Priority*. Geneva, Switzerland: WHO; 2012.

69. Dewey ME, Saz P. Dementia, cognitive impairment and mortality in persons aged 65 and over living in the community: a systematic review of the literature. *Int J Geriatr Psychiatry*. 2001;16:751–761.

70. Jotheeswaran AT, Williams JD, Prince MJ. Predictors of mortality among elderly people living in a south Indian urban community: a 10/66 Dementia Research Group prospective population-based cohort study. *BMC Public Health*. 2010;10:366.

71. Perkins AJ, Hui SL, Ogunniyi A, et al. Risk of mortality for dementia in a developing country: the Yoruba in Nigeria. *Int J Geriatr Psychiatry*. 2002;17:566–573.

72. Samba H, Guerchet M, Ndamba-Bandzouzi B, et al. Dementia-associated mortality and its predictors among older adults in sub-Saharan Africa: results from a 2-year follow-up in Congo (the EPIDEMCA-FU study). *Age Ageing*. 2016;45:681–687.

73. Paddick SM, Kisoli A, Dotchin CL, et al. Mortality rates in community-dwelling Tanzanians with dementia and mild cognitive impairment: a 4-year follow-up study. *Age Ageing*. 2015;44:636–641.

74. Larson EB, Yaffe K, Langa KM. New insights into the dementia epidemic. *N Engl J Med*. 2013;369:2275–2277.

75. Prince M, Ali GC, Guerchet M, Prina AM, Albanese E, Wu YT. Recent global trends in the prevalence and incidence of dementia, and survival with dementia. *Alzheimer's Res Ther*. 2016;8:23.

76. Wu Y-T, Fratiglioni L, Matthews FE, et al. Dementia in western Europe: epidemiological evidence and implications for policy making. *Lancet Neurol*. 2015;S1474–4422:00092–00097.

77. Gao S, Ogunniyi A, Hall KS, et al. Dementia incidence declined in African-Americans but not in Yoruba. *Alzheimer's Dement*. 2016;12:244–251.

78. Yusuf AJ, Baiyewu O, Sheikh TL, Shehu AU. Prevalence of dementia and dementia subtypes among community-dwelling elderly people in northern Nigeria. *Int Psychogeriatr*. 2011;23:379–386.

IV. FOCUS ON SPECIFIC NEUROLOGICAL SYNDROMES OR DISEASES IN TROPICAL AREAS

13

Other Neurocognitive Disorders in Tropical Health (Amyotrophic Lateral Sclerosis and Parkinson's Disease)

Benoît Marin[1,2,3], Philippe Couratier[1,2,3], Annie Lannuzel[4,5,6], and Giancarlo Logroscino[7,8]

[1]INSERM UMR 1094 NET, Limoges, France [2]University of Limoges, Limoges, France [3]CHU Limoges, Limoges, France [4]University Hospital, Pointe-à-Pitre, Guadeloupe [5]Antilles University, Pointe-à-Pitre, Guadeloupe [6]Sorbonne University, Paris, France [7]University of Bari "Aldo Moro", Bari, Italy [8]University of Bari "Aldo Moro", at "Pia Fondazione Cardinale G. Panico", Lecce, Italy

13.1 INTRODUCTION

Amyotrophic lateral sclerosis (ALS) and Parkinson's disease (PD) are age-related neurodegenerative disorders.[1,2]

ALS is a rare disease whose worldwide incidence has been estimated at 1.68/100,000 person-years of follow-up (PYFU) (US age-standardized incidence) in a recent meta-analysis.[3] ALS incidence is low before 50 years of age, increases until 70 years (peak-age), and then follows a sharp decrease. ALS is related to UMN and LMN degenerations. It is characterized by extensive paralysis and leads to death generally by respiratory failure. ALS is invariably lethal and 50% of cases

Neuroepidemiology in Tropical Health
DOI: http://dx.doi.org/10.1016/B978-0-12-804607-4.00013-7

die within 15–20 months after diagnosis (some cases die earlier, some are long progressors). To date no consistent etiologic factor has been described and apart from supportive treatments, riluzole is the only drug that was proven to extend survival by around 3 months.

PD is a debilitating neurodegenerative disorder that affects more than 4 million individuals worldwide over the age of 50, and the number of affected individuals is projected to double by the year 2030.[4] PD is characterized by progressive slowness or akinesia and rigidity, usually unilateral at onset, with or without resting tremor, which responds significantly to levodopa treatment or other kinds of dopaminergic replacement therapy. While a small percentage of PD cases and ALS cases are caused by genetic mutations, both disorders are currently viewed as multifactorial disease, resulting from the interplay of environmental factors and genetic susceptibility.

In this chapter we will present the main clinical characteristics of ALS and PD, we will focus on the variation of ALS and PD under the tropics in terms of incidence, prevalence and phenotype. Challenges and clues for a better understanding of those variations will be given.

13.2 CLINICAL FEATURES OF ALS

13.2.1 Introduction

ALS is a progressive neurodegenerative disorder that causes muscle weakness, disability and death.[5] The clinical hallmark of ALS is the combination of upper (UMN) and lower motor neuron (LMN) signs and symptoms. The UMN findings result from degeneration of frontal motor neurons located in the motor strip (Brodman area 4) and their axons. The LMN findings are a direct consequence of degeneration of LMNs in the brainstem and spinal cord producing muscle denervation.

The neuropathology of ALS is characterized by pathologic inclusions within both UMN and LMN and glia, but also in non-motor frontal and temporal cortical neurons. Inclusions stain positively for ubiquitin. A large subset also stain positively for TAR DNA-binding protein (TDP-43) and smaller subsets stain for fused in sarcoma (FUS) protein and optineurin.

While it was once presumed to be a pure motor disorder, it has become apparent that degeneration of other brain regions such as frontal and temporal cortical neurons may also occur as part of the clinicopathologic spectrum of ALS. Diagnosis is based on clinical examination and electroneuromyographic study demonstrating degeneration of LMNs and UMNs.[6] ALS

remains the most common form of motorneuron disease in adults. Differences in site of onset, speed of spread, and the degree of UMN and/or LMN dysfunction produces a disorder that is variable between individuals. Progressive muscular atrophy[7] and UMN-dominant ALS[8] have a more benign prognosis than classic ALS. In tropical countries, some atypical phenotypes have been reported such as in Asia: Madras motorneuron disease in Southern India with a typical weakness and wasting of limbs, multiple lower cranial nerve palsies and sensorineural hearing loss in young adults.[9] The difficult access to diagnosis and care in some tropical areas may explain the relative lower incidence as compared to Western countries.

13.2.2 Clinical Symptoms and Signs

The initial clinical manifestation may occur in any body segment and may manifest as UMN or LMN symptoms or signs. Asymmetric limb weakness is the most common presentation of ALS (70%). Upper extremity onset is most often characterized by hand weakness but may begin in the shoulder girdle muscles. The "split-hand syndrome" is a frequent pattern.[10] Lower-extremity onset most often begins with weakness of foot dorsiflexion while proximal pelvic girdle onset is less common. Thirty percent of patients will have onset in the bulbar segment, which most often presents with either dysarthria or dysphagia. Less common patterns include respiratory muscle weakness,[11] axial onset with head drop or weight loss with muscle atrophy, fasciculations, and cramps.[5]

Loss of UMNs results in slowness of movement and stiffness. Arm or hand UMN symptoms include poor dexterity. Leg UMN symptoms manifest as a spastic gait with poor balance and may include spontaneous leg flexor spasms and ankle clonus. Dysarthria and dysphagia are the most common bulbar UMN symptoms. Spastic dysarthria produces a characteristically strained vocal quality with slow speech. Dysphagia results from slow contraction of the swallowing muscles, which may lead to coughing and choking. Another frequent bulbar UMN symptom is the syndrome of the pseudobulbar affect manifested as inappropriate laughing, crying, or yawning. UMN bulbar dysfunction may also result in laryngospasm. Additional manifestations of UMN bulbar dysfunction may include increased masseter tone and difficulty opening the mouth. Axial UMN dysfunction may contribute to stiffness and imbalance.

Loss of LMN results in weakness, usually accompanied by atrophy and fasciculations. Muscle cramps are also common.[12] Hand weakness causes difficulty manipulating small objects. Proximal arm weakness results in difficulty elevating the arm to the level of the mouth or above the head. Foot and ankle weakness results in tripping, a slapping gait, and falling. Proximal leg weakness results in difficulty in rising from chairs, climbing stairs and getting up off the floor. Balance may also be adversely affected. Dysarthria and dysphagia can also result from LMN damage. Dysarthria may result from weakness of the tongue, lips or palate. The speech is usually slurred and may have a nasal quality. Hoarseness may be caused by associated vocal cord weakness. Dysphagia results from tongue weakness with disruption of the oral phase of swallowing or from pharyngeal constrictor weakness with disruption of the pharyngeal phase of swallowing or both. Tongue weakness may lead to pocketing of food between the cheeks and gums. Pharyngeal weakness often manifests as coughing and choking on food, liquids or secretions such as saliva or mucus. Aspiration may result. LMN weakness of the lower face may result in a poor lip seal that may contribute to drooling or sialorrhea. LMN weakness of the masseter can cause difficulty chewing. LMN weakness of the pterygoids may produce difficulty opening the mouth and moving the jaw from side to side. LMN weakness affecting the trunk and spine may result in difficulty in holding up the head and difficulty in maintaining an erect posture as well as abdominal protuberance. LMN weakness of the diaphragm produces progressive dyspnea with decreasing amounts of effort culminating in dyspnea at rest and with talking, along with reduced vocal volume. Diaphragmatic weakness may also result in orthopnea and sleep-disordered breathing.

Extraocular motor neurons are spared until very late in the disease course in patients who choose long-term mechanical ventilation.

There is a well-established link between ALS and frontotemporal executive dysfunction that may precede or follow the onset of UMN and/or LMN dysfunction.[13–16] Nevertheless, most patients with ALS do not have overt dementia, and when present, cognitive impairment is typically subtle.[17,18] In the largest report, the proportion of patients with ALS who met criteria for frontotemporal dementia (FTD) was 15%.[13] ALS with FTD may be a familial disorder. The presentation of FTD may be initially suggestive of a psychiatric disorder because of change in personality, impairment of judgment, and development of obsessional behaviors. It may also manifest as a disorder of language.

Autonomic symptoms may occur in ALS as the disease progresses. Constipation occurs frequently and is likely multifactorial. Symptoms of early satiety and bloating consistent with delayed gastric emptying also occur as the disease progresses.[19,20] Urinary urgency is common, while incontinence is uncommon.

Extrapyramidal symptoms and signs of parkinsonism may precede or follow the UMN and LMN symptoms. These extrapyramidal features may include facial masking, tremor, bradykinesia, and postural instability. At times, a supranuclear gaze abnormality occurs that is similar to that seen in progressive supranuclear palsy (PSP).[21]

Sensory symptoms may occur in 20%–30% of patients with ALS, but the sensory examination is usually normal.[22] It is not uncommon for patients with ALS, particularly those with distal limb onset of symptoms, to complain of tingling paresthesias. When queried regarding sensory loss, these patients typically will deny loss of sensation. At times, electrophysiologic studies may demonstrate reduction of amplitudes on sensory nerve conduction and or slowing of dorsal column conduction on somatosensory evoked potential testing, even in patients without sensory findings on examination.[23,24]

13.3 CLINICAL FEATURES OF PARKINSONIAN DISORDERS

13.3.1 Introduction

Parkinsonian disorders can be classified into four types: primary (idiopathic) parkinsonism, secondary (acquired, symptomatic) parkinsonism, heredodegenerative parkinsonism, and multiple system degeneration (parkinsonism plus syndromes). In Western countries, prevalence is between 200/100,000 and 300/100,000.[25] Prevalence rises with age and affects 0.4% of those over 40 years old and 1% of those over 65 years old. The mean age of onset is 57 years. It can occur in childhood or adolescence (juvenile parkinsonism). It is 1.5-times more common in men than women.[26]

PD is the second most common of the degenerative neurological disorders of the elderly. PD is a progressive neurological disorder characterized by a large number of motor and non-motor features that can impact on function to a variable degree.

13.3.2 Clinical Features

There are four cardinal features of PD that can be grouped under the acronym TRAP: *T*remor at rest, *R*igidity, *A*kinesia (or bradykinesia), and *P*ostural instability.[27] A number of rating scales are used for the evaluation of motor impairment and disability in patients with PD.[28,29] The Hoehn and Yahr scale is commonly used to compare groups of patients and to provide gross assessment of disease progression, ranging from stage 0 (no signs of disease) to stage 5 (wheelchair bound or bedridden unless assisted). The Unified

Parkinson's Disease Rating scale (UPDRS) is the most well-established scale for assessing disability and impairment.[30]

Tremor may be the first symptom noticed by the patient. When present it is a typical, low-frequency, unconscious "pill rolling" movement of the fingers, starting usually in one hand. Typically the tremor is maximal at rest, absent during sleep, and lessens with intentional movement such as reaching or grasping. It may be more noticeable with emotional stress or tiredness. Hands, arms and legs may be first affected by tremor and as the disease progresses, the jaw, tongue and eyelids become involved. In addition to rest tremor, many patients with PD also have postural tremor that is more prominent and disabling than rest tremor.[31]

In some patients, tremor is absent and only rigidity occurs. The effects of bradykinesia and rigidity are generally more disabling than tremor, impeding manual dexterity, balance and walking. As the disease progresses, movement becomes slower and more laborious (bradykinesia), with loss of spontaneous gestures.[32−34] The face develops a mask-like expressionless character. Posture becomes stooped and it becomes difficult to initiate walking, with short shuffling steps which may break into a short-stepped run to maintain balance, described as the typical "festinating gait". Impaired control of the small muscles of the hand often results in changes in the handwriting with decreasing size of the script (micrographia) and use of fewer words. The combination of these factors leads to significant impairment of activities of daily living such as dressing, bathing and using cutlery.

Patients with PD may exhibit a number of secondary motor symptoms that may impact on their functioning.[35] Because of a breakdown of the frontal lobe inhibitory mechanisms, some patients display a re-emergence of primitive reflexes: glabellar reflex, palmomental reflex, applause sign.[36] Bulbar dysfunction manifested by dysarthria, hypophonia, dysphagia, and sialorrhoea, are thought to be related to orofacial−laryngeal bradykinesia and rigidity. Speech disorders in patients with PD are characterized by monotonous, soft and breathy speech with variable rate and frequent word-finding difficulties.[37] A number of neuro-ophthalmological abnormalities may be seen including decreased blink rate, altered tear film, visual hallucinations, blepharospasm, and decreased convergence.[38]

Non-motor symptoms are a common and underappreciated features.[39] These include autonomic dysfunction, cognitive/neurobehavioral disorders, and sensory and sleep abnormalities. Autonomic failure includes orthostatic hypotension,[40] sweating,[41] sphincter, and erectile dysfunctions. A prospective study found that patients with PD are at almost sixfold increased risk for dementia.[42] PD-related dementia is also associated with a number of other neuropsychiatric comorbidities. Among 537 patients, depression (58%), apathy (54%), anxiety (49%), and hallucinations (44%) were frequently reported.[43]

Although excessive sleepiness and sleep attacks were once largely attributed to the pharmacological therapy, there are some arguments in favor of an integral part of the disease,[44,45] supported by the observation that rapid eye movement sleep behavior disorder, which occurs in approximately one-third of patients with PD, is a risk factor for the development of PD.[46] Sleep fragmentation is also frequent (>50% prevalence), but the occurrence is highly variable among patients. The sleep abnormalities observed in patients with PD may possibly be related to a 50% loss of hypocretin (orexin) neurons.[47]

Sensory symptoms such as olfactory dysfunction, pain, paresthesia, akathisia, oral pain, and genital pain are frequent but are often not recognized as parkinsonian symptoms.[48]

13.3.3 Diagnostic Criteria

PD is diagnosed on clinical criteria; there is no definitive test for diagnosis. Diagnosis is typically based on the presence of a combination of cardinal motor features, associated and exclusionary symptoms, and response to levodopa. Differentiating PD from other forms of parkinsonism can be challenging early in the course of the disease, when signs and symptoms overlap with other syndromes. Diagnostic criteria have been developed by the UK Parkinson's Disease Society Brain Bank and the National Institute of Neurological Disorders and Stroke (NINDS).[49−52] Several features, such as tremor, early gait abnormality, postural instability, pyramidal tract findings, and response to levodopa, can be used to differentiate PD from other parkinsonian disorders. Neuroimaging techniques, which are not available in low-income countries, may also be useful for differentiating PD from other parkinsonian disorders.[53]

13.4 EPIDEMIOLOGY OF ALS

13.4.1 Study of ALS in Tropical Areas, a Challenging Context

The geographical coverage of the worldwide ALS investigations is heterogeneous.[3,54,55] Most of the studies have been performed to date in temperate countries in Europe and in the USA. Conversely, few studies have been performed in the tropical areas of Africa,

Oceania, Asia and Latin America, where research focused mostly on infectious disorders and more recently on the burden of the frequent chronic disorders related to epidemiological transition.[56]

In tropical areas, the investigation of a rare neurodegenerative disorder is far from public health main concerns. Besides, the review of literature from these areas indicates that the accuracy of a large majority of the available material on ALS is questionable.[3,54] Studies are mostly hospital-based series (vs population-based), which have been described as the cause of a referral bias whose extent induces the unrepresentativeness of the included patients (that present with a younger age, with a lower proportion of bulbar onset and with a more favorable prognosis) as compared to the overall population.[1,57–59] Other issues are related in some settings to limitations to care and diagnostic facilities access. International investigations using standard methodologies are needed in tropical areas. This could lead to major discoveries in the field of ALS etiology, occurrence, and clinical presentation, prognosis, and care.

13.4.2 ALS Variability With Population's Ancestral Origin

Recent reviews support the hypothesis that ALS incidence and phenotype vary with ancestral origins.[3,54] Also, clues suggest that populations with genetic admixture display a lower occurrence of ALS.[60] Environmental factors and interaction between genes and environment could be implicated. Based on worldwide population-based data, Marin et al. reported homogeneous incidence rates in populations of European origins from Europe, Northern America, and New Zealand (pooled ALS (US standardized) incidence: 1.81 (1.66–1.97)/100,000 PYFU).[3] Differences were identified in ALS standardized incidence between Northern Europe (1.89 (1.46–2.32)/100,000 PYFU) and Eastern Asia (0.83 (0.42–1.24)/100,000 PYFU, China and Japan, $p = 0.001$) and Southern Asia (0.73 (0.58–0.89)/100,000 PYFU, Iran, $p = 0.02$). There were only two population-based studies from tropical areas (Caribbean island of Guadeloupe[61] and Hawaii[62]) included in this meta-analysis. Both reports illustrated well the heterogeneity of the disease in terms of incidence and phenotype.

Lannuzel et al.,[61] based on 63 incident ALS cases from the French overseas territory of Guadeloupe (1996–2011), displayed an overall incidence at 0.93/100,000 PYFU, but as high as 3.73/100,000 PYFU in Marie Galante, a small island of the archipelago. Around a quarter of patients had an association of ALS and atypical parkinsonism (ALS-Park). Others presented typical ALS (around 50%), bulbar form (20%), limb-onset variant (6%), or ALS with FTD (2%).

ALS-Park cases were mostly male (80% vs 23% for bulbar onset) and had a late onset (68 years vs 58 years for typical ALS, $p = 0.012$). ALS-Park patients also presented dementia (67%), falls (85%), supranuclear palsy (50%).

Based on 118 incident cases from Hawaii (1952–69), Matsumoto et al. reported a 1.50/100,000 PYFU standardized ALS incidence (1960 US population). There was heterogeneity between ethnic groups, as incidence among Filipinos was at 1.56 while incidence in Caucasians and Japanese were at 0.82 and 0.86, respectively (after standardization on the Hawaii population of 1960). Mortality data were in agreement (higher death rate in Filipinos). These data were connected by authors to the historical focus of ALS–parkinsonism–dementia complex (ALS-PDC) of the Oceanian Guam Island, as incidence of ALS among Filipino immigrants in Guam was found to be as high as for the local native Chamorros.[63] It was also hypothesized that Filipinos genetically contributed to the indigenous population from Guam.

Other arguments for ALS variability between ancestral origins are from Western multi-ethnic populations (mostly from the USA and the United Kingdom) showing that ALS incidence appears lower in the "African", "Hispanic", and "Asian" ethnic groups as compared to "Whites".[64–67] The same has been described using mortality data: lower mortality rates among "Blacks" or "non-Whites" as compared to "Whites".[68–70] The consistency of these elements would exclude under-ascertainment of cases as a possible explanation of the differences. Also, a lower access to health systems for some minorities with lower socioeconomic status needs to be discussed.[60,64,70,71]

13.4.3 ALS in Africa

Until the mid-1950s ALS was thought not to occur in Africans. The first description of two cases of ALS on this continent emerged from Kenya in 1955.[72] Since then, hospital-based studies of ALS in Africa, mostly retrospective, were published from Northern Africa (Tunisia,[73] Libya,[74] Morocco,[75,76] Sudan[77]), Eastern Africa (Kenya,[72,78] Ethiopia,[79] Zimbabwe[80,81]), Middle Africa (Cameroon,[82] Gabon[83]), Western Africa (Senegal,[84–87] Nigeria,[88] Ivory Coast[89]), Southern Africa (South Africa[90]) (Table 13.1).

Only one report focused on incidence of ALS in Africa (Libya). While crude incidence was low (0.89/100,000 PYFU), after standardization on the US population (2.03/100,000 PYFU), incidence was congruent with most Western data.[74]

The sample size of the cases reports from Africa ranged from two to around 300 patients. A few of

TABLE 13.1 Studies Reporting Phenotype of Amyotrophic Lateral Sclerosis in Africa

Subcontinent	Country	First author	Study period	Design	Number of ALS cases	Male:female sex ratio	Mean age ± SD (years)	Diagnostic delay (months)
Northern Africa	Morroco	Imounan	2008–12	A	60	1.5	50.0 ± 11.7	
	Morocco	Bougteba et al.	1986–2003	A	276†	2.0#	46.0#	
	Tunisia	Ben Hamida et al.	1974–80	A	102 (82 cl; 20 j)*	2.7; 1 j	53.3 cl; 21.3 j	55.3 cl; 14.0 j
	Libya	Radhakrishnan et al.	1980–85	A	23	2.3	50.9 ± 2.3	42.0
	Sudan	Abdulla et al.	1993–98	A	28	1.8	40.1	
Eastern Africa	Kenya	Harries et al.	1954	A	2	NC	26 and 30	
	Kenya	Adam et al.	1978–88	A	46	2.8	r: 13–80	
	Ethiopia	Tekle-Haimanot et al.	1986–88	B	3	2.0	28, 38, and 42	
	Zimbabwe	Wall et al.	1967–71	A	13	3.3	36.0	17.0
	Zimbabwe	Mielke et al.	ND	ND	13 (7b; 6w)	ND	37.0 b; 68.0 w	
Middle Africa	Cameroon	Kengne et al.	1993–2001	A	10	4.0	50.9 ± 3.3	
	Gabon	Le Bigot et al.	1980–85	A	2	ND	ND	
	Senegal	Jacquin-Cotton et al.	1960–69	A	18	8.0	r: 25–70	
Western Africa	Senegal	Collomb et al.	1960–68	A	18	17.0	47.7	24.1
	Senegal	Ndiaye et al.	1960–85	A	74 (64 cl; 10 j)†	4.0#	44.0#	
	Senegal	Sene-Diouf et al.	1993–2000	A	33	1.4	50.0**	r: 6.0–60.0
	Nigeria	Osuntokun et al.	1958–73	C	92	3.0	37.6 ± 1.5	
	Nigeria	Imam et al.	1980–99	A	16	15.0	38.6	11.5
	Ivory Coast	Piquemal et al.	1971–80	A	30 (15 cl; 15 j)‡	3.0 cl; 1.0 j	47.6 cl; 21.4 j	16.9 cl; 12.1 j
Southern Africa	South Africa	Cosnett et al.	<1989	A	86 (59 b; 16 w; 9 i; 2 c)	3.2 (1.6 b; 1.0 w; 8.0 i)	47.4 b; 54.0 w; 54.0 i	

Design: A, retrospective; B, door-to-door survey; C, prospective.
ALS, amyotrophic lateral sclerosis; b, black; c, colored; cl, classic ALS; i, Indian; j, juvenile ALS; NC, calculated (two males); ND, not determined; r, range; SD, standard deviation; w, white.
Juvenile form defined as amyotrophic lateral sclerosis/motor neuron disease onset before <30 years (*); <25 years (†); <40 years (‡); # absence of data for classic form and juvenile; ** median age.

these descriptions used the original or revised El Escorial criteria (EEC)[91,92] for ALS diagnosis. This is partly related to the fact that some reports were published before the EEC era or due to difficulties for electroneuromyography availability.

The main epidemiologic characteristics of ALS from Africa are the following: (1) a low mean age at time of onset of the classical ALS, in between 40 and 50 years; this is far lower than described in populations of European origin (65 years or more[54]) and is related to the difference in terms of life expectancy and of demographic structures of the populations (2%–4% of Africans aged 65 years and over versus 15%–20% in European and US populations);[93] (2) a median male:female sex ratio of 2.75; this contrasts with the usual data of around 1.3 from western series;[54] (3) a wide variation of diagnosis delay (between 1 and 4.5 years vs around 1 year or less in European series[54]); (4) absence or low percentage of familial ALS cases reported versus around 5% in European or US registers;[54,94] (5) reports of some cases of juvenile ALS (cut off used for juvenile definition 25, 30, or 40 years[73,75,86,89]). Age of juvenile ALS occurrence is around 20 years of age, with a sex ratio close to 1.

Clinical and prognosis data are quite rare: (1) symptomatic weakness of limb is a frequent initial complaint; (2) the proportion of bulbar onset (20%) appears lower than usually described in European populations (30%);[73,77,85,87,95] (3) some articles reported atypical signs (neurosensorial disturbance, sensory changes);[77,81] (4) prognosis or prognosis factors have not been described in a reliable way due to difficulties in performing the follow-up of patients;[96] (5) the standard of care (occidental and traditional care) has not been described; (6) for juvenile ALS cases, despite classic ALS, diagnosis delay appears lower. Onset is mostly spinal and the survival while poorly studied could be longer than classic ALS.

In order to provide more data about ALS phenotype in Africa, an initiative is currently underway: Tropals projets (http://www.tropals.unilim.fr/). This is a multicentre investigation of ALS in Africa, with the collection of demographic and clinical data using homogeneous methodology. Other initiatives, such as descriptive population-based data, etiological studies, and prognosis studies are needed because to date, they have been overlooked.

13.4.4 ALS in Oceania

Following the investigations run in the Pacific Marianna Island, especially in Guam (but also in Rota, Tinian, Saipan Island), one of the most famous and controversial etiological hypotheses of ALS emerged. In those areas an ALS-PDC (in local language "lytico"), which presents similarly to ALS, occurred at 50-times the incidence seen worldwide in the 1950s.[97] Another neurodegenerative disorder characterized by the association of parkinsonism and dementia ("bodig") was also identified.[98] Those neurodegenerative disorders run in families but with different clinical presentations. The hypothesis involves the exposure of Chamorro indigenous populations to L-BMAA,[99] a toxin supposed to be produced by cyanobacteria, through their dietary habits. In Guam, L-BMAA was present in cycas seeds (used by Chamoro for flour production) and in the meat of flying foxes (that ate cycas seeds and were eaten by Chamoro). L-BMAA was hypothesized to accumulate through the alimentary chain (biomagnification).[100] Also the toxin was found in the brain of Chamorro subjects who died from ALS.[99] In vitro and in vivo experiments illustrated the neuropathological properties of L-BMAA. Incidence of ALS reached a peak of, respectively, 73/100,000 and 41/100,000 PYFU in male and female in 1950–54, which progressively decreased, reaching 1/100,000 and 2/100,000, respectively, in 1995–99[97] along with the progressive change of local lifestyle related to the Westernization of the

Island. An ALS-PDC was also described among Auyu and Jakai people in a limited area of West New Guinea.[101,102] Incidence data from populations of European origins located in Oceania (New Zealand,[103] Australia[104]) are in the high range of incidence published in European countries.

13.4.5 ALS in Asia

As compared to western population-based incidence data, those from Asia suggest a lower occurrence of ALS in Hong Kong,[105,106] Taiwan,[107] Japan (Hokkaido Island)[108] with crude incidences of, respectively, 0.6, 0.3, 0.5, 0.7/100,000 PFYU. This could be related to a lower prevalence of familial ALS,[54] a lower frequency of ALS genetic variants (e.g., C9orf72 expansion[109]), difference in environment risk factors, impact of ancestral origins or admixture and to low life expectancies in low- and middle-income countries of this area.[110] Other data from Japan, based on medico-administrative data found an incidence of 2.3/100,000 PYFU.[111]

Sex-ratio of ALS patients appears higher in population-based studies from Asia as compared to Caucasians[54] (1.5 in Eastern Asia, 1.7 in Western Asia (Israel), 2.0 in Southern Asia (Iran)). In some studies, sex ratio was as high as 3.0 (India,[9] Japan[112]). Difficulty of case ascertainment in females has to be considered as potential explanation. Mean age at onset varies in between 46 years (India[9]) and 61 years (Japan[112]).

A population-based study reported survival times from onset in Japan of 28 months,[108] while some hospital-based studies in Japan, India, Malaysia, and China reported higher survival times of 48, 44.9, 71, and 114.8 months, respectively.[113–116] This longer survivorship pattern could be related to a selection bias linked to the design of the study. It can also be related to the clinical characteristics of ALS (less pejorative) and to an earlier age at onset. Genetic background could also directly influence the progression of ALS. Survival can also be highly influenced by nutritional and respiratory management. For example, tracheostomy is performed in about 30% of Japanese patients,[117] while in the US and Europe proportions range between 0% and 10%.[118]

Along with classical ALS, some specific clinical characteristics have been described among Asian patients: (1) juvenile muscular atrophy of distal upper extremity (India, Japan, China)[119] (also known as Hirayama disease or monomelic atrophy), and (2) Madras motor neuron disease (South-India).[120] It is not clear if these diseases are part of the ALS spectrum or are different diseases. Juvenile muscular atrophy of distal upper extremity is characterized by a slow onset between 15 and 20 years of age, mostly in male subjects, with a wasting and weakness confined to one limb (upper or lower limb). The progression of the disease is slow with a plateau of progression after 5 years.[121–123] Madras motor neuron disease was identified in South India. It occurs also in young patients that present with muscle wasting and weakness in the distal limb, facial and bulbar muscles.[120] Unlike Hirayama disease, there is a cranial nerve involvement, particularly hearing loss.

The ALS-PDC type described in Guam was also identified in the Kii peninsula of Honshu Island[112] (55/100,000 PYFU in 1948–67). Unlike the Guam focus of ALS, which totally disappeared, the incidence of ALS remains higher in Kozagawa and Koza areas in Wakayama prefecture (around 5/100,000 PYFU) than in other regions of Japan.[124] Researchers hypothesized that a higher proportion of familial ALS cases and genetic variants in this specific area could be an explanation.

13.4.6 ALS in Latin America

The reports on ALS epidemiology from South America are mainly clinical series and so far there is only one published population-based study from Uruguay.

Clinical series select younger cases as suggested in a series of 227 cases from Rio de Janeiro,[125] with a mean age at diagnosis of 54 years and survival time of 49 months. Most patients were white (71%), while black patients were 16% and mixed (one black parent and one white parent) patients were 13%. In a similar clinical series from the Ramos Mejía Hospital in Buenos Aires based on 187 patients, many subjects originate from rural areas (16%) and occupied as handy workers.[126] Another clinical series from South Brazil with 251 cases, show a strong selection in favor of spinal (88%) compared to bulbar (9.6%), with young age of onset at 54 years. Interestingly the interval first symptom-diagnosis is very long (18 months). In a series from Fortaleza (Brasil) of 87 cases, the age of onset was particularly young (42 years) with 19% with a typical onset with only two familial cases.[127]

Occasional reports describe isolated cases in native populations like Kaiapo-Xycrin, an isolated tribe living in Eastern Amazonia along a river. The authors suppose that this isolated group of cases may have some connection with eating polluted fish from a river or eating wild animals.[128]

The first national ALS survey was conducted in 1998 in Brazil involving in case identification through about 2500 neurologists. The idea was to identify cases through the neurologists taking care of them. In the survey 443 ALS cases were identified (86% definite ALS according to EEC). 70% were spinal and 6%

familiar. A large percentage of cases (60%) were identified through neurologists in private practice, using data from a health maintenance organization from a network of Italian hospitals.

Two population studies were conducted in metropolitan areas. The first study was conducted in Buenos Aires where the crude and the adjusted (adjusted to the general Buenos Aires population) incidence rates were 3.2/100,000 and 2.2/100,000, respectively, with mean age at diagnosis of 72 years.[129] The point prevalence was 8.9/100,000. The second study was conducted in the city of Porto Alegre, Brazil where the prevalence was about 5/100,000, with a slight predominance in males, peaking at age 70–79 years. This was estimated in the year 2010 from 70 ALS patients alive on the prevalence day.[130] These two studies show rates quite close to the rates found in epidemiological studies in Europe with the incidence peak after age 70 years.[131]

The only national study on ALS incidence has been conducted in Uruguay over 2 years in 2002–03, in a population of 3.2 million people.[132] Based on the inception of 86 cases the incidence was double in men (1.94) compared to women (0.84). The incidence calculated with capture recapture methodology was 1.4 and age of onset was relatively young (59 years). Socioeconomic status drove in this study an earlier referral and diagnosis. It confirmed that, in a population of European descent, with a valid national health system and with the presence of diffuse neurological care, the rates are similar to Europe.[132]

Few studies reported prognosis of ALS in a population-based setting in South America. In a study comparing survivorship in Europe and Latin America, prognosis was better in the Limousin region (France) compared to Uruguay (28 vs 19 months). This was not explained by differences in the diagnostic delay (10 months in Uruguay and 9 months in Limousin) but might be related to the slow referral to a multidisciplinary referral center in Uruguay and probably to the extreme difference in percentage of cases with diagnostic certainty at diagnosis as based on EEC (in Uruguay 80% are definite vs 10% in Limousin).[133] The presence of numerous definite cases strengthen the idea that when the study was conducted, the cases arrived in an advanced state of progression to the attention of the neurologists who finally made the diagnosis.

Mortality studies are particularly important in epidemiology of ALS when properly conducted descriptive studies are not possible or rare.[134]

In a mortality study in Sao Paolo, based on 326 deaths, the mortality rate was low, ranging from 0.4 in 2002 to 0.8 in 2006, with the peak at age 60–79 years, mainly whites, and with a similar rate in men and women.[135] The mortality rate in Chile is 1.1/100,000, the highest rate in the Austral area, where a population of European descent is present[136] and in subjects over age 64 years. The only study with the goal of describing ALS mortality in an ethnically mixed population was conducted in Cuba.[60] The study was conducted in a national central Statistics Office over a 6-year period. The authors identified 432 ALS cases with a mean age at death of 64 years with a total mortality rate of 0.83/100,000. The mortality rate was about half in the mixed population (0.55) compared to white (0.9) and blacks (0.9). Data from Cuba are unlikely to be confounded by access to healthcare or low disease ascertainment, as the Cuban health system is well developed, freely accessible, and keeps excellent records. This study, together with several previous studies in admixed populations, gave rise to the hypothesis that admixed populations are characterized by a lower risk of developing ALS. This hypothesis is currently being tested in the first multinational population based study conducted in Chile, Cuba, and Uruguay named LAENALS.

There has only been one study on ALS risk factors conducted in Latin America. In this case–control study conducted in Brazil with 367 cases and more than 4000 controls,[137] there was no change in risk due to magnetic field.

13.5 EPIDEMIOLOGY OF PD

Even based on established diagnostic criteria, the diagnosis of PD requires special attention and expertise. The distinction between PD and other forms of parkinsonism, atypical or secondary, is not easy to establish, and proves erroneous in about 10% of cases even with experts in the field of movement disorders.[138,139] Consequently measuring the prevalence and incidence of PD needs an appropriate health system, which can be lacking in some tropical countries. This part will focus on PD frequency in tropical countries (incidence or prevalence) and on the influence of environmental exposures.

13.5.1 Prevalence of Parkinsonism in Tropical Countries

The prevalence of PD rises with age.[2] Overall, pooled worldwide prevalence of PD is estimated at 315/100,000 persons and incidence at 36.5/100,000 PYFU in females and 65.5/100,000 PYFU in males.[2] Epidemiology of PD is not homogeneous. The incidence rates vary between countries from 1.5/100,000 to 20/100,000 per year.[140] Prevalence rates in Africa

(77/100,000) and in Asia (337/100,000) are lower as compared to 1046/100,000 in South America and 1398/100,000 in North America/Europe/Australia.[2] PD is 1.5-times more frequent in men than in women.[141] Male-to-female ratios increase with age.[26] Epidemiology of PD in tropical countries has been partially studied. Data are available in India, Thailand, some sub-Saharian African, and South-central American countries and in Cuba.

The prevalence of PD in sub-Saharan Africa is lower than in US and Europe. The crude prevalence of PD varies from 7/100,000 to 40/100,000.[142–144] Door-to-door studies found comparable prevalence rates in Eastern and Western Africa. In Eastern Africa, a door-to-door study in rural Tanzania estimated the crude prevalence of PD at 20/100,000. It was higher in men (30/100,000) than in women (11/100,000). The direct age-standardized prevalence (using the UK population as a reference) was 64/100,000 (men), 20/100,000 (women), and 40/100,000 (both genders combined).[145] In Nigeria, the crude prevalence of PD estimated using a door-to-door study was 10/100,000.[146] These data appear fivefold lower than prevalence in Blacks from Copiah County, Mississippi, whose USA age-adjusted prevalence was at 67/100,000 (as compared to 341/100,000 overall in the USA).[147]

From India, Bharucha et al. reported the prevalence of PD in the Parsi community of Bombay at 328.3/100,000 in the 1980s[148] with the use of specific criteria and with a door-to-door approach. Another work from India estimated the prevalence at 52.85/100,000 through a door-to-door study (2003–07) and the average annual incidence rate was 5.71/100,000 PYFU.[140] The prevalence of the non-slum population showed a significant increasing rate with advancing age above 62.5 years compared with the slum population. This paper also reported that not smoking tobacco was a significant factor decreasing risk (adjusted odds ratio (OR) 0.16, 95% confidence interval (CI) 0.03–0.75). Indian researchers also postulated that patients with early-onset PD might be more prone to complex depression and dementia.[149]

The average annual incidence rate was lower than in North America (20/100,000 PYFY), many European countries (19/100,000 PYFY), Japan (15/100,000 PYFY), Australia (7/100,000 PYFY), and Taiwan (10/100,000 PYFY), but similar to Italia (10/100,000 PYFY) and Libya (4.5/100,000 PYFY) and higher than in China (1.5/100,000 PYFY). Age-specific average annual incidence rates were also lower in comparison with similar studies in the above-defined populations. There was no significant difference in sex-specific incidence. In Thailand, a nationwide PD Registry from 2008 to 2011 allowed the crude estimation of age-adjusted prevalence of 95.3/100,000 and 424.6/100,000, respectively.[150]

In the Caribbean a door-to-door population study estimated the prevalence of PD at 135/100,000 in an urban area of Havana city in Cuba.[151] In Brazil, in a cohort of individuals aged 64 years or over, the prevalence was similar to that observed in elderly people from door-to-door surveys in other American, European, and Eastern countries.[152]

Most epidemiological data show that the prevalence of PD is lower in most tropical countries in which epidemiological studies have been conducted (India, Nigeria, Tanzania, Thailand, Cuba). In contrast, in Brazil in a population of people age of 64 years and more, the prevalence was similar to Europe and the US.

Prevalence data lack the ability to contrast the influence of incidence and of survival. Furthermore, because PD occurs in the elderly, a difference in life expectancy time can be implicated in the variation of prevalence data. Age-standardized prevalence rates from India, Tanzania, and Nigeria confirmed a lower prevalence of PD in these areas as compared to the US or Europe. Incident data from India found an average annual incidence 20-times lower than in North America and many European countries. These are strong arguments for a lower frequency of the disease in these countries.

The major problems remain the low frequency of PD and the difficulty in establishing diagnosis. Limited access to specialized care or inadequate knowledge of the disease by the population[153] may also influence these patterns. Reliable studies using gold standard methodology (door-to-door epidemiological studies)[145,146,154] and registries[150] are key to better comprehending PD prevalence and incidence heterogeneity.

13.5.2 Hypothesis for the Lower Frequency of PD in Tropical Countries

13.5.2.1 Racial or Ethnic Differences?

Epidemiological studies have revealed that older age and male gender are associated with an increased risk of developing PD. Ethnicity may also influence the frequency of the disease. Several studies have shown a difference in disease incidence between different ethnic groups in the USA with lower age- and gender-standardized prevalence among Blacks (10.2, 95% CI: 6.4, 14.0) followed by Asians (11.3, 95% CI: 7.2, 15.3), Hispanics (16.6, 95% CI: 12.0, 21.3), and non-Hispanic Whites (13.6, 95% CI: 11.5, 15.7).[155] The prevalence of PD medication use, across the national REGARDS cohort of 30,000 persons over age 45 was lower among African-Americans (0.51%) than among Whites (0.97%; OR: 1.90; 95% CI: 1.31–2.74).[156] There was no

association with income, education level, or geographic region of residence.[156] Racial difference in the susceptibility to develop PD may play a role in the low frequency of PD in tropical areas. Additional factors such as environmental factors must be involved as the age-adjusted prevalence rate in Nigeria remains fivefold lower than the prevalence among Blacks in the USA.[147] Besides, Ragothaman et al. identified a lower occurrence of PD in an admixed population of European and Indian origin as compared to Indians and postulated that genetic admixture could lower the risk of PD.[157]

13.5.3 Higher Proportion of Atypical Parkinsonism?

PD belongs to a group of neurodegenerative disorders that includes atypical parkinsonism, PD being considered as a "typical" parkinsonian disorder. Within the spectrum of diseases that constitute the syndrome of atypical parkinsonism, akinesia and rigidity become rapidly severe and characteristically involve the axis of the body. Shortly after the onset, other signs become prominent, leaving parkinsonism in the background: supranuclear oculomotor palsy, early falls and cognitive frontal dysfunction will suggest PSP; dysautonomia and cerebellar signs, multiple system atrophy (MSA); unilateral apraxia, corticobasal degeneration (CBD); dementia of the cortico-subcortical type with early hallucinations, Lewy body disease (LBD). Pathologically, PD, dementia with Lewy bodies (DLB), and MSA are characterized by abnormal accumulation of alphasynuclein, a protein that aggregates to various structures to form Lewy bodies; while in PSP and CBD another protein, tau protein accumulates in brain tissue in neurofibrillary tangles. In Europe and North America about 70% of parkinsonian patients have PD, 30% suffer from one form of atypical parkinsonism.[158,159]

In Guadeloupe, a French Caribbean island with 420,000 inhabitants, there is an abnormally high frequency of atypical parkinsonism. Only one-third of the 360 patients included in three successive case–control studies from 1999 to 2005, mostly of Afro-Caribbean origin, had idiopathic PD.[160,161] Atypical parkinsonism in Guadeloupe is characterized by a unique combination of levodopa-resistant parkinsonism, fronto-subcortical dementia (90%–100%), hallucinations (59%), dysautonomia (50%) and myoclonus. Half of patients have supranuclear gaze abnormalities as observed in PSP but the majority of them did do not match classical PSP due to the presence of signs considered to be signs of exclusion criteria (myoclonus, hallucination, and dysautonomia). The two subgroups of atypical parkinsonism taken together (Gd-PSP and Gd-PDC for Guadeloupean PDC depends on the presence of oculomotor signs), represented two-thirds of all cases of parkinsonism, an exceptionally high percentage. Neuropathogical and biochemical features of Gd-PSP were compatible with the diagnosis of PSP,[160,162] although some unusual pathological features were noted, particularly a widespread distribution of lesions out of midbrain and subcortical regions and a marked astrocytic involvement. The focus of Guadeloupe clinically resembles the PDC described on Guam, where a high prevalence of atypical parkinsonism has been reported since the Second World War, including one-third of PSP.[163] Neuropathological characteristics in particular the biochemical characteristics of the tau protein, differ in Guam PDC and in Gd-PSP. The tau pathology of Guam PDC shares similarities with Alzheimer's disease but not with PSP. Atypical parkinsonism in Guadeloupe could be a variant of PSP[164] whose geographical distribution remains to be defined.

Some clinical reports indicate that among parkinsonism, PD could be underrepresented in other tropical areas. In sub-Saharan Africa a detailed report published in 1979 identified PD for only 38% (83/217 patients) of Nigerians with parkinsonism.[165] In the same time period, Cosnett reported that Parkinsonism occurred with normal frequency, but "true idiopathic" PD was rare in South Africa.[166] Thirty years later, a clinical study did not confirm an over-representation of atypical parkinsonism in Nigerians. The proportion of PD was of 83% and 79% of PD in two distinct hospital cohorts.[167,168] Patient characteristics changed between 1979 and 2010: the age at onset increased from 55.6 to 61.5 years, the clear male preponderance persisted but to a lesser extent (3.3 vs 4.5).[168] As both studies included patients from the same geographical region, the changes in patient demographic and clinical characteristics could reflect the influence of environmental factors.

13.5.4 Environmental Exposures?

Environmental exposures have long been considered as a possible cause of, or risk factor for, PD. The observations in the 1980s, by Langston and Tetrud at the Santa Clara Valley Medical Center, of young individuals who developed an acute parkinsonian syndrome, led to the identification of the neurotoxin, MPTP.[169,170] Further investigation revealed that MPTP is metabolized to MPP$^+$, which is then transported by the dopamine transporter into neurons and, ultimately, poisons mitochondrial function.[171] Individuals with inadvertent exposures to MPTP developed acute loss of dopaminergic neurons and developed parkinsonian motor signs and symptoms. Although this classic example demonstrates that environmental exposures can lead to parkinsonian syndromes, no one thinks that MPTP exposure causes PD. On the other hand, this finding raises the possibility that environmental exposures could be associated with the enhancement of neurodegeneration that either alone, or in combination with genetic predispositions, creates the syndrome known as PD.

Since then many epidemiological studies found higher risk of developing PD in individuals exposed to pesticides.[172] For example, in India, Sanyal et al. identified, using a case—control study, that pesticide exposure and rural living were independently associated with PD, respectively, OR: 17.1 (95% CI: 5.0—58.8) and 4.0 (95% CI: 2.5—6.5).[173] Exposure to lead, manganese and welding as well as rural living are also considered risk factors for PD while others factors such as caffeine consumption and smoking may be protective.[174] There is little epidemiological data specific to the tropics on the subject. In Thailand the prevalence of PD was higher (126.83/100,000) in urban areas than in rural areas (90.82/100,000) ($p < 0.001$). Preliminary regional comparisons revealed a higher prevalence of PD in residents of the central plain valley of Thailand, an area with a large amount of pesticide use.[150]

Some experimental results suggest that dietary habits could have an influence on the epidemiology of PD, its occurrence, and the ratio between PD and atypical parkinsonism. Various phenolic compounds, such as curcumin from turmeric, have shown antioxidative and anti-inflammatory effects. Curcumin prevents aggregation of alpha alpha-synuclein in animal models of PD and possibly prevents cell death.[175,176] It is possible that in Indians the consumption of curcumin daily in their diet from early childhood confers a protection against PD.[140] In the French West Indian island of Guadeloupe, clinical and experimental data support the hypothesis that the exposure to an alimentary toxin may be implicated in the over-representation of atypical parkinsonism compared to PD. Two case—control studies[161,177] showed that patients with atypical parkinsonism consumed significantly more fruit and infusions or decoctions of leaves from plants of the Annonaceae family, particularly *Annona muricata* L. (soursop, guanabana, graviola, or corossol) than patients with PD or control subjects. The leaves of these plants are used in traditional Creole medicine from early childhood to old age, sometimes daily, for cardiac and digestive problems, for sedative purposes, or to maintain general health. Biochemical and experimental studies in cell culture and animal models have identified the putative toxin as the mitochondrial complex I inhibitor annonacin, which is the major acetogenin, a family of lipid-derived molecules specific to Annonaceae, in *A. muricata*.[178] High concentrations of annonacin are present in the fruit or aqueous extracts of the leaves of *A. muricata*, and it can cross the blood–brain barrier since it was detected in brain parenchyma of rats treated chronically with the molecule, and induced neurodegeneration of basal ganglia in these animals, similar to that observed in atypical parkinsonism.[179,180] Interestingly a similar clinical entity has also been associated with Annonaceae consumption in patients of Caribbean origin living in London[181] and in PD patients from New Caledonia, a French Western Pacific island.[182]

Finally the changes in clinical profile of Nigerian parkinsonians within 30 years, if it is confirmed, may result from modification of environmental factors. Analytic epidemiological studies focused on environmental factors and changes within the 30 past years would also be helpful to explore this hypothesis.

13.6 CONCLUSION

The general impression is that ALS and PD occur less commonly in most tropical countries than in Europe or North America. It remains unclear whether this is related to systematic errors (including referral bias related to a hospital-based inclusion of cases), variation in methods, different age structure and life expectancy, impact of environment, influence of genetic background (ancestry, variants) or interaction between genes and environment.

Researchers have to face these uncertainties in order to unravel a clear picture of ALS and PD. Wide population-based studies investigating incident cases of neurodegenerative disorders in various continents and countries are key in this context. Uniform methods among centers would be important for comparability. Studies should use parkinsonism-specific screening instruments, clearly defined and up to date diagnostic criteria for ALS, parkinsonism, and PD. The phenotypes of ALS and PD (including atypical forms) could be studied as well. Such an approach would require international consortia and agencies.

References

1. Logroscino G, Tortelli R, Rizzo G, Marin B, Preux PM, Malaspina A. Amyotrophic lateral sclerosis: an aging-related disease. *Curr Geriatr Rep*. 2015;4:142−153.
2. Pringsheim T, Fiest K, Jette N. The international incidence and prevalence of neurologic conditions: how common are they? *Neurology*. 2014;83(18):1661−1664.
3. Marin B, Boumediene F, Logroscino G, Couratier P, Babron MC, Leutenegger AL, et al. Variation in worldwide incidence of amyotrophic lateral sclerosis: a meta-analysis. *Int J Epidemiol*. 2016;:pii: dyw061.
4. Dorsey ER, Constantinescu R, Thompson JP, Biglan KM, Holloway RG, Kieburtz K, et al. Projected number of people with Parkinson disease in the most populous nations, 2005 through 2030. *Neurology*. 2007;68(5):384−386.
5. Mitsumoto H, Chad AD, Pioro EP. *Amyotrophic Lateral Sclerosis*. Philadelphia, PA: Davis Company; 1998:480 p.
6. Rowland LP. Diagnosis of amyotrophic lateral sclerosis. *J Neurol Sci*. 1998;160(Suppl 1):S6−S24.
7. Kim WK, Liu X, Sandner J, Pasmantier M, Andrews J, Rowland LP, et al. Study of 962 patients indicates progressive muscular atrophy is a form of ALS. *Neurology*. 2009;73(20):1686−1692.
8. Tartaglia MC, Rowe A, Findlater K, Orange JB, Grace G, Strong MJ. Differentiation between primary lateral sclerosis and amyotrophic lateral sclerosis: examination of symptoms and signs at disease onset and during follow-up. *Arch Neurol*. 2007;64(2):232−236.
9. Nalini A, Thennarasu K, Yamini BK, Shivashankar D, Krishna N. Madras motor neuron disease (MMND): clinical description and survival pattern of 116 patients from Southern India seen over 36 years (1971−2007). *J Neurol Sci*. 2008;269(1−2):65−73.
10. Wilbourn AJ. The "split hand syndrome". *Muscle Nerve*. 2000; 23(1):138.
11. Shoesmith CL, Findlater K, Rowe A, Strong MJ. Prognosis of amyotrophic lateral sclerosis with respiratory onset. *J Neurol Neurosurg Psychiatry*. 2007;78(6):629−631.
12. Caress JB, Ciarlone SL, Sullivan EA, Griffin LP, Cartwright MS. Natural history of muscle cramps in amyotrophic lateral sclerosis. *Muscle Nerve*. 2016;53(4):513−517.
13. Ringholz GM, Appel SH, Bradshaw M, Cooke NA, Mosnik DM, Schulz PE. Prevalence and patterns of cognitive impairment in sporadic ALS. *Neurology*. 2005;65(4):586−590.
14. Murphy JM, Henry RG, Langmore S, Kramer JH, Miller BL, Lomen-Hoerth C. Continuum of frontal lobe impairment in amyotrophic lateral sclerosis. *Arch Neurol*. 2007;64(4):530−534.
15. Beeldman E, Raaphorst J, Klein Twennaar M, de Visser M, Schmand BA, de Haan RJ. The cognitive profile of ALS: a systematic review and meta-analysis update. *J Neurol Neurosurg Psychiatry*. 2016;87(6):611−619.
16. Woolley SC, Strong MJ. Frontotemporal dysfunction and dementia in amyotrophic lateral sclerosis. *Neurol Clin*. 2015;33(4):787−805.
17. Lomen-Hoerth C, Murphy J, Langmore S, Kramer JH, Olney RK, Miller B. Are amyotrophic lateral sclerosis patients cognitively normal? *Neurology*. 2003;60(7):1094−1097.
18. Massman PJ, Sims J, Cooke N, Haverkamp LJ, Appel V, Appel SH. Prevalence and correlates of neuropsychological deficits in amyotrophic lateral sclerosis. *J Neurol Neurosurg Psychiatry*. 1996;61(5):450−455.
19. Toepfer M, Folwaczny C, Klauser A, Riepl RL, Muller-Felber W, Pongratz D. Gastrointestinal dysfunction in amyotrophic lateral sclerosis. *Amyotroph Lateral Scler Other Motor Neuron Disord*. 1999;1(1):15−19.
20. Toepfer M, Folwaczny C, Lochmuller H, Schroeder M, Riepl RL, Pongratz D, et al. Noninvasive (13)C-octanoic acid breath test shows delayed gastric emptying in patients with amyotrophic lateral sclerosis. *Digestion*. 1999;60(6):567−571.
21. Geser F, Martinez-Lage M, Robinson J, Uryu K, Neumann M, Brandmeir NJ, et al. Clinical and pathological continuum of multisystem TDP-43 proteinopathies. *Arch Neurol*. 2009;66(2):180−189.
22. Hammad M, Silva A, Glass J, Sladky JT, Benatar M. Clinical, electrophysiologic, and pathologic evidence for sensory abnormalities in ALS. *Neurology*. 2007;69(24):2236−2242.
23. Pugdahl K, Fuglsang-Frederiksen A, de Carvalho M, Johnsen B, Fawcett PR, Labarre-Vila A, et al. Generalised sensory system abnormalities in amyotrophic lateral sclerosis: a European multicentre study. *J Neurol Neurosurg Psychiatry*. 2007;78(7):746−749.

24. Amoiridis G, Tsimoulis D, Ameridou I. Clinical, electrophysiologic, and pathologic evidence for sensory abnormalities in ALS. *Neurology.* 2008;71(10):779.

25. Fletcher N. Movement disorders. In: Donaghy M, ed. *Brain's Diseases of the Nervous System.* 11th ed. Oxford: Oxford University Press; 2001:1015–1096.

26. Moisan F, Kab S, Mohamed F, Canonico M, Le Guern M, Quintin C, et al. Parkinson disease male-to-female ratios increase with age: French nationwide study and meta-analysis. *J Neurol Neurosurg Psychiatry.* 2016;87(9):952–957.

27. Jankovic J. Pathophysiology and assessment of parkinsonian symptoms and signs. In: Pahwa R, Lyons K, Koller WC, eds. *Handbook of Parkinson's Disease.* New York: Taylor and Francis Group; 2007:79–104.

28. Ramaker C, Marinus J, Stiggelbout AM, Van Hilten BJ. Systematic evaluation of rating scales for impairment and disability in Parkinson's disease. *Mov Disord.* 2002;17(5):867–876.

29. Ebersbach G, Baas H, Csoti I, Mungersdorf M, Deuschl G. Scales in Parkinson's disease. *J Neurol.* 2006;253(Suppl 4): IV32–IV35.

30. Goetz CG, Fahn S, Martinez-Martin P, Poewe W, Sampaio C, Stebbins GT, et al. Movement Disorder Society-sponsored revision of the Unified Parkinson's Disease Rating Scale (MDS-UPDRS): process, format, and clinimetric testing plan. *Mov Disord.* 2007;22(1):41–47.

31. Jankovic J, Schwartz KS, Ondo W. Re-emergent tremor of Parkinson's disease. *J Neurol Neurosurg Psychiatry.* 1999;67 (5):646–650.

32. Broussolle E, Krack P, Thobois S, Xie-Brustolin J, Pollak P, Goetz CG. Contribution of Jules Froment to the study of parkinsonian rigidity. *Mov Disord.* 2007;22(7):909–914.

33. Riley D, Lang AE, Blair RD, Birnbaum A, Reid B. Frozen shoulder and other shoulder disturbances in Parkinson's disease. *J Neurol Neurosurg Psychiatry.* 1989;52(1):63–66.

34. de Lau LM, Koudstaal PJ, Hofman A, Breteler MM. Subjective complaints precede Parkinson disease: the Rotterdam study. *Arch Neurol.* 2006;63(3):362–365.

35. Singh R, Pentland B, Hunter J, Provan F. Parkinson's disease and driving ability. *J Neurol Neurosurg Psychiatry.* 2007;78 (4):363–366.

36. Brodsky H, Dat Vuong K, Thomas M, Jankovic J. Glabellar and palmomental reflexes in Parkinsonian disorders. *Neurology.* 2004;63(6):1096–1098.

37. Critchley EM. Speech disorders of Parkinsonism: a review. *J Neurol Neurosurg Psychiatry.* 1981;44(9):751–758.

38. Biousse V, Skibell BC, Watts RL, Loupe DN, Drews-Botsch C, Newman NJ. Ophthalmologic features of Parkinson's disease. *Neurology.* 2004;62(2):177–180.

39. Zesiewicz TA, Sullivan KL, Hauser RA. Nonmotor symptoms of Parkinson's disease. *Expert Rev Neurother.* 2006;6(12): 1811–1822.

40. Senard JM, Rai S, Lapeyre-Mestre M, Brefel C, Rascol O, Rascol A, et al. Prevalence of orthostatic hypotension in Parkinson's disease. *J Neurol Neurosurg Psychiatry.* 1997;63(5):584–589.

41. Pursiainen V, Haapaniemi TH, Korpelainen JT, Sotaniemi KA, Myllyla VV. Sweating in Parkinsonian patients with wearing-off. *Mov Disord.* 2007;22(6):828–832.

42. Aarsland D, Andersen K, Larsen JP, Lolk A, Nielsen H, Kragh-Sorensen P. Risk of dementia in Parkinson's disease: a community-based, prospective study. *Neurology.* 2001;56(6):730–736.

43. Aarsland D, Bronnick K, Ehrt U, De Deyn PP, Tekin S, Emre M, et al. Neuropsychiatric symptoms in patients with Parkinson's disease and dementia: frequency, profile and associated care giver stress. *J Neurol Neurosurg Psychiatry.* 2007;78(1):36–42.

44. Ondo WG, Dat Vuong K, Khan H, Atassi F, Kwak C, Jankovic J. Daytime sleepiness and other sleep disorders in Parkinson's disease. *Neurology.* 2001;57(8):1392–1396.

45. Gjerstad MD, Alves G, Wentzel-Larsen T, Aarsland D, Larsen JP. Excessive daytime sleepiness in Parkinson disease: is it the drugs or the disease? *Neurology.* 2006;67(5):853–858.

46. Gagnon JF, Postuma RB, Mazza S, Doyon J, Montplaisir J. Rapid-eye-movement sleep behaviour disorder and neurodegenerative diseases. *Lancet Neurol.* 2006;5(5):424–432.

47. Fronczek R, Overeem S, Lee SY, Hegeman IM, van Pelt J, van Duinen SG, et al. Hypocretin (orexin) loss in Parkinson's disease. *Brain.* 2007;130(Pt 6):1577–1585.

48. Ford B, Louis ED, Greene P, Fahn S. Oral and genital pain syndromes in Parkinson's disease. *Mov Disord.* 1996;11(4):421–426.

49. Tolosa E, Wenning G, Poewe W. The diagnosis of Parkinson's disease. *Lancet Neurol.* 2006;5(1):75–86.

50. Gelb DJ, Oliver E, Gilman S. Diagnostic criteria for Parkinson disease. *Arch Neurol.* 1999;56(1):33–39.

51. de Rijk MC, Rocca WA, Anderson DW, Melcon MO, Breteler MM, Maraganore DM. A population perspective on diagnostic criteria for Parkinson's disease. *Neurology.* 1997;48(5):1277–1281.

52. Hughes AJ, Ben-Shlomo Y, Daniel SE, Lees AJ. What features improve the accuracy of clinical diagnosis in Parkinson's disease: a clinicopathologic study. *Neurology.* 1992;42(6):1142–1146.

53. Piccini P, Brooks DJ. New developments of brain imaging for Parkinson's disease and related disorders. *Mov Disord.* 2006;21 (12):2035–2041.

54. Marin B, Logroscino G, Boumediene F, Labrunie A, Couratier P, Babron MC, et al. Clinical and demographic factors and outcome of amyotrophic lateral sclerosis in relation to population ancestral origin. *Eur J Epidemiol.* 2015;31(3):229–245.

55. Chio A, Logroscino G, Traynor BJ, Collins J, Simeone JC, Goldstein LA, et al. Global epidemiology of amyotrophic lateral sclerosis: a systematic review of the published literature. *Neuroepidemiology.* 2013;41(2):118–130.

56. Omran AR. The epidemiologic transition: a theory of the epidemiology of population change. *Milbank Q.* 2005;83(4):731–757.

57. Lee JR, Annegers JF, Appel SH. Prognosis of amyotrophic lateral sclerosis and the effect of referral selection. *J Neurol Sci.* 1995;132 (2):207–215.

58. Sorenson EJ, Mandrekar J, Crum B, Stevens JC. Effect of referral bias on assessing survival in ALS. *Neurology.* 2007;68(8):600–602.

59. Logroscino G, Traynor BJ, Hardiman O, Chio A, Couratier P, Mitchell JD, et al. Descriptive epidemiology of amyotrophic lateral sclerosis: new evidence and unsolved issues. *J Neurol Neurosurg Psychiatry.* 2008;79(1):6–11.

60. Zaldivar T, Gutierrez J, Lara G, Carbonara M, Logroscino G, Hardiman O. Reduced frequency of ALS in an ethnically mixed population: a population-based mortality study. *Neurology.* 2009;72(19):1640–1645.

61. Lannuzel A, Mecharles S, Tressieres B, Demoly A, Alhendi R, Hedreville-Tablon MA, et al. Clinical varieties and epidemiological aspects of amyotrophic lateral sclerosis in the Caribbean island of Guadeloupe: a new focus of ALS associated with Parkinsonism. *Amyotroph Lateral Scler Frontotemporal Degener.* 2015;16(3–4):216–223.

62. Matsumoto N, Worth RM, Kurland LT, Okazaki H. Epidemiologic study of amyotrophic lateral sclerosis in Hawaii. Identification of high incidence among Filipino men. *Neurology.* 1972;22(9):934–940.

63. Garruto RM, Gajdusek DC, Chen KM. Amyotrophic lateral sclerosis and parkinsonism-dementia among Filipino migrants to Guam. *Ann Neurol.* 1981;10(4):341–350.

64. Rechtman L, Jordan H, Wagner L, Horton DK, Kaye W. Racial and ethnic differences among amyotrophic lateral sclerosis cases in the United States. *Amyotroph Lateral Scler Frontotemporal Degener.* 2015;16(1–2):65–71.

65. Annegers JF, Appel S, Lee JRJ, Perkins P. Incidence and prevalence of amyotrophic lateral sclerosis in Harris County, Texas, 1985–1988. *Arch Neurol.* 1991;48(6):589–593.

66. McGuire V, Longstreth Jr. WT, Koepsell TD, van Belle G. Incidence of amyotrophic lateral sclerosis in three counties in western Washington state. *Neurology*. 1996;47(2):571–573.

67. Kahana E, Zilber N. Changes in the incidence of amyotrophic lateral sclerosis in Israel. *Arch Neurol*. 1984;41(2):157–160.

68. Cronin S, Hardiman O, Traynor BJ. Ethnic variation in the incidence of ALS: a systematic review. *Neurology*. 2007;68(13):1002–1007.

69. McCluskey KLM. Racial disparity in mortality from ALS/MND in US African Americans. *Amyotroph Lateral Scler Other Motor Neuron Disord*. 2004;5:73–78.

70. Leone M, Chandra V, Schoenberg BS. Motor neuron disease in the United States, 1971 and 1973–1978: patterns of mortality and associated conditions at the time of death. *Neurology*. 1987;37(8):1339–1343.

71. Lasser KE, Himmelstein DU, Woolhandler S. Access to care, health status, and health disparities in the United States and Canada: results of a cross-national population-based survey. *Am J Public Health*. 2006;96(7):1300–1307.

72. Harries JR. Amyotrophic lateral sclerosis in Africans. *East Afr Med J*. 1955;32(8):333–335.

73. Ben Hamida M, Hentati F. [Charcot's disease and juvenile amyotrophic lateral sclerosis]. *Rev Neurol (Paris)*. 1984;140(3):202–206.

74. Radhakrishnan K, Ashok PP, Sridharan R, Mousa ME. Descriptive epidemiology of motor neuron disease in Benghazi, Libya. *Neuroepidemiology*. 1986;5(1):47–54.

75. Bougteba A, Basir A, Birouk N, Belaidi H, Kably B, OUazzani R. Sclérose latérale amyotrophique au Maroc. Etude de 276 cas. *Rev Neurol (Paris)*. 2005;161(Suppl 4):S97.

76. Imounan F, Lahbouje S, Hsaini Y, Regragui W, Ait Ben Haddou el H, Benomar A, et al. [Clinical and environmental aspects of amyotrophic lateral sclerosis in moroccan population: a study of 60 cases]. *Tunis Med*. 2015;93(6):365–370.

77. Abdulla MN, Sokrab TE, el Tahir A, Siddig HE, Ali ME. Motor neurone disease in the tropics: findings from Sudan. *East Afr Med J*. 1997;74(1):46–48.

78. Adam AM. Unusual form of motor neuron disease in Kenya. *East Afr Med J*. 1992;69(2):55–57.

79. Tekle-Haimanot R, Abebe M, Gebre-Mariam A, Forsgren L, Heijbel J, Holmgren G, et al. Community-based study of neurological disorders in rural central Ethiopia. *Neuroepidemiology*. 1990;9(5):263–277.

80. Mielke J, Adamolekun B, eds. *Motor Neuron Disease in Zimbabwe*. Durban, South Africa: Pan African Association of Neurological Sciences; 1996.

81. Wall DW, Gelfand M. Motor neuron disease in Rhodesian Africans. *Brain*. 1972;95(3):517–520.

82. Kengne AP, Dzudie A, Dongmo L. Epidemiological features of degenerative brain diseases as they occurred in Yaounde referral hospitals over a 9-year period. *Neuroepidemiology*. 2006;27(4):208–211.

83. Le Bigot P. Profil épidémiologique des affections neurologiques au Gabon. In: AUPELF-UREF, ed. *Neurologie Tropicale*. Paris: John Libbey Eurotext; 1993:17–21.

84. Jacquin-Cotton L, Dumas M, Girard PL. Les paraplégies au Sénégal. *Bull Soc Méd Afr Noire Lgue Fr*. 1970;15(2):206–220.

85. Collomb H, Virieu R, Dumas M, Lemercier G. Maladie de Charcot et syndromes de sclérose latérale amyotrophique au Sénégal. Etude clinique de 27 observations. *Bull Soc Méd Afr Noire Lgue Fr*. 1968;13(4):785–804.

86. Ndiaye IP, Ndiaye M, Gueye M, Ndiaye MB, eds. *Les maladies dégénératives de la moelle*. Abidjan, Côte d'Ivoire: Pan African Association of Neurological Sciences; 1986.

87. Sene Diouf F, Ndiaye M, Toure K, Ndao AK, Thiam A, Diop AG, et al. Aspects cliniques et épidémiologiques de la sclérose latérale amyotrophique à la clinique neurologique de Dakar. *Dakar Méd*. 2004;49(3):167–171.

88. Osuntokun BO, Adeuja AO, Bademosi O. The prognosis of motor neuron disease in Nigerian africans. A prospective study of 92 patients. *Brain*. 1974;97(2):385–394.

89. Piquemal M, Beugre K, Boa Yapo F, Giordano C. Etude clinique de 30 observations de syndrome de sclérose latérale amyotrophique observés en Côte d'Ivoire. *Afri J Neuro Sci*. 1982;1:31–40.

90. Cosnett JE, Bill PL, Bhigjee AI. Motor neuron disease in blacks. Epidemiological observations in Natal. *S Afr Med J*. 1989;76(4):155–157.

91. Brooks BR. El Escorial World Federation of Neurology criteria for the diagnosis of amyotrophic lateral sclerosis. Subcommittee on Motor Neuron Diseases/Amyotrophic Lateral Sclerosis of the World Federation of Neurology Research Group on Neuromuscular Diseases and the El Escorial "Clinical limits of amyotrophic lateral sclerosis" workshop contributors. *J Neurol Sci*. 1994;124(Suppl.):96–107.

92. Brooks BR. Problems in shortening the time to confirmation of ALS diagnosis: lessons from the 1st Consensus Conference, Chicago, May 1998. *Amyotroph Lateral Scler Other Motor Neuron Disord*. 2000;1(Suppl 1):S3–S7.

93. Demographic Yearbook. DYB Annual Issues [Internet] (cited July 21, 2014).

94. Byrne S, Walsh C, Lynch C, Bede P, Elamin M, Kenna K, et al. Rate of familial amyotrophic lateral sclerosis: a systematic review and meta-analysis. *J Neurol Neurosurg Psychiatry*. 2011;82(6):623–627.

95. Imam I, Ogunniyi V. What is happening to motor neuron disease in Nigeria? *Ann Afr Med*. 2004;3(1):1–3.

96. Marin B, Kacem I, Diagana M, Boulesteix M, Gouider R, Preux PM, et al. Juvenile and adult-onset ALS/MND among Africans: incidence, phenotype, survival: a review. *Amyotroph Lateral Scler*. 2012;13(3):276–283.

97. Plato CC, Garruto RM, Galasko D, Craig UK, Plato M, Gamst A, et al. Amyotrophic lateral sclerosis and parkinsonism-dementia complex of Guam: changing incidence rates during the past 60 years. *Am J Epidemiol*. 2003;157(2):149–157.

98. Hirano A, Kurland LT, Krooth RS, Lessell S. Parkinsonism-dementia complex, an endemic disease on the island of Guam. I. Clinical features. *Brain*. 1961;84:642–661.

99. Murch SJ, Cox PA, Banack SA, Steele JC, Sacks OW. Occurrence of beta-methylamino-l-alanine (BMAA) in ALS/PDC patients from Guam. *Acta Neurol Scand*. 2004;110(4):267–269.

100. Cox PA, Banack SA, Murch SJ. Biomagnification of cyanobacterial neurotoxins and neurodegenerative disease among the Chamorro people of Guam. *Proc Natl Acad Sci USA*. 2003;100(23):13380–13383.

101. Garruto R, Yase Y. Neurodegenerative disorders of the Western Pacific: the search for mechanisms of pathogenesis. *Trends Neurosci*. 1986;9(8):368–374.

102. Gajdusek DC, Salazar AM. Amyotrophic lateral sclerosis and parkinsonian syndromes in high incidence among the Auyu and Jakai people of West New Guinea. *Neurology*. 1982;32(2):107–126.

103. Murphy M, Quinn S, Young J, Parkin P, Taylor B. Increasing incidence of ALS in Canterbury, New Zealand: a 22-year study. *Neurology*. 2008;71(23):1889–1895.

104. Talman P, Forbes A, Mathers S. Clinical phenotypes and natural progression for motor neuron disease: analysis from an Australian database. *Amyotroph Lateral Scler*. 2009;10(2):79–84.

105. Fong GC, Cheng TS, Lam K, Cheng WK, Mok KY, Cheung CM, et al. An epidemiological study of motor neuron disease in Hong Kong. *Amyotroph Lateral Scler Other Motor Neuron Disord*. 2005;6(3):164–168.

106. Fong KY, Yu YL, Chan YW, Kay R, Chan J, Yang Z, et al. Motor neuron disease in Hong Kong Chinese: epidemiology and clinical picture. *Neuroepidemiology.* 1996;15(5):239−245.

107. Lai CH, Tseng HF. Epidemiology and medical expenses of motor neuron diseases in Taiwan. *Neuroepidemiology.* 2008;31 (3):159−166.

108. Okumura H. Epidemiological and clinical patterns of western pacific amyotrophic lateral sclerosis (ALS) in Guam and sporadic ALS in Rochester, Minnesota, U.S.A. and Hokkaido, Japan: a comparative study. *Hokkaido Igaku Zasshi.* 2003;78 (3):187−195.

109. Al-Chalabi A, Hardiman O. The epidemiology of ALS: a conspiracy of genes, environment and time. *Nat Rev Neurol.* 2013;9 (11):617−628.

110. Shahrizaila N, Sobue G, Kuwabara S, Kim SH, Birks C, Fan DS, et al. Amyotrophic lateral sclerosis and motor neuron syndromes in Asia. *J Neurol Neurosurg Psychiatry.* 2016;87 (8):821−830.

111. Doi Y, Atsuta N, Sobue G, Morita M, Nakano I. Prevalence and incidence of amyotrophic lateral sclerosis in Japan. *J Epidemiol.* 2014;24(6):494−499.

112. Yoshida S, Uebayashi Y, Kihira T, Kohmoto J, Wakayama I, Taguchi S, et al. Epidemiology of motor neuron disease in the Kii Peninsula of Japan, 1989−1993: active or disappearing focus? *J Neurol Sci.* 1998;155(2):146−155.

113. Watanabe H, Atsuta N, Nakamura R, Hirakawa A, Watanabe H, Ito M, et al. Factors affecting longitudinal functional decline and survival in amyotrophic lateral sclerosis patients. *Amyotroph Lateral Scler Frontotemporal Degener.* 2015;16(3−4):230−236.

114. Goh KJ, Tian S, Shahrizaila N, Ng CW, Tan CT. Survival and prognostic factors of motor neuron disease in a multi-ethnic Asian population. *Amyotroph Lateral Scler.* 2011;12(2):124−129.

115. Chen L, Zhang B, Chen R, Tang L, Liu R, Yang Y, et al. Natural history and clinical features of sporadic amyotrophic lateral sclerosis in China. *J Neurol Neurosurg Psychiatry.* 2015;86 (10):1075−1081.

116. Nalini A, Thennarasu K, Gourie-Devi M, Shenoy S, Kulshreshtha D. Clinical characteristics and survival pattern of 1153 patients with amyotrophic lateral sclerosis: experience over 30 years from India. *J Neurol Sci.* 2008;272(1−2):60−70.

117. Atsuta N, Watanabe H, Ito M, Tanaka F, Tamakoshi A, Nakano I, et al. Age at onset influences on wide-ranged clinical features of sporadic amyotrophic lateral sclerosis. *J Neurol Sci.* 2009;276 (1−2):163−169.

118. Rabkin J, Ogino M, Goetz R, McElhiney M, Hupf J, Heitzman D, et al. Japanese and American ALS patient preferences regarding TIV (tracheostomy with invasive ventilation): a cross-national survey. *Amyotroph Lateral Scler Frontotemporal Degener.* 2014;15 (3−4):185−191.

119. Hirayama K, Tsubaki T, Toyokura Y, Okinaka S. Juvenile muscular atrophy of unilateral upper extremity. *Neurology.* 1963;13:373−380.

120. Meenakshisundaram E, Jagannathan K, Ramamurthi B. Clinical pattern of motor neuron disease seen in younger age groups in Madras. *Neurol India.* 1970;18(Suppl 1):109.

121. Gourie-Devi M, Suresh TG, Shankar SK. Monomelic amyotrophy. *Arch Neurol.* 1984;41(4):388−394.

122. Tashiro K, Kikuchi S, Itoyama Y, Tokumaru Y, Sobue G, Mukai E, et al. Nationwide survey of juvenile muscular atrophy of distal upper extremity (Hirayama disease) in Japan. *Amyotroph Lateral Scler.* 2006;7(1):38−45.

123. Zhou B, Chen L, Fan D, Zhou D. Clinical features of Hirayama disease in mainland China. *Amyotroph Lateral Scler.* 2010;11(1-2): 133−139.

124. Kihira T, Yoshida S, Hironishi M, Miwa H, Okamato K, Kondo T. Changes in the incidence of amyotrophic lateral sclerosis in

Wakayama, Japan. *Amyotroph Lateral Scler Other Motor Neuron Disord.* 2005;6(3):155−163.

125. Loureiro MP, Gress CH, Thuler LC, Alvarenga RM, Lima JM. Clinical aspects of amyotrophic lateral sclerosis in Rio de Janeiro/Brazil. *J Neurol Sci.* 2012;316(1−2):61−66.

126. Bettini M, Gargiulo-Monachelli GM, Rodriguez G, Rey RC, Peralta LM, Sica RE. Epidemiology of amyotrophic lateral sclerosis patients in a centre in Buenos Aires. *Arq Neuropsiquiatr.* 2011;69(6):867−870.

127. de Castro-Costa CM, Oria RB, Machado-Filho JA, Franco MT, Diniz DL, Giffoni SD, et al. Amyotrophic lateral sclerosis. Clinical analysis of 78 cases from Fortaleza (northeastern Brazil). *Arq Neuropsiquiatr.* 1999;57(3b):761−774.

128. Godeiro-Junior C, Vieira-Filho JP, Felicio AC, Oliveira AS. Amyotrophic lateral sclerosis in a Brazilian Kayapo-Xikrin native. *Arq Neuropsiquiatr.* 2008;66(3b):749−751.

129. Bettini M, Vicens J, Giunta DH, Rugiero M, Cristiano E. Incidence and prevalence of amyotrophic lateral sclerosis in an HMO of Buenos Aires, Argentina. *Amyotroph Lateral Scler Frontotemporal Degener.* 2013;14(7−8):598−603.

130. Linden-Junior E, Becker J, Schestatsky P, Rotta FT, Marrone CD, Gomes I. Prevalence of amyotrophic lateral sclerosis in the city of Porto Alegre, in Southern Brazil. *Arq Neuropsiquiatr.* 2013;71(12):959−962.

131. Logroscino G, Traynor BJ, Hardiman O, Chio A, Mitchell D, Swingler RJ, et al. Incidence of amyotrophic lateral sclerosis in Europe. *J Neurol Neurosurg Psychiatry.* 2010;81(4):385−390.

132. Vazquez MC, Ketzoian C, Legnani C, Rega I, Sanchez N, Perna A, et al. Incidence and prevalence of amyotrophic lateral sclerosis in Uruguay: a population-based study. *Neuroepidemiology.* 2008;30(2):105−111.

133. Gil J, Vazquez MC, Ketzoian C, Perna A, Marin B, Preux PM, et al. Prognosis of ALS: comparing data from the Limousin referral centre, France, and a Uruguayan population. *Amyotroph Lateral Scler.* 2009;10(5−6):355−360.

134. Marin B, Couratier P, Preux PM, Logroscino G. Can mortality data be used to estimate amyotrophic lateral sclerosis incidence? *Neuroepidemiology.* 2011;36(1):29−38.

135. Matos SE, Conde MT, Favero FM, Taniguchi M, Quadros AA, Fontes SV, et al. Mortality rates due to amyotrophic lateral sclerosis in Sao Paulo City from 2002 to 2006. *Arq Neuropsiquiatr.* 2011;69(6):861−866.

136. Valenzuela D, Zitko P, Lillo P. Amyotrophic lateral sclerosis mortality rates in Chile: a population based study (1994−2010). *Amyotroph Lateral Scler Frontotemporal Degener.* 2015;16 (5−6):372−377.

137. Marcilio I, Gouveia N, Pereira Filho ML, Kheifets L. Adult mortality from leukemia, brain cancer, amyotrophic lateral sclerosis and magnetic fields from power lines: a case−control study in Brazil. *Rev Bras Epidemiol.* 2011;14(4):580−588.

138. Schrag A, Ben-Shlomo Y, Quinn N. How valid is the clinical diagnosis of Parkinson's disease in the community? *J Neurol Neurosurg Psychiatry.* 2002;73(5):529−534.

139. Rizzo G, Copetti M, Arcuti S, Martino D, Fontana A, Logroscino G. Accuracy of clinical diagnosis of Parkinson disease: a systematic review and meta-analysis. *Neurology.* 2016;86(6):566−576.

140. Das SK, Misra AK, Ray BK, Hazra A, Ghosal MK, Chaudhuri A, et al. Epidemiology of Parkinson disease in the city of Kolkata, India: a community-based study. *Neurology.* 2010;75 (15):1362−1369.

141. Wooten GF, Currie LJ, Bovbjerg VE, Lee JK, Patrie J. Are men at greater risk for Parkinson's disease than women? *J Neurol Neurosurg Psychiatry.* 2004;75(4):637−639.

142. Blanckenberg J, Bardien S, Glanzmann B, Okubadejo NU, Carr JA. The prevalence and genetics of Parkinson's disease in sub-Saharan Africans. *J Neurol Sci.* 2013;335(1−2):22−25.

IV. FOCUS ON SPECIFIC NEUROLOGICAL SYNDROMES OR DISEASES IN TROPICAL AREAS

143. Dotchin CL, Msuya O, Walker RW. The challenge of Parkinson's disease management in Africa. *Age Ageing*. 2007;36 (2):122–127.

144. Okubadejo NU, Bower JH, Rocca WA, Maraganore DM. Parkinson's disease in Africa: a systematic review of epidemiologic and genetic studies. *Mov Disord*. 2006;21(12):2150–2156.

145. Dotchin C, Msuya O, Kissima J, Massawe J, Mhina A, Moshy A, et al. The prevalence of Parkinson's disease in rural Tanzania. *Mov Disord*. 2008;23(11):1567–1672.

146. Osuntokun BO, Adeuja AO, Schoenberg BS, Bademosi O, Nottidge VA, Olumide AO, et al. Neurological disorders in Nigerian Africans: a community-based study. *Acta Neurol Scand*. 1987;75(1):13–21.

147. Schoenberg BS, Osuntokun BO, Adeuja AO, Bademosi O, Nottidge V, Anderson DW, et al. Comparison of the prevalence of Parkinson's disease in black populations in the rural United States and in rural Nigeria: door-to-door community studies. *Neurology*. 1988;38(4):645–646.

148. Bharucha NE, Bharucha EP, Bharucha AE, Bhise AV, Schoenberg BS. Prevalence of Parkinson's disease in the Parsi community of Bombay, India. *Arch Neurol*. 1988;45 (12):1321–1323.

149. Sanyal J, Banerjee TK, Rao VR. Dementia and cognitive impairment in patients with Parkinson's disease from India: a 7-year prospective study. *Am J Alzheimers Dis Other Demen*. 2014;29(7):630–636.

150. Bhidayasiri R, Wannachai N, Limpabandhu S, Choeytim S, Suchonwanich Y, Tananyakul S, et al. A national registry to determine the distribution and prevalence of Parkinson's disease in Thailand: implications of urbanization and pesticides as risk factors for Parkinson's disease. *Neuroepidemiology*. 2011;37 (3-4):222–230.

151. Giroud Benitez JL, Collado-Mesa F, Esteban EM. [Prevalence of Parkinson disease in an urban area of the Ciudad de La Habana province, Cuba. Door-to-door population study]. *Neurologia*. 2000;15(7):269–273.

152. Barbosa MT, Caramelli P, Maia DP, Cunningham MC, Guerra HL, Lima-Costa MF, et al. Parkinsonism and Parkinson's disease in the elderly: a community-based survey in Brazil (the Bambui study). *Mov Disord*. 2006;21(6):800–808.

153. Kaddumukasa M, Kakooza A, Kaddumukasa MN, Ddumba E, Mugenyi L, Sajatovic M, et al. Knowledge and attitudes of Parkinson's disease in rural and urban Mukono district, Uganda: a cross-sectional, community-based study. *Parkinsons Dis*. 2015;2015:196150.

154. Osuntokun BO, Schoenberg BS, Nottidge VA. Research protocol for measuring the prevalence of neurological disorders in developing countries: results of a pilot study in Nigeria. *Neuroepidemiology*. 1982;1:143–153.

155. Van Den Eeden SK, Tanner CM, Bernstein AL, Fross RD, Leimpeter A, Bloch DA, et al. Incidence of Parkinson's disease: variation by age, gender, and race/ethnicity. *Am J Epidemiol*. 2003;157(11):1015–1022.

156. Yacoubian TA, Howard G, Kissela B, Sands CD, Standaert DG. Racial differences in Parkinson's disease medication use in the reasons for geographic and racial differences in stroke cohort: a cross-sectional study. *Neuroepidemiology*. 2009;33(4):329–334.

157. Ragothaman M, Murgod UA, Gururaj G, Kumaraswamy SD, Muthane U. Lower risk of Parkinson's disease in an admixed population of European and Indian origins. *Mov Disord*. 2003;18(8):912–914.

158. de Rijk MC, Tzourio C, Breteler MM, Dartigues JF, Amaducci L, Lopez-Pousa S, et al. Prevalence of parkinsonism and Parkinson's disease in Europe: the EUROPARKINSON Collaborative Study. European Community Concerted Action on the Epidemiology of Parkinson's disease. *J Neurol Neurosurg Psychiatry*. 1997;62(1):10–15.

159. Caslake R, Taylor K, Scott N, Gordon J, Harris C, Wilde K, et al. Age-, gender-, and socioeconomic status-specific incidence of Parkinson's disease and parkinsonism in northeast Scotland: the PINE study. *Parkinsonism Relat Disord*. 2013;19(5):515–521.

160. Caparros-Lefebvre D, Sergeant N, Lees A, Camuzat A, Daniel S, Lannuzel A, et al. Guadeloupean parkinsonism: a cluster of progressive supranuclear palsy-like tauopathy. *Brain*. 2002;125 (Pt 4):801–811.

161. Lannuzel A, Hoglinger GU, Verhaeghe S, Gire L, Belson S, Escobar-Khondiker M, et al. Atypical parkinsonism in Guadeloupe: a common risk factor for two closely related phenotypes? *Brain*. 2007;130(Pt 3):816–827.

162. Litvan I, Agid Y, Calne D, Campbell G, Dubois B, Duvoisin RC, et al. Clinical research criteria for the diagnosis of progressive supranuclear palsy (Steele-Richardson-Olszewski syndrome): report of the NINDS-SPSP international workshop. *Neurology*. 1996;47(1):1–9.

163. Caparros-Lefebvre D, Lees AJ. Atypical unclassifiable parkinsonism on Guadeloupe: an environmental toxic hypothesis. *Mov Disord*. 2005;20(Suppl 12):S114–S118.

164. Apartis E, Gaymard B, Verhaeghe S, Roze E, Vidailhet M, Lannuzel A. Predominant cortical dysfunction in Guadeloupean parkinsonism. *Brain*. 2008;131(Pt 10):2701–2709.

165. Osuntokun BO, Bademosi O. Parkinsonism in the Nigerian African: a prospective study of 217 patients. *East Afr Med J*. 1979;56(11):597–607.

166. Cosnett JE. Neurological diseases in Natal. In: Spillane JD, ed. *Tropical Neurology*. London: Oxford University Press; 1973:259–272.

167. Femi OL, Ibrahim A, Aliyu S. Clinical profile of parkinsonian disorders in the tropics: experience at Kano, northwestern Nigeria. *J Neurosci Rural Pract*. 2012;3(3):237–241.

168. Okubadejo NU, Ojo OO, Oshinaike OO. Clinical profile of parkinsonism and Parkinson's disease in Lagos, Southwestern Nigeria. *BMC Neurol*. 2010;10:1.

169. Langston JW, Forno LS, Tetrud J, Reeves AG, Kaplan JA, Karluk D. Evidence of active nerve cell degeneration in the substantia nigra of humans years after 1-methyl-4-phenyl-1,2,3,6-tetrahydropyridine exposure. *Ann Neurol*. 1999;46(4):598–605.

170. Langston JW, Ballard P, Tetrud JW, Irwin I. Chronic Parkinsonism in humans due to a product of meperidine-analog synthesis. *Science*. 1983;219(4587):979–980.

171. Przedborski S, Vila M. The 1-methyl-4-phenyl-1,2,3,6-tetrahydropyridine mouse model: a tool to explore the pathogenesis of Parkinson's disease. *Ann NY Acad Sci*. 2003;991:189–198.

172. Hancock DB, Martin ER, Mayhew GM, Stajich JM, Jewett R, Stacy MA, et al. Pesticide exposure and risk of Parkinson's disease: a family-based case–control study. *BMC Neurol*. 2008;8:6.

173. Sanyal J, Chakraborty DP, Sarkar B, Banerjee TK, Mukherjee SC, Ray BC, et al. Environmental and familial risk factors of Parkinsons disease: case–control study. *Can J Neurol Sci*. 2010;37(5):637–642.

174. Elbaz A, Tranchant C. Epidemiologic studies of environmental exposures in Parkinson's disease. *J Neurol Sci*. 2007;262 (1–2):37–44.

175. Pandey N, Strider J, Nolan WC, Yan SX, Galvin JE. Curcumin inhibits aggregation of alpha-synuclein. *Acta Neuropathol*. 2008;115(4):479–489.

176. Spinelli KJ, Osterberg VR, Meshul CK, Soumyanath A, Unni VK. Curcumin treatment improves motor behavior in α-synuclein transgenic mice. *PLoS One*. 2015;10(6):e0128510.

177. Caparros-Lefebvre D, Elbaz A. Possible relation of atypical parkinsonism in the French West Indies with consumption of

tropical plants: a case–control study. Caribbean Parkinsonism Study Group. *Lancet.* 1999;354(9175):281–286.

178. Lannuzel A, Michel PP, Hoglinger GU, Champy P, Jousset A, Medja F, et al. The mitochondrial complex I inhibitor annonacin is toxic to mesencephalic dopaminergic neurons by impairment of energy metabolism. *Neuroscience.* 2003;121(2):287–296.

179. Champy P, Hoglinger GU, Feger J, Gleye C, Hocquemiller R, Laurens A, et al. Annonacin, a lipophilic inhibitor of mitochondrial complex I, induces nigral and striatal neurodegeneration in rats: possible relevance for atypical parkinsonism in Guadeloupe. *J Neurochem.* 2004;88(1):63–69.

180. Champy P, Melot A, Guerineau Eng V, Gleye C, Fall D, Hoglinger GU, et al. Quantification of acetogenins in *Annona muricata* linked to atypical parkinsonism in guadeloupe. *Mov Disord.* 2005;20(12):1629–1633.

181. Chaudhuri KR, Hu MT, Brooks DJ. Atypical parkinsonism in Afro-Caribbean and Indian origin immigrants to the UK. *Mov Disord.* 2000;15(1):18–23.

182. Angibaud G, Gaultier C, Rascol O. Atypical parkinsonism and Annonaceae consumption in New Caledonia. *Mov Disord.* 2004;19(5):603–604.

14

Vascular Disorders

Thierry Adoukonou[1] and Philippe Lacroix[2]

[1]University of Parakou, Parakou, Benin [2]INSERM UMR 1094 NET, Limoges, France

14.1 INTRODUCTION

In a very near future vascular disorders will be one of the most important public health issues in the tropical area. Despite the remaining burden of infectious and other transmitted diseases in sub-Saharan Africa (SSA), according to the Global Burden of Diseases (GBD), the non-transmitted diseases, particularly cardiovascular disorders, become the first cause of mortality and disability.[1] Prevalences of the main types of vascular disorders such as stroke, coronary artery disease (CAD), and peripheral artery disease (PAD) are increasing. Approximately 1.06 million of deaths are attributable to cardiac and cerebrovascular disease (CVD). By 2030, non-transmissible conditions in general, CAD and CVD in particular, will increase. They are becoming the first and third causes of mortality in low- and middle-income countries (LMICs).[1] All these diseases share common risk factors which are more and more prevalent in this area.

These estimations are supported by (1) the increasing life expectancy from 51 years in 1990 to more than 65 years in 2020 in women and approximately 60 years in men, (2) the better control of infectious disease and malnutrition; and (3) the growing burden of vascular risk factors.[2] SSA is undergoing an epidemiological transition.

The burden of stroke in SSA will be displayed in the first part of the document, then the related risk factors will be analyzed.

14.2 BURDEN OF STROKE IN SSA

14.2.1 Overview

Stroke is the major cause of adult disability and the second cause of mortality in the world.[3] Recent studies stressed the growing burden of stroke in LMICs, particularly in SSA.[4-6]

A shortage of stroke specialists and rudimentary investigations impact negatively on the prognosis of stroke in SSA.

However, epidemiological data from SAA suffer from important limits.[7,8] First of all, in order to evaluate accurately the burden of stroke in the community, the source population has to be well defined.[9] Very few high-quality censuses have been conducted in SAA. For data collection, specific tools adapted and validated to the population are necessary. Most of them were established from high-income countries' surveys. Validations of these tools were rarely conducted in the SAA population. Often modern imaging technics (computed tomography (CT)-scan or magnetic resonance imaging (MRI)) useful for describing the type of stroke are not available. Neurologists and even general physicians are lacking. Prospective surveys, useful in evaluating the exact incidence are limited. Identification of prevalent or incident cases in the community as well as stroke mortality cases needs to be validated. Thus, community-based surveys are scarce and data are often of intermediate quality. Most data are hospital based. It is meaningful to keep in mind that a high number of cases are not admitted to hospital. Therefore the true prevalence and incidence might be underestimated. In this context only the combination of community-based and hospital-based studies can describe the real burden of stroke in SSA.

The World Health Organization (WHO) defined stroke as "rapidly developing signs of focal (or global) disturbance of cerebral function, leading to death or lasting longer than 24 hours, with no apparent cause other than vascular".[10] Transient ischemic attack (TIA) and other ischemic stroke with recovery of the neurological deficit within 24 hours are excluded in this definition. However many TIA patients have cerebral infarction imaged on MRI. Nevertheless, considering the lack of availability and affordability of CT-scans elsewhere in Africa, this definition remains useful for epidemiological studies despite its limitations. Even if in high-income countries, stroke definition includes MRI abnormalities, recent data from GBD demonstrate an increase in the incidence of stroke in LMICs compared to high-income countries where the incidence is globally stable.[11] The mortality rate is extremely high in SSA. In a recent meta-analysis, the prevalences of stroke survivors were of 21.2/1000 (95% confidence interval (CI) 13.7–30.29) in Latin America and the Caribbean, and 3.5/1000 (95% CI: 1.9–5.7) in SSA.[12]

14.2.2 Epidemiology

14.2.2.1 Prevalence

The first epidemiological studies on stroke in SSA were hospital based.[13] Several studies were later conducted in the community often by door-to-door methods. The WHO definition of stroke was usually used. General characteristics of the populations were variable, particularly ages. Sometimes CT-scans were performed to ensure diagnosis. Prevalence is one of the best measures of the total burden of stroke in any population. It provides information about the number of people who survived a stroke. However, reliable estimates of stroke prevalence are difficult to obtain. For the best estimation we need information on population such as the size, the age distribution and other demographic data. Considering the increasing frequency of stroke with age, standardization to age is useful. This was not realized in most studies. Prevalences of stroke in various SSA settings are displayed in Table 14.1. The prevalence ranged from 15/100,000 persons in Ethiopia to 851/100,000 in Nigeria.[13–23] In other parts of the world stroke prevalence ranged from 500/100,000 to 1000/100,000; thus the prevalence of stroke in SSA is low.

TABLE 14.1 Prevalence of Stroke in Various Settings in Africa

Year (Reference)	Setting	Size	Definition of stroke	Methods	Number of cases	Prevalence[a]	Age-standardized
1982 [13]	Aiyete (Nigeria)	903	WHO	Door-to-door	4	440	
1982 [14]	Igo-Ora (Nigeria)	18954	WHO	Door-to-door	10	53	
1987 [15]	Udo (Nigeria)	2925	WHO	Door-to-door	2	68	
1986 [16]	Ethiopia	60820	WHO	Door-to-door	9	15	
2001 [17]	Agincourt (South Africa)	42378	WHO	Demographic surveillance	103	243	300[b] 200[c]
2007 [18]	Lagos (Nigeria)	13127	WHO	Door-to-door	15	114	
2010 [19]	Cotonou (Benin)	15155	WHO + CT-scan	Door-to-door	70	462	770[b]
2013 [20]	Hai (Tanzania)		WHO	Demographic surveillance	108		
2014 [21]	Anambra (Nigeria)	6150	WHO	Door-to-door	10	163	
2014 [22]	Delta Niger (Nigeria)	1057	WHO	Door-to-door	9	851	1230[b]
2015 [23]	Kwara (Nigeria)	12992	WHO	Door-to-door	17	131	

[a]per 100,000.
[b]World Health Organization (WHO) world population.
[c]Segi world population. [70]

A door-to-door method was used in all studies. In two studies, respectively, in Tanzania and in South Africa a demographic surveillance method with yearly census was conducted. In the first study the Hai population in a rural district of Tanzania was included, and a demographic surveillance was conducted.[18] It was the largest project on the prevalence of disabling hemiplegic stroke. The second survey by the South Africa Stroke Prevention Initiative (SASPI) was conducted in Agincourt, a rural district in South Africa.[15]

14.2.2.2 Incidence

Incidence is considered as one of the most relevant epidemiological indicators. Incidence studies can adopt many forms but require a rigorous methodology. The best option remains the register of stroke that combines population data and hospital data. However this kind of study is very expensive and restrictive to scale in SSA. Criteria for stroke incidence studies were defined by Sudlow and Warlow[24] and updated in 2004[9] (see Box 14.1). These criteria ensure not only the quality of the data but also facilitate comparisons between studies.

The first study on incidence of stroke in SSA took place in Nigeria in 1973.[25] The incidence was of 15/100,000 persons per year. Later other studies took place in SSA and the reported incidences ranged from 15/100,000 to 149/100,000 persons.[25–30] Other studies used weaker diagnostic criteria and were limited to hospital-based data. Incidences in various SSA settings are displayed in Table 14.2.

Worth noting is that in Africa most strokes occur at ages <60 years; conversely, the mean age is close to 65 in Western Europe and North America.

14.2.2.3 Mortality

Mortality data are the second important indicator after the incidence data. In SSA a reliable epidemiological study on stroke mortality is difficult to conduct.[9,24] We need two sources to register the mortality data: the death certificate and a verbal autopsy. Those data are not available everywhere in tropical areas. If available, the data are often unreliable. The real burden of stroke cannot be estimated from isolated hospital data. In the community, many deaths were not notified. Moreover, data from verbal autopsy are limited if they are collected several weeks or months after the event. The awareness of stroke symptoms is very poor in SSA. Then it is difficult retrospectively to attribute the death to stroke as far as people did not know those symptoms. Another issue is the lack of reliable demographic data. The registries of death in Africa are scarce and

---

BOX 14.1

UPDATED CRITERIA FOR IDEAL STUDY OF STROKE INCIDENCE AND/OR MORTALITY

Standard definition

– WHO definition of stroke.
– At least 80% verification by CT or MRI of diagnosis of ischemic, intracerebral hemorrhage, and subarachnoid hemorrhage.
– Classification of ischemic stroke into subtypes (e.g., large-artery disease, small-artery disease, cardioembolic, other) if possible.
– First ever in a lifetime and recurrent stroke (separately and combined).

Standard methods

– Complete population-based case ascertainment, based on multiple overlapping sources of information: hospitals (including admissions for acute vascular problems and cerebrovascular imaging studies and/or interventions), outpatient clinics (including regular checking or general practitioners' databases), death certificates.

– Prospective study design, ideally with "not pursuit" or cases.
– Large, well-defined stable population.
– Follow-up or patients vital census for at least 1 month.
– Reliable method for estimating denominator (census data not more than 5 years old).

Standard data presentation

– Complete calendar years data or data: not more than 5 years of data averaged together.
– Men and women presented separately.
– Recommended reporting of age specific estimates within standard mild decade age bands (e.g., 45–54 years).
– 95% CIs around rates.

Source: Feigin V, Hoorn SV. How to study stroke incidence? Lancet 2004;363:1920.

TABLE 14.2   Community Incidence Studies in Various Setting in Sub-Saharan Africa

Year (Reference)	Setting	Size	Definition of stroke	Methods	Type of stroke	Number of cases	Incidence[a]	Age-standardized[b]
1973 [25]	Ibadan	803098	WHO	DDSS	All	318	15	
1997 [26]	Harare	887768	WHO	Hospital	First ever	275	31	68
1986 [27]	Pretoria	114931	HCSR criteria and CT-scan (79%)	Hospital[c]	All	116	101	
2007 [28]	Maputo		WHO, Autopsy CT-scan (92.3%)	Hospital[c]	All	651	148.7	260.1
2010 [29]	Dar es Salam	56517	WHO	DDSS	All	183	107.9	315.6
2010 [29]	Hai	159814	WHO	DDSS	All	453	94.5	108.6
2013 [30]	Lagos	750.000	WHO	Hospital[c]	First ever	189	25.2	54.08

[a]Incidence rate per 100,000 person-years.
[b]Segi World population.
[c]Hospital: hospital-based study.
DDSS, door-to-door survey or demographic system surveillance.

TABLE 14.3   Stroke Mortality in Various Settings in Africa

Years (Reference)	Setting	Size	Stroke mortality[a]	Part of mortality (%)
1975–1980 [31]	Accra (Ghana)	4075		8
1992–1995 [32]	Agincourt (South Africa)	63000	127	6
1992–1995 [33]	Dar Es Salam (Tanzania)	65826	158	6.1
1992–1995 [33]	Hai (Tanzania)	142414	165	8.8
1992–1995 [33]	Morogoro (Tanzania)	99672	82	2.5

[a]per 100,000 inhabitants.

the reliability is fragmentary. Then it is very difficult to estimate the real mortality of stroke in Africa. Nevertheless by combining verbal autopsy, hospital data and other sources of notification, stroke mortality can be evaluated. Mortality rates varied from 82 to more than 200 per 100,000 inhabitants.[31–33] The available data are summarized in Table 14.3.

14.2.2.4 Stroke Case-Fatality

In SSA among stroke patients, many died at home or before hospital admission. In these countries, acute management is lacking. The case fatality is very high in LMICs compared to those reported in the high-income countries. One-month case-fatality rates ranged from 20% to more than 50% in the tropics.[34–36] After 7 years of follow-up in Tanzania, 82.3% of the stroke subjects were dead.[37] Many factors influenced the mortality in the acute phase of stroke. Good management can prevent most of them. Early detection and management of these complications such as pneumonia, venous thrombo-embolism, urinary tract infection, cerebral edema, and early recurrence of stroke is of major importance. Nevertheless, information about the

trend and the temporal census of the case-fatality are useful.

14.2.2.5 Stroke Subtypes

The CVD events are strokes, cerebral venous thrombosis and subarachnoid hemorrhage. Stroke are commonly ischemic or hemorrhagic. Mechanisms and management of both are different. Many scales or scores were developed in order to distinguish hemorrhagic from thrombotic strokes (Siriraj Stroke Scale, Guy's Hospital) but their accuracy is limited. According to a recent meta-analysis, they cannot help clinicians in practice.[38] Imaging techniques such as CT-scan and MRI allow an accurate diagnosis. MRI is the best imaging strategy. It cannot be routinely used in tropical areas because of its availability and affordability. Nevertheless the available data indicates that the ischemic stroke accounted for 40%–70% and hemorrhage for 21%–60% in SSA.[39] Some recent surveys illustrate a switch from a hemorrhagic predominance to an ischemic one.

According to the Trial of ORG 10172 in Acute Stroke Treatment (TOAST)[40] classification of stroke and on the basis of available data[41,42] the etiologies of stroke in the black population may be:

1. small vessel disease due to hypertension and diabetes mellitus;
2. undefined cause (lack of investigation);
3. cardio embolic;
4. atherosclerosis;
5. other causes (vasculitis, sickle cell disease, etc.).

Vasculitis due to HIV infection may be an important part of ischemic stroke where the HIV infection is prevalent particularly among young patients. In fact, due to the growing prevalence of vascular risk factors in SSA, the role of atherosclerosis will be important in the coming years. Current large comprehensive studies will contribute to a better understanding of stroke etiologies in this setting.[43] Regarding the hemorrhagic strokes, about 70%–80% can be explained by hypertension with small-vessel disease involvement causing small lipohyalinotic aneurysms and their rupture. The location of the hematoma (deep location) and the past history of hypertension or the other damage of this disease (cardiac, eye, kidney, etc.) can help to ascertain diagnosis.

14.2.3 Comments

Considering the strict application of the incidence or mortality criteria difficulties, especially in some tropical areas, the WHO proposed a new approach in three steps in order to obtain reliable epidemiological data in developing countries.[44] This approach may be generalized. The first step is the identification of hospital cases. The second is the identification of the fatality cases outside the hospital, and the third is the

---

BOX 14.2

ROSIER SCALE PROFORMA

Assessment	Date: I__I__I/I__I__I/I__I__I__I__I		Time: I__I__I/I__I__I
Symptom onset	Date: I__I__I/I__I__I/I__I__I__I__I		Time: I__I__I/I__I__I

GCS: E I__I V I__I M I__I   BP: I__I__I__I/I__I__I__I mmHg   BG: I__I__I,I__Immol/l

If BG<3.5mmol/l treat urgently and reassess once blood glucose normal

Has there been loss of consciousness or syncope?	I__I (Yes= -1, No=0)
Has there been seizure activity?	I__I (Yes=-1, No=0)
Is there a NEW ACUTE onset (or on awakening from sleep)	
Asymmetric facial weakness	I__I (Yes=+1, No=0)
Asymmetric arm weakness	I__I (Yes=+1, No=0)
Asymmetric leg weakness	I__I (Yes=+1, No=0)
Speech disturbance	I__I (Yes=+1, No=0)
Visual field deficit	I__I (Yes=+1, No=0)
TOTAL SCORE	I__I (-2 to +5)

Provisional diagnosis (Score>0 predicts Stroke, Score≤0 Non-stroke)

I__I Stroke   I__I Non-stroke (specify)..

BG, blood glucose; BP, blood pressure; E, eye; GCS, Glasgow Coma Scale; M, motor; V, voice.

Source: *Nor AM, Davis J, Sen B, et al. The Recognition of Stroke in the Emergency Room (ROSIER) scale: development and validation of a stroke recognition instrument. Lancet Neurol. 2005;4(11):727–734.*

---

identification of non-fatality cases in community. This step-wise approach documents the initial outcomes of stroke patients: events in hospital, fatal events in community, and non-fatal events in community. The stroke surveillance system begins with cases admitted to hospital: this group is easily identified, and patients are followed until discharge or death. The second level consists of identifying and validating the diagnoses of fatal stroke events for patients not admitted to hospital, i.e., the fatal events in the community. The third step represents non-fatal, non-hospitalized events. The optimal stroke surveillance system requires collection of data from all three steps.

However, in resource-limited countries due to the unavailability and inaccessibility of modern imaging techniques, it is difficult to identify hospital cases. Furthermore it is necessary to ensure that stroke cases are properly confirmed at least clinically. The WHO definition of stroke is very useful for epidemiological studies but non-applicable for hospital cases requiring treatment which depends on the subtype. Appropriate identification is sometimes difficult and differential diagnoses have to be excluded especially in case of neurological disorders of sudden onset. Seizure, hypoglycemia, confusion and other differential diagnoses might be confusing. Then a good emergency clinical diagnosis by non-neurologist health workers coupled with the CT-scan could optimize the management. Lack of neurologists in those areas should encourage the development of simple diagnostic tools. Several tools have been developed for this purpose.[45−47] The Rosier scale[48] (see Box 14.2) seems easier and is useful in the emergency room, and has a good accuracy. This tool evaluates several items with a total score from −2 to +5. A score >0 predicts stroke with a sensitivity of 92%, a specificity of 86%, a negative predictive value of 88% and a negative predictive value of 91%.

Importantly, in SSA the poor awareness of stroke signs and definition amongst health workers[49] and in the community[50] is a strong limitation for diagnosis and management. The analysis of the epidemiological studies conducted in SSA ought to integrate these limitations.

14.3 RISK FACTORS

Africa is facing an epidemiological transition, characterized by an increasing burden of non-communicable disorders (NCD) including hypertension, diabetes mellitus and dyslipidemia. The rapid urbanization results in lifestyle changes, particularly in tobacco smoking, overweight and obesity driven by changes in food supply, alcohol consumption and decreased physical activity. The influence of all these risk factors was extensively documented in high-income countries' populations. Most of them are considered as stroke risk factors. Longitudinal surveys conducted in high-income countries' communities illustrated this relationship. Large prospective epidemiological studies are very scarce in SSA.[51] Most data come from cross-sectional surveys particularly in hospital setting. The finding in high-income countries cannot always be generalized to the SSA population. Even in high-income countries, data from epidemiological studies demonstrate a wide variation in the prevalence of exposure to risk factors and events outcome.[52] Regarding SSA, the prevalence of the stroke risk factors was documented but often the level of the relationship was not weighed. In addition to modifiable risk factors, NCD might influence the stroke outcome, particularly cardiac abnormalities (atrial fibrillation, valvular diseases, heart failure). These diseases share most of the risk factors for stroke. Few risk factors seem to be more specific to SSA population. HIV status was independently associated with stroke risk in SSA.[53] Stroke is also a recognized complication of sickle cell disease, a most prevalent condition in the African population. Genetic risk factors more prevalent in African people may contribute to stroke incidence. Very few are monogenic such as sickle cell disease. In many cases, strokes result from complex multifactorial genetic factors. Few genetic variants related to hypertension were described in the African population.[54]

Thus, the underlying factors for stroke in SSA are complex. The epidemiological characteristics of these factors in this setting will be discussed.

Hypertension is highly prevalent in the SSA community. In the STEPS (STEPwise approach to noncommunicable disease risk factor surveillance) surveys, more than 25% of the subjects describe this condition.[55−57] In Benin, similar prevalences were reported in urban and in rural areas.[55] Differences were minimal between male and female individuals.[51,58] The prevalence of hypertension was steadily increasing with age up to 16%, 26%, 35%, and 44% at mean ages of 30, 40, 50, and 60 years, respectively.[59] Only 27% out of those with hypertension were aware of the condition, less rarely treated and very few were controlled. In SSA the available evidence is that hypertension remains the dominant causal risk factor regardless of the stroke type. The relationship was documented in community studies; data from Nigeria and Tanzania showed that >80% of the stroke index cases had hypertension.[23,29] In the hospital setting, the prevalence of hypertension was up to 68% in the Johannesburg hospital stroke register.[42] In the Interstroke study[60] O'Donnell analyzed the risk of stroke associated with hypertension according the region. Compared with other countries, the risk was the highest in Africa: 4.96 (3.11−7.91). Maredza reported similar data in the Agincourt health and demographic surveillance system (HDSS), 38% of the stroke burden was due to hypertension.[61] These relationships illustrate the potentially dramatic impact of hypertension detection and treatment. It is estimated that up to one-half of the strokes might be prevented through risk factors detection and prevention.

Overall prevalence rate of diabetes mellitus in Africa is up to 2%−3%.[55,62] A survey conducted in Tanzania[53] showed that, in the community, 11% of the stroke index cases were diabetics. In the Johannesburg hospital registry diabetes prevalence was up to 15%.[42] The urbanization and the aging of the SSA populations would result in changes in dietary patterns that may explain the increasing prevalences.

In the INTERSTROKE study[60] the risk of stroke associated with smoking was lowest in Africa, at 2.18 (1.07−4.43). The burden of stroke due to smoking seems to be limited in SSA, particularly in the black population. In the Johannesburg stroke register[42] only 22% of the white patients were in the never smoking group; conversely, 65% of the black patients were in the same group. Residency might influence the smoking habits; in Malawi and Benin the prevalence of daily tobacco smoking was higher in rural than in urban areas.[55,57]

Few studies highlighted the substantial burden of stroke attributable to elevated body mass index (BMI) and abdominal obesity in SSA. The overall risk of stroke related to waist to hip ratio in the INTERSTROKE study[60] was at 1.73 (0.99−3.02) in Africa but at 2.70 (1.95−3.74) in the female group; conversely the risk was lower in the male group at 1.25 (0.99−1.59). In the Agincourt HDSS,[61] excess BMI was considered to be responsible for 20% of the stroke burden (3.5% males; 16% females). This risk was concentrated amongst the females and the youngest age groups. Overweight and obesity were more prevalent in urban than in rural populations.[55,56] Elevated cholesterol level appeared more prevalent among high educational and urban populations.[63]

Stroke has been reported to be associated with HIV infection in European cohorts. Ischemic stroke is the predominant type. Prospective studies assessing the role of HIV infection in stroke risk in ASS are lacking. In a retrospective case−control study conducted in South Africa,[64] the infection was associated with a trend toward an increased risk of stroke. However, later a relationship between acquired immunodeficiency syndrome and stroke risk was shown in the Baltimore-Washington Cooperative Young Stroke Study.[65] The adjusted relative risk was 9.1 for cerebral infarction (95% CI, 3.4−24.6) and 12.7 for intracerebral hemorrhage (95% CI, 4.0−40.0). In 2013 a study conducted in Tanzania[53] showed that the HIV infection was an independent risk factor of stroke (odds ratio (OR): 4.20, 95% CI: 1.56−11.30). The other factors were the hypertension, smoking status, previous stroke, TIA or myocardial infarction and a high ratio of total to high-density lipoprotein cholesterol ratio. None or low alcohol consumption was protective.

Stroke is a well-known complication of sickle cell disease. This monogenic disorder is highly prevalent in SSA; the gene carrier rate was estimated to be up to 40% in Nigeria.[66] Amongst 5721 Nigerian sickle cell patients registered in clinics, prevalence of stroke was 12.4/1000.[67] Other mono- or multigenic disorders may influence the stroke epidemiology in SSA, through their impact on usual risk factors such as hypertension, diabetes, and dyslipidemia.

Other known stroke risk factors include heart diseases. The shift from hemorrhagic toward ischemic stroke reflects the growing burden of these disorders. In Tanzania[53] 14%−20% of the stroke cases versus 3%−4% in the control group presented a history of previous cardiac events. In high-income countries, atrial fibrillation (AF) is the most common cardiac arrhythmia associated with stroke. The prevalence of AF in the SSA community was poorly documented. In older Tanzanian subjects (≥60 years old)[68] the prevalence was 0.67% but the related 1-year mortality rate was extremely high (50%−66%). AF might be related to hypertension and heart failure. Additionally in ASS, rheumatic heart disease due to undertreated streptococcal infection may contribute to the burden of AF.

Stroke risk may be increased by sociocultural factors. Poverty and lack of infrastructure for healthcare have been documented as risk factors for stroke and were common in SSA.

14.4 THE GAPS

The burden of stroke is growing in SSA, this rise is driven substantially by the influence of common risk factors. Without further investments in identification of subjects at risk and prevention, the stroke incidence will increase. In order to identify the subjects at risk of cardiovascular events, particularly stroke, tools dedicated to this population have to be developed and validated. Cardiovascular risk scales have been developed from high-income country cohorts such as the Framingham study or the SCORE project. The transportability of such models in SSA was not studied and most of them included biological data not available in low- or middle-income countries. The main risk factors for stroke, particularly hypertension, are not only undetected but even if detected the effective management is suboptimal. Alternative approaches are required to control these treatable risk factors. The interventions have to be sustainable, cost effective, and acceptable to the local population. In low-resource settings, telemedicine and mHealth interventions[69] might be relevant for the clinical approaches for hypertension and diabetes. Community health workers can be integrated in such interventions and have to be included in educational program delivery on tobacco use, unhealthy diet and low activity.

References

1. Mathers CD, Loncar D. Projections of global mortality and burden of disease from 2002 to 2030. *PLoS Med.* 2006;3(11):e442.
2. Murray CJ, Lopez AD. Alternative projections of mortality and disability by cause 1990−2020: Global Burden of Disease Study. *Lancet.* 1997;349(9064):1498−1504.
3. Lopez AD, Mathers CD, Ezzati M, Jamison DT, Murray CJ. Global and regional burden of disease and risk factors, 2001: systematic analysis of population health data. *Lancet.* 2006;367 (9524):1747−1757.
4. Lemogoum D, Degaute JP, Bovet P. Stroke prevention, treatment, and rehabilitation in sub-Saharan Africa. *Am J Prev Med.* 2005;29(5 Suppl 1):95−101.
5. Joubert J, Prentice LF, Moulin T, Liaw ST, Joubert LB, Preux PM, et al. Stroke in rural areas and small communities. *Stroke.* 2008;39(6):1920−1928.
6. Brainin M, Teuschl Y, Kalra L. Acute treatment and long-term management of stroke in developing countries. *Lancet Neurol.* 2007;6(6):553−561.
7. Bharucha N, Odermatt P, Preux PM. Methodological difficulties in the conduct of neuroepidemiological studies in low- and middle-income countries. *Neuroepidemiology.* 2014;42(1):7−15.
8. Preux PM, Ratsimbazafy V, Bhalla D, Ngoungou E, Quet F, Druet-Cabanac M. Méthodologie des études neuroépidémiologiques dans les pays tropicaux : un challenge ? *Rev Neurol (Paris).* 2012;168(3):211−215.
9. Feigin V, Hoorn SV. How to study stroke incidence? *Lancet.* 2004;363:1920.
10. Aho K, Harmsen P, Hatano S, Marquardsen J, Smirnov VE, et al. Cerebrovascular disease in the community: results of a WHO Collaborative Study. *Bull World Health Organ.* 1980;58:113−130.
11. Feigin VL, Forouzanfar MH, Krishnamurthi R, et al. Global and regional burden of stroke during 1990−2010: findings from the Global Burden of Disease Study 2010. *Lancet.* 2014;383 (9913):245−254.
12. Ezejimofor MC, Chen YF, Kandala NB, et al. Stroke survivors in low- and middle-income countries: a meta-analysis of prevalence and secular trends. *J Neurol Sci.* 2016;364:68−76.
13. Osuntokun BO, Schoenberg BS, Nottidge VA, et al. Research protocol for measuring the prevalence of neurologic disorders in developing countries. *Neuroepidemiology.* 1982;1(3):143−153.
14. Osuntokun BO, Adeuja AO, Schoenberg BS, et al. Neurological disorders in Nigerian Africans: a community based study. *Acta Neurol Scand.* 1987;75:13−21.
15. Longe AC, Osuntokun BO. Prevalence of neurological disorders in Udo, a rural community in Southern Nigeria. *Trop Geogr Med.* 1989;41:36−40.
16. Tekle-Haimanot R, Abebe M, Gebre-Mariam A, Forsgren L, Heijbel J, et al. Community-based study of neurological disorders in rural central Ethiopia. *Neuroepidemiology.* 1990;9:263−277.

17. Connor M. Prevalence of stroke survivors in rural South Africa: results from the Southern Africa Stroke Prevention Initiative (SASPI) Agincourt field site. *Stroke.* 2004;35(3):627−632.
18. Danesi M, Okubadejo N, Ojini F. Prevalence of stroke in an urban, mixed-income community in Lagos, Nigeria. *Neuroepidemiology.* 2007;28:216−223.
19. Cossi M-J, Gobron C, Preux P-M, Niama D, Chabriat H, et al. Stroke: prevalence and disability in Cotonou, Benin. *Cerebrovasc Dis.* 2012;33:166−172.
20. Dewhurst F, Dewhurst MJ, Gray WK, Aris E, Orega G, et al. The prevalence of neurological disorders in older people in Tanzania. *Acta Neurol Scand.* 2013;127(3):198−207.
21. Enwereji KO, Nwosu MC, Ogunniyi A, Nwani PO, Asomugha A, Enwereji EE. Epidemiology of stroke in a rural community in southeastern Nigeria. *Vasc Health Risk Manag.* 2014;10:375−388.
22. Onwuchekwa AC, Tobin-West C, Babatunde S. Prevalence and risk factors for stroke in an adult population in a rural community in the Niger Delta, south-south Nigeria. *J Stroke Cerebrovasc Dis.* 2014;23(3):505−510.
23. Sanya EO, Desalu OO, Adepoju F, Aderibigbe SA, Shittu A, Olaosebikan O. Prevalence of stroke in three semi-urban communities in middle-belt region of Nigeria: a door to door survey. *Pan Afr Med J.* 2015;20:33.
24. Sudlow CL, Warlow CP. Comparing study incidence worldwide: what makes studies comparable? *Stroke.* 1996;27:550−558.
25. Osuntokun BO, Bademosi O, Akinkugbe OO, Oyediran AB, Carlisle R. Incidence of stroke in an African City: results from the Stroke Registry at Ibadan, Nigeria, 1973−1975. *Stroke.* 1979;10:205−207.
26. Matenga J. Stroke incidence rates among black residents of Harare—a prospective community-based study. *S Afr Med J.* 1997;87:606−609.
27. Rosman KD. The epidemiology of stroke in an urban black population. *Stroke.* 1986;17:667−669.
28. Damasceno A, Gomes J, Azevedo A, Carrilho C, Lobo V, et al. An epidemiological study of stroke hospitalizations in Maputo, Mozambique: a high burden of disease in a resource-poor country. *Stroke.* 2010;41:2463−2469.
29. Walker R, Whiting D, Unwin N, Mugusi F, Swai M, et al. Stroke incidence in rural and urban Tanzania: a prospective, community-based study. *Lancet Neurol.* 2010;9:786−792.
30. Danesi MA, Okubadejo NU, Ojini FI, Ojo OO. Incidence and 30-day case fatality rate of first-ever stroke in urban Nigeria: the prospective community based Epidemiology of Stroke in Lagos (EPISIL) phase II results. *J Neurol Sci.* 2013;331(1−2):43−47.
31. Chukwumeka AC, Pobee JOM, Larbi E, et al. Are cardiovascular diseases important causes of mortality in Africa? Results of a WHO/UGMS cardiovascular disease follow-up study over a fi veyear period in a suburb of Accra. *Trop Cardiol.* 1982;8:105−109.
32. Kahn K, Tollman SM. Stroke in rural South Africa—contributing to the little known about a big problem. *S Afr Med J.* 1999;89:63−65.
33. Walker RW, McLarty DG, Kitange HM, et al. Stroke mortality in urban and rural Tanzania. *Lancet.* 2000;355:1684−1687.
34. Sene Diouf F, Basse AM, Toure K, et al. Prognosis of stroke in department of neurology of Dakar. *Dakar Med.* 2006;51(1):17−21.
35. Wahab KW, Okubadejo NU, Ojini FI, Danesi MA. Predictors of short-term intra-hospital case fatality following first-ever acute ischaemic stroke in Nigerians. *J Coll Physicians Surg Pak.* 2008;18 (12):755−758.
36. Longo-Mbenza B, Lelo Tshinkwela M, Mbuilu Pukuta J. Rates and predictors of stroke-associated case fatality in black Central African patients. *Cardiovasc J Afr.* 2008;19(2):72−76.
37. Walker RW, Wakefield K, Gray WK, Jusabani A, Swai M, Mugusi F. Case-fatality and disability in the Tanzanian Stroke Incidence Project cohort. *Acta Neurol Scand.* 2016;133(1):49−54.

38. Mwita CC, Kajia D, Gwer S, Etyang A, Newton CR. Accuracy of clinical stroke scores for distinguishing stroke subtypes in resource poor settings: a systematic review of diagnostic test accuracy. *J Neurosci Rural Pract*. 2014;5(4):330–339.

39. Adoukonou TA, Vallat JM, Joubert J, et al. Prise en charge des accidents vasculaires cérébraux en Afrique subsaharienne. *Rev Neurol (Paris)*. 2010;166(11):882–893.

40. Adams Jr. HP, Bendixen BH, Kappelle LJ, Biller J, Love BB, Gordon DL, et al. Classification of subtype of acute ischemic stroke. Definitions for use in a multicenter clinical trial. TOAST. Trial of Org 10172 in Acute Stroke Treatment. *Stroke*. 1993;24(1):35–41.

41. Markus HS, Khan U, Birns J, et al. Differences in stroke subtypes between black and white patients with stroke: the South London Ethnicity and Stroke Study. *Circulation*. 2007;116(19):2157–2164.

42. Connor MD, Modi G, Warlow CP. Differences in the nature of stroke in a multiethnic urban South African population: the Johannesburg hospital stroke register. *Stroke*. 2009;40(2):355–362.

43. Akpalu A, Sarfo FS, Ovbiagele B, et al. Phenotyping stroke in sub-Saharan Africa: Stroke Investigative Research and Education Network (SIREN) Phenomics Protocol. *Neuroepidemiology*. 2015;45(2):73–82.

44. World Health Organization. *WHO STEPS Stroke Manual: The WHO STEPwise Approach to Stroke Surveillance*. Geneva, Switzerland: WHO; 2006.

45. Harbison J, Hossain O, Jenkinson D, Davis J, Louw SJ, Ford GA. Diagnostic accuracy of stroke referrals from primary care, emergency room physicians, and ambulance staff using the face arm speech test. *Stroke*. 2003;34:71–76.

46. Kothari RU, Pancioli A, Liu T, Brott T, Broderick J. Cincinnati prehospital stroke scale: reproducibility and validity. *Ann Emerg Med*. 1999;33:373–378.

47. Kidwell CS, Starkman S, Eckstein M, Weems K, Saver J. Identifying stroke in the field-prospective validation of the Los Angeles prehospital stroke screen (LAPSS). *Stroke*. 2000;31:71–76.

48. Nor AM, Davis J, Sen B, et al. The Recognition of Stroke in the Emergency Room (ROSIER) scale: development and validation of a stroke recognition instrument. *Lancet Neurol*. 2005;4(11):727–734.

49. Akinyemi RO, Ogah OS, Ogundipe RF, et al. Knowledge and perception of stroke amongst hospital workers in an African community. *Eur J Neurol*. 2009;16(9):998–1003.

50. Cossi MJ, Preux PM, Chabriat H, Gobron C, Houinato D. Knowledge of stroke among an urban population in Cotonou (Benin). *Neuroepidemiology*. 2012;38(3):172–178.

51. Dalal S, Holmes MD, Laurence C, et al. Feasibility of a large cohort study in sub-Saharan Africa assessed through a four-country study. *Glob Health Action*. 2015;8:27422.

52. Vikhireva O, Pajak A, Broda G, et al. SCORE performance in Central and Eastern Europe and former Soviet Union: MONICA and HAPIEE results. *Eur Heart J*. 2014;35:571–577.

53. Walker RW, Jusabani A, Aris E, et al. Stroke risk factors in an incident population in urban and rural Tanzania: a prospective, community-based, case-control study. *Lancet Glob Health*. 2013;1:e282–e288.

54. Akinyemi RO, Ovbiagele B, Akpalu A, et al. Stroke genomics in people of African ancestry: charting new paths. *Cardiovasc J Afr*. 2015;26(2 Suppl 1):S39–S49.

55. Houehanou YC, Lacroix P, Mizehoun GC, et al. Magnitude of cardiovascular risk factors in rural and urban areas in Benin: findings from a nationwide steps survey. *PLoS One*. 2015;10(5):e0126441.

56. Nakibuuka J, Sajatovic M, Nankabirwa J, et al. Stroke-risk factors differ between rural and urban communities: population survey in Central Uganda. *Neuroepidemiology*. 2015;44:156–165.

57. Msyamboza KP, Kathyola D, Dzowela T, et al. The burden of hypertension and its risk factors in Malawi: nationwide population-based STEPS survey. *Int Health*. 2012;4:246–252.

58. Addo J, Smeeth L, Leon DA. Hypertension in sub-Saharan Africa: a systematic review. *Hypertension*. 2007;50:1012–1018.

59. Atakle F, Erqou S, Kaptoge S, et al. Burden of undiagnosed hypertension in sub-Saharan Africa: a systematic review and meta-analysis. *Hypertension*. 2015;65:291–298.

60. O'Donnell MJ, Xavier D, Liu L, et al. Risk factors for ischemic and intracerebral haemorrhagic stroke in 22 countries (the INTERSTROKE study): a case-control study. *Lancet*. 2010;376:112–123.

61. Maredza M, Bertram MY, Gómez-Olivé XF, et al. Burden of stroke attributable to selected lifestyle risk factors in rural South Africa. *BMC Public Health*. 2016;16:143.

62. Owolabi MO, Akarolo-Anthony S, Akinyemi R, et al. The burden of stroke in Africa: a glance at the present and a glimpse into the future. *Cardiovasc J Afr*. 2015;26:S27–S38.

63. BeLue R, Okoror TA, Iwelunmor J, et al. An overview of cardiovascular risk factor burden in sub-Saharan African countries: a socio-cultural perspective. *Glob Health*. 2009;5:10.

64. Hoffmann M, Berger JR, Nath A, et al. Cerebrovascular disease in young, HIV-infected, black Africans in the KwaZulu Natal province of South Africa. *J Neurovirol*. 2000;6:229–236.

65. Cole JW, Pinto AN, Hebel JR, et al. Acquired immunodeficiency syndrome and the risk of stroke. *Stroke*. 2004;35:51–56.

66. Okpala I. Epidemiology, genetics and pathophysiology of SCD. In: Okpala I, ed. *Practical Management of Haemoglobinopathies*. Oxford: Blackwell Publishing; 2004:20–25.

67. Jude MA, Aliyu GN, Nalado AM, Garba KU, et al. Stroke prevalence amongst sickle cell disease patients in Nigeria: a multicentre study. *Afr Health Sci*. 2014;14(2):446–452.

68. Dewhurst MJ, Adams PC, Gray WK, et al. Strikingly low prevalence of atrial fibrillation in elderly Tanzanians. *J Am Geriatr Soc*. 2012;60:1135–1140.

69. Hall CS, Fottrell E, Sophia Wilkinson S, et al. Assessing the impact of mHealth interventions in low- and middle-income countries—what has been shown to work?. *Glob Health Action*. 2014;7:25606.

70. Mitsuo Segi, Susumu Fujisaku, Minoru Kurihara, Yoko Narai, Kyoko Sasajima. The Age-adjusted Death Rates for Malignant Neoplasms in Some Selected Sites in 23 Countries in 1954–1955 and their Geographical Correlation. *Tohoku J. Exper. Med*. 1960;72:91–103.

CHAPTER

15

Neuromuscular Disorders in Tropical Areas

Stéphane Mathis[1], Laurent Magy[2], and Jean-Michel Vallat[2]

[1]CHU Bordeaux, Bordeaux, France [2]CHU Limoges, Limoges, France

15.1 INTRODUCTION

Neuromuscular disorders (NMDs) constitute a complex group of heterogeneous diseases that may be acquired or inherited: they are usually characterized by a primary or secondary involvement of skeletal muscle (muscle weakness and wasting), and can be broadly subdivided into disorders mainly affecting the anterior horn cell, the peripheral nerve, the neuromuscular junction, and the muscle. Epidemiological studies in sub-Saharan Africa (SSA) are quite rare; in 1987, in Nigeria, Osuntokun et al. reported the following causes of NMD (in cases/100,000 people): 253 for polyneuropathies (185 "tropical ataxic neuropathies", 42

idiopathic neuropathies, 10 hereditary neuropathies, 10 diabetic neuropathies, and 5 other nutritional neuropathies), 15 for mononeuropathies (10 lepromatous neuropathies and 5 multiplex mononeuritis), 15 motorneuron diseases, 10 poliomyelitis, 10 myopathies (5 pyomyositis and 5 muscular dystrophies), and 5 post-polio syndromes.[1]

The World Health Organization (WHO) has identified 17 neglected tropical diseases (mostly infectious diseases), with potential central and peripheral neurological complications. However, while the WHO currently focuses on these 17 priority diseases, the list of conditions eligible to be considered as neglected tropical diseases is much longer:[2] in tropical areas, NMDs

Neuroepidemiology in Tropical Health
DOI: http://dx.doi.org/10.1016/B978-0-12-804607-4.00015-0

195

may be also due to other conditions such as toxic causes and malnutrition.

15.2 INFECTIOUS NMDS

15.2.1 Bacterial Infections

15.2.1.1 Leprosy

Leprosy classicaly causes "mononeuritis multiplex" resulting in autonomic, sensory, and motor neuropathy. Sensory loss is the most constant finding in lepromatous neuropathy, sometimes with an "insular pattern". It may be associated with anhydrosis and vasomotor areflexia.[3] Motor disturbances, amyotrophy, plantar ulcers, and trophic changes are usually a late event of the disease, but facial palsy is a classical sign of the disease.[3] Nerve biopsy is an important means of diagnosing this disease in pure neural forms where skin lesions are not clinically or histologically definitive.[4] Multi-drug therapy came into vogue in the 1970s and was popularized by the WHO in the 1980s, resulting in a rapid decline in the new case detection rates: in early cases, standard therapy recommended by the WHO includes rifampicin, dapsone, and clofamidine for 6–12 months, depending on the form of the disease.

15.2.1.2 Buruli Ulcer

Buruli ulcer, or *Mycobacterium ulcerans* disease, represents one of the neglected tropical diseases for which the WHO stressed the need to improve treatment. It occurs in scattered foci around the world in riverine areas with humid and hot climates (as well as sometimes in temperate climates): it is endemic in areas in West Africa.[5] This chronic skin disease typically presents with non-ulcerated lesions, papules, nodules, plaques or edema, or undermined ulcers.

A major characteristic is painlessness, and this is the reason why its diagnosis may be underestimated and it sometimes leads to severe sequellae such as limb amputation (as it may be observed in diabetic neuropathy). This phenomenon is probably induced by mycolactone (a toxic lipid producing by *M. ulcerans*): a study has shown that injection of mycolactone in footpads of mice gives hyper- then hypo-esthesia, associated with skin and peripheral nerve damage.[6]

15.2.1.3 Tuberculosis

Tuberculosis may induce neurological manifestations, mainly central nervous system (CNS) complications. Peripheral nervous system (PNS) involvement is rare and controversial:[7] neuropathy can occur as a complication of a tuberculous meningitis (spinal nerve roots and cranial nerves palsies), or rarely as a

manifestation of a nerve vasculitis; peripheral nerve involvement may also result from long-lasting increased intracranial pressure (hydrocephalus for cranial nerves) or a direct compression (carpal tunnel syndrome may result from a tuberculous tenosynovitis); direct infiltration (tuberculous granulomata) of the peripheral nerve has been exceptionnally reported.[8] In tuberculosis, neuropathies may be induced by the neurotoxic effect of some antituberculous medications, such as the combination of isoniazide and pyridoxine (causing a vitamin B6 deficiency).[8] The tuberculosis/ human immunodeficiency virus (HIV) co-infection also represents a serious public health problem.

Finally, very rarely, tuberculous pyomyositis may be observed; it has a different clinical presentation from the classical pyogenic pyomyositis: the onset may be insidious with an indolent presentation and nonspecific symptoms (fever, night sweats, malaise, and weight loss).[9]

15.2.1.4 Brucellosis

Neurobrucellosis represents 7% of the cases of brucellosis, affecting the CNS but also the PNS (acute or chronic demyelinating polyradiculoneuropathy).[10] Myalgias are present in 36.1% of cases of brucellosis,[11] but direct involvement of muscles has exceptionally been described.[12]

15.2.1.5 Tropical Pyomyositis

This entity (also known as "myositis tropicans", "tropical skeletal muscle abscess", or "tropical myositis") corresponds to suppuration within skeletal muscles manifesting as single or multiple abscesses: intermuscular abscesses, abscesses extending into muscles from adjoining tissues, and those secondary to previous septicemia are not considered as tropical pyomyositis. The bacteria that most commonly cause tropical myositis are *Staphylococcus aureus* (90% in tropical areas, vs 75% in temperate zones), Group A *Streptococcus* (in 1%–5% of cases), more rarely *Pneumococcus*, *Streptococcus* from of other groups, *Haemophilus*, and gram-negative bacilli.[13]

The maximum incidence of this pyogenic myositis is observed in persons aged from 10 to 40 years.[13] Classically, three successive stages are observed: (1) the invasive stage (subacute onset of variable fever, painful firm swelling, and sometimes erythema); (2) the suppurative stage (appearing during the second or third weeks, with high spiky fever); and (3) the late stage (corresponding to the dissemination of the infection when the disease is not treated: bacteriemia and septicemia, septic shock, etc.).[13]

15.2.1.6 Leptospirosis and Other Spirochetes

Among the potential complications of leptospirosis, neurological manifestations (neuroleptosirosis) usually consist of aseptic meningitis; neuromuscular involvement may also be observed: acute inflammatory polyradiculopathy (Guillain–Barré syndrome (GBS)-like), polyneuropathy, mononeuritis, neuralgia, autonomic lability, polymyositis, and facial (or other cranial nerve) palsy.[14]

"Cat scratch disease" is caused by *Bartonella henselae*, following a bite, lick or scratch by a cat (less commonly a dog). Acute encephalitis is the commonest neurological manifestation, but PNS disorders have been reported: cranial nerve palsies, peripheral nerve palsies, sensory neuropathy, meningoradiculitis, and chronic inflammatory demyelinating polyradiculoneuropathy (CIDP).[15]

Human granulocytic ehrlichiosis may be a rare cause of peripheral neuropathy and brachial plexopathy.[16]

15.2.1.7 Rickettsiosis

Cranial nerve palsies have been reported in cases of meningitis with *Ricketssia typhi*.[17]

"Scrub typhus" is another rickettiossis caused by *Orientia tsutsugamushi* (vector: trombiculid mite larvae), rarely giving brachial plexopathy.[18]

"African tick bite fever" is caused by *Rickettsia africae* (vector: cattle tick of the *Amblyomma* genus): a few cases of subacute neuropathy (radiating pain, paresthesia and/or motor weakness of extremities, hemifacial pain and paresthesia, and one case of unilateral sensorineural hearing loss) have been reported.[19]

"Rocky montain spotted fever" (observed in Central and South America) is caused by *Rickettsia rickettsii* (vector: *Dermacentor variabilis* and other American ticks) may give cranial nerve palsies, peripheral neuropathy and sensorimotor hearing loss (sometimes with long-term sequellae).[20]

"Mediterranean spotted fever" (observed in India and Africa) is caused by *Ricketsia conorii* (vector: ticks of the *Rhipicephalus* genus). Neurological complications are not rare, mainly involving CNS: PNS involvements are rare, comprising meningoradiculitis, GBS, and cranial nerve palsies.[21]

15.2.1.8 Diphtheria

Neurological manifestations occur in 15% of the cases, usually after a severe pharyngeal infection. Diphtheritic polyneuropathy is a severe acute polyneuropathy mimicking GBS but with a higher occurrence of bulbar (98%) and respiratory (30%) involvement, as well as autonomic disturbances in almost all patients.[22] The hallmark sign of diphtheritic

polyneuropathy is the early presence of bulbar and respiratory involvement (initially without severe limb involvement), then an evolution lasting more than 4 weeks; myocarditis and pulmonary complications represent the major cause of death. The administration of diphtheritic antitoxins is a therapeutic option in this disease (with endotracheal ventilation if necessary), but seems ineffective if administered after the second day of diphtheritic symptoms.[22]

15.2.1.9 Botulism

Botulism may cause a wide spectrum of clinical manifestations ranging from mild hypotonia to a combination of bilateral cranial nerve palsies, flaccid paralysis, and diaphragmatic weakness; clinical presentation is often more severe in type A than in type B botulism; treatment is based on the administration of intravenous botulism immunoglobulins and supportive care.[23]

15.2.1.10 Tetanus

Tetanus is due to powerful exotoxins secreted by spores (*Clostridium tetani*) widely distributed in the soil (introduced into human tissue under anaerobic conditions): tetanolysin damages otherwise viable tissue, whereas tetanospasmin causes the neurological manifestations; tetanospamin binds to local nerve terminals throughout the body, then it is internalized and transported by retrograde axoplasmic flow to the cell body (via motor, sensory, and autonomic nerves), and finally moves to the brainstem by retrograde axonal transport and transsynaptic spread.[24] Tetanus can be localized at the site of injury, but generalized tetanus is the commonest form (trismus, muscle rigidity, and reflex spasms). Its treatment is based on antibiotherapy and the administration of antitetanus serum; however, tetanus is preventable by vaccination.[24]

15.2.1.11 Typhoid Fever

PNS invovement is rare, with only seven cases observed in a series of 959 patients with typhoid fever:[25] "typhoid polyneuritis" is a sensorimotor polyneuropathy; the sensory signs (paresthesia and acute paroxystic pain) usually precede motor weakness (commonly initially localized in the lower limbs, more marked at the extremities and in the extensors);[7] autonomic neuropathy is exceptionnal;[26] motor neuron disease has been observed in one patient.[25] Its evolution is usually favorable (completely recovery after several months), but post-typhoid neuralgia have been described.[7]

15.2.2 Viral Infections

15.2.2.1 Human Immunodeficiency Virus

Distal symmetric polyneuropathy (DSP) represents the most common neurological complication of HIV, estimated to be present in more than 50% of patients with advanced disease; its clinical presentation is similar to other forms of DSP: symmetric and predominantly distal sensory loss, sometimes with numbness, tightness, burning, or hyperelgesia of the feet. Its treatment is currently focused on the management of the neuropathic pain.[27] Other neuropathic conditions also occur in HIV infection, such as inflammatory demyelinating polyneuropathies, mononeuropathies (entrapment neuropathies are common in HIV patients), multiplex mononeuropathies, autonomic neuropathies, and polyradiculopathy. The causes of these neuropathies also depend of the context: inflammatory neuropathies usualy occur at higher CD4 counts, neuropathies due to opportunistic infections occur at lower CD4 counts, and some neuropathies may be due to a co-infection (syphilis, varicella-zoster virus (VZV), or tuberculosis).[27]

Muscular disorders are rare in patients with HIV, but many forms have been reported (symptoms ranging from myalgias and asymptomatic elevation of creatine kinase (CK) to rhabdomyolysis). The most common form is the HIV-associated polymyositis, quite similar to polymyositis observed in HIV-negative patients: it may occur at all stages of the disease as a slowly progressive proximal and symmetric weakness of the limbs: its treatment is based on corticotherapy and immunomodulatory therapies. Toxic myopathy may also be induced by some antiretrovirals.[27]

Other NMDs are rare. Motoneuron disorder may be observed in HIV patients, the differences with the HIV-negative patients being a younger age of onset, a more rapidly progressive course, and a response to antiretrovirals. Myasthenia gravis (MG) has been reported in the course of the disease, without a clear causal relationship. Finally, the rare diffuse infiltrative lymphocytosis syndrome (a Sjögren-like syndrome) may induce a peripheral neuropathy and rarely a muscular involvement.[27]

15.2.2.2 The Neurotropic Herpes Viruses

Primary infection with VZV (varicella, or chickenpox) may cause acute inflammatory demyelinating polyneuropathy (AIDP); reactivation of VZV may induce AIDP, as well as peripheral facial nerve (or other cranial nerve) palsy, focal motor or diaphragmatic weakness, radiculopathy, sphincter dysfunction, and post-herpetic neuralgia;[28] rarely, myositis is associated with viral infection such as VZV.[29]

Meningitis is the most frequent neurological complication of herpes simples virus-2 (HSV-2) infection, and may be sometimes complicated by lumbosacral radiculopathy or sphincter dysfunction (sometimes isolated);[30] AIDP has rarely been reported in a context of HSV-2 infection.[31]

15.2.2.3 Cytomegalovirus

Cytomegalovirus (CMV) neuropathy is a treatable neuropathy occuring at a late stage of immunodeppression (such as acquired immune deficiency syndrome (AIDS)). This neuropathy is often associated with retinitis or with other symptoms due to CMV infection.[32] Two disctint clinical patterns of this neuropathy have been observed: polyradiculopathy and multifocal neuropathy, sometimes with a mixture of both; usually, cerebrospinal fluid (CSF) abnormalities are found, with pleiocytosis, decreased CSF glucose, and high levels of protein. It is important to make the diagnosis of CMV neuropathy because it is accessible to a specific treatment by ganciclovir or foscarnet.[3] Moreover, after *Campylobacter jejuni*, CMV is the most frequent pathogen (13%−22%) inducing GBS.[33]

Rarely, myositis and rhabdomyolysis are associated with CMV infection.[34]

15.2.2.4 Human T-lymphotropic Virus 1

Polyneuropathy may be associated with tropical spastic paraparesis (TSP).[35] Polymyositis and body inclusion myositis have been associated with human T-lymphotropic virus 1 (HTLV1) infection, mostly in a context of TSP: it is characterized by myalgias and raised concentrations of CK in the serum.[35]

Motoneuron disorder has been reported in HTLV1 infection; the differences with a classical amyotrophic lateral sclerosis (ALS) are a slower progression, early bladder dysfunction, possible sensory signs, and sometimes a positive response to steroids.[35]

15.2.2.5 The Hepatitis Viruses

Cases of peripheral neuropathy, MG, and myopathy have been occasionnally reported in hepatitis A virus (HAV) infection.[36]

PNS involvement after hepatitis B virus (HBV) infection consists in neuropathy due to HBV-related polyarteritis nodosa, mononeuropathy due to HBV-related non-panvasculitis, GBS, as well as chronic neuropathies (CIDP, neuropathy simplex/multiplex); inflammatory myopathy is extremly rarely reported.[37]

Hepatitis C virus (HCV) is known to induce sensorimotor peripheral mononeuropathy or polyneuropathy in the context of mixed cryoglobulinemia (found to up to 30% of HCV-positive cryoglobulinemic patients);[38] MG and myopathy have also rarely been reported.[39]

PNS involvement is a rare complication of hepatitis E virus (HEV) infection, mainly being GBS and brachial neuritis (rarely cranial nerve palsies)[40] or acute ataxic neuropathy.[41] Few cases of myositis induced by HEV have been reported.[42]

The use of interferon as a treatment of hepatitis may induce peripheral neuropathy. Finally, various NMDs have been described after vaccination against HAV (GBS, inflammatory myopathy, and macrophagic myofasciitis)[36] and HBV (GBS, myositis, MG),[43] even if epidemiological surveys have failed to establish unequivocal causality between the hepatitis vaccine and the development of these autoimmune disorders.

15.2.2.6 Arboviruses

Dengue virus-related PNS involvement is due to an autoimmune reaction to the virus, inducing GBS or Miller Fisher syndromes (representing 30% of the neurological complications of Dengue virus infection); these forms are similar to the classical GBS. Focal neuropathies (sometimes affecting the cranial nerves) may occur in Dengue virus infection, as well as benign myositis (by direct viral invasion or immune-mediated damage of muscle fibers).[44]

GBS has also been less frequently reported with other arboviruses such as Zika virus,[45] West Nile virus,[46] chikungunya,[47] and Japanese encephalitis virus.[48]

15.2.2.7 Poliovirus

Acute anterior poliomyelitis (polio) is due to poliovirus infection; it is both sporadic and epidemic, often found in isolated areas, most common in summer, and customarily attacks infants under 3 years of age.[49] The cardinal symptom is motor weakness that may come on the first day (sometimes until the seventh day). It begins with full intensity and, when discovered, is at its maximum; usually it is a lower extremity paralysis (but muscles elsewhere may be affected). Spinal or brain foci of irritation determine the seat of paralysis. Reflexes are lost (not dependent upon the muscles paralyzed). Electromyography confirmed a diffuse neurogenic pattern. Since its launch in 1988, the "Global Polio Eradication Initiative" (vaccination) spearheaded interruption of indigenous wild poliovirus transmission of all three serotypes in all but three countries (Afghanistan, Pakistan, Nigeria) by 2013;[50] however, as long as wild polioviruses circulate anywhere, then can cause outbreaks in previously polio-free areas that (for political or other reasons) do not maintain high population immunity through intense vaccination, as in Nigeria in 2003.[51]

Post-polio syndrome is a condition that can affect polio survivors years after recovery from an initial paralytic attack by the poliovirus. It is characterized by progressive or new muscle weakness or decreased muscle endurance in muscles that were previously affected by the poliovirus infection; other symptoms may include generalized fatigue and pain.[52]

15.2.2.8 Rabies

This infection induced by the rabies virus genotype 1 is endemic worldwide, but is mainly observed in Asia and Africa; it is commonly associated with dogs bites, but also with bats bites in America.

Two-thirds of the patients infected with dog rabies virus present the classical "furious rabies": sensory loss (sensory neuronopathy) and focal neuropathic pain, fluctuating consciousness and psychiatric disturbance, phobic spasms, before becoming comatose. The remaining third develops the "paralytic rabies": sensory loss (sensory neuronopathy) and focal neuropathic pain, then ascending motor weakness (predominantly involving proximal and facial musculature) with electrophysiological evidence of peripheral demyelination and axonopathy, but with preserved consciousness.[53] Rabies associated with bat rabies virus variants presents atypical features (focal brainstem signs, myoclonus, hemichorea, or Horner's syndrome).[53]

Altough it can be treated through the timely administration of post-exposure prophylaxis, approximatively 59,000 human deaths per year are caused worldwide by rabies.[54]

15.2.2.9 Rubella

PNS involvement is rare, manifesting as GBS and brachial neuritis: it is usually painful, with signs of demyelination on electrophysiological study[7]. GBS has also been observed after vaccination against rubella virus.[55]

15.2.2.10 Influenza

In a Mexican study, among the potential neurological complications of H1N1 influenza, PNS involvement appears to be the most frequent (GBS or cranial neuropathies).[56] GBS may occur after influenza infection as well as after seasonal influenza vaccination.[57]

15.2.3 Parasitic and Fungal Infections

15.2.3.1 American Trypanosomiasis

Acute Chagas' disease (CD) gives various symptoms (fever, malaise, etc.), and leads to chronic stages (two-thirds being asymptomatic); the PNS may be involved with autonomic neuropathies.[58] Of 511 patients with CD, signs of neuropathy have been found in 52 patients.[59]

Muscular pain and weakness are frequent symptoms in chagasic patients. American trypanosomiasis may lead to acute myositis: muscle biopsies have shown histological and ultrastructural changes in myofibers.[60]

15.2.3.2 Trichinellosis (or Trichinosis)

Muscle involvement is characterized by myalgias and proximal weakness (more intense a few weeks after the primo-infection), sometimes contractures; it is usually associated with skin abnormalities mimicking dermatomyositis.

Neurological complications are variable and called "neurotrichinellosis"; most of them affect the CNS, but cranial nerve palsies and polyneuropathy have been reported.[61]

15.2.3.3 Cysticercosis

Most of the neurological manifestations are due to brain or spinal cord lesions. Rarely, radiculopathy has been observed in spinal cysticercosis.[62]

Rarely, symptomatic skeletal muscle involvement has been reported: the clinical manifestations are variable, ranging from the asymptomatic form to mild tenderness and either muscle atrophy or hypertrophy; muscle tomodensitometry and magnetic resonance imaging (MRI) may help with diagnosis, showing calcifications, scolex, and cysts.[63]

15.2.3.4 Echinococcosis

Spinal echinococcosis may give compression of the rootlets; cases of mononeuropathy by compression of the nerve by an adjacent hydatid cyst are exceptional;[64] an immune-mediated mechanism has been proposed in one case of polyneuropathy in the context of hydatidosis.[65]

15.2.3.5 Schistosomiasis (or Bilharziasis)

Neuroschistosomiasis consists in a severe complication of the disease, but granuloma (around Schistosoma eggs) mainly develop in the brain and/or in the spinal cord. Schistosoma eggs may also involve skeletal muscles: motor weakness (with focal or generalized amyotrophy) has been observed;[66] neuritis has been reported in some cases;[67] recurrent lumbosacral and brachial plexopathies have been observed after Schistosoma japonica infection.[68]

15.2.3.6 Malaria

NMDs are rare complications of Plasmodium infections: 11 cases of GBS have been reported following malaria (half of the patients infected by Plasmodium falciparum have developed respiratory complications, whereas none of those with Plasmodium vivax infection have presented such troubles).[69] Periodic paralysis has been reported in one patient, probably caused by the transient rise of plasma potassium described during febrile episodes of malaria (in a patient with some genetic predisposition).[70]

Many drugs are used in treatment and prophylaxis of malaria, particularly quinolines and related drugs. Several cases of reversible "neuromyopathy" occuring after chloroquine treatment or prophylaxis have been described since the 1960s: in such cases, microscopical studies of peripheral nerves show segmental demyelination/remyelination, as well as cytoplasmic lipidic inclusions in Schwann cells, perineurium, and endoneurium (perineurial calcifications and curvilinear profiles, in perineurium and Schwann cells, may be observed).[71]

15.2.3.7 Gnathosomiasis

One of the main manifestations of neurognathomiasis are radicular pains and headaches; radiculitis may be associated with focal motor weakness due to the involvement of cranial nerves or spinal roots; multiple cranial nerve palsies may be observed in patients with rhombencephalitis.[72]

15.2.3.8 Angiostrongyliasis

The main clinical feature of angiostrongyliasis is eosinophilic meningoencephalitis; other manifestations are radiculitis (radicular pain), cranial nerve palsies, myelitis, and cerebellar ataxia, rarely respiratory failure and coma.[73]

15.2.3.9 Dracunculiasis

The Dracunculiasis worm may die into the subcutaneous tissues (especially in case of trauma or its injudicious extraction), leading to cellulitis or chronic abscess (depending on the "rupture" of the worm): rarely, entrapment neuropathy may be induced by abscessed and calcified aberrant worm in subcutaneous tissues.[74]

15.2.3.10 Infection by Haycocknema perplexum

This parasite is able to complete its life cycle within human muscle, giving myositis: muscle biopsy shows endomysial aggregates of lymphocytes, histiocytes, and eosinophils (associated with actively degenerating/regenerating fibers); parasites are visible within myofibers (female worms, with or without eggs, containing retractile granules in their rectal region; rarely, larvae are seen invaded myofibers).[75]

15.2.3.11 Coenuriasis

Human infections of coenuriasis are characterized by subcutaneous cysts, but muscle and CNS may also be involved.[76]

15.2.3.12 *Toxoplasmosis*

In immunocompromised patients, PNS is rarely involved: few cases of GBS have been reported.[77] Polymyositis has also been described.[78]

15.2.3.13 *Onchocerciasis*

A case of median nerve compression has been reported in onchocerciasis.[79]

15.2.3.14 *Giardiasis*

Two cases of polyneuropathy have been reported in the context of giardiasis, probably as a consequence of the malabsorptive syndrome caused by the parasite.[80]

15.2.3.15 *Bothriocephalus* Infection

Bothriocephalus infection causes "megaloblastic bothriocephalus anemia" due to vitamine B12 deficiency (rarely to folic acid deficiency),[81] then potentially sensory polyneuropathy.

15.2.3.16 *Leishmaniasis*

Neurological manifestations are not common in leishmaniasis. "Neuritis" (inflammatory cells with the perineurium, as well as in dermal nerves) has been demonstrated in both human and experimental cutaneous leishmaniasis.[82]

15.2.3.17 *Sarcocystosis*

Sarcocystidae are known to produce muscular infestations in humans: all patients develop fever and myalgia within 17 days; muscle biopsy may show inflammation (sometimes mild to moderate) and sarcocysts within muscle fibers (detected by light microscopy).[83]

15.2.3.18 *Microsporidial Infections*

Microsporidia represent a group of obligate intracellular organisms spreading among hosts through a spore stage. The microsporidian *Anncaliia algerae* is an emerging human pathogen causing various manifestations in immunocompromised hosts; it may induce severe myositis, also reported after infection with *Anncaliia vesicularum* and *Anncallia connori*, as well as other microsporidia such as *Trachipleistophora*, *Pleistophora*, and *Tubulinosema* species.[84]

Anncaliia algerae myositis is an uncommon infection with high case-fatality rate. It causes fever, weight loss, fatigue, generalized muscle weakness and pain, dysphagia, glossitis, peripheral edema, and diarrhea. Muscle biopsy shows necrotic myocytes containing spores: ovoid spores are observed after silver stains (Warthin-Starry) and by electron microscopy (with a possibility to differentiate *Anncaliia* species from other microsporidia). Treatment is based on the reduction of immunosuppression, measures to prevent complications, and albendazole.[84]

15.3 MALNUTRITION AND MICRONUTRIENT DEFICIENCIES

Underfeeding (a serious problem in some tropical areas) leads to metabolic response and hypovitaminosis; malnutrition may also be caused by chronic disease, such as parasitic infection of the digestive tract (*Giardia* and *Bothriocephalus* infections).

15.3.1 Vitamin B1 (Thiamin) Deficiency

It causes *beriberi*, which is a Hindustani word meaning "unsteady gait"; it develops when the consumer absorbs less than 0.4 mg of thiamine daily for a fairly long period, with a fat-free diet.[85] Beriberi polyneuropathy represents a classical model of polneuropathy due to undernourishment/malnutrition; its classical presentation is a symmetrical sensorimotor neuropathy (affecting the lower limbs) with edema of the feet and cardiac enlargement.[86] Rarely, thiamine deficiency gives acute motor weakness mimicking GBS. Both acute and chronic PNS involvement may improve after thiamine administration.[87]

15.3.2 Vitamin B2 (Riboflavin) Deficiency

Vitamin B2 (riboflavin) deficiency leads to angular stomatitis, seborrheal dermatitis of the face, and blepharal eczema. This hypovitaminosis plays a role in causing pellagra polyneuropathy and complex malabsorption syndromes with neuropathy.[85]

15.3.3 Vitamin B3 (Niacin) Deficiency

Pellagra is due to deficit of niacin causing pathognomonic skin abnormalities, CNS/PNS involvement, and chronic diarrhea. The first manifestations of pellagra polyneuropathy are paresthesia and "burning sensation" in the feet, the hands, and sometimes the face (with normal deep tendon reflexes in most of the cases), before a distal symmetrical motor weakness of the lower limbs appears (with a decrease of motor nerve conduction velocity).[85]

15.3.4 Vitamin B5 (Pantothenic Acid) Deficiency

Vitamin B5 (pantothenic acid) deficiency is characterized by symptoms of peripheral neurological excitation (sensory overexcitability of the feet, and spontaneous painful "burning feet") and predominant sensory polyneuropathy affecting the lower limbs.[85]

15.3.5 Vitamin B6 (Pyridoxine) Deficiency

Vitamin B6 (pyridoxine) deficiency causes a predominantly sensory polyneuropathy, with pathological signs (sural nerve biopsy) close to those of thiamin deficiency.[85]

15.3.6 Vitamin B8 (Biotin) Deficiency

Vitamin B8 (biotin) deficiency brings on clinical symptoms similar to those of vitamin B5 deficiency.[85]

15.3.7 Vitamin B12 (Cobalamin) Deficiency

Polyneuropathy is an established complication of cobalamin deficiency (sometimes its first symptom). It is characterized by less pain and lower limb weakness than patients with idiopathic polyneuropathy, with more likely concomitant involvement of the lower and upper extremities (sometimes with onset in the hands).[88] Symptom duration appears to be a major determinant for the extent of neurological improvement after treatment (cyanocobalamin): it is possible that this replacement therapy needs to be started early for significant improvement in PNS symptoms.[88]

15.3.8 Copper Deficiency

Copper has specific functions in metabolic processes, neurotransmitters, brain peptide synthesis, and oxydative defense: alterations of its homeostasis result in increased production of free radicals and oxydative damage leading to neurological disorders. The main neurological complication of copper deficiency is progressive myelopathy with ataxic gait and neuropathy: conversely to cobalamin deficiency, PNS involvement is rarely inaugural, corresponding to a mixed, pure sensory, or pure motor axonal peripheral neuropathy; copper administration may improve the symptoms.[89]

15.3.9 Tropical Sprue

Tropical sprue (TS), endemic in some areas such as the Carribean (sporadic in South America, Africa, and India), is believed to be uncommon currently in spite of contrary evidence. Pathogenesis of TS is multifactorial: gut dysbiosis (nowadays, no specific pathogen has been found, but patients frequently have coliform colonization of the proximal small bowel), muccosal disaccharidase deficiency, deficiency of vitamin B12 and folates, and colonic dysfunction.[90] Patients with TS present with chronic diarrhea, anorexia, weight loss, (megaloblastic) anemia, fatigue, and sometimes neuropathy (in part due to the vitamin deficiency). TS responds to treatment with antibiotics (tetracycline) and folic acid.[90]

15.3.10 Tropical Ataxic Neuropathy

Tropical ataxic neuropathies have been known of for more than a century, since the first report by Strachan.[91] They constitute a heterogeneous group of neuromyelopathies of obscure etiology that occurs in geographical isolates, in tropical regions of Africa, Asia, and Latin America.[92] Usually most of the patients come from a rural background.

Clinically, tropical neuropathies are broadly categorized into tropical ataxic neuropathy (TAN: prominent sensory ataxia and symmetrical sensory polyneuropathy, most of the patients complaining of "burning feet"), sensory-motor polyneuropathy and TSP (prominent spastic gait disturbance with minimal sensory deficit).[92] A majority of patients have gradual onset of symptoms, while acute or subacute onset rarely occurs. Pathological reports are quite rare in the literature:[93] microscopic signs consist of severe axonal involvement affecting both unmyelinated and myelinated fibers. It is uncertain whether the clinical heterogeneity of tropical myeloneuropathies is related to differences in the etiopathogenesis, to genetic predispositions, or to a combination of these factors. The early descriptions attributed tropical neuromyelopathies to a nutritional deficiency resulting from combined poor dietary intake and malabsorption; later reports incriminated cassava neurotoxicity.[94]

Some authors have proposed that TAN may be explained by vitamin deficiencies, but studies have shown that serum levels of niacin, folic acid, pyridoxine, cobalamin, and panthothenic acid are normal in patients with TAN; however, there is some evidence implicating thiamin deficiency in TAN so that a long-term thiamin supplementation program for susceptible individuals in endemic areas may be effective in the control and eventual eradication of the disease.[95]

15.4 TOXIC CAUSES

Toxic neuropathies and myopathies are caused by drugs, industrial and environnemental agents, and heavy metals (Table 15.1). In some cases, a nerve injury may also be the consequence of a local compression (hemorrhage in the context of anticoagulant treatment) or the local injection of a cytotoxic agent such

TABLE 15.1   Main Drugs/Substances Inducing Neuropathies and Myopathies

	Paresthesiae	Sensory polyneuropathy	Distal sensorimotor polyneuropathy	Predominantly motor polyneuropathy	Myopathy/rhabdomyolysis
Antineoplastic drugs	Cytarabine (Ara-C)	Thalidomide Cisplatin (and platinum drugs) Procarbazine Nitrofurazone Taxanes Cytarabine (Ara-C)• Etoposide	Vincristine/vinca alkaloids Podophyllum Chlorambucil Thalidomide Taxanes Cytarabine (Ara-C)• Etoposide	Vincristine/vinca alkaloids Sulfonamides Amphotericin Cytosine arabinoside• Suramin•	Vincristine (AM) Hydroxyurea□
Antimicrobial drugs	Colistin Streptomycin Nalidixic acid	Ethionamide Choramphenicol Thiamphenicol Diamines Metronidazole Nitrofurantoin NRTI	Isoniazid Ethambutol Streptomycin Nitrofurantoin Clioquinol Metrodinazole Chloroquine/ hydroxychloroquine Anti-retrovirals (stavudine, didanosine, zalcitabine)	Nitrofurantoin Suramin•	Ofloxacin/levofloxacin Daptomycin Anti-retrovirals (zidovudine, tenofovir/ abicavir, raltegravir) (MM) Chloroquine/hydroxychloroquine Amphotericin (HM)
Cardiovascular drugs	Propanolol		Amiodarone• Perhexiline maleate• Hydrallazine Disopyramide Clofibrate	Amiodarone• Perhexilline maleate	Cholesterol-lowering agents† Amiodarone Labetolol† Proprolol† Diuretics (HM)
Hypnotics and psychotropics	Phenelzine		Methaqualone Glutethymide Amitriptyline	Imipramine	Tricyclicic antidepressants Other antidepressants (venlafaxine, sertraline, escitalopram) Antipsychotic (risperidone, haloperidol, aripriprazole, olanzapine, quetiapine, clozapine) Depakote
Antirheumatic drugs			Indomethacin Colchicine Chloroquine Phenylbutazone Cyclosporine D-penicillamine•	Tacrolimus• Cyclosporine D-penicillamine•	Colchicine (AM) Cyclosporine† D-penicillamine□ Corticosteroids (HM)
Other drugs	Sulthiame Chlorpropamide Methysergide Topiramate	Calcium carbimide Sulfoxone Ergotamine Propylthiouracil Pyridoxine (vitamin B6) Podophyllin Almitrine Interferon alpha Disulfiram Clioquinol	Phenytoin Disulfiram Carbutamide Tolbutamide Chlorpropamide Methimazole Lithium•	Dapsone Exitotoxin• Saxitoxin• Cimetidine	Aminocaproic acid Antihistamines Interferon alpha□ Lithium Phenytoin□ Lamotrigine□ L-tryptophan□ Cimetidine□ Imatinib□ Laxatives (HM) Omeprazole Isoretinoin Finasteride Emetine Nondepolarizing neuromuscular blocking agents

(Continued)

IV. FOCUS ON SPECIFIC NEUROLOGICAL SYNDROMES OR DISEASES IN TROPICAL AREAS

TABLE 15.1   (Continued)

	Paresthesiae	Sensory polyneuropathy	Distal sensorimotor polyneuropathy	Predominantly motor polyneuropathy	Myopathy/rhabdomyolysis
Industrial and environmental agents		Alcohol Acrylamide• Ethylene oxyde• Vinyl benzene (styrene)•	Alcohol Organophosphates Vinyl benzene (styrene)•	Organophosphates Hexacarbons (glue sniffer)• Carbon disulfide• Dichloro-diphenyl-trichloroethane (DDT)	Alcohol[†] or HM Toluene (HM) Liquorice (HM)
Heavy metals		Thallium	Mercury Gold•	Lead Arsenic•	

For neuropathies: some drug-induced neuropathies may sometimes be associated with a conduction slowing on ENMG (•).
For myopathy/rhabdomyolysis: †, necrotizing myopathy; ¤, inflammatory myopathy.
AM, antimicrotubular myopathy; HM, hypokaliemic myopathy; MM, mitochondrial myopathy; NRTI, nucleoside analog reverse-transcriptase inhibitors.

as sciatic nerve injury after intramuscular injection in the gluteal muscle (quinine, etc.). Various mechanisms may explain the occurrence of toxic myopathies (depending on the drug): necrotizing myopathy, drug-induced autophagic lysosomal myopathy, antimicrotubular myopathy, mitochondrial myopathy, inflammatory myopathy, hypokaliemic myopathy, or critical illness myopathy.[96] Moreover, MG may be exacerbated by many drugs, such as antimalarials and antibiotics (Table 15.2).

15.4.1 NMDs Caused by Envenimation

15.4.1.1 Spider Bites

Some neuromuscular manifestations may occur after *Latrodectus* species bites (generalized myalgia, fasciculations, and paresthesias), but after other symptoms of lactrodectism (intense pain at the site of the bite, then arterial hypertension, profuse diaphoresis, nausea/vomiting, chest and abdominal pain, rarely myocarditis and acute renal failure).[97]

Funnel-web spiders (*Hexathelidae*) envenoming causes hyperstimulation of postsynaptic receptors (through the action of δ-atracotoxins) resulting in sensory disturbance and muscle paralysis, as well as autonomic dysfunction.[97]

15.4.1.2 Scorpion Stings

After a scorpion bite, systemic manifestations occur in a delay from 10 minutes to 24 hours, consisting of muscle weakness (with generalized rigidity, increased muscle stretch reflexes, and fasciculations through disturbance of the neuromuscular transmission), with also cardiac manifestations (cardiac arrythmia, myocarditis, and pulmonary edema) and CNS involvement (seizures, tremors, and cerebrovascular complications); in some cases, rhabdomyolysis may be observed.[97]

15.4.1.3 Snake Bites

Damage of the PNS is more frequent after the bite of elapids (occuring in 2%–12%), rarely after the bite of vipers. Muscle weakness (induced by a pre- or postsynaptic blockade of the neuromuscular junction) is caused by the effect of α-neurotoxin and β-neurotoxin (starting from minutes to a few hours after the inoculation of venom). Usually, the first neuromuscular signs are ptosis and external ophthalomplegia, then other muscles are affected in the following hours (facial muscles, muscles of deglutition, vocal cords, neck muscles, intercostal muscles, and the diaphragm), before the installation of a flaccid quadriparesis (the nerve conduction velocity is initially most often normal). In most cases, weakness is reversible and patients show complete recovery after a few days, provided that proper therapy (including mechanical ventilation) had been promptly started.[97]

Generalized myokimia (syndrome of continuous and spontaneous muscular activity) may be observed after the bite of some rattlesnake species, due to a particular neurotoxin acting on voltage-gated potassium channels.[97]

Acute polyneuropathy (including some cases of typical GBS) have been rarely observed, probably induced by an immune-mediated reaction to the venom itself (or to antivenom or toxoid administration).[97]

15.4.1.4 Marine Poisoning

Marine poisoning (usually giving gastrointestinal and neurological symptoms) results from the ingestion of marine animals (an important part of human diet worldwide) containing toxic substances; its incidence is beyond 1200/100,000 people per year in the Pacific area. Marine toxins affect voltage-gated sodium channels in myelinated and unmyelinated peripheral nerves.[98]

TABLE 15.2 Drugs That may Induce or Exacerbate Myasthenia Gravis

	Drugs contraindicated in MG	Drugs to be used with caution in MG
Antibiotics	Aminoglycosides Tetracycline Colistine Telithromycine Polymyxin A and B	Aminoglycosides (topical treatment) Lincomycine Clindamycine Neomycine Fluoroquinolones Thyrothricine Bacitracine Erythromycine
Antimalarials	Quinine/chloroquine Mefloquine	–
Cardiovascular drugs	Quinidine Procainamide Beta-blockers Ajmaline	Lidocaine Calcium antagonist drugs Furosemide (hypokaliemia) Statines Trimetaphan
Drugs used during anesthesia	Non-depolarizing agents Flunitrazepam Midazolam Halothane Succinylcholine	Barbiturates Ketamin Propanidid
Psychotropics	–	Chlorpromazine Tricyclic antidepressants (high dose) Benzodiazepine (high dose) Carbamates Zopiclone Zolpidem
Anticonvulsivants	Phenytoin	Carbamazepine (high dose) Gabapentin Ethosuximide
Antirheumatics drugs	D-penicillamine	Colchicine
Analgesics	–	Morphine
Antineoplastic drugs	Cisplatin	–
Other drugs	Magnesium sulfates Dantrolen Baclofen Trimethadione Halofantrine Oxybutynine Chlorproethazine Tiopronin Botulinic toxin Quinquina	Corticosteroids Lithium carbonate Fluphenazine Calcium injection Phenothiazine Interferon apha Myorelaxant drugs Pyritinol Liquorice (hypokaliemia) Riluzole Glatiramer acetate Gadopentetate dimeglumine (gadolinium) Emetine Antitetanus immunoglobulins Nicotine patch Mydriatics

MG, myasthenia gravis.

Ciguatera (endemic throughout tropical and subtropical areas) is the most common cause of marine poisoning, caused by the ingestion of ciguatoxins and their metabolites (resulting from the transformation of gambierotoxins produced by the plankton dinoflagellates *Gambierdiscus toxicus*) accumulated in some tropical fishes. It is rarely fatal, usually causing moderate neurological (myalgia, "electric-shock"-like shooting

IV. FOCUS ON SPECIFIC NEUROLOGICAL SYNDROMES OR DISEASES IN TROPICAL AREAS

pains, cold allodynia, pruritis, and ataxia) and gastrointestinal (vomiting, diarrhea, and abdominal cramps) manifestations, 1–48 hours after ingestion; the polyneuropathy is a sensory length-dependent neuropathy involving large and small fibers (but with a prominent small-fiber dysfunction).[98]

Puffer fish poisoning (mainly occuring in Southeast Asia) is due to the consumption of some fishes, such as puffer fish fillet ("fugu"), containing tetrodotoxin (also present in xanthid crabs, horseshoe crabs, and other fish species). The symptoms (that seem to depend on the amount ingested) are first sensory disturbances (perioral and distal limb numbnesses and paresthesia) and gastrointestinal manifestations; after more important ingestion, weakness appears on skeletal muscles (with also bulbar signs), before ataxia; in the more severe form of the disease, there is a generalized flaccid paralysis with respiratory failure, aphonia, and fixed dilated pupils. No antidote is available, with only the need for supportive care.[98]

Paralytic shellfish poisoning is due to the consumption of some bivalve molluscs that may become toxic for humans during certain seasons (worldwide, but mainly in temperate areas); it is due to the presence of the curarizing agent saxitoxin (induced by microalgae dinoflagellate). The symptoms are sensory disturbances (paresthesias) and sometimes muscular paralysis; saxitoxin, similar to tetrodotoxin (but far more potent), is able to block tetrodotoxin-sensitive sodium channels disrupting nerve conduction and resulting in motor and sensory nerves abnormalities. The first symptoms (occuring 30 minutes to 3 hours after the ingestion) are paresthesia (tingling and numbness of the tongue and the lips, then the face and the neck, and finally fingers and toes) with a sensation of "dizziness"; then arm and leg weakness occurs, with generalized paresthesia, ataxia, and headache. The effects usually resolve in 2–3 days, but the more severe cases may be fatal within 12 hours. A variant, "neurotoxic shellfish poisoning" (caused by brevetoxins) is limited to the west coast of Florida, North Carolina, and New Zealand: it is clinically similar to ciguatera.[98]

Clupeotoxism, resulting from the ingestion of plankton-eating fish (sardines and herrings), has been reported in Carribean and Indo-Pacific regions; its cause is not well known, but it could be due to the palytoxin. Neurological symptoms are dilated pupils, paresthesia, muscle cramps and paralysis, back pain, then coma (with a high fatality rate).[98]

15.4.1.5 Plant Sources and Other Animal Poisons

As a way of protection (against microorganisms, insects and other animals, or even other plants), many plant species are able to synthesize and accumulate toxic compounds. Traditional societies have used these toxic substances as arrow and dart poisons.

Alkaloids (from Strychnos species) are able to induce two different types of toxic mechanisms: the first one is a tetanizing activity caused by strychnine; the other one is a paralyzing action due to a series of quaternary alkaloids included in curare poisons.[99] Indeed, toxins of many animals have been used for arrow and dart poison, especially for their neuromuscular effects. Dendrobatidea, a variety of frogs, secrete toxins (in their skin) ranking among the most powerful poisons (batrachotoxin and homobatrachotoxin) that are steroidal alkaloids (probably by accumulation through the consumption of some insects containing these toxins): these substances induce irreversible depolarization of nerves and muscles, fibrillation, cardiac arrythmia and possibly cardiac failure. The presence of batrachotoxin alkaloids has also been reported in skin and feathers of toxic passerine birds from New Guinea, probably due to the consumption of a variety of beetle (Melyridae).[99]

15.5 OTHER ACQUIRED NMDS

15.5.1 Diabetes Mellitus

PNS involvement includes diffuse (diabetic sensorimotor polyneuropathy, autonomic neuropathy, small-fiber neuropathy) and focal (carpal tunnel syndrome, entrapment neuropathies, cranial nerve palsies, lumbosacral radiculoplexopathy of Bruns–Garland, femoral neuropathy, and thoracoabdominal neuropathy) neuropathies. If diabetes mellitus (DM) may affect skeletal muscles, clinical myopathy is not typical of patients with DM; however, a myopathy found in a context of diabetes needs to be investigated for other causes of myopathy such as hereditary mitochondrial disorders.[100]

Soft tissue infections are common in DM patients, especially in Africa where hand infections are reported more and more, giving a "diabetic hand" in part due to "peripheral neuropathy ulcerations", ulnar nerve compression, and carpal tunnel syndrome (as well as limited joint mobility, Dupuytren's contracture, trigger fingers, skin and nail pathology, etc.).[101]

15.5.2 Immune-Mediated NMDs

15.5.2.1 Guillain–Barré Syndrome

Guillain-Barré syndrome (GBS) is one of the most severe immune-mediated neuropathies, leading to flaccid paralysis in less than 4 weeks in most cases; some patients may have high level of protein in the cerebrospinal fluid (without cells).[102] Most of the cases recover in weeks or months, but a significant number keep motor sequelae. It is usually secondary to an infection, such as *C. jejuni* and CMV.

In tropical countries, GBS has been described following other infections such as arbovirus infection or malaria. Treatment is based on intravenous immunoglobulins (IgIV) or plasma exchange (PLEX), as well as intensive care support, if necessary.[102]

15.5.2.2 CIDP

CIDP is a chronic acquired disorder, which usually presents as a progressive or relapsing sensory and motor, proximal and distal neuropathy with generalized areflexia, developing over at least 8 weeks; it is a rare disorder that can occur at any age, with a global prevalence of ~1/100,000 to 10/100,000. Its treatment is mainly based on IgIV, steroids or PLEX, but other immunsuppressant drugs have been used in refractory forms.[103]

Other classical immune-mediated neuropathies are multifocal motor neuropathies with conduction blocks, demyelinating neuropathies with monoclonal gammopathy (with or without anti-myelin associated glycoprotein (MAG) antibodies) and paraneoplastic neuropathies.

15.5.2.3 Myasthenia Gravis

MG is an uncommon worldwide autoimmune NMD caused by acetylcholine receptor antibodies at the neuromuscular junction. The hallmark of the disease is a fatigable muscle weakness, usually with ophthalomolgical (ptosis, diplopia) and/or bulbar (dysphagia) manifestations at onset. Another form (rare) is due to the presence of anti-MuSK (anti-Muscle Specific Kinase) antibodies. The treatment is based on cholinesterase inhibitors and immunosuppressants, and sometimes thymectomy; IgIV and PLEX may be used in myasthenic crisis. It is important to know that MG may be exacerbated by many common drugs (including antimalarial drugs) (Table 15.2).

15.5.3 Systemic Diseases

Vasculitic neuropathies may be due to many systemic diseases (rheumatoid athritis, systemic lupus erythematosus, sarcoidosis, etc.) but seem to be uncommon in Africa.[104] Indeed, some differences have been also noted between European populations and some populations living in tropical areas; for example, Polynesians and Maoris are generally not affected by "European autoimmune diseases" (multiple sclerosis, etc.), but develop specific autoimmune disorders.[105]

Sarcoidosis is a chronic multisystemic granulomatous disease of unknown cause (where lungs, eyes, and skin are more frequently affected). Its frequency is lower in African people, but with a high degree of severity.[106] Neuromuscular manifestations (occuring in about 5%–15%) are various and may be isolated (making its diagnosis sometimes difficult). Proximal muscles of the limbs are frequently involved, with three forms described (acute, chronic, and nodular): muscle biopsy may help to make the diagnosis by showing granuloma (inflammatory cells around noncaseating necrotic tissue).[107] Most of sarcoid polyneuropathies are sensorimotor polyneuropathies or pure motor polyneuropathies; however, small fiber neuropathy, lumbosacral plexopathy, and cranial nerve palsies are sometimes reported.[108]

Inflammatory myopathies (myositis) are characterized by subacute motor weakness (mainly proximal) of the four limbs (with myalgia and usually high levels of CK in the serum); skin abnormalities may be observed in dermatomyositis. Inflammatory myopathies are infrequent, mostly affecting people aged from 20 to 40 years in Africa; they are sometimes associated with neoplasia, and their treatment is based on corticosteroids.[109] In the Western world, the classification of myositis is now presented in four subtypes according to the ages of the patients, the clinical presentation and the type of microscopical lesions observed in the muscle biopsy: dermatomyositis, polymyositis, necrotizing autoimmune myositis, and inclusion body myositis.[95] At present, there is no report suggesting that, in tropical countries, such entities may be individualized in this way.

15.6 HEREDITARY NMDS

Many hereditary NMDs have been described worlwide, affecting PNS, muscle, motor neurons or neuromuscular junctions.

The growing number of various implicated genes and mutations implicated in hereditary neuropathies makes their classification more and more complex and difficult.[110] An increased risk of recessive disorders is considered to be the main cause of a higher risk of birth defects among children with consanguineously related parents: consanguineous marriage is customary in the Middle East and parts of South Asia, Zoroastrians, some Jewish communities, and many tribes in SSA and South East Asia.[111] We can give the example of the recessive autosomal form of Charcot–Marie–Tooth disease due to mutations of the *MED25* (mediator complex subunit 25) gene found in Costa Rican patients (but, in this case, the mutation is probably due to Spanish ancestry of immigrants in the 16th century).[112]

Inherited myopathies seem to be less frequent in black populations. For example, in South Africa, the prevalence was 1/250,000 for Duchenne's muscular dystrophy (vs in 1/3500 in high-income countries), and 1/750,000 for Becker's muscular dystrophy (vs 1/35,000−40,000 in high-income countries); indeed, myotonic dystrophy (prevalence of 1/10,000 in high-income countries) is considered to be uncommon in black populations.[109]

15.7 IDIOPATHIC NEUROPATHIES

"Idiopathic neuropathies" are defined as neuropathies without cause found despite investigations: 20%−30% of total neuropathies are said to be idiopathic,[113] but this proportion could be more important in some areas (>50% in Africa).[104] However, with new thorough investigations and tests, a cause can be found in most of cases: impaired glucose metabolism in 25.4%, CIDP in 20%, monoclonal gammopathy in 20%, and various other causes.[113]

ABBREVIATIONS

AIDP acute inflammatory demyelinating polyneuropathy
AIDS acquired immune deficiency syndrome
ALS amyotrophic lateral sclerosis
CD Chagas' disease
CIDP chronic inflammatory demyelinating polyneuropathy
CK creatine kinase
CMV cytomegalovirus
CNS central nervous system
CSF cerebrospinal fluid
DENV Dengue virus
DM diabetes mellitus
DSP distal symmetric polyneuropathy
GBS Guillain-Barré syndrome
HAV hepatitis A virus
HBV hepatitis B virus
HCV hepatitis C virus
HEV hepatitis E virus
HIV human immunodeficiency virus
HTLV human T-lymphotrophic virus
IVIg intravenous immunoglobulins
MAG myelin-associated glycoprotein
MED25 mediator complex subunit 25
MG myasthenia gravis
NCV nerve conduction velocity
NMD neuromuscular disorders
MRI magnetic resonance imaging
PLEX plasma exchange
PNS peripheral nervous system
polio acute anterio poliomyelitis
SSA Sub-Saharan Africa
TAN tropical ataxic neuropathy
TS tropical sprue
TSP tropical spastic paraparesis
VZV varicella-zoster virus

WHO World Healt Organization

References

1. Osuntokun BO, Adeuja AO, Schoenberg BS, Bademosi O, Nottidge VA, Olumide AO, et al. Neurological disorders in Nigerian Africans: a community-based study. *Acta Neurol Scand.* 1987;75:13−21.
2. Jannin J, Gabrielli AF. Neurological aspects of neglected tropical diseases: an unrecognized burden. *Handb Clin Neurol.* 2013;114:3−8.
3. Said G. Infectious neuropathies. *Neurol Clin.* 2007;25:115−137.
4 Chimelli L, Vallat JM. Infectious and tropical neuropathies. In: Vallat JM, Weis J, Gray F, Keohane K, eds. *Peripheral Nerve Disorders: Pathology and Genetics.* 1st ed. Chichester, UK: John Wiley & Sons, Ltd; 2014:196−209.
5. de Zeeuw J, Alferink M, Barogui YT, Sopoh G, Phillips RO, van der Werf TS, et al. Assessment and treatment of pain during treatment of Buruli ulcer. *PLoS Negl Trop Dis.* 2015;9:e0004076.
6. En J, Goto M, Nakanaga K, Higashi M, Ishii N, Saito H, et al. Mycolactone is responsible for the painlessness of *Mycobacterium ulcerans* infection (Buruli ulcer) in a murine study. *Infect Immun.* 2008;76:2002−2007.
7. Boudouresques J, Khalil R, Vigouroux A, Daniel F, Gosset A. Infectious diseases of nerves. In: Vinken PJ, Bruyn GW, eds. *Handbook of Clinical Neurology.* New York: North Holland Publishing Company; 1970:473−494.
8. Orrell RW, King RH, Bowler JV, Ginsberg L. Peripheral nerve granuloma in a patient with tuberculosis. *J Neurol Neurosurg Psychiatry.* 2002;73:769−771.
9. Simopoulou T, Varna A, Dailiana Z, Katsiari C, Alexiou I, Basdekis G, et al. Tuberculous pyomyositis: a re-emerging entity of many faces. *Clin Rheumatol.* 2016;35(4):1105−1110.
10. Kutlu G, Ertem GT, Coskun O, Ergun U, Gomceli YB, Tulek N, et al. Brucella: a cause of peripheral neuropathy. *Eur Neurol.* 2009;61:33−38.
11. Buzgan T, Karahocagil MK, Irmak H, Baran AI, Karsen H, Evirgen O, et al. Clinical manifestations and complications in 1028 cases of brucellosis: a retrospective evaluation and review of the literature. *Int J Infect Dis.* 2010;14:e469−e478.
12. Naha K, Karanth S, Dasari S, Prabhu M. Polymyositis-like syndrome with rhabdomyolysis in association with brucellosis. *Asian Pac J Trop Med.* 2012;5:755−756.
13. Chauhan S, Jain S, Varma S, Chauhan SS. Tropical pyomyositis (myositis tropicans): current perspective. *Postgrad Med J.* 2004;80:267−270.
14. Panicker JN, Mammachan R, Jayakumar RV. Primary neuroleptospirosis. *Postgrad Med J.* 2001;77:589−590.
15. Massei F, Gori L, Macchia P, Maggiore G. The expanded spectrum of bartonellosis in children. *Infect Dis Clin North Am.* 2005;19:691−711.
16. Horowitz HW, Marks SJ, Weintraub M, Dumler JS. Brachial plexopathy associated with human granulocytic ehrlichiosis. *Neurology.* 1996;46:1026−1029.
17. Moy WL, Ooi ST. Abducens nerve palsy and meningitis by *Rickettsia typhi*. *Am J Trop Med Hyg.* 2015;92:620−624.
18. Ting KS, Lin JC, Chang MK. Brachial plexus neuropathy associated with scrub typhus: report of a case. *J Formos Med Assoc.* 1992;91:110−112.
19. Jensenius M, Fournier PE, Fladby T, Hellum KB, Hagen T, Prio T, et al. Sub-acute neuropathy in patients with African tick bite fever. *Scand J Infect Dis.* 2006;38:114−118.
20. Archibald LK, Sexton DJ. Long-term sequelae of Rocky Mountain spotted fever. *Clin Infect Dis.* 1995;20:1122−1125.

21. Alioua Z, Bourazza A, Lamsyah H, Erragragui Y, Boudi O, Karouach K, et al. Manifestations neurologiques de la fièvre boutonneuse méditerranéenne: à propos de quatre observations. *Rev Med Interne*. 2003;24:824–829.

22. Logina I, Donaghy M. Diphtheritic polyneuropathy: a clinical study and comparison with Guillain-Barre syndrome. *J Neurol Neurosurg Psychiatry*. 1999;67:433–438.

23. Rosow LK, Strober JB. Infant botulism: review and clinical update. *Pediatr Neurol*. 2015;52:487–492.

24. Murthy JM, Dastur FD, Khadilkar SV, Kochar DK. Rabies, tetanus, leprosy, and malaria. *Handb Clin Neurol*. 2014;121:1501–1520.

25. Osuntokun BO, Bademosi O, Ogunremi K, Wright SG. Neuropsychiatric manifestations of typhoid fever in 959 patients. *Arch Neurol*. 1972;27:7–13.

26. Blumenfeld AM, Sieling WL. Acute autonomic neuropathy with *Salmonella typhi* infection. A case report. *S Afr Med J*. 1987;71:532–533.

27. Robinson-Papp J, Simpson DM. Neuromuscular diseases associated with HIV-1 infection. *Muscle Nerve*. 2009;40:1043–1053.

28. Steiner I, Kennedy PG, Pachner AR. The neurotropic herpes viruses: herpes simplex and varicella-zoster. *Lancet Neurol*. 2007;6:1015–1028.

29. Joseph TP, Chand RP, Tariq SM, Johnston WJ, Muirhead D, Buhl L. Acute proximal myopathy due to herpes zoster. *J R Soc Med*. 1993;86:360.

30. Léger JM, Blétry O, Karabinis A, Brunet P, Godeau P, Cukier J. Rétention vésicale complète à début brutal secondaire a une infection génitale herpétique. *J Urol (Paris)*. 1982;88:281–283.

31. Cosson A, Tatu L, Decavel P, Parratte B, Rumbach L, Monnier G. Polyradiculoneuropathie aiguë au cours d'une primo-infection à virus Herpes simplex. *Rev Neurol (Paris)*. 2002;158:833–835.

32. Said G, Lacroix C, Chemouilli P, Goulon-Goeau C, Roullet E, Penaud D, et al. Cytomegalovirus neuropathy in acquired immunodeficiency syndrome: a clinical and pathological study. *Ann Neurol*. 1991;29:139–146.

33. Boucquey D, Sindic CJ, Lamy M, Delmee M, Tomasi JP, Laterre EC. Clinical and serological studies in a series of 45 patients with Guillain–Barré syndrome. *J Neurol Sci*. 1991;104:56–63.

34. Sato K, Yoneda M, Hayashi K, Nakagawa H, Higuchi I, Kuriyama M. A steroid-responsive case of severe rhabdomyolysis associated with cytomegalovirus infection. *Rinsho Shinkeigaku*. 2006;46:312–316.

35. Araujo AQ. Update on neurological manifestations of HTLV-1 infection. *Curr Infect Dis Rep*. 2015;17:459.

36. Stubgen JP. Neuromuscular complications of hepatitis A virus infection and vaccines. *J Neurol Sci*. 2011;300:2–8.

37. Stubgen JP. Neuromuscular disorders associated with hepatitis B virus infection. *J Clin Neuromuscul Dis*. 2011;13:26–37.

38. Adinolfi LE, Nevola R, Lus G, Restivo L, Guerrera B, Romano C, et al. Chronic hepatitis C virus infection and neurological and psychiatric disorders: an overview. *World J Gastroenterol*. 2015;21:2269–2280.

39. Stubgen JP. Neuromuscular diseases associated with chronic hepatitis C virus infection. *J Clin Neuromuscul Dis*. 2011;13:14–25.

40. Cheung MC, Maguire J, Carey I, Wendon J, Agarwal K. Review of the neurological manifestations of hepatitis E infection. *Ann Hepatol*. 2012;11:618–622.

41. Bruffaerts R, Yuki N, Damme PV, Moortele MV, Wautier M, Lagrou K, et al. Acute ataxic neuropathy associated with hepatitis E virus infection. *Muscle Nerve*. 2015;52:464–465.

42. Mengel AM, Stenzel W, Meisel A, Buning C. Hepatitis E induced severe myositis. *Muscle Nerve*. 2016;53(2):317–320.

43. Stubgen JP. Neuromuscular disorders associated with Hepatitis B vaccination. *J Neurol Sci*. 2010;292:1–4.

44. Puccioni-Sohler M, Rosadas C, Cabral-Castro MJ. Neurological complications in dengue infection: a review for clinical practice. *Arq Neuropsiquiatr*. 2013;71:667–671.

45. Oehler E, Watrin L, Larre P, Leparc-Goffart I, Lastere S, Valour F, et al. Zika virus infection complicated by Guillain-Barre syndrome--case report, French Polynesia, December 2013. *Euro Surveill*. 2014;19(9):pii: 20720.

46. Asnis DS, Conetta R, Teixeira AA, Waldman G, Sampson BA. The West Nile Virus outbreak of 1999 in New York: the Flushing Hospital experience. *Clin Infect Dis*. 2000;30:413–418.

47. Das T, Jaffar-Bandjee MC, Hoarau JJ, Krejbich Trotot P, Denizot M, Lee-Pat-Yuen G, et al. Chikungunya fever: CNS infection and pathologies of a re-emerging arbovirus. *Prog Neurobiol*. 2010;91:121–129.

48. Ravi V, Taly AB, Shankar SK, Shenoy PK, Desai A, Nagaraja D, et al. Association of Japanese encephalitis virus infection with Guillain-Barre syndrome in endemic areas of south India. *Acta Neurol Scand*. 1994;90:67–72.

49. Wilson WH. Acute anterior poliomyelitis. *J Natl Med Assoc*. 1925;17:64–65.

50. Duintjer Tebbens RJ, Pallansch MA, Cochi SL, Wassilak SG, Thompson KM. An economic analysis of poliovirus risk management policy options for 2013–2052. *BMC Infect Dis*. 2015;15:389.

51. Samba E, Nkrumah F, Leke R. Getting polio eradication back on track in Nigeria. *N Engl J Med*. 2004;350:645–646.

52. Koopman FS, Beelen A, Gilhus NE, de Visser M, Nollet F. Treatment for postpolio syndrome. *Cochrane Database Syst Rev*. 2015;5:CD007818.

53. Hemachudha T, Ugolini G, Wacharapluesadee S, Sungkarat W, Shuangshoti S, Laothamatas J. Human rabies: neuropathogenesis, diagnosis, and management. *Lancet Neurol*. 2013;12:498–513.

54. Hampson K, Cleaveland S, Briggs D. Evaluation of cost-effective strategies for rabies post-exposure vaccination in low-income countries. *PLoS Negl Trop Dis*. 2011;5:e982.

55. Cusi MG, Bianchi S, Santini L, Donati D, Valassina M, Valensin PE, et al. Peripheral neuropathy associated with anti-myelin basic protein antibodies in a woman vaccinated with rubella virus vaccine. *J Neurovirol*. 1999;5:209–214.

56. Cardenas G, Soto-Hernandez JL, Diaz-Alba A, Ugalde Y, Merida-Puga J, Rosetti M, et al. Neurological events related to influenza A (H1N1) pdm09. *Influenza Other Respir Viruses*. 2014;8:339–346.

57. Kwong JC, Vasa PP, Campitelli MA, Hawken S, Wilson K, Rosella LC, et al. Risk of Guillain-Barre syndrome after seasonal influenza vaccination and influenza health-care encounters: a self-controlled study. *Lancet Infect Dis*. 2013;13:769–776.

58. Nunes MC, Dones W, Morillo CA, Encina JJ, Ribeiro AL. Chagas disease: an overview of clinical and epidemiological aspects. *J Am Coll Cardiol*. 2013;62:767–776.

59. Genovese O, Ballario C, Storino R, Segura E, Sica RE. Clinical manifestations of peripheral nervous system involvement in human chronic Chagas disease. *Arq Neuropsiquiatr*. 1996;54:190–196.

60. Maldonado IR, Ferreira ML, Camargos ER, Chiari E, Machado CR. Skeletal muscle regeneration and *Trypanosoma cruzi*-induced myositis in rats. *Histol Histopathol*. 2004;19:85–93.

61. Taratuto AL, Venturiello SM. Trichinosis. *Brain Pathol*. 1997;7:663–672.

62. Hawk MW, Shahlaie K, Kim KD, Theis JH. Neurocysticercosis: a review. *Surg Neurol*. 2005;63:123–132:discussion 32.

63. Mishra P, Pandey D, Tripathi BN. Cysticercosis of Soleus muscle presenting as isolated calf pain. *J Clin Orthop Trauma*. 2015;6:39–41.

IV. FOCUS ON SPECIFIC NEUROLOGICAL SYNDROMES OR DISEASES IN TROPICAL AREAS

64. Combalia A, Sastre-Solsona S. Hydatid cyst of gluteus muscle. Two cases. Review of the literature. *Joint Bone Spine.* 2005;72:430–432.

65. Paolino E, Granieri E, Tugnoli V. Peripheral neuropathy as a manifestation of hydatid disease. A case report. *Acta Neurol (Napoli).* 1987;9:256–262.

66. Mannsour SE, Bauman PM, Reese HH, Otto GF. Myopathy in mice experimentally infected with *Schistosoma mansoni. Trans R Soc Trop Med Hyg.* 1965;59:87–89.

67. Mostafa M, Habib MA, Abdel-Moneim S, Abdel-Hamid T, Sherif M, Basmy K, et al. Peripheral polyneuropathy in bilharziasis. *J Egypt Med Assoc.* 1972;55:44–59.

68. Marra TA. Recurrent lumbosacral and brachial plexopathy associated with schistosomiasis. *Arch Neurol.* 1983;40:586–587.

69. Sithinamsuwan P, Sinsawaiwong S, Limapichart K. Guillain-Barre's syndrome associated with *Plasmodium falciparum* malaria: role of plasma exchange. *J Med Assoc Thai.* 2001;84:1212–1216.

70. Senanayake N, Wimalawansa SJ. Periodic paralysis complicating malaria. *Postgrad Med J.* 1981;57:273–274.

71. Tégner R, Tomé FM, Godeau P, Lhermitte F, Fardeau M. Morphological study of peripheral nerve changes induced by chloroquine treatment. *Acta Neuropathol.* 1988;75:253–260.

72. Boongird P, Phuapradit P, Siridej N, Chirachariyavej T, Chuahirun S, Vejjajiva A. Neurological manifestations of gnathostomiasis. *J Neurol Sci.* 1977;31:279–291.

73. Martins YC, Tanowitz HB, Kazacos KR. Central nervous system manifestations of Angiostrongylus cantonensis infection. *Acta Trop.* 2015;141:46–53.

74. Balasubramanian V, Ramamurthi B. An unusual location of guineaworm infestation. Case report. *J Neurosurg.* 1965;23:537–538.

75. McKelvie P, Reardon K, Bond K, Spratt DM, Gangell A, Zochling J, et al. A further patient with parasitic myositis due to *Haycocknema perplexum*, a rare entity. *J Clin Neurosci.* 2013;20:1019–1022.

76. Webman RB, Gilman RH. Coenuriasis. In: Magill AJ, Ryan ET, Hill D, Solomon T, eds. *Hunter's Tropical Medicine Emerging Infectious Diseases.* 9th ed. Philadelphia, PA: W. B. Saunders; 2013:921–922.

77. Bossi P, Caumes E, Paris L, Darde ML, Bricaire F. *Toxoplasma gondii*-associated Guillain-Barre syndrome in an immunocompetent patient. *J Clin Microbiol.* 1998;36:3724–3725.

78. Montoya JG, Liesenfeld O. Toxoplasmosis. *Lancet.* 2004;363:1965–1976.

79. Simmons EH, Van Peteghem K, Trammell TR. Onchocerciasis of the flexor compartment of the forearm: a case report. *J Hand Surg Am.* 1980;5:502–504.

80. Bassett ML, Danta G, Cook TA. Giardiasis and peripheral neuropathy. *Br Med J.* 1978;2:19.

81. Odeberg B, Hansen HA, Lanner LO. Megaloblastic bothriocephalus anemia mainly due to folic acid deficiency. *Acta Med Scand.* 1963;174:155–162.

82. Satti MB, el-Hassan AM, al-Gindan Y, Osman MA, al-Sohaibani MO. Peripheral neural involvement in cutaneous leishmaniasis. A pathologic study of human and experimental animal lesions. *Int J Dermatol.* 1989;28:243–247.

83. Italiano CM, Wong KT, AbuBakar S, Lau YL, Ramli N, Syed Omar SF, et al. Sarcocystis nesbitti causes acute, relapsing febrile myositis with a high attack rate: description of a large outbreak of muscular sarcocystosis in Pangkor Island, Malaysia, 2012. *PLoS Negl Trop Dis.* 2014;8:e2876.

84. Watts MR, Chan RC, Cheong EY, Brammah S, Clezy KR, Tong C, et al. *Anncaliia algerae* microsporidial myositis. *Emerg Infect Dis.* 2014;20:185–191.

85. Erbslöh F, Abel M. Deficiency neuropathies. In: Vinken PJ, Bruyn GW, eds. *Handbook of Clinical Neurology.* Amsterdam: North Holland Publishing Company; 1970:558–663.

86. Ohnishi A, Tsuji S, Igisu H, Murai Y, Goto I, Kuroiwa Y, et al. Beriberi neuropathy. Morphometric study of sural nerve. *J Neurol Sci.* 1980;45:177–190.

87. Koike H, Ito S, Morozumi S, Kawagashira Y, Iijima M, Hattori N, et al. Rapidly developing weakness mimicking Guillain-Barre syndrome in beriberi neuropathy: two case reports. *Nutrition.* 2008;24:776–780.

88. Saperstein DS, Wolfe GI, Gronseth GS, Nations SP, Herbelin LL, Bryan WW, et al. Challenges in the identification of cobalamin-deficiency polyneuropathy. *Arch Neurol.* 2003;60:1296–1301.

89. Zara G, Grassivaro F, Brocadello F, Manara R, Pesenti FF. Case of sensory ataxic ganglionopathy-myelopathy in copper deficiency. *J Neurol Sci.* 2009;277:184–186.

90. Ghoshal UC, Srivastava D, Verma A, Ghoshal U. Tropical sprue in 2014: the new face of an old disease. *Curr Gastroenterol Rep.* 2014;16:391.

91. Strachan H. On a form of multiple neuritisprevalent in the West Indies. *Practitioner.* 1897;59:477–484.

92. Roman GC, Spencer PS, Schoenberg BS. Tropical myeloneuropathies: the hidden endemias. *Neurology.* 1985;35:1158–1170.

93. Vallat JM, Dumas M, Giordano C, Piquemal M, Sonan T. Histologic study of peripheral nerve biopsy from 20 patients with tropical ataxic neuropathy. *Neurology.* 1987;35:255.

94. Oluwole OS, Onabolu AO, Cotgreave IA, Rosling H, Persson A, Link H. Incidence of endemic ataxic polyneuropathy and its relation to exposure to cyanide in a Nigerian community. *J Neurol Neurosurg Psychiatry.* 2003;74:1417–1422.

95. Dalakas MC. Inflammatory muscle diseases. *N Engl J Med.* 2015;372:1734–1747.

96. Pasnoor M, Barohn RJ, Dimachkie MM. Toxic myopathies. *Neurol Clin.* 2014;32:647–670:viii.

97. Del Brutto OH. Neurological effects of venomous bites and stings: snakes, spiders, and scorpions. *Handb Clin Neurol.* 2013;114:349–368.

98. Isbister GK, Kiernan MC. Neurotoxic marine poisoning. *Lancet Neurol.* 2005;4:219–228.

99. Philippe G, Angenot L. Recent developments in the field of arrow and dart poisons. *J Ethnopharmacol.* 2005;100:85–91.

100. Bril V. Neuromuscular complications of diabetes mellitus. *Continuum (Minneap Minn).* 2014;20:531–544.

101. Papanas N, Maltezos E. The diabetic hand: a forgotten complication? *J Diabetes Complications.* 2010;24:154–162.

102. Magy L, Mathis S, Vallat JM. Diagnostic and therapeutic challenges in chronic inflammatory demyelinating polyneuropathy and other immune-mediated neuropathies. *Curr Opin Crit Care.* 2011;17:101–105.

103. Vallat JM, Sommer C, Magy L. Chronic inflammatory demyelinating polyradiculoneuropathy: diagnostic and therapeutic challenges for a treatable condition. *Lancet Neurol.* 2010;9:402–412.

104. Howlett WP. Disorders of peripheral nerves. In: Howlett WP, ed. *Neurology in Africa Clinical Skills and Neurological Disorders.* 1st ed. Cambridge: Cambridge University Press; 2012:259–282.

105. Edinur HA, Dunn PP, Hammond L, Selwyn C, Brescia P, Askar M, et al. HLA and MICA polymorphism in Polynesians and New Zealand Maori: implications for ancestry and health. *Hum Immunol.* 2013;74:1119–1129.

106. Coquart N, Cadelis G, Tressieres B, Cordel N. Epidemiology of sarcoidosis in Afro-Caribbean people: a 7-year retrospective study in Guadeloupe. *Int J Dermatol.* 2015;54:188–192.

107. Matsuo M, Ehara S, Tamakawa Y, Chida E, Nishida J, Sugai T. Muscular sarcoidosis. *Skeletal Radiol.* 1995;24:535–537.

108. Kerasnoudis A, Woitalla D, Gold R, Pitarokoili K, Yoon MS. Sarcoid neuropathy: correlation of nerve ultrasound, electrophysiological and clinical findings. *J Neurol Sci.* 2014;347:129−136.

109. Howlett WP. Myopathies and myasthenia gravis. In: Howlett WP, ed. *Neurology in Africa Clinical Skills and Neurological Sisorders.* 1st ed. Cambridge: Cambridge University Press; 2012:311−323.

110. Vallat JM, Goizet C, Magy L, Mathis S. Too many numbers and complexity: time to update the classifications of neurogenetic disorders? *J Med Genet.* 2016;53(10):647−650.

111. Modell B, Darr A. Science and society: genetic counselling and customary consanguineous marriage. *Nat Rev Genet.* 2002;3:225−229.

112. Tazir M, Bellatache M, Nouioua S, Vallat JM. Autosomal recessive Charcot-Marie-Tooth disease: from genes to phenotypes. *J Peripher Nerv Syst.* 2013;18:113−129.

113. Farhad K, Traub R, Ruzhansky KM, Brannagan III TH. Causes of neuropathy in patients referred as "idiopathic neuropathy". *Muscle Nerve.* 2015.

CHAPTER

16

Headaches in Tropical Areas

Dismand Houinato[1,2] and Athanase Millogo[2,3]

[1]University of Abomey-Calavi, Cotonou, Bénin [2]University of Limoges, Limoges, France
[3]CHU, Bobo-Dioulasso, Burkina Faso

16.1 INTRODUCTION

Cephalalgia, widely referred to as headache, is one of the most commonly reported symptoms prompting patients to consult general practitioners and specialists, in particular neurologists.[1] Manifestations are generally subjective and can be recognized only when expressed by the patient. The most common functional sign, pain, is rather difficult to measure.[2] The pain is most often attributable to tissue damage leading to stimulation of nociceptive peripheral receptors in an intact nervous system. It may also be secondary to a lesion or abnormal activation of the sensory pathways of the central or peripheral nervous system.[2] The overall prevalence of headache varies from 35% to almost 100%.[3] According to a World Health Organization (WHO) study, nearly half of the adult population had at least one headache during the year prior to being asked.[4] Approximately 1.7%–4.0% of the global adult population is affected by headache for at least 15 days per month.[4]

Based on a systematic review of the literature, we present evidence on headaches in tropical areas. We searched the following databases: PUBMED, HINARI, SCIENCEDIRECT, INDIANJPAIN, and AJN. We collected all the original articles that used the International Headache Society (HIS) classification and were published between 1989 and 2016.

16.2 EPIDEMIOLOGICAL ASPECTS

Most studies have focused on primary headaches including migraine and tension headaches. However, prevalence data exist for all types of headache. Table 16.1 summarizes the prevalences found in the literature.[1,5−19] The overall prevalence of headache varies between 27.5% and 96.1%. The prevalence of migraine varies from 3.3% to 64.0% and that of tension headaches between 27% and 40%.[5,6]

Almost all studies report that female patients predominate. Table 16.2 presents prevalences according to gender.[1,5−9,14,15,17−22] All age groups are represented. Distribution varies according to population. In 2007, Belo et al. found a prevalence of 57.4% in subjects aged 10–14 years and 54.4% in those aged 15–19 years.[5]

TABLE 16.1 Prevalence of Headache in Tropical Areas

Target	Authors (location, date)	Number of subjects	Prevalence of		
			Headache (%)	Migraine (%)	Tension headache (%)
General population	Andriantseheno et al. (Madagascar, 2005)	496		19.0	
	Lavados (Chile, 1997)	1385		39.7	26.9
	Junior et al. (Brazil, 2009)	1605	65.4		
	Dominges et al. (Brazil, 2009)	102	60.7	38.8	
	Rao et al. (India, 2012)	2329	63.6	25.2	
	Phanthumchinda et al. (Thailand, 1989)	540		29.1	
	Jeongwook et al. (South Korea, 2014)	1507	1.8		
	Zarka et al. (India, 2016)	6960	66.2		
	Houinato et al. (Benin, 2010)	1113		3.3	
	Mengistu et al. (Ethiopia, 2013)	1105	96.1		
	Winkler et al. (Tanzania, 2010)	7412		4.3	
Schools, students, workplace	Belo et al. (Togo, 2009)	171	41.9	39.8	41.5
	Maiga et al. (Mali, 2011)	733	86.5	20.0	38.0
	Ojini et al. (Nigeria, 2009)	376	46.0	6.4	18.1
	Adoukonou et al. (Benin, 2009)	336		11.3	
	Adoukonou et al. (Benin, 2009)	938		8.9	24.8

TABLE 16.2 Prevalence of Headache According to Gender

Target	Authors (location, date)	Number of subjects	Prevalence in	
			Males (%)	Females (%)
General population	Andriantseheno et al. (Madagascar, 2005)	496	9.4	26.8
	Lavados (Chile, 1997)	1385	46.7	53.3
	Junior et al. (Brazil, 2009)	1605	60.9	69.5
	Dominges et al. (Brazil, 2009)	102	51.5	65.2
	Félicio et al. (Brazil, 2006)	3328	20.0	80.0
	Karen et al. (Brazil, 2013)	2500	14.0	86.0
	Jeongwook et al. (South Korea, 2014)	1507	1.4	2.3
	Adoukonou et al. (Benin, 2009)	938	7.2	20.8
	Houinato et al. (Benin, 2010)	1113	2.2	4.0
	Mengistu et al. (Ethiopia, 2013)	1105	95.9	98.6
	Stark et al. (Australia, 2007)	5663	6.1	14.9
Schools, students, workplace	Belo et al. (Togo, 2009)	171	41.8	43.7
	Maiga et al. (Mali, 2011)	733	6.4	23.0
	Ojini et al. (Nigeria, 2009)	376	20.5	30.1
	Adoukonou et al. (Benin, 2009)	336	6.8	18.3

TABLE 16.3 Distribution of Headache According to Age

Authors (location, date)	≤ 29 years (%)	30−39 years (%)	40−49 years (%)	≥ 50 years (%)
Adoukonou et al. (Benin, 2009)	6.5	7.7	11.3	8.6
Jeongwook et al. (South Korea, 2014)	1.7	1.2	1.2	2.2
Lavados (Chile, 1997)	34.7	21.8	15.0	28.2

Table 16.3 shows the prevalence of headache according to age reported by few authors.[8,12,19]

Common triggering factors in the literature were difficulty falling asleep and menstruation[14,15,23] varying from 36.6% to 63.2% and from 2.6% to 7.2%, respectively. Other factors include noise and anxiety.[15] Their frequency varies between 12% and 71%.

16.3 IMPACT

The impact of headaches is assessed in terms of loss of productivity, particularly lost working time, and via Migraine Disability Assessment (MIDAS) and Headache Impact Test (HIT-6) scores. Loss of performance was often noted in studies evaluating the impact of headache as estimated either in school or vocational absenteeism, or in lost working days. The loss of productivity and absenteeism were more frequent in migraine patients than in people suffering from tension headaches. However, the burden of these headaches varied greatly among the populations studied. In 2009, Maiga et al.[6] found that 29.9% of migraine patients in Mali had a frequent drop in concentration, with absenteeism of less than 15 days in 58.0% and more than 15 days in 4.7%. Mengistu and Alemayehu[15] in 2007 in Ethiopia found that 34.8% of primary headache cases lost an average of 13.8 working days per year and this was higher with migraine (about 16 days) and lower with tension headaches (11 days). Amayo et al.[24] in Kenya in 1994 noted that 22.6% of cephalalgic students had to miss course days due to headache with an average of 3.2−3.4 days lost. Absenteeism was more frequent in students with migraine (31.3%) than in those with tension headaches (15.5%). Lavados et al.[8] in Chile in 1997 reported that 34.6% of headache patients reported absences at work and 26.0% were unable to perform activities of daily living. Other authors have used the HIT and MIDAS criteria.[5,25] The majority of patients had a low MIDAS score, which contrasted with a HIT-6 score of 60 or more in 30.9%−48.6%. This can be explained by the fact that these two scores assess the impact of headaches in different ways.

16.4 CLINICAL ASPECTS

16.4.1 Migraine

The reported ratio of migraine with and without aura varies. In the tropics, there appears to be a higher prevalence of migraine without aura. Indeed, the proportions vary from 16.7% to 42.1% for migraine with aura.[25] Only two authors, Adeney et al.[25] in Peru and Berk et al.[26] in South Africa reported proportions of migraine with aura in more than half of all patients, 51.1% and 59.5%, respectively.

The duration of migraine headache varies from one study to the next, but most authors agree on a duration of headache of less than 24 hours. The proportion of this duration ranges from 63.7% to 82.0%.

Pulsatile pain is by far the most frequent form ranging from 62.5% to 92.9%. However, some authors report a low proportion of this type of pain: Mehta et al.[27] in India and Zahid et al.[28] in Pakistan reported 31.5% and 6.2%, respectively.

Among the symptoms associated with migraine, photophobia is by far the most prevalent, varying from 50% to 100% depending on the author. A single author reports a low frequency of photophobia.

Information on migraine headaches is summarized in Table 16.4[5−7,13,16,17,25−30] and Table 16.5 reports the signs associated with migraine.[5,6,16,25,26,28,29]

Migraine headache may be associated with other disorders, such as cysticercosis, which was reported by Cruz et al.[31] in 1995 in Ecuador, where the authors showed that migraine was significantly more present in neurocysticercosis. Conditions such as depression, anxiety and stress were also associated with these conditions, as was alcohol consumption. Table 16.6 summarizes the different prevalences of headache in patients with other conditions reported by authors.[25,30−33]

16.4.2 Tension Headaches

Table 16.7 shows the proportion of tension headaches and the characteristics studied by the authors.[7,23,26,34,35]

TABLE 16.4 Characteristics of Migraine Headache

Authors (location)	Aura (%)	Pulsatile pain (%)	Intensity			Duration	
			Minimal	Moderate	Severe	<24 h	≥ 24 h
Zahid (Pakistan)	38.3	16.7	24.7	48.1	27.2	—	—
Mehta (India)	20.7	31.5	—	—	—	—	—
Adeney (Peru)	51.1	92.9	—	—	—	100.0	0.0
Ezeala-Adikaibe (Nigeria)	28.9	—	—	—	—	—	—
Ojini (Nigeria)	16.7	62.5	41.0	49.7	9.3	94.8	5.2
Lavados (Chile)	48.5	—	7.9	45.5	46.5	—	—
Maiga (Mali)	18.0	97.6	11.8	43.2	37.8	3.9	96.1
Berk (South Africa)	59.5	—	5.0	38.0	73.0	84.0	16.0
Adoukonou (Benin)	42.1	—	—	—	—	76.3	27.7
Kowacs (Brazil)	27.2	90.9	—	45.5	54.5	63.7	18.2
Andriantseheno (Madagascar)	—	89.5	—	—	—		
Dent (Tanzania)			5.7	65.9	28.5	51.3	33.3
Houinato (Benin)	—	—	—	—	24.3	59.5	12.6

TABLE 16.5 Associated Signs With Migraine

Authors (location)	Nausea (%)	Vomitting (%)	Photophobia (%)	Phonophobia (%)
Kowacs et al. (Brazil)	63.6	36.4	—	100.0
Berk et al. (South Africa)	82.0	27.0	77.0	73.0
Maiga et al. (Mali)	38.6		67.7	97.6
Andriantseheno et al. (Madagascar)	89.5		85.0	
Ojini et al. (Nigeria)	47.8		91.7	
Adeney et al. (Brazil)	78.6		50.0	
Zahid et al. (Pakistan)	4.9	4.9	12.4	7.4

16.4.3 Secondary Headache

These headaches were found not only in tropical infectious pathology (malaria and salmonellosis) but also in various pathologies such as stroke and arterial hypertension. Only very few studies have addressed specifically these headache etiologies. They were generally conducted in hospitals. In Cameroon, Mapoure et al.[36] reported a prevalence of 43.5% of secondary headache, with 33.5% of infectious causes. Maiga et al.[5] in Mali found that 42% of symptomatic headache was related to an infectious disease. In Togo, Belo et al.[4] found that secondary headaches occurred in 7.6% of the student population with headache.

Some authors address headaches within studies focused on other diseases, and in these cases headaches are described as symptoms revealing the disease. Thus we can identify two groups of studies:

- Group 1: headache studies, exploring the etiology of headache.
- Group 2: studies describing headache as a reason for consultation or symptom.

Table 16.8 summarizes the etiologies of headaches in both groups.[4,19,36–50]

16.4.3.1 Headache of Ophthalmological Origin

In ametropia, the prevalence of headache is around 30%. More than 98% had settled in a progressive mode, as in the work of Ebana et al.[38] In this series, the characteristics of these headaches were as follows:

TABLE 16.6   Prevalence of Migraine and Tension Headache in Subjects With Other Disorders

Authors (date)	Target	Headache types	Associated diseases	Prevalence (%)
Cruz (1995)	General population	Migraine	Cysticercosis	33.3
Cruz (1995)		Tension headache	Neurocysticercosis	16.7
Gelaye (2013)	Workplace	Migraine	Moderate depression	4.7
			Severe depression	1.4
			Moderate anxiety	5.6
			Severe anxiety	3.8
			Moderate stress	2.8
			Severe stress	2.8
Soares (2013)	University	Migraine	Irritable bowel	19.3
Ezeala-Akaibe (2012)	General population	Migraine	High blood pressure	9.9
			Diabetis	2.2
			Epilepsy	33.3
			Stroke	5.6
			Alcohol	6.0
			Tobacco	6.6
		Tension headache	High blood pressure	20.4
			Diabetes	4.6
			Epilepsy	16.7
			Stroke	27.8
			Alcohol	16.3
			Tobacco	15.1
Adeney (2006)	Parturients	Migraine	High blood pressure	7.1
			Asthma	7.1
			Fibroma	21.4
			Dysthyroïdia	7.1
			Depression	28.6

TABLE 16.7   Characteristics of Tension Headaches

Authors (location, date)	Proportion		Intensity			Frequency		Mean duration
	Episodic (%)	Chronic (%)	Minimal (%)	Moderate (%)	Severe (%)			
Fogang (Sénégal, 2013)	55.0	45.0						12.6 days ± 7.4
Amayo et al.						43.0% (1/month)	20.6 (>4/months)	
						36.3% (2−3/months)		
Berk et al.	90.4	9.6	38.1	49.0	12.6	1j: 41%	5−10j: 0%	
						5j: 13%	>10j: 89%	
Lavados			38.1	49.0	12.6	<5: 70.4%		
Da Silva Junior	39.3	42.9						

TABLE 16.8   Prevalence of Secondary Headaches

Disorders	Group 1 Author, location	Group 1 Prevalence (%)	Group 2 Author, location	Group 2 Prevalence (%)
High blood pressure	Bigal, Brazil	11.1	Makani, Bassakouahou, Congo	45.5
			Kaba, Guinea	25.0
			Diao, Senegal	27.3
Stroke	Felicio, Brazil	0.3	Bamouni, Burkina Faso	49.2
	Mapoure, Cameroon	9.5	–	–
Ophtalmology	Ebana, Cameroon	30.6	–	–
	Belo, Togo	3.5	–	–
Otorhinolaryngology	Antoniuk, Brazil	7.5	–	–
	Bigal, Brazil	7.8	–	–
	Mapouré, Cameroon	3.8	–	–
	Belo, Togo	6.0	–	–
Stomatology	Belo, Togo	2.4	–	–
Malaria	Mapoure, Cameroon	3.4	Menan, Ivory Coast	53.7
	Belo, Togo	10.8		
Acute infections	Bigal, Brazil	15.7	Dovonou, Benin	94.8 (salmonellosis)
	Felicio, Brazil	0.8		
Intracranial infections	Mapoure, Cameroon	9.7	Kadjo, Ivory Coast	58.4 (toxoplasmosis)
	Bigal, Brazil	0.4	Smadja, Martinique	71.0 (toxoplasmosis)
	–	–	Sakho, Senegal	53.6 (brain abcesses)
Cranial traumas	Bigal,	1.1	N'Gbesso, Ivory Coast	87.9
	Mapoure, Cameroon	3.2	–	–
Intracranial tumors	Mapoure, Cameroon	3.2	–	–
Cough	Belo, Togo	4.8	–	–
Psychiatric disorders	Felicio, Brazil	0.2	Mouanga, Congo	38.1
	Mapoure, Cameroon	0.7	–	–
	Mouanga, Congo	8.4	–	–

- triggering factors such as fixation, light, reading: 92.8%.
- spontaneous triggering: 7.2%.
- localization in the frontal region in 59.5% of patients, in the temporo-parietal region in 19.6% and diffuse in 20.9%.
- Stopping visual effort and rest relieved or suppressed pain in 2/3 of the patients.
- For the others, remission occurred spontaneously or after an analgesic.
- Intermittent headache in 71.5% of cases

16.4.3.2 Headache in Intracranial Infection

Headaches in intracranial infections are frequent in tropical areas and generally associated with other signs of the concerned infection, thus explaining their coexistence in the context of an intracranial hypertension syndrome. Hospital series report headaches in the context such as bacterial meningitis,[37] cerebral abscesses,[49] or cerebral toxoplasmosis.[47,4,9] These authors describe headache in intracranial infections as one of the most common signs. In cerebral toxoplasmosis, for example, in HIV-infected patients, headache was present in 58.4% of the patients after fever according to Kadjo et al.[47] Smadja et al.[48] found a higher proportion of headache (71%) in toxoplasmosis. In their study of cerebral abscesses in Senegal, Sakho et al.[49] found that headaches were the third (53.8%) clinical manifestation after hyperthermia (74.3%) and motor deficit (56.4%)

and were more present than convulsions, vomiting and signs of meningeal irritation.

16.4.3.3 Work-Related Headache

In their study of health problems among workers in Democratic Republic of Congo, Wangata et al.[51] found a prevalence of headache of 74.7% in drivers versus 86.5% in support staff. It should be noted that this work concerned the staff in an informal workplace and therefore taking questionable charge of their health.

16.5 CONCLUSION

Headache is relatively common in the tropics, but the studies that relate to it mainly focus on migraine. Hospital series consider only the causes of secondary headaches falling within the scope of general medical and surgical pathology. The causes of headache falling within the specific framework of tropical pathology are very little studied. The 2000 edition of the WHO's Global Burden of Disease study listed migraine as the 19th cause of disability in the world, for 1.4% of all Years Living with Incapacity. This study has been cited several times since then and has drawn the attention of public authorities to headaches and their acceptance as a public health problem. This study found that although headache studies are numerous, they refer principally to migraine headaches and tension headaches. The few studies that address secondary headaches are those performed in hospitals and in specialized services. This is a barrier to extrapolation of the results to the general population.

References

1. Headache Classification Subcommittee of the International Headache Society. The international classification of headache disorders, 2nd edition. *Cephalalgia.* 2005;24:1–150.
2. Clavelou P, Nathalie G. Epidémiologie des céphalées et impact socioéconomique. In: Bousser MG, Ducros A, Massiou H, eds. *Migraine et céphalées.* 1st ed. France: Doin; 2005.
3. Castaigne A, Godeau B, Lejone J-L, Schaeffer A. *Sémiologie médicale : initiation à la physiopathologie.* 3rd ed. Rueil-Malmaison: Novartis Pharma SA; 1993.
4. Organisation Mondiale de la Santé. *OMS Atlas des céphalées.* <http://www.who.int/mental_health/management/atlas_headache_disorders/en/>. Accessed 20.04.16.
5. Belo M, Assogba K, Awidina-Ama A, et al. Céphalées et qualité de vie en milieu scolaire à Lomé, Togo. *Afr J Neurol Sci.* 2009;28:29–33.
6. Maiga Y, Boubacar S, Kanikomo D, et al. La migraine en milieu scolaire à Gao au Mali. *Afr J Neurol Sci.* 2011;30:49–53.
7. Andriantseheno L, Rafidison J, Andriantseheno O. Prévalence de la migraine à Madagascar : résultats d'une enquête menée dans une population générale. *Afr J Neurol Sci.* 2005;24:13–15.
8. Lavados P, Tenhamm E. Epidemiology of migraine headache in Santiago, Chile: a prevalence study. *Cephalalgia.* 1997;17:771–777.
9. Domingues R, Cezar P, Filho J, et al. Prevalence and impact of headache and migraine among Brazilian Tupiniquim natives. *Arq Neuropsiquiatr.* 2009;67:413–415.
10. Rao G, Kulkarni G, Gururaj G. The burden of headache disorders in India: methodology and questionnaire validation for a community-based survey in Karnataka State. *J Headache Pain.* 2012;13:543–550.
11. Phanthumchinda K, Sithi-Amorn C. Prevalence and clinical features of migraine: a community survey in Bangkok, Thailand. *Headache.* 1989;29:594–597.
12. Jeong-Wook P, Heui-Soo M, Jae-Moon K, Kwang-Soo L, Min Kyung C. Chronic daily headache in Korea: prevalence, clinical characteristics, medical consultation and management. *J Clin Neurol.* 2014;10:236–243.
13. Zarka A, Parvaiz A, Irfan I. Prevalence of headache in Kashmire Valley, India. *Neurol Asia.* 2016;21:145–153.
14. Houinato D, Adoukonou T, Ntsiba F, Adjien C, Avode D, Preux PM. Prevalence of migraine in a rural community in south Benin. *Cephalalgia.* 2010;30:62–67.
15. Mengistu G, Alemayehu S. Prevalence and burden of primary headache disorders among a local community in Addis Ababa, Ethiopia. *J Headache Pain.* 2013;14:30.
16. Winkler A, Dent W, Stelzhammer B, et al. Prevalence of migraine headache in a rural area of northern Tanzania: a community-based door-to-door survey. *Cephalalgia.* 2010;30:582–592.
17. Ojini F, Okubadejo N, Danesi M. Prevalence and clinical characteristics of headache in medical students of the University of Lagos, Nigeria. *Cephalalgia.* 2009;29:472–477.
18. Adoukonou T, Houinato D, Kankouan J, et al. Migraine among university students in Bénin (Benin). *Headache.* 2009;49:887–893.
19. Adoukonou T, Houinato D, Adjien C, Gnonlonfoun D, Avode G, Preux PM. Prévalence de la migraine dans une population de travailleurs à Bénin au Bénin. *Afr J Neurol Sci.* 2009;28:16–24.
20. Felicio A, Bichuetti D, Dos Santos W, De Oliveira G, Marin L, De Souza C. Epidemiology of primary and secondary headaches in a Brazilian tertiary-care center. *Arq Neuropsiquiatr.* 2006;64:41–44.
21. Karen S, Eckeli A, Dach F, Jose G. Comorbidities, medications and depressive symptoms in patients with restless legs syndrome and migraine. *Arq Neuropsiquiatr.* 2013;71:87–91.
22. Stark R, Valenti L, Miller C. Management of migraine in australian general practice. *Med J Aust.* 2007;187:142–146.
23. Bousser MG, Massiou H. Céphalées: classification et démarche diagnostique. In: Bousser MG, Ducros A, Massiou H, eds. *Migraine et céphalées.* France: Doin; 2005.
24. Amayo E, Jowi J, Njeru E. Migraine headaches in a group of medical students at the Kenyatta National Hospital, Nairobi. *East Afr Med J.* 1996;73:594–597.
25. Adeney KL, Flores JL, Perez JC, Sanchez SE, Williams MA. Prevalence and correlates of migraine among women attending a prenatal care clinic in Lima, Peru. *Cephalalgia.* 2006;26:1089–1096.
26. Berk M, Fritz VU, Schofield G. Patterns of headache in panic disorder: a survey of members of the South African Panic Disorder Support Group. *S Afr Psychiatry Rev.* 2004;7:28–30.
27. Mehta S. Study of various social and demographic variables associated with primary headache disorders in 500 school-going children of central India. *J Pediatr Neurosci.* 2015;10:13–17.
28. Zahid M, Sthanadar A, Kaleem M, et al. Prevalence and perceptions about migraine among students and patients in Khyber Pakhtunkhwa Province, Pakistan. *Adv Biosci Biotech.* 2014;5:508–516.
29. Kowacs PA, Piovesan EJ, Lange MC, et al. Prevalence and clinical features of migraine in a population of visually impaired subjects in Curitiba, Brazil. *Cephalalgia.* 2001;21:900–905.

30. Ezeala-Adikaibe A, Stella E, Ikenna O, Ifeoma U. Frequency and pattern of headache among medical students at Enugu, South East Nigeria. *Niger J Med*. 2012;21:205−208.

31. Cruz ME, Cruz I, Preux PM, Schantz P, Dumas M. Headache and cysticercosis in Ecuador, South America. *Headache*. 1995;35:93−97.

32. Gelaye B, Peterlin BL, Lemma S, Tesfaye M, Berhane Y, Williams MA. Migraine and psychiatric comorbidities among sub-Saharan African adults. *Headache*. 2013;53:310−321.

33. Soares RLS, Moreira-Filho PF, Maneschy CP, Breijã JF, Schmidte NM. The prevalence and clinical characteristics of primary headache in irritable bowel syndrome: a subgroup of the functional somatic syndromes. *Arq Gastroenterol*. 2013;50:281−284.

34. Fogang FY, Touré K, Naeije G, Ndiaye M, Gallo DA, Ndiaye MM. L'attention sélective en période non douloureuse des étudiants souffrant de céphalées primaires : une étude cas-témoins. *Afr J Neurol Sci*. 2013;32:36−44.

35. Da Silva-Júnior AA, Faleiros BE, Dos Santos TM, Gómez RS, Teixeira L. Relative frequency of headache types. *Arq Neuropsiquiatr*. 2010;68:878−881.

36. Mapoure NY, Gnonlonfoun D, Ossou-Guiet PM, Mbatchou Ngahane BH, Njimah Amadou N, Motah M. Etiologies des céphalées en milieu hospitalier tertiaire en Afrique subsaharienne. *Bénin Méd*. 2012;51:17−23.

37. Bigal ME, Bordini CA, Speciali JG. Etiology and distribution of headaches in two Brazilian primary care units. *Headache*. 2000;40:241−247.

38. Ebana Mvogo C, Ellong A, Bella AL, Luma H, Nyame Dipepa E. Les céphalées dans les amétropies. *Méd Afr Noire*. 2007;54:156−160.

39. Antoniuk S, Kozak M, Michelon L, Netto MM. Prevalence of headache in children of a school from Curitiba, Brazil, comparing data obtained from children and parents. *Arq Neuropsiquiatr*. 1998;56:726−733.

40. Mouanga AM, Missontsa DA. Profil socio démographique et clinique des patients observés en consultation externe de psychiatrie au Congo Brazzaville. *Méd Afr Noire*. 2005; 52:279−282.

41. Makani Bassakouahou JK, Ikama MS, Ellenga Mbolla BF, et al. Profil des patients admis pour hypertension dans le service de cardiologie et médecine interne du centre hospitalier et universitaire de Brazzaville (Congo). *Méd Afr Noire*. 2015;62:598−602.

42. Kaba ML, Camara B, Bah AO, Kourouma ML, Toure YI, Simon P. Prévalence de l'hypertension artérielle en milieu scolaire à Conakry. *Méd Afr Noire*. 2008;55:557−560.

43. Diao M, Ba FG, Kane AD, et al. Urgence hypertensive: aspects cliniques et évolutifs. *Méd Afr Noire*. 2013;60:231−236.

44. Bamouni YA, Lougue-Sorgho CL, Cisse R, Zanga SN, Tapsoba TL. Aspects épidémiologiques, cliniques et évolutifs des accidents vasculaires cérébraux ischémiques au C.H.U.Y.O. de Ouagadougou. *Méd Afr Noire*. 2006;53:349−355.

45. Menan EIH, Yavo W, Oga SSA, Kassi RR, Evi JB. Diagnostic clinique présomptif du paludisme : part réelle de la maladie. *Méd Afr Noire*. 2007;54:139−144.

46. Dovonou AC, Adoukonou TA, Sanni A, Gandaho P. Aspects épidémiologique, clinique, thérapeutique et évolutif des salmonelloses majeures dans le service de médecine interne du CHDU/Borgou au nord du Bénin. *Méd Afr Noire*. 2011;58:527−532.

47. Kadjo AK, Ouattara B, Kra O, Sanogo S, Yao H, Niamkey EK. Toxoplasmose cérébrale chez le sidéen dans le service de médecine interne du chu de Treichville. *Méd Afr Noire*. 2007;54:13−16.

48. Smad JAD, Fournerie P, Cabre P, Cabie A, Olindo S. Efficacité et bonne tolérance du cotrimoxazole comme traitement de la toxoplasmose cérébrale au cours du SIDA. *Presse Med*. 1998;27:1315−1320.

49. Sakho Y, Zabsonre S, Gaye M, et al. Approche diagnostique et thérapeutique de l'abcès du cerveau au Sénégal à propos d'une série de 39 cas. *Méd Afr Noire*. 2012;59:73−82.

50. N'Gbesso RD, N'Goan-Domoua AM, Ould Beddi M, Yoman AMF, Keita AK. Traumatismes crâniens en Côte d'ivoire : Evaluation TDM de 297 cas. *Méd Afr Noire*. 2004;51:595−601.

51. Wangata J, De Brouwer C. Condition de travail et prévalences des problèmes de santé chez les travailleurs du secteur informel dans le transport urbain à Kinshasa, République Démocratique du Congo. *Méd Afr Noire*. 2014;61:319−329.

17

Neuropsychiatric Disorders and Addictions

Jean-Pierre Clément[1,2,3] *and Philippe Nubukpo*[1,2,3]

[1]INSERM UMR 1094 NET, Limoges, France [2]University of Limoges, Limoges, France
[3]Hospital Center Esquirol, Limoges, France

17.1 INTRODUCTION

Substance use disorders (SUDs) and psychiatric disorders (PDs) cause a significant burden of disease in low- and middle-income countries. Comorbid psychopathology and longer duration of untreated PDs or SUDs can negatively affect treatment outcomes and quality of life in these regions of the world. While 14% of the global burden of disease is attributed to these disorders, most of the people affected—75% in many low-income countries—do not have access to the treatment they need.

According to the World Health Organization (WHO), 75%–85% of people with serious mental disorders receive no treatment in countries with low or intermediate income versus 35%–50% in high-income countries. In sub-Saharan Africa (SSA), only 1% of national budgets for health is allocated to mental health, while the WHO estimates that it represents 13% of the global burden of disease. In this area of the world, the situation of mental health is aggravated by the devastating effects of conflict, violence against women, and the poor health of children.

In Section 17.2 we describe the situation of psychiatry and mental health in SSA, on epidemiological aspects, then we focus on barriers, and we shape possible solutions. In Section 17.3 we describe the situation of addiction in SSA from history, to solutions facing this new expanding public health matter, through epidemiology and analyzing causes.

17.2 PSYCHIATRY AND MENTAL HEALTH IN SSA

Often derived from colonization, operative mental health and psychiatric care in most of the countries of SSA have changed little over the years, and often involve poor child health.

17.2.1 Epidemiology in Mental Health in SSA

17.2.1.1 Prevalence of Mental Illness and Associated Factors

A recent meta-analysis reported that, compared to other regions of the world, apart from North and

Neuroepidemiology in Tropical Health
DOI: http://dx.doi.org/10.1016/B978-0-12-804607-4.00017-4

South East Asia, 1-year prevalence rates of PDs were low in SSA, whereas English-speaking counties returned the highest lifetime prevalence estimates.[1]

Depression in SSA has some specificities which can hide the classic symptomatology and compromise the diagnosis. Ignoring this fact can lead to simplistic ethnocentrism. From a clinical standpoint, depressive illness characterized by somatic manifestations, delusions of persecution, and anxiety are increasingly uncommon. As African societies modernize, these traditional forms are being gradually supplanted by states with symptoms and prognoses more like those observed in industrialized countries. Hybrid depressive syndromes are now the most widespread. Epidemiologically, the notion widely held only a few decades ago that depression is a rare occurrence in Africa has now been dispelled. Many studies have been conducted to determine the exact incidence, age distribution, and sex ratio, but more precise data are still needed. This investigation will require improvements in screening and diagnostic methods which must be not only suitable for clinical use but also adaptable to local conditions. Treatment facilities are different in urban and rural areas but care is often dispensed in unconventional settings and may be combined with traditional methods. Drug availability is limited by problems involving supply and cost.[2]

Some authors hypothesized that "westernalization" (considered as the loss at an individual level of traditional ways of life, working habits, cultural patterns and languages in favor of different attitudes influenced by western culture) may represent a risk factor for depressive illness. It has been hypothesized that rapid changes in the social organization tend to exacerbate attitudes of "compulsive hyper-responsibilization", a cognitive set of basic assumptions which may be considered at the same time both as a product of "westernalization" at an individual level and a risk factor for depression.[3]

In richer countries, three-times as many men die of suicide than women do, but in low- and middle-income countries the male-to-female ratio is much lower at 1.5 men to each woman. Globally, suicides account for 50% of all violent deaths in men and 71% in women.[4]

There have been few epidemiological surveys to establish the prevalence and associated risk factors of psychosis in SSA. A descriptive and analytical cross-sectional study on the epidemiological aspects of schizophrenia in the health area of Ouidah—Kpomassè—Tori-Bossito (Benin) was carried out in 2013 through the MINI tool (Mini International Neuropsychiatric Interview). The prevalence of schizophrenia was 1.1%. It was significantly higher in individuals with a family history of schizophrenia ($p = 0.04$), the unemployed ($p < 0.0001$), and the separated/divorced/widowed ($p = 0.003$).[5]

Prevalence of schizophrenia found in other settings in SSA are different. The prevalence of psychotic symptoms in urban Tanzania within a random sample of 899 adults aged 15—59 years surveyed using the Psychosis Screening Questionnaire (PSQ), was that 3.9% of respondents reported one or more psychotic symptoms in the previous year. Significantly higher rates of symptoms were found in those who had recently experienced two or more stressful life events, those with common mental diseases and people who had used cannabis in the preceding year.[6]

Jenkins et al.[7] reported, from a population-based epidemiological survey in rural Kenya (Maseno, Kisumu District of Nyanza province), using the PSQ, a prevalence of single psychotic symptoms in 8% of the adult population, but only 0.6% had two symptoms and none had three or more psychotic symptoms in this sample size. Psychotic symptoms were evenly distributed across this relatively poor rural population and were significantly associated to a lesser extent with poor physical health and housing type.

Schizophrenia's cases are difficult to recognize (mainly the less severe form in the general population), and some think that this can explain the low prevalence usually found in low-income countries.

As a platform for INTREPID (a program for research into psychoses in India, Nigeria, and Trinidad), some authors sought to establish comprehensive systems for detecting representative samples of cases of psychosis by mapping and seeking to engage all professional and folk (traditional) providers and potential key informants in defined catchment areas. They used a combination of official sources, local knowledge of principal investigators, and snowballing techniques. Ibadan (Nigeria) has the most extensive folk (traditional) sector. They identified and engaged in their detection system: (1) all mental health services professional at each site (in- and out-patient services: Chengalpet, 6; Ibadan, 3; Trinidad, 5); (2) a wide range of folk providers (Chengalpet, three major healing sites; Ibadan, 19 healers; Trinidad, 12 healers); and (3) a number of key informants, depending on need (Chengalpet, 361; Ibadan, 54; Trinidad, 1).

They concluded that there is a necessity to develop tailored systems for the detection of representative samples of cases with untreated and first-episode psychosis as a basis for robust, comparative epidemiological studies.[8]

17.2.1.2 Women's Mental Health

A mental health epidemiological survey was conducted in a demographic surveillance site of a Kenyan household population in 2013 to test the hypothesis that the prevalence of psychotic symptoms would be similar to that found in an earlier sample drawn from the same sample frame in 2004, using the same overall methodology and instruments. This 2013 study found that the prevalence of one or more psychotic symptoms was 13.9%, and 3.8% with two or more symptoms, while the 2004 study had found that the prevalence of single psychotic symptoms in rural Kenya was 8% of the adult population, but only 0.6% had two symptoms, and none had three or more psychotic symptoms. This change was accounted for by a striking increase in psychotic symptoms in women (17.8% in 2013 compared with 6.9% in 2004, $p < 0.001$), whereas there was no significant change in men (10.6% in 2013 compared with 9.4% in 2004, $p = 0.58$).[9]

Mental disorders represent 9% of the disease burden in Ghana. Women are more affected by common mental disorders, and are underrepresented in treatment settings.[10] The authors used the SF-36 and K6 forms and four psychosis questions to examine physical and social correlates of mental illness in 2814 adult women living in Accra, Ghana, as part of a larger cross-sectional population-based survey of women's health.

Low levels of education, poverty and unemployment are negatively associated with mental health; marriage is neither good nor bad for mental health. Physical ill health is also associated with mental distress. No association was found between mental distress and religion or ethnicity. Some additional risk factors were significant for one, but not both of the outcome variables. Only 0.4% of women reported seeing a mental health professional in the previous year, whereas 58.6% had visited a health center.[10]

In areas of conflict, common sexual assaults are a factor in mental illness among women and girls.[9]

17.2.1.3 Childhood Mental Health

Published data from Africa are limited on childhood mental health. From a review including nine studies with four studies coming from South Africa, two each from Democratic Republic of Congo (DRC) and Nigeria, respectively, and one from Ethiopia, the authors found that the prevalence of attention deficit hyperactivity disorders (ADHD) varied with rates of between 5.4% and 8.7%, among school children, 1.5% among children from the general population between 45.5% and 100% among special populations of children with possible organic brain pathology. Common associated comorbid conditions were oppositional defiant disorder, conduct disorder, as well as anxiety/depressive symptoms.[11]

Children are paying a heavy price in wars and conflict and the mental health of child soldiers is a real public health problem.

17.2.1.4 Mental Health in Prisons

There is a high prevalence of mental disorders among prisoners in the prison population in SSA. In Durban, South Africa, one of the largest prisons in the Southern hemisphere, 193 prisoners were interviewed using the Mini Neuro-psychiatric Interview. The study demonstrated that 55.4% of prisoners had an Axis 1 disorder; the most common disorder was substance and alcohol use disorders (42%); 23.3% of prisoners were diagnosed with current psychotic, bipolar, depressive and anxiety disorders; 46.1% were diagnosed with antisocial personality disorder. The majority of these prisoners were untreated in prison, related to no detection of the mental disorder.[12]

17.2.1.5 Consequences of Mental-Health Diseases and Mental Quality of Life

A review paper in Nigeria reported a poor quality of life in Nigerian patients suffering from PDs; this finding was associated with illness-related factors such as comorbid medical problems, presence of anxiety and depressive symptoms and no adherence to medications.[13]

Fekadu et al.[14] followed-up 919 adults (from 68,378 screened) with the Severe Mental Illness instrument (SMI) over 10 years. In total, 121 patients (13.2%) died. The overall standardized mortality ratios (SMR) were twice that of the general population, and were higher for men and people with schizophrenia. Patients died about three decades prematurely, mainly from infectious causes (49.6%). Suicide, accidents, and homicide were also common causes of death.

The authors suggested that premature death and mortality related to self-harm should be considered in the estimation of the global burden of disease by the SMI.[14]

Dementia is associated with increased mortality risk. Between 2012 and 2014, a longitudinal population-based cohort study carried out to determine dementia prevalence in central Africa throughout the EPIDEMCA (Epidemiology of Dementia in Central Africa) program, assessed dementia-related mortality among Congolese older people after 2 years of follow-up. Older participants were traced and interviewed in rural and urban Congo annually between 2012 and 2014. DSM-IV and NINCDS-ADRDA criteria were required for dementia diagnosis. Data on vital status were collected throughout the follow-up. Cox proportional hazards model was used to assess the link between baseline dementia diagnosis and mortality risk. After 2 years of follow-up, 101/910 (9.8%) participants had died. Compared with participants with normal cognition, patients with dementia had 2.5-times higher mortality risk (hazard ratio (HR) = 2.53, 95% confidence interval (CI) = 1.42−4.49, p = 0.001). These results highlight the need for targeted health policies and strategies for dementia care in SSA.[15]

17.2.2 Support for Mental Illness and Promotion of Mental Health

17.2.2.1 Obstacles

Support for patients suffering from mental illness in most countries in SSA face obstacles that can be summarized as follows: a deficit of trained human resources, inadequate prevention, screening and diagnosis of mental illness, insufficient informational support of people with mental illness and their families, and insufficient education of civil society on mental health problems.

For example, in Togo, a west African low-income country, non-communicable diseases (NCDs) have a relative importance due to changes in lifestyle and dietary habits, including smoking, excessive alcohol consumption, the use of drugs and other psychoactive substances, obesity, etc. The health system of Togo is relatively well provided with infrastructure. Nevertheless, geographical, economic and social inequalities in the supply and access to essential care persist. In this country, only three psychiatrists (including a young psychiatrist working simultaneously with the Togolese army health services) are working on mental health in Togo. For example, Benin has more than 17 psychiatrists, Niger has its school of psychiatric nursing which has specialized for 4 years. Structures dedicated to mental health, centered on the capital, Lomé, or nearby are pathetic compared to the needs. There is a huge gap between the needs and the means available to deal with the significant burden of neurological and behavioral disorders which, according to the latest estimates, would represent 12.3% of the total burden of disease.

According to the WHO,[16] persons with mental illness, and in particular those who are placed in institutions, have higher rates of mortality from cancer and heart disease than the general population. None of the 35 nurses, auxiliaries or aides in charge of the Psychiatric Hospital of Zebe (the first psychiatric hospital in the country) patients received a diploma in psychiatry. If any serious diseases or injuries are treated on the physical plane (amputation, human immunodeficiency virus (HIV)), little or no consideration is given to the mental dimension in the medical management of the patient.

Due to a lack of medical specialists in psychiatry, psychologists—within regional hospitals—are currently in charge of the mentally ill and prescribe medicines!

The attitude of the population towards mental illness in Togo (as in many countries in Africa) is still strongly influenced by traditional and supernatural beliefs. This belief system often leads to answers unsuitable or even harmful to mental illness, to stigmatize people who have the disease, and to hesitation or to delay in seeking appropriate care for these problems.

A number of influencers and decision makers (including encountered health professionals) believe that mental illness is mostly incurable or, in any case, does not respond to conventional medical practices. Where they exist, the legal provisions relating to mental illness are often outdated and must be revised. Mental health services are not so far integrated into primary health care, as recommended by all WHO studies. This is the picture of the situation of mental health in most of countries in SSA.

17.2.2.2 Solutions for More Effectively Dealing With Mental Health in SSA

Solutions to the above problems could be based on the following three actions: creating a partnership of public−private−civil society, the availability and accessibility of drugs, and combating the stigma (professional training, increasing the capacity of civil society, advocacy for the mental health component, etc.). At an international level, some tools have been created. For example, WHO's comprehensive mental health action plan 2013−20[17] was adopted by the 66th World Health Assembly. The four major objectives of the action plan are: (1) to strengthen effective leadership and governance for mental health; (2) to provide comprehensive, integrated and responsive mental health and social care services in community-based settings; (3) to implement strategies for promotion and prevention in mental health; and (4) to strengthen information systems, evidence and research for mental health. Moreover, the WHO Mental Health Gap Action Program (mhGAP)[18] aims at scaling up services for mental, neurological and SUDs for countries especially with low- and middle-income. The program asserts that with proper care, psychosocial assistance and medication, tens of millions could be treated for depression, schizophrenia, and epilepsy, prevented from suicide and begin to lead normal lives—even where resources are scarce.

Finally, the WHO Assessment Instrument for Mental Health Systems (WHO-AIMS) is a new WHO tool for collecting essential information on the mental health system of a country or region. The goal of collecting this information is to improve mental health systems and to provide a baseline for monitoring the change.[19]

For example, in this context, in Togo, great efforts have been made in integrated strategic planning to combat NCDs (2012–15): the creation of a National Program for Mental Health with the appointment of three focal points, the creation of a master's degree in mental health for the formation of middle management professionals in mental health across the country.

17.3 ADDICTIONS IN SSA

Many drugs, natural or not, can change behavior. Many of them are voluntarily consumed in a traditional way, in almost all societies. In 1986, a report by the Regional Office of the WHO in Brazzaville noted a general increase in the drug problem in Africa. Important regions have become dependent on income that comes from the cultivation of marijuana. The report also highlighted the excessive availability and use of psychotropic substances without prescription in many African countries. This was in addition to the already widespread use of legal drugs like alcohol and tobacco.

SSA is now an important link in drug dealing. Early preventive policies can prevent its spread. It is particularly important that the links between certain addictions and HIV/AIDS are established. The co-occurrence in the same individual, of a drug-abuse-associated disorder and another psychiatric disorder is also frequent.

According to the WHO, alcohol represents the third risk factor of morbidity and mortality in the world. Damage due to harmful use of alcohol are important. Alcohol is a major determinant for NCDs and neuro-psychiatric disorders.[20,21] Every year, 2.5 million deaths have their origins in Alcohol Use Disorder (AUD); among these are direct mortality caused by roads accidents, hetero-aggressive and auto-aggressive behavior (suicide) as well as deferred mortality (cancers, cardiovascular diseases, hepatic diseases, etc.). AUD constitutes the primary cause of cognitive impairment among people aged 60 and less.[22,23] Alcohol dependency is a well-spread disorder; the prevalence during a lifetime is estimated at 7%–12.5% in the majority of Western countries.[24]

17.3.1 History and Generality of Addictions in SSA

In West Africa, there is a traditional spread of cannabis crops. In 1990, according to Interpol, Africa was involved at three levels: production and cannabis trafficking, transit of opiates, trafficking and importation of psychotropic drugs.

17.3.1.1 The Local Production of Cannabis

Cannabis grows wild in most African countries. In some countries, it is cultivated by the rural population for local consumption and export.

17.3.1.2 Local Production of Heroin and Cocaine

There is a culture of poppies in the North of Nigeria, in Kano region, but it is a marginal activity.[25] A large number of capitals of West Africa are used as places of transit for heroin (e.g., air transport via Lagos to European capitals). The transport of heroin is by "mail" or "mules" who conceal the product in body cavities. Cocaine trafficking is also a transit traffic controlled by the same traffickers of heroin.[25] The important role of the West Africa in the European market for cocaine and heroin has been underlined.[26] This traffic is known to have experienced a meteoric progression because smuggling networks are very organized on the continent.

The West African cocaine seizures increased between 2002 (95 kg) and 2007 (6.5 tons out of global traffic estimated at 40 tons or US$1.8 billion). Cocaine is transported by commercial flights and trans-Saharan land routes to the EU. Cocaine use is rising quickly in West Africa (Burkina, Guinea, Nigeria, Ghana, Senegal, Sierra Leone, Togo) and southern Africa (Angola, Mozambique, South Africa). Since 1987, the African continent is suffering from a traffic of psychotropic drugs, often diverted from licit markets from Europe and India, and coming also from illicit productions. The sale in SSA is favored by the absence of legislation imposing orders. In Nigeria in 1990, 70% of available drugs came from counterfeiting. In East Africa, large quantities of Methaqualone (mandrax*) and phenobarbital are produced.[27]

17.3.2 Epidemiology of Addictions in SSA

17.3.2.1 Prevalence of Addictions in SSA

From a literature review, and true personal experiences, we provide an overview of the epidemiological situation concerning the problem of addiction in SSA. The prevalences are highly variable depending on the country and on the basis of the substances. Cocaine use in SSA has reached a prevalence of 7.6% (i.e., more than 1.1 million people; 14.5 million consumers worldwide). In Ghana, the abuse of cannabis is greater than 13% for individuals aged 15–65 years (3.8% in the world). The share of methamphetamine in applications for care related to narcotics in South Africa increased between 1996 (1%) and 2007 (41%) (UNODC). In Togo, drug abuse accounted for 5% of hospitalizations in 1990 in the Neuropsychiatry Unit at the University Hospital in Lomé, the capital. According to the results of a survey conducted among the military, college students, transport drivers, apprentice hairdressers, and dressmakers' apprentices in Lomé, people who were addicted to drugs (17.2%), consumed especially cannabis and amphetamines.[27] The figures for South Africa are even higher.[28]

17.3.2.1.1 PREVALENCE OF ADDICTION TO INTRAVENOUS DRUG MISUSE IN SSA

Estimations indicate that about 3 million drug-injecting users live in SSA with 200,000 in Kenya, and at least 250,000 in South Africa. They are generally victims of exclusion and repression.[29] In Togo, the frequency of drug injection among students from Lomé was 6.1%.[27] A recent study confirms that in West Africa, drug injection has become more and more frequent, and this may get worse against the efforts of the administrative authorities who are struggling to implement programs to reduce the risks in general and more specifically in the population with HIV.[30]

17.3.2.1.2 DRUG ABUSE AND HIV

In many low or intermediate income countries (LICs) in SSA, HIV prevalence is higher among intravenous drugs users (from 26% to 88%) compared to the rest of the population (<1%).[29] Officially, according to the UNODC, only three African countries reported the presence of injecting drug users (IDU) on their territory: South Africa, Kenya and Mauritius.[30]

Authors have shown from a review of the literature that, among drug injecters in 343 areas of the world including in SSA (four countries included in the review), the reduction of infection of hepatitis C virus below a certain threshold (30%), could reduce the risk of HIV infection.[31] In many LICs of SSA, HIV prevalence is higher among intravenous drug users than in the general population. Prevalences are estimated to be between 68% and 88% in Kenya; 58% in women and 27% in men in Tanzania; 28% in South Africa; 26% in Zanzibar; while the prevalence in the general population is <1%.[29]

17.3.2.2 Specificity of the Addictions in Women in SSA

There is a significant problem in South Africa related to the phenomenon of "women who drink in the kitchen cupboard", which is very common in the suburbs of Johannesburg. These are women who hide their consumption of alcohol from their spouses. Fetal alcohol syndrome (FAS) is also a major public health problem in South Africa. Indeed, the Cape Province

has the highest prevalence of FAS in the world: 40.5−46.4/1000 children.[32]

17.3.2.3 Prevalence of Addictions in Childhood and Adolescents in SSA

This is a frequent and growing phenomenon (see Table 17.1). This increase relates both to tobacco and alcohol and cannabis. The prevalence are increasing: 5.2 to 48%; the age of onset is usually between 10 and 15 years. In order of importance, products used by teenagers are alcohol, tobacco, and cannabis (the third most used substance in East Africa).

17.3.2.4 Factors Associated With Addictions in SSA

In general, it is observed that prevalence of addictions is increasing among adolescents.

17.3.2.4.1 FACTORS ASSOCIATED WITH SMOKING IN SSA

Regarding smoking, some authors, in a review of the literature from 54 articles, found that 14 of the 48 countries in SSA show a male predominance with an increased risk in the 30−49 years age group, while younger and older people smoke less. This seemed to have no connection with the social environment. Smoking started in late adolescence or in early adulthood. In Benin, a low level of education and living in rural areas have been found to be factors associated with smoking.[37]

TABLE 17.1 Prevalence of Addiction Among Adolescents in SSA

Country	Tobacco (%)	Alcohol (%)	Cannabis (%)	Authors
South Africa			2−9	Peltzer et al. [28]
Kenya		5.2	1.7	Ndetei et al. [33]
Senegal		15	70	D'Hondt and Vandewiele [34]
Nigeria	14.3	31.6	4.1	Igwe and Ojinnaka [35]
	18.1			
Benin	4			Houinato [36]
Niger	22			
Burkina-Faso	20.3			
Gabon	10.9			
Uganda	5.6			

17.3.2.4.2 FACTORS ASSOCIATED WITH ALCOHOLISM IN SSA

A review of the literature on alcoholism in SSA shows that the most frequently associated factors are comorbidity with tobacco consumption and socioeconomic environment. Thus, a case–control study in two suburbs of the capital of Tanzania, Dar es Salaam (middle class Ilala, and Saba, a shantytown) including 899 adults aged 15–59 years, showed that the prevalence of alcoholism and current smoking was higher in the poor suburb of Saba.[38] Comorbidity with HIV was also found to be a factor associated with alcoholism. From a review of the literature (21 publications from the East or the South of Africa), authors found a statistically significant positive correlation between alcoholism and HIV infection or its conversion to AIDS. There is a higher risk of HIV infection during the consumption of alcohol in sexual contexts.[39]

17.3.2.4.3 FACTORS ASSOCIATED WITH INTRAVENOUS DRUG INJECTION IN SSA

Among the associated factors, special mention should be made of stigma and the role of psychological trauma. Thus, intravenous drug users are generally victims of exclusion and repression, and among intravenous-substance-using women in Tanzania, a history of sexual assault in childhood was found.[40]

17.3.2.4.4 FACTORS ASSOCIATED WITH ADDICTION IN TEENAGERS IN SSA

A case–control study among 900 teenagers in 29 secondary schools of Enugu in Nigeria, revealed in some drug abusers ($n = 290$) a significant social dysfunction in 24.1% of abusers (vs 10.7% of controls); 18.6% of abusers (vs 7.7% of controls) had a ≥ 50 scale depression; the depression score was more severe in poly-abusers.[35] Another study in 939 students in 39 schools in the city of Cape Town (South Africa) showed a statistically significant link between post-traumatic stress disorder and addiction.[41]

17.3.2.4.5 FACTORS ASSOCIATED WITH ADDICTIONS IN WOMEN SSA

The most significant factors associated with FAS are the low socioeconomic status of women, ignorance of the community regarding the effects of alcohol on pregnancy, and the cultural context marked by the "Dop System". The "Dop System", invented by the settlers in the 17th century, is a mode of payment of farm workers with bread, wine and tobacco, and promotes alcohol comsumption.[42] Indeed in the Cape province in South Africa, women are agricultural workers (30%); they have two- to three-times more seasonal jobs than men. Fifty percent of pregnant women in the region of the Western Cape drink alcohol compared to 34% of pregnant women in the metropolitan areas of South Africa. Women whose children have FAS, drink 12.6 standard drinks per week versus 2.4 in controls. Among the reasons for this increased alcohol consumption is the stress of having a violent and alcoholic husband. Women do not seek help often enough for their drinking because they are afraid of the stigma, and of being rejected by their loved ones; they also face losing their children, either through their inability to look after them or through having them taken away. O'Connell et al.[43] described 619 pregnant black women in the suburbs of the city of Cape Town, and the factors associated with FAS were the following: youth, celibacy, active smoking, living conditions, long delay between beginning of gestation and diagnosis of pregnancy, domestic violence, high prevalence of depressive symptoms.

17.3.2.5 The Consequences of Addictions in SSA

Estimations indicate that smoking is responsible for five million deaths each year of which two million are in Africa. By 2030, this figure will reach 10 million deaths per year with 70% of these deaths in developing countries. As a result of the consequences of alcoholism in SSA, according to the WHO, the harmful use of alcohol is the fifth risk factor by order of importance in premature death and disability in the world. It is the leading cause of death in low-mortality developing countries. Alcohol is responsible for 2.2% of all deaths in SSA; 2.5% of all disability-adjusted life years are related to alcohol.

According to the report of the world situation with regard to alcohol, alcohol consumption is showing a global downward trend in the WHO African Region (AFR); however, the alcohol-attributable mortality is not the lowest in that region.[20,44] The role of alcohol in industrial accidents and in work-related deaths has been noted.[45]

In order to reduce the burden of harmful use of alcohol in the world, the WHO developed a world strategy including the epidemiologic surveillance of alcohol consumption.[44]

Greater economic wealth is broadly associated with higher levels of consumption and lower abstention rates. Moreover, a rise in alcohol consumption is expected to increase the alcohol-attributable burden of disease in developing economies.[46]

Schneider et al.,[47] using the WHO comparative risk assessment (CRA) methodology, in adults aged 15 years and above, found that alcohol harm accounted for an estimated 7.1% of all deaths and 7.0% of total disability-adjusted life years in 2000. Injuries and cardiovascular incidents ranked first and second in terms of attributable deaths, respectively. Top rankings for overall attributable burden were interpersonal violence, neuropsychiatric conditions and road traffic injuries. In a multi-disease community-based screening campaign for hypertension, diabetes, and HIV in Uganda including 1245 women and 1007 men, Kotwani et al.[48] reported a prevalence of hypertension of 14.6%; they found a strong link with modifiable factors including overweight/obesity and alcoholism: 24.1% men consumed ≥ 10 alcoholic drinks per month.

A high prevalence of peripheral artery disease (PAD) has been found in the older general population of two countries of Central Africa (Congo and the Central African Republic).[15]

In South Africa, within the fishing communities, alcohol consumption seems to be a real risk factor for HIV contamination. Fishermen who drank two or more times a week were 7.9-times more likely to have had transactional sex compared with those who never drank alcohol.[49]

One study, in Zambia,[50] also found a correlation between alcohol dependence disorders in persons receiving treatment for HIV and tuberculosis (TB) at 16 primary healthcare centers along with some other factors. Factors associated with alcohol dependence disorder in men included being single, divorced or widowed compared with being married and working.

Anti-Retroviral Treatments (ART) initiation may be an opportune time to implement interventions for alcohol consumption and other health behaviors.[51]

In Togo, a cross-sectional, descriptive and analytic study using the STEP-Wise approach was performed between December 1, 2010 and January 23, 2011, across the five regions of Togo, a low-income country in West Africa with a population of 6,306,000 inhabitants. Cluster sampling methods were used to select 4800 people aged 15–64 years. Results showed a young population with a male predominance. About one-third of respondents were alcohol abstainers the majority being women; it was the same proportion for current drinkers (daily consumption). Concerning Total Alcohol Consumption, the reported daily average consumption of alcohol was of 13 pure alcohol grams for men and 9 pure alcohol grams for women. The maximum mean number of heavy drinking days, during the last 30 days was more important in men (3 days) and concerned 37.5% of drinking men. The prevalence of current smokers was 6.8% with a male predominance (12.4%). In the Togo STEP-Wise study, 6.8% of the sample were current smokers, with a male predominance (12.4%); about three-quarters were daily smokers. Harmful use of alcohol was higher among tobacco users (5.9%) than tobacco non-users (2.8%) ($p < 10^{-6}$).[52]

In Benin, a neighboring country, according to a STEP-Wise study in 2008 on 6758 people aged 25–64

years, a tobacco use prevalence of 17.1% (95% CI = 15.9–18.3) was reported. Old age, male gender, Yom ethnicity, living in rural setting, low instruction level and county of living were significantly associated with tobacco use.[37] In Ghana, a prevalence of 7.6% was reported in adults.[53]

Some women interviewed during a STEP study in Togo were pregnant. Although the burden of alcohol use during pregnancy was a significant problem, limited data currently exist for the majority of SSA countries. Furthermore, significant variation likely exists within various populations. From a review of 12 selected studies,[54] it was estimated that the prevalence of alcohol use during pregnancy ranged from 2.2% to 87% in SSA and alcohol use is currently associated with domestic violence, including sexual domestic violence.[55]

17.3.2.6 The Causes of Addiction in SSA

All of the causes of addiction can be summed up by the poverty in SSA. In addition, high-income countries, looking for opportunities for licit drugs (tobacco and alcohol), try to circumvent the tightening of regulations.

17.3.2.6.1 CAUSES OF ALCOHOLISM IN SSA

Contrary to what one might believe, the alcohol consumed significantly in Africa is not imported; it is from non-industrial local production. However, in the 1990s, a shift from the use of traditional products to the consumption of industrial products was observed in Africa. Propagation of breweries can be linked to their probable role in economics or even politics in many countries.[25] Thus, in Burkina Faso in the 1980s, the manufacture of beer had tripled in 5 years.[56]

17.3.2.6.2 CAUSES OF THE TOBACCO EPIDEMIC IN DEVELOPING COUNTRIES

To respond to the decline in sales in developed countries, tobacco companies have begun to turn to foreign markets in developing countries; in these, the tobacco sales increase is three-times faster than elsewhere. The US tobacco industry controls 85% of the world production of tobacco leaves, and three of the six multinational corporations that dominate the tobacco industry in the world are located in the USA (Philip Morris, RJ Reynolds and American Brands). For 40% of world production between 1954 and 1980, tobacco met the conditions to be included in the aid program "Food for Peace". So each year, the Ministry of agriculture expedited the supply of millions of dollars-worth of tobacco with food to countries suffering from hunger.[57]

In developing countries, tobacco firms Fund aggressive advertising, sports and artistic events, and free entries into nightclubs. In Kenya, there is a cigarette brand called "life and sportsman" which presents itself as a key to success. Cigarettes sold in developing countries contain more tar and nicotine than those in developed countries. In the developing countries which produce tobacco, 71% of the total land area are devoted to the cultivation of this plant.

17.3.2.6.3 CAUSES OF ILLICIT DRUG ADDICTION IN SSA

The minimum gap between the average income of a agricultural worker in the sub-Saharan area and the average income of a farmer in the sub-Saharan area who produces cannabis has been evaluated from 1 to 10 in 1989. Cannabis would require equal production, $\frac{1}{4}$ of working time and 1/5 of the land necessary for the average of other cultures.[25]

17.3.2.6.4 CAUSES OF TRAFFICKING OF DRUGS IN SSA

The traffic of drugs in SSA is partly a response to a shortage in the economy and the absence of legislation on orders. In Nigeria in 1990, 70% of available drugs came from counterfeiting.[25,57] The illicit production of psychotropic drugs escape checks by the International Vienna Convention of 1971 on psychotropic substances, as well as the laws of the country of importation. This illicit production comes from traffic with some countries: amphetamines (produced in Bulgaria), pemoline (produced in Yugoslavia), ephedrine (produced in Germany, United Kingdom, China, and Asia).

The countries involved in the importation and redistribution are Nigeria, Togo, and Guinea, in West Africa, and Kenya, Zambia, Zimbabwe in East, and South Africa. Ephedrine and pseudoephedrine, are used to develop amphetamine-type stimulants and methamphetamine (crystal, and MMDA better known as Ecstasy). In 2007, 23 tons of pseudoephedrine were intercepted in the DRC. Nigeria was suspected to be implicated in exportation of $KMnO_4$ (potassium permanganate used in the manufacture of cocaine) to South America.[25,27]

17.3.3 Solutions to Addiction in SSA

17.3.3.1 Fighting Tobacco Control in SSA

Only a few African countries (South Africa, Botswana, Mali, Mauritius) have adopted comprehensive regulatory frameworks to fight against the use of tobacco, including tax policy, bans on advertising, restrictions on consumption, judgment and effective education programs. An international legal instrument, was adopted in 2003 to fight against tobacco (convention by WHO), which advocates three ways to fight tobacco taxation, the dissemination of research and the facilitation of access to alternative treatments. In West Africa, there are six signatories: Benin, Burkina Faso, Republic of Ivory Coast, Guinea, Mali, Niger, Senegal.[58]

17.3.3.2 The Fight Against Alcoholism

Controls of State structure are different depending on the country. When there is a legislative or religious ban, alcohol is introduced by smuggling.[25] At its 63rd meeting on May 20, 2010, the WHO through 193 member countries, adopted a text recommending that nations raise the price of alcohol and encourage a better control of the advertisments and sponsorships promoted by the alcohol industry.

17.3.3.3 The Fight Against Drug Abuse and Trafficking of Psychotropic Drugs

Although the WHO encourages the implementation of substitution programs, these are overpriced and are not yet available in most countries in SSA. The American program PEPFAR (President Emergency Plan for AIDS Relief), which has made available antiretrovirals to 2.1 million HIV-positive people in SSA, has unfortunately not yet had an effect on intravenous drug users.[29] In this respect, South Africa seems to be a model country. Indeed, there are national recommendations for the treatment of opioid addiction by substituting with methadone and buprenorphine HD, although they are not yet popular enough.[59]

In Bela-Bela and in greater Pretoria, education programs on substance abuse linked with HIV/AIDS exist to address students and community workers; their mid-term evaluation is not yet conclusive.[60] Prevention and risk reduction policy is desirable. Finally, strong legislation in matters of importance associated with a real drug policy is essential.

17.4 CONCLUSION

In sub-Saharan Africa, mental health remains the poor child of the health Policy. Women and children are paying a heavy price. Poverty, war, stigma, and magical beliefs remain the main factors associated with psychopathology, which resembles that observed elsewhere in the world. The WHO has made many efforts to encourage States to develop community and integrated mental health policies.

Addictions in SSA are characterized by disparate prevalence figures that call for studies with a common methodology. There is an increase in problems related to smoking and alcoholism in women, especially in the region of Cape Town, South Africa. It should also be noted that there is an increase in adolescent addictions. Intravenous drug abuse is also increasing. There are clearly links between addictions and HIV, with an urgency for prevention. Psychiatric co-morbidities with SUDs have not been studied extensively enough. Support is still in its infancy. All of these findings argue for the need to develop teaching about addiction in SSA and the importance of improving addiction medicine policy in each country.

References

1. Steel ZI, Marnane C, Iranpour C, et al. The global prevalence of common mental disorders: a systematic review and meta-analysis 1980—2013. *Int J Epidemiol*. 2014;43:476—493.

2. Perez SI, Junod A. La dépression en Afrique sub-saharienne. *Med Trop*. 1998;58:168—176.

3. Carta MG, Coppo P, Reda MA, Hardoy MC, Carpiniello B. Depression and social change. From transcultural psychiatry to a constructivist model. *Epidemiol Psychiatr Soc*. 2001;10:46—58.

4. World Health Organization. *Preventing Suicide: A Global Imperative*. Geneva, Switzerland: WHO; 2014:141 p.

5. Ezin Houngbé J, Gansou MG, Agongbonou R, et al. Prévalence de la schizophrénie au Bénin. *Rev Epidémiol Santé Publique*. 2014;62:S226.

6. Jenkins R, Mbatia J, Singleton N, White B. Prevalence of psychotic symptoms and their risk factors in urban Tanzania. *Int J Environ Res Public Health*. 2010;7:2514—2525.

7. Jenkins R, Njenga F, Okonji M, et al. Psychotic symptoms in Kenya: prevalence, risk factors, and relationship with common mental disorders. *Int J Environ Res Public Health*. 2012;9:1748—1756.

8. Morgan C, Hibben M, Esan O, et al. Searching for psychosis: INTREPID (1): systems for detecting untreated and first-episode cases of psychosis in diverse settings. *Soc Psychiatry Psychiatr Epidemiol*. 2015;50:879—893.

9. Jenkins R, Othieno C, Ongeri L, et al. Adult psychotic symptoms, their associated risk factors and changes in prevalence in men and women over a decade in a poor rural district of Kenya. *Int J Environ Res Public Health*. 2015;12:5310—5328.

10. De Menil V, Osei A, Douptcheva N, Hill AG, Yaro P, De-Graft Aikins A. Symptoms of common mental disorders and their correlates among women in Accra, Ghana: a population-based survey. *Ghana Med J*. 2012;46:95—103.

11. Bakare MO. Attention deficit hyperactivity symptoms and disorder (ADHD) among African children: a review of epidemiology and co-morbidities. *Afr J Psychiatry (Johannesburg)*. 2012;15:358—361.

12. Naidoo S1, Mkize DL. Prevalence of mental disorders in a prison population in Durban, South Africa. *Afr J Psychiatry (Johannesburg)*. 2012;15:30—35.

13. Aloba O, Fatoye O, Mapayi B, Akinsulore S. A review of quality of life studies in Nigerian patients with psychiatric disorders. *Afr J Psychiatry (Johannesburg)*. 2013;16:333—337.

14. Fekadu A, Medhin G, Kebede D, et al. Excess mortality in severe mental illness: 10-year population-based cohort study in rural Ethiopia. *Br J Psychiatry*. 2015;206:289—296.

15. Samba H, Guerchet M, Ndamba-Bandzouzi B, et al. Dementia-associated mortality and its predictors among older adults in sub-Saharan Africa: results from a 2-year follow-up in Congo (the EPIDEMCA-FU study). *Age Ageing*. 2016;45:681—687.

16. World Health Organization. *Mental Health and Development: Targeting People with Mental Health Conditions as a Vulnerable Group*. Geneva, Switzerland: WHO; 2010:108 p.

17. World Health Organization. *Mental Health Action Plan 2013-2020. 1. Mental Health. 2. Mental Disorders—Prevention and Control*. Geneva, Switzerland: WHO; 2013:50 p.

18. World Health Organization. *MhGAP Intervention Guide for Mental, Neurological and Substance Use Disorders in Non-specialized Health Settings: Mental Health Gap Action Programme (mhGAP)*. Ginebra: OMS; 2010:126 p.

19. World Health Organization. *Mental Health Systems in Selected Low- and Middle-Income Countries: A WHO-AIMS Cross National Analysis*. Geneva, Switzerland/Herndon, VA: WHO/Stylus Pub., LLC; 2009:108 p.

20. World Health Organization. *Global Status Report on Alcohol and Health*. Geneva, Switzerland: WHO; 2014:<http://www.who.int/substance_abuse/publications/global_alcohol_report/en/>.

21. Pierucci-Lagha A, Derouesné C. Alcool et vieillissement. *Geriatr Psychol Neuropsychiatr Vieil*. 2003;1:234—249.

22. Ménecier P, Afifi A, Menecier-Ossia L, et al. Alcool et démence: des relations complexes. *Rev Gériatrie*. 2006;31:11—18.

23. Nubukpo P, Laot L, Clément JP. Addictions de la personne âgée. *Geriatr Psychol Neuropsychiatr Vieil*. 2012;10:315—324.

24. Pirkola SP, Poikolainen K, Lönnqvist JK. Currently active and remitted alcohol dependence in a nationwide adult general population—results from the Finnish Health 2000 study. *Alcohol Alcohol*. 2006;41:315—320.

25. Cesoni ML. La route des drogues: explorations en Afrique sub-saharienne. *Tiers Mondes*. 1992;33(131):645—671.

26. Naudé PF. Le continent de tous les trafics. Jeune Afrique du 4/09/2008. www.jeuneafrique.com.

27. Nubukpo P. Les addictions en Afrique Subsaharienne. *Courr Addict*. 2011;4:27—30.

28. Peltzer K, Ramlagan S, Johnson BD, Phaswana-Mafuya N. Illicit drug use and treatment in South Africa: a review. *Subst Use Misuse*. 2010;45:2221—2243.

29. Nabel EG, Stevens S, Smith R. Lives to save: PEPFAR, HIV, and injecting drug use in Africa. *Lancet*. 2009;373(13):2006—2007.

30. Raguin G, Leprêtre A, Ba I, et al. Usage de drogues et VIH en Afrique de l'Ouest : un tabou et une épidémie négligée. *Transcriptase*. 2010;143:15—17.

31. Vickerman P, Hickman M, May M, Kretzschmar M, Wiessing L. Can hepatitis C virus prevalence be used as a measure of injection-related human immunodeficiency virus risk in populations of injecting drug users? An ecological analysis. *Addiction*. 2010;105:311—318.

32. Pretorius L, Naidoo A, Reddy SP. "Kitchen cupboard drinking": a review of South African women's secretive alcohol addiction, treatment history, and barriers to accessing treatment. *Soc Work Public Health*. 2009;24:89—99.

33. Ndetei DM, Khasakhala LI, Mutiso V, Ongecha-Owuor FA, Kokonya DA. Drug use in a rural secondary school in Kenya. *Subst Abus*. 2010;31:170—173.

34. D'Hondt W, Vanderwiele M. Use of drug among Senegalese school going adolescents. *J Youth Adolesc*. 1984;3:253—266.

35. Igwe WC, Ojinnaka NC. Mental health of adolescents who abuse psychoactive substances in Enugu, Nigeria—a cross-sectional study. *Ital J Pediatr*. 2010;10:36—53.

36. Houinato D. *Prévalence du tabagisme et de la consommation nocive d'alcool en population générale au Bénin en*. Thèse Médecine; 2008.

37. Nubukpo P, Gbary AR, Ouendo Em, et al. Le tabagisme en population générale au Bénin (Afrique de l'Ouest). *Alcoologie et Addictologie*. 2012;34:265—272.

38. Mbatia J, Jenkins R, Singleton N, White B. Prevalence of alcohol consumption and hazardous drinking, tobacco and drug use in urban Tanzania, and their associated risk factors. *Int J Environ Res Public Health*. 2009;6:1991—2006.

39. Pithey A, Parry C. Descriptive systematic review of sub-Saharan African studies on the association between alcohol use and HIV infection. *Sahara J.* 2009;6:155–169.

40. McCurdy SA, Ross MW, Williams ML, Kilonzo GP, Leshabari MT. Flashblood: blood sharing among female injecting drug users in Tanzania. *Addiction.* 2010;105:1062–1070.

41. Saban A, Flisher AJ, Distiller G. Association between psychopathology and substance use among school-going adolescents in Cape Town, South Africa. *J Psychoactive Drugs.* 2010;42:467–476.

42. Parry CD. Alcohol policy in South Africa: a review of policy development processes between 1994 and 2009. *Addiction.* 2010;105:1340–1345.

43. O'Connell R, Chishinga N, Kinyanda E, et al. Prevalence and correlates of alcohol dependence disorder among TB and HIV infected patients in Zambia. *PLoS One.* 2013;8:e74406.

44. World Health Organisation. *Global Status Report on Alcohol and Health.* Geneva, Switzerland: WHO; 2010:<http://www.who.int/substance_abuse/publications/global_alcohol_report>.

45. Anderson P, Moller L, Gadea G, et al. *Alcohol in the European Union: Consumption, Harm and Policy Approach.* Copenhagen, Denmark: WHO Regional Office for Europe; 2012:p. 101 <http://www.euro.who.int/>.

46. Grittner U, Kuntsche S, Gmel G, Bloomfield K. Alcohol consumption and social inequality at the individual and country levels results from an international study. *Eur J Public Health.* 2013;23:332–339.

47. Schneider M, Norman R, Parry C, et al. Estimating the burden of disease attributable to alcohol use in South Africa in 2000. *S Afr Med J.* 2007;97:664–672.

48. Kotwani P, Kwarisiima D, Clark TD, et al. Epidemiology and awareness of hypertension in a rural Ugandan community: a cross-sectional study. *BMC Public Health.* 2013;13:1151.

49. Tumwesigye NM, Atuyambe L, Wanyenze RK, et al. Alcohol consumption and risky sexual behaviour in the fishing communities. *BMC Public Health.* 2012;12:1069.

50. O'Connor MJ, Tomlinson M, Leroux IM, Stewart J, Greco E, Rotheram-Borus MJ. Predictors of alcohol use prior to pregnancy recognition among township women in Cape Town, South Africa. *Soc Sci Med.* 2011;72:83–90.

51. Santos Santos G-M, Emenyonu N-I, Bajunirwe F, et al. Self-reported alcohol abstinence associated with ART initiation among HIV-infected persons in rural Uganda. *Drug Alcohol Depend.* 2014;134:151–157.

52. Agoudavi H, Dalmay F, Legleye S, Kumako K, Preux PM, Nubukpo P. Epidemiology of Alcohol Use Disorder (AUD) in Togo general population. *Addict Behav Rep.* 2015;2:1–5.

53. Yawson AE, Akosua Baddo A, Hagan-Seneadza NA, et al. Tobacco use in older adults in Ghana: sociodemographic characteristics, health risks and subjective wellbeing. *BMC Public Health.* 2013;13:979.

54. Culley CL, Ramsey DT, Mugyenyi G, et al. Alcohol exposure among pregnant women in sub-saharian Africa: a systematic review. *J Popul Ther Clin Pharmacol.* 2013;20:321–333.

55. Semahegn A, Belachew T, Abdulahi M. Domestic violence and its predictors among married women in reproductive age in Fagitalekoma Woreda, Awi zone, Amhara regional state, North Western Ethiopia. *Reprod Health.* 2013;10:63.

56. Treillon R, Gattegno I. Canettes contre calebasses ou une comparaison économique des filières bière industrielle, bière artisanale au Burkina Faso. In: Bricas N, et al. , eds. *Nourrir les villes en Afrique Subsaharienne.* Paris: L'Harmattan; 1985:279.

57. Barry M. Le tiers monde malade du tabac. *La Recherche.* 1991;236:1190–1193.

58. Da Costa e Silva VL. Health consequences of the tobacco epidemic in West African French-speaking countries and current tobacco control. *Promot Educ.* 2005;(Suppl 4):7–12.54.

59. Weich L, Perkel C, van Zyl N, Rataemane ST, Naidoo L. Medical management of opioid dependence in South Africa. *S Afr Med J.* 2008;98:280–283.

60. Perngparn U, Assanangkornchai S, Pilley C, Aramrattana A. Drug and alcohol services in middle-income countries. *Curr Opin Psychiatry.* 2008;21:229–233.

Further Reading

Townsend L, Flisher AJ, Gilreath T, King G. A systematic literature review of tobacco use among adults 15 years and older in sub-Saharan Africa. *Drug Alcohol Depend.* 2006;84:14–27.

18

Neurological Syndromes or Diseases Caused by Parasites in Tropical Areas

Francisco J. Carod-Artal[1,2], Hector H. Garcia[3,4],
Andrea S. Winkler[5,6], and Daniel Ajzenberg[7]

[1]Raigmore Hospital, Inverness, United Kingdom [2]International University of Catalonia, Barcelona, Spain [3]Cayetano Heredia University, Lima, Peru [4]National Institute of Neurological Sciences, Lima, Peru [5]University of Oslo, Oslo, Norway [6]Technical University of Munich, Munich, Germany [7]University of Limoges, Inserm UMR 1094 NET, Limoges, France

18.1 INTRODUCTION

Most infectious and parasitic diseases in the tropics are considered neglected tropical diseases. They affect poor resource populations mostly in rural areas and urban slums in Latin-America, Africa and Asia. Parasitic diseases are an unrecognized burden in the tropics. However, the involvement of the central nervous system (CNS) in parasite diseases is the most common cause of disability in these regions. Epilepsy and disability due to cerebral and spinal cord syndromes may affect young adults and cause long-term sequelae.[1] Recently, it has been estimated that a group of 12 diseases (Chagas disease, African human trypanosomiasis, schistosomiasis, leishmaniasis, lymphatic filariasis, onchocercosis,

ascariasis, trichuriasis, hookworn disease, trachoma, dracunculiasis, and leprosy) accounted for about 56.6 million disability-adjusted life years lost and 534,000 deaths.[2] In addition, epilepsy is also a silent epidemic in the tropics and prevalence of seizures could be at least three-times the prevalence observed in developed countries. Neurocysticercosis (NCC), cerebral malaria, bacterial meningitis, and other parasite infections of the CNS are the most common causes of seizures in the tropics.[1,3−7]

In this chapter, clinical and epidemiological data about the most common parasite infections involving the CNS in tropical areas, including NCC, trypanosomiasis, cerebral malaria, schistosomiasis, toxoplasmosis, gnathostomiasis, and angiostrongylosis will be reviewed. The nodding syndrome (NS) and neurologic infections caused by free living amebae are also included in this review because these diseases are considered to be emerging.

18.2 CYSTICERCOSIS

Neurologic infection by the larvae of the pork tapeworm *Taenia solium* (NCC) is a very common cause of seizures and other neurological manifestations in most developing countries around the world.[8,9] The cycle of *T. solium* is typical of complex parasites and includes a definitive host (humans) harboring the adult tapeworm in the intestine, and an intermediate host (pigs) harboring the cystic larvae or cysticerci in the tissues. The cycle is initiated when humans ingest improperly cooked pork, and the larvae evaginate by action of intestinal enzymes and bile, and anchor in the intestinal mucosa using the cephalic fixation organs (hooks and suckers). From there, the adult tapeworm develops by forming units or proglottids from the neck region. Distal proglottids are gravid and contain thousands of infective eggs that are released to the environment with stools of the human tapeworm carrier. Then the cycle is closed when the pig ingests the eggs by accessing human feces in places with poor sanitation, and the embryos in the eggs are released and cross the intestinal mucosa, reach the circulatory system, and are transported to all tissues where they establish as cysticerci. Unfortunately, humans may also act as intermediate hosts when they ingest eggs via fecal oral contamination.[10]

Taenia solium is endemic in most of Latin-America, most of Africa, and parts of Asia including China, the Indian subcontinent and some countries in Southeast Asia. Active transmission does not seem to occur in the UK, Europe or the USA/Canada, although clinical cases are seen because of immigration and travel. Similarly, transmission does not occur in Muslim countries but cases may still occur in immigrants.[8,11,12] Epidemiological data from multiple serological studies[13–15] and a few community-based studies using neuroimaging[16–20] suggest that human exposure to *T. solium* in endemic regions is a very common event, and marks of previous infection (cerebral calcifications) can be found in 10%–20% of unselected individuals in these areas. Most individuals infected with NCC in community settings seem to be asymptomatic, although mild neurological disease cannot be ruled out. A minority of NCC-infected individuals will develop symptoms, particularly late-onset seizures, headache, intracranial hypertension, and less frequently motor or cognitive deficits. The clinical expression of NCC differs depending on the stage of parasites, its number, size, location, and the immune response of the host.[8]

Most of the literature classifies cases of NCC based on the location and stage of parasites. A major distinctive factor is whether the parasites are in the brain parenchyma or outside (subarachnoid space and ventricles). Intraparenchymal NCC presents with seizures and epilepsy and follows a more benign prognosis. Cerebral parasites establish as viable cysts without an evident inflammatory reaction from the host. At some point the host's immune system detects the parasites and launches an inflammatory response with cellular infiltration that ends up destroying the cyst. During this stage (degenerating cysts), perilesional inflammation and edema result in more symptoms. Finally the inflammation subsides and the cyst remnants are cleared or, most commonly, a calcified scar remains.[21] Conversely, cysts in the ventricles or the subarachnoid space most frequently grow and infiltrate, resulting in blockage of cerebrospinal fluid (CSF) circulation and hydrocephalus, or in clusters of cysts that occupy space and produce mass effects. Parasite growth is more notable when the lesions are located in the Sylvian fissure or the basal cysterns.[22]

A table of diagnostic criteria and degrees of diagnostic certainty was developed in 1996 and updated in 2001.[23] This set of criteria is a very useful instrument to homogenize criteria and make cases series comparable across regions. Most clinicians experienced with NCC will focus on neuroimaging as the primary line of diagnosis, and rely on serology for confirmatory purposes or to clarify cases where imaging is unclear. Either computed tomography (CT) or magnetic resonance imaging (MRI) is useful to diagnose NCC. MRI provides better imaging definition and is more sensitive to demonstrate small lesions, ventricular lesions, lesions near the skull, or signs of perilesional inflammation. On the other hand, CT is much more sensitive to detect calcified cysts. Imaging of multiple cystic lesions with scolex is characteristic of NCC.[23] Immunodiagnosis of cysticercosis is more frequently oriented to antibody detection, and the test of choice is the enzyme-linked immunoelectrotransfer blot assay (EITB, western blot) using purified parasite glycoprotein antigens. This test is highly sensitive and specific, although the background rates of seropositivity in endemic regions may affect its positive predictive value.[24] Antigen detection (using monoclonal antibody-based enzyme-linked immunosorbent assay (ELISA)) is less sensitive but has the advantage of demonstrating the presence of live parasites and thus serves to orientate the therapeutic approach.[25]

The management of NCC implies both symptomatic and specific measures. Symptomatic management (including analgesics, antiepileptic drugs, anti-inflammatory agents, or measures to control intracranial hypertension) is the primary line of work, and should be appropriately established and effective before considering the use of antiparasitic agents.[26] In most patients, destroying live parasites by using antiparasitic agents is of benefit for their long-term evolution.[27,28] This process, however, may cause an increase of perilesional inflammation in the initial days of therapy with associated neurological symptoms.[26] The available agents are albendazole (commonly used at 15 mg/kg per day, for 8–30 days) and praziquantel (commonly used at 50 mg/kg per day, for 15 days). Combined therapy with both albendazole and praziquantel is more effective than albendazole alone in patients with multiple cysts.[28,29] Antiparasitic agents should be given with concomitant steroid coverage to modulate the acute inflammatory reaction caused by the attack on the parasite and the subsequent antigen liberation resulting in local inflammation.[26] Caution should be exercised when using antiparasitic agents in patients with heavier parasite loads because of the risk of diffuse inflammation and severe intracranial hypertension.[30]

The *T. solium* taeniasis/cysticercosis complex is considered eradicable on the basis of having a single definitive host, a known intermediate host closing the cycle, no wild vectors, and available therapeutic agents. After multiple initial attempts for control at small scale, the Cysticercosis Working Group in Peru reported in 2016 a large elimination program that interrupted transmission in a large endemic area.[31] The strategy was based on mass treatment of humans with niclosamide (followed by coproantigen stool examinations[32] to identify tapeworm carriers and confirm that they were cured), mass treatment of pigs with oxfendazole,[33] and pig vaccination with the TSOL18 antigen.[34] This program was complex and labor-intensive, however it provides proof of concept that transmission can be interrupted, opening the possibility for wider elimination and eventual eradication of the disease.

18.3 AMERICAN TRYPANOSOMIASIS

American trypanosomiasis, also called Chagas disease, is a parasitic disease caused by the flagellate protozoan *Trypanosoma cruzi*. Chagas disease has been for decades a neglected tropical disease in Latin-America, and nowadays it has become a global health problem. A wild vector-borne transmission occurs predominantly in rural endemic areas where the disease is transmitted to humans by the contaminated feces of blood-sucking triatomine bugs. Although more than 130 triatomine species have been described to be potential *T. cruzi* vectors, the main ones are *Triatoma infestans*, *Rhodnius prolixus*, and *Panstrongylus megistus*.[35] A large-scale rural-to-urban migration occurred in the last half century and, as a consequence, non-vector transmission through infected blood transfusion, congenital infection and organ transplantation predominates in urban areas. Oral transmission outbreaks have been reported in the Amazon basin and Venezuela.[36]

Geographically, Chagas disease is endemic in 21 Latin-American countries and spread from north of Argentina and Chile, Bolivia, Brazil, Venezuela, Colombia, and the Caribbean region to Mexico and southern Texas. According to the World Health Organization (WHO), between 8 and 14 million people may be infected with *T. cruzi*, including 2 million infected women of fertile age.[37] However the true prevalence of chronic Chagas disease may be underestimated as many chagasic people do not know they harbor *T. cruzi* infection. Around 20% of the Latin-American population (100 million people) is living in areas which are triatominae habitats, and more than 25 million people are at risk of the disease.[37] The annual burden has been estimated as $627.5 million in healthcare and more than 806,000 disability-adjusted life-years.[38] The number of acute cases decreased from 700,000 in 1990 to approximately 41,000 in 2006, and the number of annual deaths were below 14,000 after the implementation of entomological surveillance and vector control programs.[37]

A number of hyperendemic areas have been reported in several countries. Although the prevalence of Chagas disease in Brazil is around 4%, the infection rate in the northeast region may range between 5% and 13%. The prevalence of *T. cruzi*-infected people in Bolivia is near 7% and in Mexico ranges between 1% and 6%.[35] The appearance of secondary peridomestic vectors and the resistance to pyrethroid insecticides are additional difficulties in controlling the disease in highly endemic areas. The Chaco region and the Bolivian Highlands remain at risk for *Triatoma* spp. infestation, and a high prevalence of infected pregnant women has been found.[39]

Congenital transmission from chagasic pregnant women who have the chronic or indeterminate form of the disease is an important concern. Vertical transmission rate may range from 5.6% (Uruguay) and 19% (Northern Chile and Argentina) to 40% in some areas of southern Bolivia. Between 2% and 10% of babies may acquire the congenital infection.[40] Symptomatic infection may happen in half of newborns who have the congenital infection, with a mortality rate of 2%–13% if left untreated.[40] Compulsory screening for *T. cruzi* infection in most Latin-American blood banks has reduced dramatically the rate of blood transfusion-transmitted cases. However, the rate of seropositive donors in Brazilian and Argentinean blood banks is still high (1.3% and 4%, respectively).[41]

Chagas disease is considered an emerging disease in non-endemic countries as a consequence of the growing migration flows from Latin-America to non-endemic countries. Cases have been reported in Europe, North America, Australia, and Japan. Migration of *T. cruzi* infected patients has spread Chagas disease by means of non-vector-borne transmission, mainly through blood transfusion and congenital transmission. Prevalence of Chagas disease among immigrants in North America and Europe may range between 2% and 5.2%.[35] A least 300,000 *T. cruzi* infected immigrants may be currently living in the United States and approximately 65,000 will develop the chronic form of the disease.[42] Nevertheless, the estimated prevalence data may not reflect the real burden of Chagas disease in non-endemic countries because illegal or undocumented immigrants may have not been included in these studies. The number of *T. cruzi* infected immigrants in Europe may exceed 120,000 with more than 4300 laboratory-confirmed cases in France, Belgium, Spain, Italy, Switzerland, and the United Kingdom. However, the index of underdiagnosis is very high (around 95%).[43] Screening of blood supply for Chagas disease is now mandatory in North America and most European countries.

Primary chagasic infection usually occurs during childhood and most acute cases are asymptomatic but severe cases of myocarditis and encephalitis may occur in children. An indeterminate stage follows and may last years. Nevertheless, between one-third and half of *T. cruzi*-infected patients will develop cardiac, digestive, cardio-digestive, or neurological complications along their lives.[37] At least one-third of chagasic patients will suffer from a chronic progressive cardiomyopathy. Heart failure, cardiac arrhythmias, conduction disorders, sudden death, and cardioembolism are common complications.[44] The chronic gastrointestinal form may cause megaoesophagus and megacolon.

Chagas disease is an independent risk factor for ischemic stroke and a major cause of cardioembolic stroke. Factors that predispose to stroke include cardiac arrhythmias, conduction defects, left ventricle apical aneurysm, mural thrombus, severe heart failure and left ventricle dysfunction. More than 70% of ischemic stroke chagasic patients have electrocardiography abnormalities and an apical aneurysm can be found in 40% of them.[45] Ischemic stroke may also be the first manifestation of Chagas disease in patients with asymptomatic *T. cruzi* infection and/or with mild left ventricle dysfunction. Around one-third of chagasic patients who suffer a stroke have an asymptomatic *T. cruzi* infection, and cardiac damage can also be detected at this stage.[46]

The aging of the chagasic population and the adoption of urban habits (sedentary style of life, smoking, and obesity) may induce mixed forms of cardiomyopathy and an increased risk of ischemic stroke. Vascular risk factors may also explain the appearance of non-cardioembolic stroke subtypes, and the detection of large-artery ischemic stroke and lacunar infarctions in some patients. More than 20% of chagasic stroke patients may suffer stroke recurrence during the follow-up. The 10-year accumulative risk of death in *T. cruzi* infected patients is two-times higher (4.8%) as compared to non-infected ischemic stroke patients.[47]

In Central Brazil, the prevalence of *T. cruzi* infection in a hospital case-series study of 478 ischemic stroke patients was 20%.[45] Population studies showed that around 5.4% of *T. cruzi* infected people in the community had suffered an ischemic stroke.[48] Chagasic cardiomyopathy harbors a double risk of provoking cardioembolic stroke as compared to other types of cardiomyopathy.[49] The magnitude of the problem may be greater as many patients who suffered a stroke did not know they were infected by *T. cruzi*. In another Brazilian study, 40% of stroke patients were diagnosed as having Chagas disease after suffering their first ischemic stroke.[50]

Acute reactivation of the indeterminate or chronic chagasic stages has been described in immune-suppressed individuals suffering from AIDS, on chronic use of corticoids or under other immunosuppressive therapies. Neurological involvement occurs in at least 85% of cases. Encephalitis, meningoencephalitis and space-occupying lesions called chagomas are common forms of presentation. Cerebral reactivation of Chagas disease is considered an AIDS-defining opportunistic infection in Latin-America.

Due to the dynamic changes in the epidemiology of Chagas disease, several recommendations have arisen to policy makers.[51] The screening for Chagas disease among ischemic stroke patients in endemic regions is recommended. There is a need for active surveillance programs among asymptomatic *T. cruzi* infected individuals. From a public health perspective, it is also necessary to develop measures to prevent congenital and blood-transfusion *T. cruzi* transmission in both endemic and non-endemic countries.

18.4 AFRICAN TRYPANOSOMIASIS

Human African trypanosomiasis (HAT), also called sleeping sickness, is a parasitic disease caused by the protozoan *Trypanosoma brucei*, and transmitted to human beings by the blood-sucking tsetse fly (genus: *Glossina*). The two human pathogen subspecies, *Trypanosoma brucei gambiense* and *Trypanosoma brucei rhodesiense*, are morphologically indistinguishable. *Trypanosoma brucei gambiense* represents 98% of reported cases and causes the Western and Central sub-Saharan form of HAT, which leads to a chronic infection that may last for years. In Eastern and southern Africa *T. b. rhodesiense* is considered a zoonosis with occasional presence in humans causing an acute infection that may be fatal within weeks or months.[52]

HAT is considered a neglected tropical disease affecting poor populations in rural areas and is endemic in 36 countries in sub-Saharan Africa (SSA). A patchy geographical distribution has been noted. Twenty-four countries are endemic for *gambiense* HAT, and the 95% of cases were reported in only five countries: Democratic Republic of Congo (78%), Central African Republic (8%), South Sudan (4%), Chad (3.5%), and Angola (3%). More than 2 million people live at very high risk for western HAT as defined by an average annual number of cases of at least 1/1000 inhabitants.[53] Tsetse fly infestation may cover about 10 million km^2 of African landmass, and the total endemic area is around 1.308 million km^2. Approximately 70 million people are at risk of contracting the disease in SSA. The 80% are at risk of *T. b. gambiense* HAT.[54]

In the last decades the improved control and surveillance system resulted in a sustained reduction in the number of annual new cases, which fell below 10,000 per year in 2012.[53] Nevertheless, underdetection is a big problem and approximately 50% of cases may not be reported. Around 20,000 people across Africa may be infected with HAT.

18.5 MALARIA

Malaria is a protozoan disease transmitted by Anopheles mosquitoes in poor, tropical and subtropical areas of the world and is one of the leading causes of death throughout humanity, especially in children. The highest burden of the disease is in SSA where young children under 5 years and pregnant women are the most vulnerable.[55] The immensity of this burden is difficult to imagine because morbidity, disability, and mortality directly and indirectly caused by malaria are also accompanied by dramatic social and economic effects.[56–58] Among the five species of *Plasmodium* causing malaria in human beings, *Plasmodium falciparum* causes the majority of infections in Africa and almost all deaths, so we will focus our overview on *falciparum* malaria in SSA.

The importance of *P. falciparum* in human evolutionary history is so high that malaria is the infectious disease that had the strongest selective pressure on the genome of populations from Africa where the disease is highly endemic.[59] The most famous example of genetic adaptation is the HbS mutation in the β-globin gene. Although the HbS mutation can be fatal in persons who are homozygous by causing sickle cell anemia, it was maintained at high frequency in the genome of populations from malaria-endemic regions because it is associated with an 89% reduced risk for cerebral malaria and severe malarial anemia in persons who are heterozygous for this mutation.[60,61] There are many other examples of natural selection due to malaria, including blood group O, other hemoglobin variants such as HbC, or a novel locus close to a cluster of glycophorin genes involved in erythrocyte invasion by *P. falciparum*, which are associated with resistance to malaria although they are less strongly protective than HbS against severe malaria.[61,62] Other mutations that result in reduced enzyme activity of the glucose-6 phosphate dehydrogenase or low sensitivity to bitter-tasting substances are often-quoted examples of mutations supposed to be selected by malaria pressure on African populations but their protective effect against severe malaria was not tested or was not obvious in large multi-center case—control studies.[59,61]

Plasmodium falciparum is one of the most pathogenic infectious agents for humans, and it is of great interest to find out since when humans have had to face this deadly cohabitation. There is good evidence that modern humans in SSA were already infected by *P. falciparum* when they started their exodus out of Africa less than 62,000—95,000 years ago.[63,64] This indicates that *P. falciparum* originated in SSA and colonized the world by following human migrations, but how and when was it introduced into humans? This question has raised considerable debate in the literature with basically two different theories.[65] The co-speciation scenario favors an ancient origin: a common ancestor of *P. falciparum* and *Plasmodium reichenowi*, a parasite of the chimpanzee, was already present in the ancestor of humans and chimpanzees and diverged at the same time as their respective hosts between 4 and 7 million years ago. The host-switch scenario favors a more recent origin: a *P. falciparum*-like parasite was present in African great apes and was transferred into humans in whom it became *P. falciparum*. This scenario echoes the cross-species transmissions of human immunodeficiency virus (HIV) from mangabey and chimpanzee to humans in Africa.[66] The Bonobo, Chimpanzee, and gorilla, were all nominated in the literature to receive the prize of the origin of *P. falciparum* but there is now convincing data that *P. falciparum* is of gorilla origin and emerged in humans following a single cross-species transmission event.[67] However it is still impossible to know when this event occurred.[65,67] The good news is that great apes are not reservoirs of *Plasmodium*, including *P. falciparum*, for humans because *P. falciparum*-like and other *Plasmodium* species that infect present-day great apes do not infect humans.[68]

The public health problem of malaria is so big that it was central to seven out of the eight Millenium Development goals (MDGs) at the 2000 United Nations general Assembly.[69] In 1998, WHO and other partners created the Roll Back Malaria (RBM) Partnership as a coordinated plan to reduce malaria cases from 2000 levels by 75% and malaria deaths to near zero in 2015.[70] Reliable estimates of malaria cases are needed to see the real effect of these efforts against malaria since the Millenium declaration in 2000. Unfortunately, because this disease occurs in resource-poor settings with unreliable national health records and where many sick children are not brought to health facilities, there is a massive uncertainty in estimates of malaria cases.[58] According to the WHO in the malaria world report 2015,[69] there was a total of 218 million cases of malaria in 2013 but the confidence interval (CI) is so large that this estimate could be anywhere between 148 and 302 million. In other words, the amount of uncertainty in annual malaria cases is larger than the population of the ninth most populous country in the world, the Russian federation. According to this report, 86% of malaria cases were in SSA in 2013 and the four countries with the higher number of cases were Nigeria, Democratic Republic of the Congo, India, and Mozambique with 59, 21, 17, and 9 million cases, respectively.[69] By using a different approach, the estimation of malaria cases in 2013 by the Global burden of disease (GBD) study was much lower with 165 million cases and the CI was also large (95–284 million).[71] Another striking difference with WHO estimates is that only 56% of malaria cases were in SSA because India had a total of 60 million cases in the GBD study 2013.[71] The good news, even if the objective of RBM to reduce malaria cases from 2000 levels by 75% was not achieved, is that malaria control interventions with insecticide-treated bednets, indoor residual spraying, and prompt treatment of malaria cases with artemisinin-based combination therapy, are a success in SSA because the incidence of clinical disease fell by 40% between 2000 and 2015.[72]

According to the WHO, the commonest and most important complications of *P. falciparum* infection in children are cerebral malaria, severe anemia, respiratory distress/metabolic acidosis, and hypoglycemia.[73] In certain African countries, case fatality rates following treatment are as high as 10% for severe malarial anemia but may exceed 25% for cerebral malaria.[61] Data from a large randomized trial comparing Artesunate with quinine in African children with severe *falciparum* malaria showed that the three most frequent and independent predictors of death were acidosis indicated by a large base deficit, cerebral malaria, and renal impairment, with a mortality of 43% when the three are combined.[74] Even more than

for case estimates, the annual number of deaths due to malaria is also highly debated. The majority of malaria deaths occurs at home[58] and mortality statistics rely on verbal autopsy which is not as reliable as post-mortem autopsy.[75] The other matter of debate is that only direct deaths are estimated and not the indirect ones such as infant and fetal mortality caused by malaria in pregnancy.[55,57] In the world malaria report 2015, there were 839,000 (95% CI 653,000–1,099,000) and 450,000 (95% CI 240,000–660,000) malaria deaths in 2000 and 2013, respectively, which represents a 46% decline.[69] The GBD estimates are much higher with malaria deaths peaking at 1.2 million (95% CI 1.1–1.4 million) in 2004 and having a 29% decline to 855,000 (95% CI 703,000–1,032,000) in 2013.[71] Even if the GBD estimates and other studies suggest that the death rate attributed to malaria is largely underestimated in India,[76] the WHO and GBD reports were in agreement to conclude that the majority of deaths in 2013 were in SSA, 89% and 83%, respectively, and that malaria deaths were concentrated in children younger than 5 years.[69,71] Although uncertain, these numbers prove that the RBM objective to reduce malaria deaths to near zero by 2015 was largely utopic but the good news is that malaria is no longer the leading cause of death among children in SSA.[69]

Among children who survive cerebral malaria, up to 25% have long-term neurocognitive deficits (visual/hearing/cognitive/language impairment, ataxia, hemiparesis, motor deficit, dysphasia, behavioral/learning difficulties, and epilepsy) and 10% show evidence of mental health disorders.[57,77] Considering that cerebral malaria might be one of the more common causes of epilepsy in malaria-endemic regions, it is clear that the burden of neurological sequelae related to cerebral malaria is largely underestimated but is also difficult to estimate because diagnosis of cerebral malaria is challenging in malaria-endemic regions.[78] Cerebral malaria is characterized anatomopathologically by the sequestration of infected red cells in the cerebral microvasculature and clinically by impaired consciousness or coma with repetitive and focal seizures. The standard clinical case definition for cerebral malaria in children is a Blantyre Coma score of ≤2 with asexual *P. falciparum* parasitemia and no other evident etiology for coma.[73] Because asymptomatic parasitemia may reach 80% in highly endemic countries, this clinical definition is likely to overestimate the number of cerebral malaria cases as evidenced in autopsy studies.[79] The distinction between children with cerebral malaria from those with non-malarial coma is therefore difficult but can be greatly improved when a funduscopic examination is performed to look for malaria retinopathy.[78] Among the classical causes of coma in African children, acute bacterial meningitis is a leading one,[80] but viruses, especially adenoviruses, are also important even in children who had the retinopathy thought to be characteristic of cerebral malaria.[81]

There is hope that malaria can be controlled successfully even in the dramatic situation of SSA and even if the emergence of resistance to artemisinin in South East Asia is of major concern[55] but this ambitious goal will never be achieved without interventions to support socioeconomic development of SSA.[82]

18.6 SCHISTOSOMIASIS

Schistosomiasis, also called bilharziasis, is a helminthic parasitic disease caused by blood-dwelling flukes of the genus *Schistosoma*. At least five species of this trematode worm are known to infect humans: *Schistosoma mansoni*, *Schistosoma haematobium*, *Schistosoma japonicum*, *Schistosoma mekongi*, and *Schistosoma intercalatum*. *Schistosoma mansoni* is the most prevalent species and is endemic in 54 countries including South America, the Caribbean, Middle East, and Africa. *Schistosoma japonicum* is endemic in China, Japan, and the Philippines, and *S. haematobium* is endemic in many African and Middle Eastern countries. There is a large overlap of *S. mansoni*- and *S. haematobium*-endemic regions in Africa, and as a consequence many people may be at increased risk of co-infection by both species.

Schistosomiasis is endemic in 78 countries. The WHO estimated that at least 260 million people worldwide were infected and required preventive treatment, including approximately 120 million with clinical symptoms, and 800 million were are at risk of infection in tropical and subtropical regions, with 85% in African countries.[83] Approximately 1 million people are infected by *S. japonicum* in China, and 60 million are at risk of infection. The burden of the disease, 70 million disability-adjusted life-years, is also significant. However a substantial decline in mortality and morbidity occurred in recent years, due to the widespread use of antischistosomal drugs in endemic countries. The WHO calculated that the number of people treated for schistosomiasis rose from 12.4 million in 2006 to more than 40 million in 2013.[83]

In the last decade, demographic, environmental, and climatic factors favored schistosomiasis radiation to new geographical areas. High mobility of populations, socioeconomic factors, poor hygiene conditions, and construction of dams facilitated the spread to endemic areas. In endemic areas, infection rate may be increased by co-infection with HIV. Cases of schistosomiasis have been reported in Europe due to immigration and tourism travel to endemic areas.

Neuroschistosomiasis, the infection of the CNS by *Schistosoma* spp., can provoke severe disability. Neurological symptoms occur in cerebral and spinal neuroschistosomiasis as a consequence of the immune reaction around the *Schistosoma* eggs deposited in the CNS. The size and morphology of eggs may play a role in the pathogenesis of neuroschistosomiasis. The egg characteristics may explain the increased proportion of *S. mansoni* and *S. haematobium* infection in spinal schistosomiasis, because their eggs are larger and are retained more frequently in the spinal cord. *Schistosoma japonicum* eggs are smaller and reach the brain more easily and cause seizures as the main neurological manifestations.[84] Acute transverse myelitis, myeloradiculopathy, and cauda equine syndrome are common in spinal neuroschistosomiasis. Epilepsy, headache, and focal deficits are frequent in cerebral schistosomiasis. Hydrocephalus and intracranial hypertension may occur in tumor-like and cerebellar neuroschistosomiasis. Schistosomiasis is an under-recognized cause of myelopathy in tropical regions. In Central Brazil, approximately 6% of patients admitted with non-traumatic myelopathy were due to *S. mansoni* spinal neuroschistosomiasis.[85]

18.7 TOXOPLASMOSIS

Toxoplasmosis is a food-borne disease caused by *Toxoplasma gondii*, a ubiquitous protozoan parasite that is virtually able to infect all mammals and birds.[86] While usually asymptomatic, *T. gondii* infection can cause severe disease including encephalitis, retinochoroiditis, or multivisceral dissemination especially in congenitally infected fetuses and newborns and in immunocompromised patients, particularly those with AIDS.[86–89] *Toxoplasma gondii* prevalence has significantly decreased in the US and Europe during the past decades, but remains high (>50%) in many tropical countries.[86,90]

The highest burden of toxoplasmosis is clearly in tropical South America because (1) incidence and severity of congenital toxoplasmosis are higher than elsewhere, especially in Brazil;[87,91] (2) the high prevalence of *T. gondii* in the general population maintains a high rate of toxoplasmic encephalitis (TE) in HIV-infected patients, as shown in Brazil;[92] (3) prevalence levels of ocular toxoplasmosis in the general population of Southern Brazil, Colombia, and Northeastern Argentina are so high that they are a public health issue;[89,93] (4) there is a reservoir of wild strains in the Amazonian forest that are highly pathogenic to humans, as shown in French Guiana.[88] The severity of toxoplasmosis in tropical South America is probably the consequence of an impaired host immune response to the genetically divergent and diverse strains of this area.[87,93,94]

In SSA, *T. gondii* prevalence is high and TE is a leading cause of hospitalization and mortality in patients infected with HIV.[95] The burden of toxoplasmosis is likely to be underestimated in Western Africa because indirect data suggest high incidence levels of toxoplasmic retinochoroiditis.[96] Few data are available in tropical Asia but considering that *T. gondii* seroprevalence is very low in many countries of this area, the associated burden of toxoplasmosis is supposed to be also low.[97] Notable exceptions to this general pattern could be found in the populous areas of Indonesia where *T. gondii* prevalence is above 60%[90] and of the Indian subcontinent where toxoplasmosis can apparently be a serious problem for people.[98]

18.8 GNATHOSTOMIASIS

Gnathostomiasis is a parasitic and food-borne infection caused by larvae of *Gnathostoma* spp. *Gnathostoma spinigerum* and *Gnathostoma hispidum* are the most important species that may cause human infection. Gnathostomiasis is endemic in geographical regions where people eat raw or marinated seafood, fish, shellfish eels, frogs, or snakes. Less frequent routes of transmission are the ingestion of freshwater contaminated with infected copepods and direct skin penetration in food handlers by larvae during the preparation of infected fish or frogs.[99]

Endemic gnathostomiasis foci are localized in Southeast Asia (particularly in Thailand; Cambodia, Myanmar, Laos, Vietnam, Bangladesh, Indonesia, and Malaysia), China, India, Japan, and Philippines. In recent years, gnathostomiasis has been an increasing problem in some Latin-American countries, especially Mexico, Guatemala, Peru, Ecuador, and Colombia.[99] Imported gnathostomiasis have been reported in travelers returning from endemic countries. Eating *sashimi* and *ceviche* dishes prepared with cheaper freshwater fish (such as tilapia, bass or trout) were risk factors. Autochthonous cases in non-endemic regions are likely to be increased as a result of the importation of live *Gnathostoma*-infected species.[100]

Clinical manifestations in humans can be summarized as follows: (1) cutaneous involvement caused by the invasive migratory larvae with intermittent migratory swelling and eosinophilia; (2) visceral disease (pulmonary, hepatic, gastrointestinal, and less frequently genitourinary involvement); (3) cerebral and ocular gnathostomiasis. Hemorrhagic worm tracks can be seen in the liver and brain on post-mortem studies. Cutaneous gnathostomiasis is described in endemic regions as "Yangtze River edema" (China), "migratory eosinophilic nodular paniculitis" (Latin-America), and "tuao child" (Japan).

Eosinophilic meningitis/meningoencephalitis, radiculomyelitis, and subarachnoid and intracerebral hemorrhages have been reported in neurognathostomiasis.[101] Excruciating radicular pain, headache and cranial nerve and extremities paralysis are common. Neurognathostomiasis has been associated with high case fatality rate (between 8% and 25%) and chronic disability (one-third of survivors may have long-term sequelae). However, these data were published before the introduction of ivermectin and albendazole treatments. Massive subarachnoid hemorrhage may occur if the larva burrows through a cerebral arteriole. Case series from Thailand suggest that 6% of sub-arachnoid hemorrhage in adults and 18% in children may be due to neurognathostomiasis.[102]

Eradication of gnathostomiasis is unlikely due to the global distribution of the nematode and the ethnic and culinary eating habits in some regions of the planet. Educational campaigns are needed to raise public awareness about the risk of eating raw fish or marinated fish. Adequate cooking is recommended to kill the larvae, and freezing infected fish to −20°C for 3−5 days may also be effective.

18.9 ANGIOSTRONGYLIASIS

Angiostrongylus cantonensis, a food-borne zoonotic parasite, represents the most relevant infectious cause of eosinophilic meningitis. It belongs to the group of re-emerging infections and is currently spreading globally from its former endemic regions in South east Asia (mainly Thailand, China, and Taiwan) and the Pacific islands to Australia, the Americas and also recently to parts of Africa. Increased migration and international mobility has led to increased numbers of infected travelers returning from endemic areas.[103–105]

The life cycle of this neurotropic nematode, also called rat lungworm, includes rats as definitive hosts and mollusks of various types (e.g., snails and slugs) as intermediate hosts.[104,105] Humans can become accidental hosts, mainly through consumption of raw snails, contaminated vegetable and paratenic hosts such as frogs, crabs, fish, and shrimps. Larvae are transported via the bloodstream to the CNS, where they do not transform but die, causing pronounced inflammation which, in turn, can lead to eosinophilic meningitis (often resulting in meningoencephalitis), encephalitis/encephalomyelitis (with more unfavorable outcome) and radiculitis, among others. Ocular angiostrongyliasis can occur, but is rare.[105] The main complaints are acute severe headache, signs and symptoms of meningeal irritation with or without fever, paresthesia, cranial nerve palsies, muscle weakness, orbital/retro-orbital pain, diplopia/blurred vision and rarely convulsions, reduced level of consciousness, coma and respiratory failure. Neurological disease manifests as either an acute form with an incubation period of around 2 weeks presenting with acute symptoms, as described above, and a self-limiting course of around 20 days or, rarely, as a chronic disease with persisting paresthesia, weakness and cognitive deficits.[103,105]

Most patients portray blood eosinophilia and some show moderate leukocytosis. There is a variety of immunoassays testing for antibodies and antigens with different sensitivities and specificities and in different formats, although none of the tests is available commercially.[104] In CSF samples, the majority of patients show eosinophilia with a moderately elevated cell count. Protein count is usually slightly elevated, glucose level is normal, and larvae are seldom found.[103] The development of polymerase chain reaction (PCR) assay for *A. cantonensis* currently shows promising results in CSF samples.[106] Brain MRI often shows leptomeningeal enhancement and thickening, increased signal in the basal ganglia (both on T1-weighted post-contrast images), as well as small hemorrhages seen with gradient imaging.[107] Therapy of eosinophilic meningitis consists of symptomatic treatment including repeat spinal taps to relieve intracranial pressure. Prednisolone 60 mg/kg body weight for 2 weeks reduced the percentage of patients with persistent headache and its mean duration. There was no superiority in efficacy when combining corticosteroids and antihelminthic medication.[108,109] Prophylaxis mainly consists of food safety, water sanitation, information campaigns, and personal hygiene.

18.10 NODDING SYNDROME AND ONCHOCERCIASIS

NS represents an epilepsy disorder with onset during childhood in previously healthy children, leading to physical (e.g., malnourishment, wasting, stunting, and delayed sexual development) and cognitive decline as well as behavioral problems in the majority of patients, if left untreated. It clusters in time and space in areas of South Sudan and northern Uganda (first reports in the 1990s in both sites with notable and steady increase since the early 2000s). To date high numbers are still reported from South Sudan, whereas no new cases have been reported from northern Uganda since 2013,[110] although this has never been confirmed through population-based studies. NS was first described and classified in Tanzania, where it has been reported endemic for many decades.[111,112]

The clinical core feature of NS is a paroxysmal repetitive forward head drop of variable duration. In most patients, there is an association with other seizures types, mainly generalized tonic-clonic seizures.[111,113,114] Evidence from electroencephalography suggests that NS may represent atonic seizures, although latest ictal data from well-characterized children with NS demonstrate a variety of epileptic activity patterns including atypical absences, clonic, and atonic seizures.[112,113] Comprehensive symptomatic management of affected children consisting of antiepileptic medication, nutritional diet, physiotherapy, and intellectual stimulation shows improvement in most of the children.[112,115] Up to now the cause of NS has still not been discovered despite analyses of potential infectious, nutritional, environmental, and genetic causes.[114,116] In search for the cause of NS, one infectious agent, however, stands out and has been discussed rather controversially as a culprit for NS and other forms of epilepsy. *Onchocerca volvulus*, a filarial nematode that is transmitted by the black-fly, *Simulium* spp., causes river blindness and acute as well as chronic dermatitis. There is some epidemiological evidence that communities affected with onchocerciasis have a higher prevalence of epilepsy, although so far evidence of contact of the parasite with the brain has not been established.[117] There is however the possibility of an indirect autoimmune-mediated mechanism that may affect nervous tissue by cross-reacting antibodies, thereby causing an autoimmune encephalitis/encephalopathy. Results in this direction are eagerly awaited as they may open up new avenues for causative treatment.

18.11 NEUROLOGIC INFECTIONS BY FREE-LIVING AMEBAE

Human CNS infections by free-living amebae of the genus *Acanthamoeba* (*A. castellanii*, *A. culbertsoni*, *A. polyphaga*, and other species), *Naegleria* (*N. fowleri*), and *Balamuthia* (*B. mandrillaris*) have been reported in hundreds of cases worldwide, with high mortality rates.[118] These infections produce amebic encephalitis that may be acute (*Naegleria*) or sub-acute (*Acantamoeba* and *Balamuthia*), most likely acquired via nasal invasion. *Naegleria fowleri* produces an aggressive, acute meningoencephalitis. It is more frequent in younger individuals and is strongly related to exposure to ponds, pools or other freshwater reservoirs. Deaths due to *N. fowleri* are on the rise and more cases are expected in the near future because ameba density in water supplies and recreational activity in freshwater areas will increase as a consequence of rising temperatures due to global warming.[119]

Granulomatous encephalitis by *Acanthamoeba* spp. occurs in immunosuppressed or otherwise debilitated patients and apparently manifests months after infection. It is characterized by headache, alterations of personality, that progress to intracranial hypertension, seizures, coma and death. In the early 90s *B. mandrillaris* was recognized as causing similar neurologic disease to that caused by *Acanthamoeba* sp.[120] but may affect immunocompetent individuals with a higher frequency, and may show with a cutaneous amebic lesion prior to the development of neurological symptoms.[121] CSF findings in *Acanthamoeba* or *Balamuthia* infections include mononuclear pleocytosis, increased protein, and normal glucose levels. Neuroimaging shows heterogeneous supra or infratentorial lesions that enhance peripherally after contrast injection. Directly finding the amebae in CSF samples is rare, although they may grow in CSF culture. DNA testing may also confirm the diagnosis. *Naegleria* infections differ from the other species in demonstrating polymorphonuclear pleocytosis, and amebae can be directly found in CSF samples. Diagnosis of amebic encephalitis is often delayed by lack of suspicion and poor availability of diagnostic tests. Treatment with combinations of pentamidine, fluconazole, clarithromycin, fluocytosine, sulfadiazine, or more recently miltefosine, has succeeded in very few cases.[122]

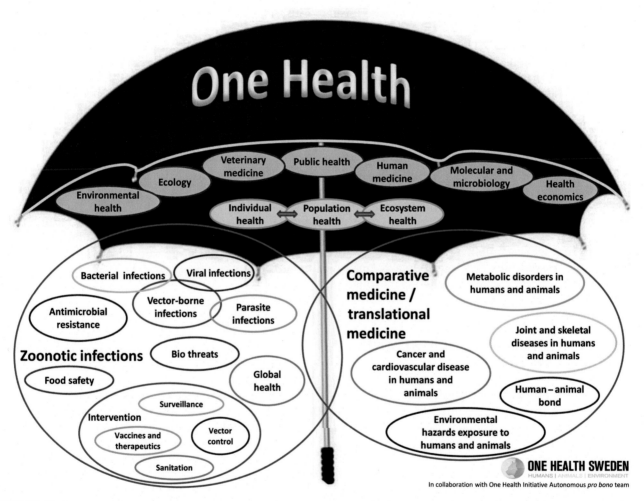

FIGURE 18.1　The "one-health umbrella" summarizing the overall idea of the "one-health" approach (developed by the networks "One Health Sweden" and "One Health Initiative Autonomous *pro bono* team" (available on www.onehealthinitiative.com)).

Multiple disciplines come together under one "umbrella" to work towards a common goal, at the same time tackling different "health domains" such as individual, population, and ecosystem health. Suitable disease complexes that are best dealt with by resorting to the "one-health" concept are infections, especially those that are vector borne, which require a multi- and cross-disciplinary approach. The same applies to interventions, such as surveillance, vector control, sanitation, food safety as well as development of vaccines, and therapeutics. Antimicrobial resistance and bio-threats also belong here. The "one-health" approach oftentimes is also used for translational medicine, mainly for non-communicable diseases that involve human and animal health alike.

18.12 FUTURE DIRECTIONS

Parasitic diseases that are often zoonotic in origin currently undergo changes in occurrence and geographic distribution due to various reasons such as climate change, increased migration of human populations, changes in dietary habits, increased global transportation of food as well as international trade and traveling. Containment as well as eradication of emerging or better re-emerging diseases need awareness creation and lobbying by key stakeholders. Many parasitic diseases with the exception of malaria have been summarized into the group of "Neglected Tropical (NTD)/Zoonotic (NZD) Diseases" which has led to increased visibility and subsequent attraction of funds. In addition, there is a relatively new movement creating excellent possibilities for scientific, capacity and financial leverage, termed the "one-health approach", of which parasitic zoonotic diseases form a major part (Fig. 18.1). The "one-health" concept is based on multi- and cross-disciplinarity involving human, animal and environmental health, amongst others, leading to a true concerted effort towards disease eradication.

References

1. Carod-Artal FJ. [Tropical causes of epilepsy]. *Rev Neurol.* 2009;49 (9):475–482.
2. Hotez PJ, Molyneux DH, Fenwick A, Ottesen E, Ehrlich Sachs S, Sachs JD. Incorporating a rapid-impact package for neglected tropical diseases with programs for HIV/AIDS, tuberculosis, and malaria. *PLoS Med.* 2006;3(5):e102.
3. Ngoungou EB, Preux PM. Cerebral malaria and epilepsy. *Epilepsia.* 2008;49(Suppl 6):19–24.
4. Quet F, Guerchet M, Pion SD, Ngoungou EB, Nicoletti A, Preux PM. Meta-analysis of the association between cysticercosis and epilepsy in Africa. *Epilepsia.* 2010;51(5):830–837.
5. Quattrocchi G, Nicoletti A, Marin B, Bruno E, Druet-Cabanac M, Preux PM. Toxocariasis and epilepsy: systematic review and meta-analysis. *PLoS Negl Trop Dis.* 2012;6(8):e1775.
6. Ba-Diop A, Marin B, Druet-Cabanac M, Ngoungou EB, Newton CR, Preux PM. Epidemiology, causes, and treatment of epilepsy in sub-Saharan Africa. *Lancet Neurol.* 2014;13(10):1029–1044.
7. Ngoungou EB, Bhalla D, Nzoghe A, Dardé ML, Preux PM. Toxoplasmosis and epilepsy--systematic review and meta analysis. *PLoS Negl Trop Dis.* 2015;9(2):e0003525.
8. Garcia HH, Nash TE, Del Brutto OH. Clinical symptoms, diagnosis, and treatment of neurocysticercosis. *Lancet Neurol.* 2014;13 (12):1202–1215.
9. Newton CR, Garcia HH. Epilepsy in poor regions of the world. *Lancet.* 2012;380(9848):1193–1201.
10. Flisser A. Taeniasis and cysticercosis due to *Taenia solium. Prog Clin Parasitol.* 1994;4:77–116.
11. Schantz PM, Wilkins PP, Tsang VCW. Immigrants, imaging and immunoblots: the emergence of neurocysticercosis as a significant public health problem. In: Scheld WM, Craig WA, Hughes JM, eds. *Emerging Infections 2.* Washington, DC: ASM Press; 1998:213–241.
12. Schantz PM. *Taenia solium* cysticercosis: an overview of global distribution and transmission. In: Singh GP, Prabhakar S, eds. *Taenia Solium Cysticercosis: From Basic to Clinical Science.* Oxon, UK: CABI Publishing; 2002:1–12.
13. Garcia HH, Gilman R, Martinez M, et al. Cysticercosis as a major cause of epilepsy in Peru. The Cysticercosis Working Group in Peru (CWG). *Lancet.* 1993;341(8839):197–200.
14. Chopra JS, Kaur U, Mahajan RC. Cysticerciasis and epilepsy: a clinical and serological study. *Trans R Soc Trop Med Hyg.* 1981;75 (4):518–520.
15. Sarti E, Schantz PM, Plancarte A, et al. Epidemiological investigation of *Taenia solium* taeniasis and cysticercosis in a rural village of Michoacan state, Mexico. *Trans R Soc Trop Med Hyg.* 1994;88(1):49–52.
16. Moyano LM, Saito M, Montano SM, et al. Neurocysticercosis as a cause of epilepsy and seizures in two community-based studies in a cysticercosis-endemic region in Peru. *PLoS Negl Trop Dis.* 2014;8(2):e2692.
17. Montano SM, Villaran MV, Ylquimiche L, et al. Neurocysticercosis: association between seizures, serology, and brain CT in rural Peru. *Neurology.* 2005;65(2):229–233.
18. Sanchez AL, Lindback J, Schantz PM, et al. A population-based, case-control study of *Taenia solium* taeniasis and cysticercosis. *Ann Trop Med Parasitol.* 1999;93(3):247–258.
19. Medina MT, Duron RM, Martinez L, et al. Prevalence, incidence, and etiology of epilepsies in rural Honduras: the Salama Study. *Epilepsia.* 2005;46(1):124–131.
20. Del Brutto OH, Santibanez R, Idrovo L, et al. Epilepsy and neurocysticercosis in Atahualpa: a door-to-door survey in rural coastal Ecuador. *Epilepsia.* 2005;46(4):583–587.
21. Escobar A. The pathology of neurocysticercosis. In: Palacios E, Rodriguez-Carbajal J, Taveras JM, eds. *Cysticercosis of the Central Nervous System.* Springfield, IL: Charles C. Thomas; 1983:27–54.
22. Fleury A, Carrillo-Mezo R, Flisser A, Sciutto E, Corona T. Subarachnoid basal neurocysticercosis: a focus on the most severe form of the disease. *Expert Rev Anti Infect Ther.* 2011;9 (1):123–133.
23. Garcia HH, Del Brutto OH. Imaging findings in neurocysticercosis. *Acta Trop.* 2003;87(1):71–78.
24. Tsang VC, Brand JA, Boyer AE. An enzyme-linked immunoelectrotransfer blot assay and glycoprotein antigens for diagnosing human cysticercosis (*Taenia solium*). *J Infect Dis.* 1989;159(1): 50–59.
25. Rodriguez S, Wilkins P, Dorny P. Immunological and molecular diagnosis of cysticercosis. *Pathog Glob Health.* 2012;106(5): 286–298.
26. Nash TE, Garcia HH. Diagnosis and treatment of neurocysticercosis. *Nat Rev Neurol.* 2011;7(10):584–594.
27. Garcia HH, Pretell EJ, Gilman RH, et al. A trial of antiparasitic treatment to reduce the rate of seizures due to cerebral cysticercosis. *N Engl J Med.* 2004;350(3):249–258.
28. Garcia HH, Gonzales I, Lescano AG, et al. Efficacy of combined antiparasitic therapy with praziquantel and albendazole for neurocysticercosis: a double-blind, randomised controlled trial. *Lancet Infect Dis.* 2014;14(8):687–695.
29. Garcia HH, Lescano AG, Gonzales I, et al. Cysticidal efficacy of combined treatment with praziquantel and albendazole for parenchymal brain cysticercosis. *Clin Infect Dis.* 2016;62(11): 1375–1379.
30. Rangel R, Torres B, Del Bruto O, Sotelo J. Cysticercotic encephalitis: a severe form in young females. *Am J Trop Med Hyg.* 1987;36(2):387–392.
31. Garcia HH, Gonzalez AE, Tsang VC, et al. Elimination of *Taenia solium* transmission in northern Peru. *N Engl J Med.* 2016;374 (24):2335–2344.

32. Bustos JA, Rodriguez S, Jimenez JA, et al. Detection of *Taenia solium* taeniasis coproantigen is an early indicator of treatment failure for taeniasis. *Clin Vaccine Immunol.* 2012;19(4):570–573.

33. Gonzales AE, Garcia HH, Gilman RH, et al. Effective, single-dose treatment or porcine cysticercosis with oxfendazole. *Am J Trop Med Hyg.* 1996;54(4):391–394.

34. Lightowlers MW. Cysticercosis and echinococcosis. *Curr Top Microbiol Immunol.* 2013;365:315–335.

35. Carod-Artal FJ. American trypanosomiasis. *Handb Clin Neurol.* 2013;114:103–123.

36. Bern C. Chagas' disease. *N Engl J Med.* 2015;373(19):1882.

37. WHO. Research priorities for Chagas disease, human African trypanosomiasis and leishmaniasis. *World Health Organ Tech Rep Ser.* 2012;(975):v–xii, 1–100.

38. Lee BY, Bacon KM, Bottazzi ME, Hotez PJ. Global economic burden of Chagas disease: a computational simulation model. *Lancet Infect Dis.* 2013;13(4):342–348.

39. Lescure FX, Le Loup G, Freilij H, et al. Chagas disease: changes in knowledge and management. *Lancet Infect Dis.* 2010;10(8):556–570.

40. Sanchez Negrette O, Mora MC, Basombrio MA. High prevalence of congenital *Trypanosoma cruzi* infection and family clustering in Salta, Argentina. *Pediatrics.* 2005;115(6):e668–e672.

41. Coura JR, Dias JC. Epidemiology, control and surveillance of Chagas disease: 100 years after its discovery. *Mem Inst Oswaldo Cruz.* 2009;104(Suppl 1):31–40.

42. Schmunis GA, Yadon ZE. Chagas disease: a Latin American health problem becoming a world health problem. *Acta Trop.* 2010;115(1–2):14–21.

43. Basile L, Jansa JM, Carlier Y, et al. Chagas disease in European countries: the challenge of a surveillance system. *Euro Surveill.* 2011;16(37):pii, 19968.

44. Carod-Artal FJ, Gascon J. Chagas disease and stroke. *Lancet Neurol.* 2010;9(5):533–542.

45. Carod-Artal FJ, Vargas AP, Horan TA, Nunes LG. Chagasic cardiomyopathy is independently associated with ischemic stroke in Chagas disease. *Stroke.* 2005;36(5):965–970.

46. Carod-Artal FJ, Vargas AP, Falcao T. Stroke in asymptomatic *Trypanosoma cruzi*-infected patients. *Cerebrovasc Dis.* 2011;31(1):24–28.

47. Lima-Costa MF, Matos DL, Ribeiro AL. Chagas disease predicts 10-year stroke mortality in community-dwelling elderly: the Bambui cohort study of aging. *Stroke.* 2010;41(11):2477–2482.

48. Lima-Costa MF, Castro-Costa E, Uchoa E, et al. A population-based study of the association between *Trypanosoma cruzi* infection and cognitive impairment in old age (the Bambui Study). *Neuroepidemiology.* 2009;32(2):122–128.

49. Oliveira-Filho J, Viana LC, Vieira-de-Melo RM, et al. Chagas disease is an independent risk factor for stroke: baseline characteristics of a Chagas disease cohort. *Stroke.* 2005;36(9):2015–2017.

50. Carod-Artal FJ, Ribeiro Lda S, Vargas AP. Awareness of stroke risk in chagasic stroke patients. *J Neurol Sci.* 2007;263(1–2):35–39.

51. Carod-Artal FJ. Policy implications of the changing epidemiology of Chagas disease and stroke. *Stroke.* 2013;44(8):2356–2360.

52. Kennedy PG. Clinical features, diagnosis, and treatment of human African trypanosomiasis (sleeping sickness). *Lancet Neurol.* 2013;12(2):186–194.

53. WHO. Control and surveillance of human African trypanosomiasis. WHO Technical Report Series 984; 2013. <http://apps.who.int/iris/bitstream/10665/95732/1/9789241209847_eng.pdf>.

54. Franco JR, Simarro PP, Diarra A, Ruiz-Postigo JA, Jannin JG. The journey towards elimination of *gambiense* human African trypanosomiasis: not far, nor easy. *Parasitology.* 2014;141(6):748–760.

55. White NJ, Pukrittayakamee S, Hien TT, Faiz MA, Mokuolu OA, Dondorp AM. Malaria. *Lancet.* 2014;383(9918):723–735.

56. Sachs J, Malaney P. The economic and social burden of malaria. *Nature.* 2002;415(6872):680–685.

57. Breman JG, Alilio MS, Mills A. Conquering the intolerable burden of malaria: what's new, what's needed: a summary. *Am J Trop Med Hyg.* 2004;71(2 Suppl):1–15.

58. Breman JG. The ears of the hippopotamus: manifestations, determinants, and estimates of the malaria burden. *Am J Trop Med Hyg.* 2001;64(1–2 Suppl):1–11.

59. Campbell MC, Tishkoff SA. African genetic diversity: implications for human demographic history, modern human origins, and complex disease mapping. *Annu Rev Genomics Hum Genet.* 2008;9:403–433.

60. Aidoo M, Terlouw DJ, Kolczak MS, et al. Protective effects of the sickle cell gene against malaria morbidity and mortality. *Lancet.* 2002;359(9314):1311–1312.

61. Malaria Genomic Epidemiology Network. Reappraisal of known malaria resistance loci in a large multicenter study. *Nat Genet.* 2014;46(11):1197–1204.

62. Malaria Genomic Epidemiology Network, Band G, Rockett KA, Spencer CC, Kwiatkowski DP. A novel locus of resistance to severe malaria in a region of ancient balancing selection. *Nature.* 2015;526(7572):253–257.

63. Tanabe K, Mita T, Jombart T, et al. *Plasmodium falciparum* accompanied the human expansion out of Africa. *Curr Biol.* 2010;20(14):1283–1289.

64. Fu Q, Mittnik A, Johnson PL, et al. A revised timescale for human evolution based on ancient mitochondrial genomes. *Curr Biol.* 2013;23(7):553–559.

65. Prugnolle F, Durand P, Ollomo B, et al. A fresh look at the origin of *Plasmodium falciparum*, the most malignant malaria agent. *PLoS Pathog.* 2011;7(2):e1001283.

66. Gao F, Bailes E, Robertson DL, et al. Origin of HIV-1 in the chimpanzee *Pan troglodytes troglodytes*. *Nature.* 1999;397(6718):436–441.

67. Liu W, Li Y, Learn GH, et al. Origin of the human malaria parasite *Plasmodium falciparum* in gorillas. *Nature.* 2010;467(7314):420–425.

68. Sundararaman SA, Liu W, Keele BF, et al. *Plasmodium falciparum*-like parasites infecting wild apes in southern Cameroon do not represent a recurrent source of human malaria. *Proc Natl Acad Sci USA.* 2013;110(17):7020–7025.

69. WHO. World malaria report 2015; 2015. <http://apps.who.int/iris/bitstream/10665/200018/1/9789241565158_eng.pdf>.

70. Partnership RBM. *The Global Malaria Action Plan for a Malaria Free World.* Geneva, Switzerland: Roll Back Malaria Partnership; 2008.

71. Murray CJ, Ortblad KF, Guinovart C, et al. Global, regional, and national incidence and mortality for HIV, tuberculosis, and malaria during 1990–2013: a systematic analysis for the Global Burden of Disease Study 2013. *Lancet.* 2014;384(9947):1005–1070.

72. Bhatt S, Weiss DJ, Cameron E, et al. The effect of malaria control on *Plasmodium falciparum* in Africa between 2000 and 2015. *Nature.* 2015;526(7572):207–211.

73. WHO. *Management of Severe Malaria: A Practical Handbook.* 3rd ed. Geneva, Switzerland: WHO; 2013:<http://www.who.int/malaria/publications/atoz/9789241548526/en/>.

74. von Seidlein L, Olaosebikan R, Hendriksen IC, et al. Predicting the clinical outcome of severe *falciparum* malaria in African children: findings from a large randomized trial. *Clin Infect Dis.* 2012;54(8):1080–1090.

75. Cox JA, Lukande RL, Lucas S, Nelson AM, Van Marck E, Colebunders R. Autopsy causes of death in HIV-positive individuals in sub-Saharan Africa and correlation with clinical diagnoses. *AIDS Rev.* 2010;12(4):183—194.

76. Dhingra N, Jha P, Sharma VP, et al. Adult and child malaria mortality in India: a nationally representative mortality survey. *Lancet.* 2010;376(9754):1768—1774.

77. Idro R, Kakooza-Mwesige A, Asea B, et al. Cerebral malaria is associated with long-term mental health disorders: a cross sectional survey of a long-term cohort. *Malar J.* 2016;15(1):184.

78. Birbeck GL, Molyneux ME, Kaplan PW, et al. Blantyre Malaria Project Epilepsy Study (BMPES) of neurological outcomes in retinopathy-positive paediatric cerebral malaria survivors: a prospective cohort study. *Lancet Neurol.* 2010;9(12):1173—1181.

79. Taylor TE, Fu WJ, Carr RA, et al. Differentiating the pathologies of cerebral malaria by postmortem parasite counts. *Nat Med.* 2004;10(2):143—145.

80. Berkley JA, Mwangi I, Mellington F, Mwarumba S, Marsh K. Cerebral malaria versus bacterial meningitis in children with impaired consciousness. *QJM.* 1999;92(3):151—157.

81. Mallewa M, Vallely P, Faragher B, et al. Viral CNS infections in children from a malaria-endemic area of Malawi: a prospective cohort study. *Lancet Glob Health.* 2013;1(3):e153—e160.

82. Tusting LS, Willey B, Lucas H, et al. Socioeconomic development as an intervention against malaria: a systematic review and meta-analysis. *Lancet.* 2013;382(9896):963—972.

83. WHO. Schistosomiasis; 2015. <http://www.who.int/mediacentre/factsheets/fs115/en/index.html>.

84. Carod Artal FJ. Cerebral and spinal schistosomiasis. *Curr Neurol Neurosci Rep.* 2012;12(6):666—674.

85. Carod Artal FJ, Vargas AP, Horan TA, Marinho PB, Coelho Costa PH. *Schistosoma mansoni* myelopathy: clinical and pathologic findings. *Neurology.* 2004;63(2):388—391.

86. Jones JL, Dubey JP. Foodborne toxoplasmosis. *Clin Infect Dis.* 2012;55(6):845—851.

87. Gilbert RE, Freeman K, Lago EG, et al. Ocular sequelae of congenital toxoplasmosis in Brazil compared with Europe. *PLoS Negl Trop Dis.* 2008;2(8):e277.

88. Demar M, Hommel D, Djossou F, et al. Acute toxoplasmoses in immunocompetent patients hospitalized in an intensive care unit in French Guiana. *Clin Microbiol Infect.* 2012;18(7):E221—E231.

89. de-la-Torre A, Lopez-Castillo CA, Gomez-Marin JE. Incidence and clinical characteristics in a Colombian cohort of ocular toxoplasmosis. *Eye (Lond).* 2009;23(5):1090—1093.

90. Pappas G, Roussos N, Falagas ME. Toxoplasmosis snapshots: global status of *Toxoplasma gondii* seroprevalence and implications for pregnancy and congenital toxoplasmosis. *Int J Parasitol.* 2009;39(12):1385—1394.

91. Torgerson PR, Mastroiacovo P. The global burden of congenital toxoplasmosis: a systematic review. *Bull World Health Organ.* 2013;91(7):501—508.

92. Coelho L, Cardoso SW, Amancio RT, et al. Trends in AIDS-defining opportunistic illnesses incidence over 25 years in Rio de Janeiro, Brazil. *PLoS One.* 2014;9(6):e98666.

93. Rudzinski M, Khoury M, Couto C, Ajzenberg D. Reactivation of ocular toxoplasmosis in non-Hispanic persons, Misiones province, Argentina. *Emerg Infect Dis.* 2016;22(5):912—913.

94. de-la-Torre A, Sauer A, Pfaff AW, et al. Severe South American ocular toxoplasmosis is associated with decreased Ifn-gamma/Il-17a and increased Il-6/Il-13 intraocular levels. *PLoS Negl Trop Dis.* 2013;7(11):e2541.

95. Lewden C, Drabo YJ, Zannou DM, et al. Disease patterns and causes of death of hospitalized HIV-positive adults in West Africa: a multicountry survey in the antiretroviral treatment era. *J Int AIDS Soc.* 2014;17:18797.

96. Gilbert RE, Stanford MR, Jackson H, Holliman RE, Sanders MD. Incidence of acute symptomatic *Toxoplasma* retinochoroiditis in south London according to country of birth. *BMJ.* 1995;310(6986):1037—1040.

97. Pengsaa K, Hattasingh W. Congenital toxoplasmosis: an uncommon disease in Thailand. *Paediatr Int Child Health.* 2015;35(1):56—60.

98. Palanisamy M, Madhavan B, Balasundaram MB, Andavar R, Venkatapathy N. Outbreak of ocular toxoplasmosis in Coimbatore, India. *Indian J Ophthalmol.* 2006;54(2):129—131.

99. Herman JS, Chiodini PL. Gnathostomiasis, another emerging imported disease. *Clin Microbiol Rev.* 2009;22(3):484—492.

100. Diaz JH. Gnathostomiasis: an emerging infection of raw fish consumers in gnathostoma nematode-endemic and nonendemic countries. *J Travel Med.* 2015;22(5):318—324.

101. Katchanov J, Sawanyawisuth K, Chotmongkoi V, Nawa Y. Neurognathostomiasis, a neglected parasitosis of the central nervous system. *Emerg Infect Dis.* 2011;17(7):1174—1180.

102. Visudhiphan P, Chiemchanya S, Somburanasin R, Dheandhanoo D. Causes of spontaneous subarachnoid hemorrhage in Thai infants and children. A study of 56 patients. *J Neurosurg.* 1980;53(2):185—187.

103. Graeff-Teixeira C, da Silva AC, Yoshimura K. Update on eosinophilic meningoencephalitis and its clinical relevance. *Clin Microbiol Rev.* 2009;22(2):322—348, Table of Contents.

104. Eamsobhana P. Eosinophilic meningitis caused by *Angiostrongylus cantonensis*—a neglected disease with escalating importance. *Trop Biomed.* 2014;31(4):569—578.

105. Martins YC, Tanowitz HB, Kazacos KR. Central nervous system manifestations of *Angiostrongylus cantonensis* infection. *Acta Trop.* 2015;141(Pt A):46—53.

106. Qvarnstrom Y, Xayavong M, da Silva AC, et al. Real-time polymerase chain reaction detection of *Angiostrongylus cantonensis* DNA in cerebrospinal fluid from patients with eosinophilic meningitis. *Am J Trop Med Hyg.* 2016;94(1):176—181.

107. Walker MD, Zunt JR. Neuroparasitic infections: nematodes. *Semin Neurol.* 2005;25(3):252—261.

108. Chotmongkol V, Sawanyawisuth K, Thavornpitak Y. Corticosteroid treatment of eosinophilic meningitis. *Clin Infect Dis.* 2000;31(3):660—662.

109. Chotmongkol V, Kittimongkolma S, Niwattayakul K, Intapan PM, Thavornpitak Y. Comparison of prednisolone plus albendazole with prednisolone alone for treatment of patients with eosinophilic meningitis. *Am J Trop Med Hyg.* 2009;81(3):443—445.

110. Burton A. Uganda: how goes the nodding syndrome war? *Lancet Neurol.* 2016;15(1):30—31.

111. Winkler AS, Friedrich K, Konig R, et al. The head nodding syndrome—clinical classification and possible causes. *Epilepsia.* 2008;49(12):2008—2015.

112. Spencer PS, Kitara DL, Gazda SK, Winkler AS. Nodding syndrome: 2015 international conference report and Gulu accord. *eNeurologicalSci.* 2016;3:80—83.

113. Sejvar JJ, Kakooza AM, Foltz JL, et al. Clinical, neurological, and electrophysiological features of nodding syndrome in Kitgum, Uganda: an observational case series. *Lancet Neurol.* 2013;12(2):166—174.

114. Dowell SF, Sejvar JJ, Riek L, et al. Nodding syndrome. *Emerg Infect Dis.* 2013;19(9):1374—1384.

115. Idro R, Namusoke H, Abbo C, et al. Patients with nodding syndrome in Uganda improve with symptomatic treatment: a cross-sectional study. *BMJ Open.* 2014;4(11):e006476.

116. Foltz JL, Makumbi I, Sejvar JJ, et al. An epidemiologic investigation of potential risk factors for nodding syndrome in Kitgum District, Uganda. *PLoS One*. 2013;8(6):e66419.

117. Winkler AS. Onchocerciasis and its potential association with epilepsy. In: Chopra J, Sawhney I, eds. *Neurology in the Tropics*. 2nd ed. New Delhi: Elsevier; 2016:315–326.

118. Visvesvara GS. Infections with free-living amebae. *Handb Clin Neurol*. 2013;114:153–168.

119. Siddiqui R, Khan NA. Primary amoebic meningoencephalitis caused by *Naegleria fowleri*: an old enemy presenting new challenges. *PLoS Negl Trop Dis*. 2014;8(8):e3017.

120. Visvesvara GS, Schuster FL, Martinez AJ. *Balamuthia mandrillaris*, N. G., N. Sp., agent of amebic meningoencephalitis in humans and other animals. *J Eukaryot Microbiol*. 1993;40(4):504–514.

121. Bravo FG, Alvarez PJ, Gotuzzo E. *Balamuthia mandrillaris* infection of the skin and central nervous system: an emerging disease of concern to many specialties in medicine. *Curr Opin Infect Dis*. 2011;24(2):112–117.

122. Martinez DY, Seas C, Bravo F, et al. Successful treatment of *Balamuthia mandrillaris* amoebic infection with extensive neurological and cutaneous involvement. *Clin Infect Dis*. 2010;51(2):e7–e11.

19

Bacterial Diseases of the Nervous System

Jean-François Faucher[1] and Marie-Cécile Ploy[2]

[1]Limoges University Medical Center, Limoges, France [2]UMR 1092 University of Limoges, Limoges, France

19.1 INTRODUCTION

Bacterial diseases of the nervous system comprise a broad range of diseases with their related pathogens. Many bacterial infections can spread to the nervous system, most of them by the hematogenous route. Some of them, like bacterial meningitis (BM), represent a high disease burden worldwide, which disproportionately affect young children and deserve vigorous interventions for prevention by vaccination. As tuberculosis (TB) is still highly prevalent in tropical areas, tuberculous meningitis (TBM) will be presented separately.

19.2 BM/ENCEPHALITIS/CEREBRAL ABSCESS (MYCOBACTERIA EXCLUDED)

BM, which is very common in many tropical areas, is a medical emergency. Large meningococcal epidemics have occurred in Brazil, China, and sub-Saharan Africa (SSA) and they still threaten populations in resource-poor settings. In addition to a high mortality rate, a high risk of neurologic sequelae is associated with BM. Neonatal meningitis and meningitis occurring in children and adults will be detailed separately.

19.2.1 Neonatal Meningitis

19.2.1.1 Epidemiology and Physiopathology

Early BM is usually part of a syndrome of sepsis neonatorum. A significant part of the mortality attributed to neonatal sepsis (which represents 7% of the mortality in under-five death worldwide[1]) is related to neonatal meningitis (NNM).

The incidence of NNM is much higher in low-income areas than in other areas, from 0.8/1000 to 6.1/1000 live births.[2,3] The epidemiology and microbiology of NNM is similar to neonatal sepsis.[4] Neonatal meningitis has a high mortality rate, up to 58% in the tropical world.[4] Mortality is higher in premature babies and specifically in low birth-weight neonates. Meningitis in neonates are divided into early- (≤7 days of life) or late- (from 7 to 90 days) onset diseases. Different bacterial species have been isolated in neonates meningitis

TABLE 19.1 Main Etiologies of Bacterial Meningitis in Neonates in the Tropical World

Bacterial species	Early- (first 7 days of life) and late- (from 7 days to 3 months) onset meningitis
Group B Streptococcus (Streptococcus agalactiae)	Both
Escherichia coli	Mostly early
Streptococcus pneumoniae	Mostly late
Neisseria meningitidis	Late
Staphylococcus sp	Late
Haemophilus influenzae	Late
Gram-negative bacteria, including Salmonella enterica	Late

TABLE 19.2 Clinical Signs of Neonatal Meningitis[2,14]

Neurologic	Aspecific
Irritability or lethargy	Fever or hypothermia
	Feeding intolerance or vomiting
	Respiratory distress
Hypotonia	Apnea
Seizures	Bradycardia
Bulging anterior fontanel	Hypotension
Nuchal rigidity	Poor perfusion
	Jaundice
	Hypo- or hyperglycemia
	Diarrhea

(Table 19.1).[2,5–8] Bacteria involved in early-onset meningitis are mainly commensal of the vaginal flora of the mothers.

Early-onset meningitis is due to bacteria transmitted from mother to newborn before or during delivery. Bloodstream infection with hematogenous spread to the meninges is the most common mechanism.

The physiopathology of late-onset meningitis is less obvious, with both vertical and horizontal transmission.

Risk factors for neonatal meningitis do not seem to differ between developing countries and the rest of the world:[6] low birth weight and prematurity, premature rupture of membranes, prolonged rupture of membranes (>18 hours), maternal colonization with Group B Streptococcus (GBS), maternal chorioamnionitis, urinary tract infection, and low socioeconomic status.

Group B Streptococcus (Streptococcus agalactiae) is a gram-positive cocci normally present in the rectovaginal flora but the prevalence differs according to the country[9] and the socio-economic status.[10] Neonates who have been colonized at birth can develop an infection (1%–3%[11]). There are 10 serotypes of Group B Streptococcus and serotypes I and III are predominant in invasive diseases in neonates and showed a high morbidity and mortality rates, with a higher risk of late-onset disease in human immunodeficiency virus (HIV)-exposed neonates.[12]

Escherichia coli is the second most prevalent bacteria found in early-onset meningitis. The mortality is higher for neonates infected with the strain expressing the K1 antigen.[13]

Salmonella species have been reported to be among the most prevalent gram-negative bacilli in reports from Africa.[7]

Contrary to what happens in more developed countries, Listeria monocytogenes is less prevalent in the tropical world, probably due to the low availability of the usual contaminated dietary sources.[4]

The likelihood of neonatal meningitis takes into account the above-mentioned risk factors and various clinical signs.[14] Clinical presentation is often subtle and indistinguishable from that of sepsis without meningitis[15] (Table 19.2).

19.2.1.2 Diagnosis

As much as possible, blood cultures and a lumbar puncture (LP) should be performed to collect the cerebrospinal fluid (CSF), but the latter is frequently deferred due to the concern of exacerbating clinical deterioration in the infant with hemodynamic instability. Approaches in which only infants with confirmed bacteremia are evaluated for meningitis will result in missed diagnoses of meningitis.[16] The current recommendation is to perform an LP (whenever possible, it should be performed prior to the administration of antibiotics) on all clinically stable infants suspected to have neonatal sepsis and who are exhibiting signs of infection.[17]

A cytologic analysis (red and white cells) of the CSF will be performed. For neonates, the normal white cell count can reach up to 30 cells/mm^3. Above this threshold, a formula should be performed to distinguish neutrophils, lymphocytes and other nuclear cells. A Gram stain should be done systematically. The other parameters to be performed are the CSF protein (normal when <0.4 g/L, but premterm neonates have higher values) and glucose (normally >50% of the glycemia) levels.

There is great difficulty in predicting the diagnosis of meningitis solely based on CSF parameters due to overlaps of values between infants with and without confirmed meningitis.[2,13] The CSF is normally sterile and the culture should be done systematically with rich media incubated for at least 3 days to confirm that bacterial culture is negative.

Bacterial antigens can also be detected directly from the CSF, as for *Streptococcus pneumoniae* for which an immunochromatographic test is available allowing detection of pneumococcal antigen in 15 minutes.

Lastly, the detection of bacterial DNA by polymerase chain reaction (PCR) should be performed either with specific PCRs for bacterial agents or with a universal PCR (targeting the 16SRNA encoding DNA) and sequencing allowing the detection of any bacterial DNA. These tests require specialized technicians and different equipments and are not easy to implement in low-income countries. In-house multiplex molecular tests have recently been described and introduced[18] with a higher detection rate of the etiological infectious agent than culture. Fully automated multiplex tests are being developed and some are already available as a matter of routine allowing the detection of different infectious agents (bacteria, virus, fungi, parasites) in one assay in less than 2 hours.[19] These tests could be performed by non-specialized technicians as the technology includes all steps from DNA extraction to amplification detection. These tests are expensive but medico-economic studies including all the costs (diagnosis, treatment, hospital length of stay, complications, etc.) should be performed to address the added-value of such tests in meningitis. Recent studies underline the potential clinical impact of the reduction of the time-to-diagnosis[20] induced by this approach.

19.2.1.3 *Treatment*

Along with supportive care, parenteral antibiotics should be administered as a matter of emergency because delays in treatment are associated with increased mortality and morbidity.[21] Empirical antibiotic therapy must be effective and achieve bactericidal activity in the CSF[22] without toxicity. The World Health Organization (WHO) recommends ampicillin plus gentamicin combination or a third-generation cephalosporin.[23]

The areas most affected by bacteria which are resistant to the first-line antibiotics are resource-poor countries,[24] where local data on antimicrobial resistance are generally lacking.[25]

The prevention of nosocomial infections in these children is a challenge, especially in low-income countries. Interventions like clustering of nursing care, implementing a simple algorithm for empirical therapy of suspected early-onset sepsis, minimal invasive care and promotion of early discharge proved to be effective for reducing the incidence of nosocomial infections in Senegal.[26]

There is evidence that mortality of neonatal meningitis is much higher in resource-poor countries (roughly 50%) than in wealthy countries (10%−15%).[5,6,27]

Survivors of neonatal meningitis are at considerable risk of long-term neurologic impairment.[28,29] Taking into account the disease burden, there are very few data on disability in survivors from resource-limited countries in the literature.[30]

In many countries, the adoption of intrapartum antibiotic prophylaxis (IAP) and universal antenatal screening for GBS colonization lead to a reduction limited to early onset-GBS infections.[31] In some areas of developing countries, early-onset GBS disease is sufficiently prevalent to justify the implementation of IAP (−60). However, it is unclear whether IAP[32] can reduce the incidence of neonatal sepsis in areas where the incidence of GBS early-onset sepsis is low.[31,33−35] Maternal immunity to the most common serotypes of GBS can be transferred passively to the infant and protect against early- and late-onset infections.[36] Surrogate markers for a prevention of early-onset GBS neonatal sepsis by vaccination have been defined and a GBS vaccine is currently in development.[37]

19.2.2 BM in Children and Adults

BM is frequent in developing countries. *Neisseria meningitidis*, *S. pneumoniae*, and *Haemophilius influenzae* serotype b (Hib) are the main pathogens. Meningococcal disease is characterized by its association with large meningitis epidemics. However, microbiological diagnosis is rarely possible and presumptive treatment is the current standard of care in most care settings. Prevention of BM by vaccination is currently a critical issue.

19.2.2.1 *Epidemiology*

Acute BM is found throughout the world, but the relative contributions of the three more common pathogens varies considerably.

19.2.2.1.1 MENINGOCOCCAL DISEASE

Neisseria meningitidis is a gram-negative cocci only found in humans. It can be found as a commensal bacteria of the oropharynx. In a study on the *N. meningitides* carriage in the African meningitis belt, the overall rate of acquisition of meningococci was 2.4% per month[38] but the average duration of carriage was relatively short (3−4 months).

Different *N. meningitidis* serogroups have been defined according to the composition of the capsule. In Africa, the serogroup A is the most prevalent. However, the global plan for vaccination in this part of the world allowed a decrease in the incidence of meningococcal meninigitis.[39] However, other serogroups then emerged, mainly C and W.[40] In the northern hemisphere, serogroups B and C predominate but serogroup Y incidence increased for few years.

In tropical areas, meningococcal disease has a very high incidence in children and adults, and this is especially true for the meningitis belt, where more than 10,000 cases of meningococcal meningitis were recorded annually. A common feature for the occurrence of frequent epidemics is the 300−1100 mm mean annual rainfall isohyets. Thus, the area where meningitis epidemics frequently occur extends beyond the meningitis belt. The largest outbreaks, as defined by a case incidence superior to 10 cases per week/100,000 inhabitants, occur during Sahelian dry season (December−May).[41] Typically, epidemics occur in 5- to 10-year cycles and spread from the East (Red Sea) to the West (Atlantic). Epidemics in tropical areas were mostly related to serogroup A meningococci.

19.2.2.1.2 STREPTOCOCCUS PNEUMONIAE

In the absence of a major meningococcal epidemic, the incidence of S. pneumoniae meningitis in the meningitis belt and elsewhere, may be higher than the meningococcal meningitis incidence for persons less than 1 and over 20 years of age.[42] Streptococcus pneumoniae (pneumococcus) is a gram-positive cocci which is a commensal of the oropharynx. The bacterial capsule is antigenic and lead to the description of 93 serotypes. The capsule is a major virulence factor allowing the bacteria to avoid the immune system and producing adhesion to epithelial cells.[43] Moreover, the pneumococcus produces a lot of enzymes as pneumolysin involved in bacterial virulence. Some serotypes are frequent as commensals of the oropharynx, especially in children, and are frequently highly resistant to antibiotics. Other serotypes, less able to colonize the oropharynx for a long time and are often more susceptible to antibiotics. In Africa, the serotype 1 is the main serotype of the pneumococcus strains isolated from CSF or blood cultures from patients with a BM.[44] Other serotypes are also described (19F, 4, 3, 9V, etc.). However, the recent introduction of conjugate pneumococcal vaccines in immunization programs might change the epidemiology of pneumococcal meningitis.[45,46] Interestingly, it has been recently shown that the serotype 1 strains circulating in Africa lack the main virulence factors of the European Serotype 1 strains[43] suggesting the specificity of the clones disseminating in Africa.

Since 2000, there have been outbreaks of pneumococcal meningitis in the African belt with cumulative attack rates of 100/100,000[47] and a mortality rate higher than for meningococcal meningitis.

19.2.2.1.3 OTHER PATHOGENS

Other pathogens can be involved in BM.

Hib is a gram-negative bacteria mainly described in children below 5 years of age in all regions of the world where the conjugate vaccine is not used routinely.

Streptococcus suis is an important zoonotic pathogen in South East Asia,[48] and by far the first cause of BM in southern Vietnam.[49]

Non-typhoidal Salmonella meningitis has been described in South Africa, where it is associated with HIV infection in adults.[8]

Listeriosis in the developed world most commonly occurs at the extremes of life, during severe immune suppression and during pregnancy. This pattern may be different in developing countries, were cases have rarely been reported.[50]

19.2.2.2 Pathophysiology

Patients usually develop a meningitis via hematogenous dissemination of the bacteria. Streptococcus pneumoniae is also responsible for meningitis via direct spread from the nasopharyngeal niche to the central nervous system (CNS). Direct entry through dural defects is possible.

Bacterial components are recognized by receptors of brain cells. This leads to the release of pro-inflammatory cytokines. This inflammatory response allows granulocytes to go through the blood−brain barrier. Lysis of the bacterial cells due to the antibiotic treatment or autolysis (S. pneumoniae) allows the release of bacterial components (lipoteichoic acids, lipopolysaccharide, peptidoglycan, etc.) increasing the inflammatory host response. Moreover, patients with deficiencies in the complement system are more susceptible to invasive diseases with capsulated bacteria.[46]

19.2.2.3 Clinical Features

BM should be promptly considered, when symptoms and signs of infection or inflammation of the meninges are found. There may be signs of raised intracranial pressure (headache, bulging fontanelle). Purpura is reported more frequently with N. meningitidis meningitis, than with meningitis of other causes. There is little information on the incidence of purpura fulminans in the meningitis belt.[51] Pneumococcal meningitis is more likely to be associated with focal signs on admission.

As large meningitis epidemics occur in poor-resource settings, surveillance in these areas is based on a simple case definition that can be implemented in any health care setting (Table 19.3, adapted from WHO/EMC/BAC/98.3).

TABLE 19.3   Standard Case Definition of Meningococcal Meningitis

1. Suspected case of acute meningitis[a]
- sudden onset of fever (>38.5°C rectal or 38.0°C axillary)
  WITH
- stiff neck

In patients under 1 year of age, a suspected case of meningitis occurs when fever is accompanied by a bulging fontanelle

2. Probable case of bacterial meningitis[b]
- suspected case of acute meningitis as defined above
  WITH
- turbid cerebrospinal fluid

3. Probable case of meningococcal meningitis[b]
- suspected case of either acute or bacterial meningitis as defined above
  WITH
- Gram stain showing gram-negative diplococcus
  OR
- ongoing epidemic
  OR
- petechial or purpural rash

4. Confirmed case[c]
- suspected or probable case as defined above
  WITH EITHER
- positive cerebrospinal fluid antigen detection for *Neisseria meningitidis*
  OR
- positive culture of cerebrospinal fluid or blood with identification of *N. meningitidis*

[a]*Often the only diagnosis that can be made in dispensaries (peripheral level of health care).*
[b]*Diagnosed in health centers where lumbar punctures and cerebrospinal fluid examination are feasible (intermediate level).*
[c]*Diagnosed in well-equipped hospitals (provincial or central level).*

The HIV status, as far as possible, should be quickly determined when the etiology has HIV infection as a risk factor (*S. pneumoniae*, non-typhoidal *Salmonella*), and also in meningitis of unknown etiology since there are specific etiologies of meningitis in these patients.

The differential diagnosis with cerebral malaria in malaria endemic areas is complicated by the fact that BM can occur in a patient who asymptomatically harbors malaria parasites, in which case meningitis can easily (malaria rapid diagnostic tests (mRDT) have been widely implemented in the recent years) be misdiagnosed as cerebral malaria. In a study performed in Kenya in children with encephalopathy, Histidin-Rich Protein 2 was detectable in the blood of children with meningitis, epilepsy and severe sepsis.[52] In children with meningitis, this pittfall is likely to delay a prompt administration of an adequate treatment. As cerebral malaria is associated with a high (partially sequestered) parasite biomass, attempts to identify a more specific marker of cerebral malaria than malaria infection, in patients with encephalopathy in malaria endemic areas are underway.[52–54]

19.2.2.4 Diagnosis

The diagnosis is identical to the one described for neonates. However, for children and adults, the white cell normal count is less than 10 white cells per mm^3

and the protein and glucose parameters are more discriminant.

When patients have petechial or purpural rash, it can be helpful to sample the petechiae either for bacterial culture or molecular diagnosis by PCR. This can also be helpful for a post-mortem diagnosis.[55]

19.2.2.5 Treatment

Whatever the care setting, a patient suspected of meningitis should be urgently referred for hospitalization. For patients with a history of clinically overt meningitis, an inappropriate delay in commencing therapy incrementally increases the risk of permanent injury.[56] Therefore, BM treatment is an emergency.

Penicillin and other β-lactams are effective against the most common pathogens and the CSF concentration tends to be close to the minimum inhibitory concentrations for moderately susceptible bacteria. Penicillin-resistant pneumococci have been reported from all parts of the world[25] and their impact on mortality has been established.[57] However, epidemiological data on the etiologies of BM and antimicrobial resistance of etiologic agents, which could inform rationale guidelines on the management of BM, are scarce or absent in many parts of the world.

Vancomycin is widely recommended when penicillin-resistant pneumococci might be present, but because it crosses the blood–brain barrier poorly it should be used in conjunction with another antimicrobial, often a cephalosporin.

Antibiotics to be recommended may vary according to national guidelines, but high-dose third-generation cephalosporins have become the cornerstone of BM therapy worldwide.

In the meningitis belt, during the meningitis epidemics, the recommend regimens depend on the age group:[58] intravenous or intramuscular ceftriaxone (100 mg/kg per day in one dose) for 7 days in children below 2 months of age, intravenous or intramuscular ceftriaxone (100 mg/kg per day in one dose) for 5 days in children from 2 months to 14 years of age, intravenous or intramuscular ceftriaxone (2 g/day in one dose) for 5 days in children above 14 years of age and in adults. Out of meningitis epidemic periods, WHO recommends 7- to 10-day treatment with intravenous or intramuscular ceftriaxone, whatever the age group. Chloramphenicol is not recommended any more.

In many settings, corticosteroids administered before or at the same time as the first dose of antibiotics are recommended in children as well as in adults on the basis of the results of randomized trials. In children, the risk of hearing loss following meningitis can be reduced with corticosteroids, as confirmed by a recent metanalysis.[59] There is a strong negative correlation between (1) severity of illness at the time of presentation, and (2) delay in the initiation of appropriate antibiotic therapy and hearing outcome.[60,61] This may explain why studies of dexamethasone therapy in children with BM from low-income countries found no beneficial effects.[59]

In adults, likewise, mortality and neurologic disabilities were reduced with corticosteroids in a double-blind trial in a resource-rich setting[62] but not confirmed in adults, 90% of whom were HIV-positive with high rates of malnutrition.[49,63] In Vietnam, however, treatment with dexamethasone was associated with reduced hearing loss in S. suis meningitis patients.[49]

19.2.2.6 Outcome

Many factors influence mortality: age,[64] underlying diseases,[65] etiologic agent[66] and clinical status at admission.[67] Mortality with S. pneumoniae meningitis is higher than with N. meningitidis meningitis, but broadly varies according to care setting, with mortality rates as high as 54% in Malawi,[68] whereas it was estimated at 17% in the Netherlands[64] and below 10% in Vietnam.[49] In Malawi, nearly 90% of patients were HIV-positive (5% of whom received antiretrovirals).

Impaired consciousness at admission and HIV infection increase the odds of death from BM in Malawian children, in whom mortality was 28%.[69]

Some data indicate that in resource-poor settings, a substantial part (10%; range = 0%–18%) of mortality occurs within weeks after discharge.[70]

Mortality in S. suis BM in Vietnam was below 5%.[71]

Global and regional risks of disabling sequelae from BM (children and adults) have been reviewed.[72] The risk of at least one major sequela (cognitive deficit, bilateral hearing loss, motor deficit, seizures, visual impairment, hydrocephalus) was 12% and of at least one minor sequela (behavioral problems, learning difficulties, unilateral hearing loss, hypotonia, diplopia) was 8%. The risk of at least one major sequela was 24% in pneumococcal meningitis, 9% in Hib, and 7% in meningococcal meningitis. The most common major sequela was hearing loss (33%), and 19% had multiple impairments. All-cause risk of a major sequela was twice as high in the African (25%) and southeast Asian regions (21%) as in the European region (9%).

In a systematic literature review on sequelae due to BM among African children,[70] about a quarter of children surviving pneumococcal meningitis and Hib meningitis had neuropsychological sequelae by the time of hospital discharge, a risk higher than in meningococcal meningitis cases (7%).

In Vietnam, 6 months after randomization, 42% of adults with definite meningitis were disabled.[49] Deafness was observed in 9% of those who had received dexamethasone. Hearing loss was a common sequel occurring in 53% of patients with S. suis BM.[73]

These data suggest that in some parts of the world, a substantial part of intellectual impairment and hearing loss is related to BM.[74] However, there is little data describing intellectual disability or hearing loss geographically and across population and age groups, with their etiologies.[75] These data are needed, especially in resource-poor countries, in order to plan for and meet the needs of patients who survive with sequelae.

19.2.2.7 Prevention

19.2.2.7.1 CHEMOPROPHYLAXIS

Chemoprophylaxis is effective in preventing subsequent cases of meningococcal disease in household contacts of a case of meningococcal disease.[76] The number needed to treat to prevent one subsequent case is 200. Drugs recommended for chemoprophylaxis are the following: rifampicin, ceftriaxone, ciprofloxacine.[77] Chemoprophylaxis is however not promoted, unlike vaccination, as a preventive tool in response to meningitis epidemics in the meningitis belt.[58]

19.2.2.7.2 VACCINATION

A high burden of sequelae could be averted by vaccination with Hib, pneumococcal, and meningococcal vaccines.[70,72]

Conjugate vaccines are immunogenic in infants and can prevent transmission by their activity on carriage.

Consequently conjugate vaccines are a more appropriate tool for effective prevention strategies than polysaccharide vaccines in SSA as well as in other parts of the world. Immunization based on a meningococcal serogroup A polysaccharide/tetanus toxoid conjugate vaccine (PsA-TT) is being implemented in countries of the African meningitis belt.[78]

Getting developing countries the best vaccines beyond protection of meningococcal A disease (other meningococcal serogroups, Hib, *S. pneumoniae*), at affordable prices is a critical global health governance issue.[79]

Immunization strategies in Southeast Asia, where *S. suis* is highly prevalent, have to be determined.[46]

19.2.3 Cerebral Abscess

Relatively few studies from poor-resource countries are available in the literature on this subject, as shown in a recent meta-analysis.[80] Although opportunistic infections related to AIDS involve more frequently parasites or fungi than bacteria, HIV screening is the rule as far as a cerebral abscess is suspected or diagnosed. In a monocentric study from Burkina Faso, where the abscess etiologic agents were isolated in 16% of patients, HIV seroprevalence was 5%.[81]

Data on cerebral abscess incidence are scarce. The incidence of cerebral abscesses was estimated to be 1/100,000 per year in a part of South Africa.[82] They arise from direct dissemination of adjacent infection (of odontogenic origin, sinusitis, otitis, meningitis), hematogenous spread, head trauma. Several bacteria may be isolated in cerebral abcesses, anaerobes (*Fusobacterium, Bacteroides, Propionibacterium acnes*), *Streptococcus* of the buccal flora, *Actinomyces* and *Nocardia, Aggregatibacter, Staphylococcus aureus*, and Enterobacteriaceae. The etiology depends on the origin of infection.[82,83] Although disseminated nocardiosis with cerebral abscess infection has been well described in immunodeficient patients including AIDS patients,[84] very few case reports have been published from Africa, where most patients with AIDS live,[85] probably due to the lack of means of diagnosis.

Mortality can be estimated at 10%−20%.[86] Although a global trend to a better prognosis with as much as 70% of patients with full recovery has been found in a broad study,[80] it may not reflect the outcome in many settings, with up to 50% of patients living with sequelae.[81]

19.3 MYCOBACTERIAL INFECTIONS

19.3.1 TB of the CNS

Three main types of CNS tuberculous infection are seen: TBM, solitary lesions in the brain or the spinal cord (tuberculoma) and spinal tuberculous arachnoiditis. These three types can overlap.[87] CNS TB by post-primary dissemination of TB is frequent in children and encountered in regions where TB incidence is high. CNS TB is also diagnosed in immunocompromised adults as a result of reactivation bacillemia.

TBM may be seen at any age, although it disproportionately affects young children.[88] It is the most common form of CNS TB, and one of the most severe forms of TB. The meningeal inflammation typically leads to a rhombencephalitis.

19.3.1.1 Epidemiology

In the USA, TBM represents 1% of all cases of TB and 5% of extrapulmonary meningitis.[89] There is a preponderance of the disease type in children and in HIV-positive individuals. All patients with any form of TB should be proposed for HIV screening. In a review devoted to CNS TB in children, CNS TB represented 10% of all TB cases.[90] TBM could be the second leading cause of meningitis in African patients living with HIV.[91] TBM among HIV-positive patients was associated with advanced HIV-infection, as evidenced by the median CD4 count at admission, significantly lower with TBM than with pulmonary TB.[92] It is therefore expected that broad access of HIV-positive patients to antiretrovirals can reduce the incidence of TBM.

19.3.1.2 Pathophysiology

Mycobacterium tuberculosis belongs to the *M. tuberculosis* complex which includes several species with nearly identical sequences. This group contains the human pathogens *M. tuberculosis, Mycobacterium africanum*, and *Mycobacterium canetti*. Five major lineages of *M. tuberculosis* have been described in the different regions of the world, lineage 1 (East Africa, Philippines, in the region of the Indian Ocean), lineage 2 (East Asia), lineage 3 (East Africa, Central Asia), lineage 4 (Europe, America, Africa) and lineage 7 (Ethiopia[93]).

The precise mechanism of how *M. tuberculosis* invades the brain is still unknown. There is probably a hematogenous dissemination that occurs early in the disease, before the control by the immune system. Indeed, patients co-infected by HIV or patients with polymorphisms of innate immune response (TLR2) are more susceptible to developing TBM. A possible role for vitamin D has been also proposed. Furthermore, it seems that some strain lineages (Indo-Oceanic or East Asian Beijing) are more susceptible to causing TBM. Lastly, the *Rv0931c* gene (also known as *pknD*), that encodes a serine/threonine protein kinase is necessary for brain endothelial invasion and thus enable bacteria to cross blood−brain barrier.[94]

19.3.1.3 Clinical Features

Patients with TBM present with a subacute illness combining fatigue and most of the times fever, meningeal signs (headache, neck stiffness, vomiting), alterations of mental status (personality change, lethargy, confusion, coma), focal signs (cranial nerves alterations, long-tract signs), and seizures. A categorization in three stages of increasing severity has been proposed:[95] (1) stage I patients are lucid with no focal neurologic signs or evidence of hydrocephalus; (2) stage II patients exhibit lethargy, confusion; they may have mild focal signs, such as cranial nerve palsy or hemiparesis; (3) stage III represents advanced illness with delirium, stupor, coma, seizures, multiple cranial nerve palsies, and/or dense hemiplegia. Abnormalities on chest radiographs may be seen in half of cases, ranging from focal lesions to a miliary pattern. Papilloedema is a common feature of raised intacranial pressure, which can complicate TBM.

There is an overlap between TBM and miliary TB. Miliary TB should therefore be suspected in case of TBM and vice versa.

Tuberculomas present like symptomatic intracranial mass lesions: headache, seizures, focal signs and signs of raised intracranial pressure.[96] Occasionally, tuberculomas can present like intramedullary mass lesions: subacute spinal cord compression with the appropriate motor and sensory findings, depending on the level of the lesion.[97]

With spinal tuberculous arachnoiditis, symptoms develop and progress slowly over weeks to months. Patients present with the subacute onset of nerve root and cord compression signs: spinal or radicular pain, hyperesthesia or paresthesias; lower motor neuron paralysis; and bladder or rectal sphincter dysfunction.[98]

Neurological symptoms can occur in the course of spinal TB, as compression of the spinal cord can complicate Pott's disease.[99,100]

19.3.1.4 Diagnosis

Diagnosis of TBM will combine CSF cytology, microscopy, culture and molecular techniques. That means that a volume of CSF >1 mL has to be sampled. The examination of the CSF in case of TBM usually reveals a clear appearance and white cells count >500 mm^{-3} with a mixture of neutrophils and lymphocytes, with less than 50% neutrophils. The protein level is increased and the ratio of CSF to plasma glucose level is less than 0.5.

The sensitivity of the CSF Ziehl—Neelsen staining and microscopy does not exceed 60% and highly depends on the volume of CSF used.

Culture must be performed but give final results in 8 weeks as *M. tuberculosis* is a slow-growing bacteria.

Commercial real-time PCR tests have been developed enabling the diagnosis of *M. tuberculosis* complex in less than 2 hours. Some studies showed that these commercial tests are less sensitive than culture but more sensitive than microscopy. As for microscopy, the sensitivity depends on the volume of CSF available. Rapid detection of multidrug resistant (MDR) strains (resistant to rifampin and isoniazid) must be performed directly from the CSF. Commercial molecular techniques allow detection of at least rifampicin or rifampicin and isoniazid resistance. Indeed, the management of patients infected with MDR strains is different than for patients infected with susceptible strains. The new cases of TB occur mainly in Asia and Africa. MDR and XDR (resistant to rifampin and isoniazid and at least one fluoroquinolone, and any of the second-line injectable treatments (amikacin, capreomycin, kanamycin)) strains are increasing in these countries. MDR strains are highly prevalent in the tropical world, especially in the BRICS countries (Brazil, Russia, India, China, South Africa). Around 3% of new cases of TB are due to MDR strains (20% for patients who have already received a treatment for TB). In central Asia, South Africa and Eastern Europe, the proportion of MDR and XDR cases increases dramatically.[101]

Few studies have analyzed the value of interferon-gamma release assay performed on CSF and concluded that sensitivity and specificity will be acceptable only for CSF volumes of 5—10 mL.

In patients with TBM, computed tomography and magnetic resonance imaging can define the presence and extent of basilar arachnoiditis, cerebral edema, infarction, and hydrocephalus.[102]

In patients with tuberculoma, LP is usually avoided because of concerns for raised intracranial pressure and risk of brainstem herniation; in the occasional reported case where CSF has been examined, the findings are normal or nonspecific.

On contrast CT imaging, early-stage lesions are low density or isodense, often with edema out of proportion to the mass effect and little encapsulation. Later-stage tuberculomas are well encapsulated, isodense or hyperdense, and have peripheral ring enhancement.[102]

19.3.1.5 Treatment

Anti-tuberculous therapy is based on a combination of four drugs. It generally combines isoniazide, rifampin, pyrazinamide, and ethambutol. Some experts advise treating TBM with a fluoroquinolone (moxifloxacin or levofloxacine),[103] instead of ethambutol which poorly penetrates into meninges. Intensive therapy is advised for 2 months, with a further continuation phase (isoniazide plus rifampin) of 10 months.

In many areas, however, treatment and prevention of TB has become more complex because of resistance to commonly used antituberculous drugs. Hence, multiresistant TBM is predictive of an increased risk of death.[104,105] Management of drug-resistant TB can be difficult and may necessitate the use of second-line drugs and/or surgical resection. Duration of therapy should be extended to 18−24 months.

It is widely acknowledged than adding glucocorticoids to anti-tuberculous therapy reduces mortality, at least in children and adults patients who are HIV negative, with no effect on the number of people who survive TBM with disabling neurological deficit.[106] Dexamethasone or prednisone should be used in this indication.

In HIV-positive patients, initiation of antiretroviral therapy (ART) within 2 weeks of antituberculous therapy was associated with increased mortality in patients with TBM.[107] Thus, for ART-naïve HIV patients with CNS TB, initiation of ART should be delayed for the first 4−8 weeks of antituberculous therapy, regardless of CD4 count.[108,109]

Patients with hydrocephalus may require surgical decompression of the ventricular system in order to manage the complications of raised intracranial pressure.[110]

Unlike other CNS mass lesions, medical management is preferred for clinical tuberculomas unless the lesion produces obstructive hydrocephalus or compression of the brainstem.

When the diagnosis is in doubt, surgery is also required to obtain tissue for culture and sensitivity studies.[110]

19.3.1.6 Outcome

In a recent meta-analysis, mortality of TBM in children is estimated at 19% and the probability of surviving without neurological sequelae at 36%, which means that the risk of neurological sequelae is slightly above 50% in survivors.[88] Mortality was estimated to be 10% in children with pooled early stages, whereas it was 34% in children with the most advanced disease stage. In a study performed in south Africa, drug resistance was more likely encountered in children with HIV, and multiresistance was independently associated with a higher risk of death.[111]

The influence of drug resistance on the risk of death in TBM is dramatic in HIV-infected adults,[112] with a mortality up to 100% with multiresistant TBM.

Mortality in adults with TBM is high, but certainly varies among care settings and populations of patients. It is estimated to be at 25% in HIV-negative adults, and 67% for HIV-positive adults.[107,113] Mortality in adults with TBM in Africa, as reflected by literature, is as high as 60% and seems to have remained stagnant over the past half century.[114]

Stroke can complicate MTB, early in the course of disease as a consequence of vasospasms. Later strokes involve proliferative intimal disease.[115] Stroke complicates approximately one-third of TBM cases.[116]

An open label randomized study in adults with TBM, on the role of aspirin showed a beneficial effect on mortality.[117] However, patients selectively received corticosteroids, which may have biased the results.[118]

19.3.1.7 Prevention

19.3.1.7.1 CHEMOPROPHYLAXIS

Treating latent *M. tuberculosis* in persons at risk for TB is recommended in low TB burden control.[119]

In high-burden countries, HIV-positive patients are clearly at risk for TB. The Temprano study, which was performed in HIV1 positive patients with a CD4+ between 200 and 800 cells/μL, has shown that immediate 6 months of isoniazide preventive therapy (IPT) and ART independently led to lower rates of severe illness than did deferred ART and no IPT, both overall and among patients with CD4+ counts of at least 500 cells per cubic millimeter.[120]

19.3.1.7.2 VACCINATION: BACILLE CALMETTE−GUÉRIN

Newborns and infants are the demographic group with greatest potential benefit from Bacille Calmette−Guérin (BCG) vaccination, and this intervention has been adopted for prevention of TB worldwide. BCG protects children both against infection and against disease.[121] In countries where the prevalence of TB is moderate to high, neonatal vaccination is recommended by the WHO and is administered routinely.

BCG vaccination is not appropriate for infants or adults with known immunodeficiency (including HIV infection) nor for infants with symptoms consistent with HIV infection in the absence of laboratory confirmation of actual HIV infection.[122]

19.3.2 Leprosy

Mycobacterium leprae, the agent of Hansen's disease, is a slow-growing mycobacterium and has a tropism for the skin, and for peripheral nerves (Schwann cells). The reservoir of leprosy is human, although anecdotal cases acquired by contact with armadillos have been reported. The main dissemination route of leprosy seems to be aerosol/droplets,[123] which occurs through repeated contact with untreated infected individuals, is favored by a long asymptomatic or paucisymptomatic incubation period (more than 5 years most of the time).

Living in close contact with a patient is therefore a major risk factor for leprosy. Multibacillary forms and multilesional forms of leprosy are more contagious than paucibacillary forms with a single lesion.[124]

19.3.2.1 Epidemiology

Leprosy is endemic in Asia, Africa, South America, Middle-East, South Pacific, and the Carribean. In 2014, 213,899 new cases were recorded. Patients from India, Brazil and Indonesia represent 81% of incident cases. The trend in terms of leprosy prevalence was marked by a dramatic decrease between 1980 and 2005, after the approval and wide use of a multidrug therapy but it has reached a plateau for 10 years (between 200,000 and 250,000 yearly cases).[125]

19.3.2.2 Pathophysiology

Mycobacterium leprae belongs to the Mycobacteriaceae family. The respiratory tract seems to be the potential portal of entry for *M. leprae* and transmission probably occurs through nasal or sputum excretions. *Mycobacterium leprae* has a very long doubling time (14 days) compared to *M. tuberculosis* (24 hours). Leprosy has thus a slow evolution with various incubation periods, from very short in young children to very long in adults (up to 30 years). The mechanisms of transmission from the latent infection to symptomatic infection are still unknown. On the other hand, host immunity plays an important role in the evolution of the disease. Tuberculoid leprosy is associated with a Th1 response and lepromatous leprosy involves a predominant Th2 response leading to an inadequate immune response to an intracellular bacterium and thus the uncontrolled multiplication of the bacteria.[126]

19.3.2.3 Clinical Features

Paucibacillary (or tuberculoid) leprosy is the most frequent form of the disease. Most of the time, only the skin (hypochromic macules with loss of sensation within the lesions) is involved, but peripheral nerve enlargement and/or tenderness, including the ulnar nerve at the elbow, median and superficial radial cutaneous nerve at the wrist, great auricular nerve in the neck, and common peroneal nerve at the popliteal fossa may be associated. Typically, sensory and/or motor loss generally occurs in the distribution of nerves in the vicinity of the skin lesion.

Mutibacillary (or lepromatous) leprosy skin lesions are usually generalized and consist of erythematous macules, papules, and/or nodules. A diffuse infiltration of the nose and ear lobes can lead to the leonine facies in extreme forms. Nerve damage may become bilateral and generalized. Commonly, the nerve trunks involved include the ulnar and median nerves (claw hand), the common peroneal nerve (foot drop), the posterior tibial nerve (claw toes and plantar insensitivity), the facial nerve (lagophthalmos), the radial cutaneous nerve, and the great auricular nerve.

Pure neurologic forms of the disease occur in less than 10% of cases.[127] They should be suspected in persons exposed to leprosy, with an asymmetrical neuropathy of unknown origin.

19.3.2.4 Diagnosis

Mycobacterium leprae cannot be cultured in vitro and the diagnosis will thus be based mainly on optical microscopy from skin lesion samples (punch biopsy). A rapid diagnosis can also be performed from a tissue fluid smear test on the earlobe. PCR methods have also been developed. Pathological analysis of the skin biopsies must also be performed. The infiltrations observed are specific to the tuberculoid or lepromatous leprosy states.

19.3.2.5 Treatment

Treatment may be prescribed in the setting of a national program, following the WHO guidelines, or as an individualized treatment.

The WHO advise a combination of dapsone plus rifampicin for 6 months in paucibacillary TB, and a combination of dapsone plus rifampicin (intermittent) plus clofazimine (intermittent) for 12 months[128] in multibacillary TB. The same drugs can be prescribed to be administered on a daily basis in individualized therapy.

Alternatives (ofloxacin, minocycline, clarithromycin) can be necessary either because of poor tolerance/toxicity, or because of (rarely observed) resistance to the standard regimen.

19.3.2.6 Outcome

Type 1 reaction is known as reversal reaction and type 2 reaction as erythema nodosum leprosum (ENL). They are defined by severe activation/reactivation of the immune and inflammatory systems and primarily affect peripheral nerves. Fifteen to 65% of leprosy patients present with irreversible nerve damage during the course of the disease, mainly during leprosy reactions.[129]

Patients can present with leprosy reactions before being diagnosed with leprosy, during treatment, or once an adequate treatment has been completed. Loss of nerve function can be dramatic.

The peak time for type 1/reversal reactions is in the first 2 months of treatment, but they can continue to occur for 12 months, and occasionally after multidrug therarpy (MDT) is completed. Corticosteroids (starting at 40−60 mg daily), prolonged for several months, are the treatment of choice for type 1 reactions.

The expected recovery rate for nerve function is 60%–70%. Recovery is less in patients with preexisting impairment of nerve function or with chronic or recurrent reactions.

ENL (type 2 reactions) occur in about 20% of lepromatous forms. Patients are febrile with erythema nodosum; other signs are iritis, neuritis, lymphadenitis, orchiitis, bone pain, dactylitis, arthritis, and proteinuria. ENL can start during the first or second year of antimicrobial therapy and can relapse intermittently over several years. Thalidomide (400 mg daily) is better than steroids in controlling ENL and is the drug of choice when not contraindicated. Clofazimine has a useful anti-inflammatory effect in ENL and can be used (300 mg daily) for several months.

The disabilities associated with leprosy are secondary to nerve damage. A patient with an anesthetic hand or foot needs to understand the importance of protection when undertaking potentially dangerous tasks, and inspection for trauma.

Plantar ulcers should be treated with rest; weight-bearing should not be permitted until the ulcer has healed. More complicated ulcers require surgical management.

Likewise, surgery can be indicated in the following situations: contractures of hands, and feet, foot drop, lagophthalmos, entropion, and ectropion.

19.3.2.7 Prevention

BCG gives variable protective efficacy against leprosy in different countries, ranging from 34% to 80%.[129] Rifampicin has been proposed as a post-exposure prophylaxis among contacts in non-endemic areas,[129] but it's efficacy is unknown.

19.4 OTHER BACTERIAL INFECTIONS INVOLVING THE CNS

19.4.1 Neurosyphilis

In the context of syphilis resurgence in rich-resource countries, much remains to be known on the consequences of the disease in poor-resource countries. Neurosyphilis is a well-known and possibly severe form of syphilis, for which a CSF analysis is mandatory in order to establish a diagnosis. Thus, in poor-resource countries, limited access to care is an obstacle to a proper estimation of neurosyphilis incidence. The rapid expansion of antibiotics consumption in poor-resource countries, prescribed or purchased for any reason, may also interfere in the disease in an unknown manner, since the agent of neurosyphilis is sensitive to many widely available antibiotics.

19.4.1.1 Epidemiology

In 2012, an estimated 5.6 million new cases of syphilis occurred among 15- to 49-year-olds worldwide. There are an estimated 18 million prevalent cases of syphilis.[130]

Syphilis is transmitted through sexual contact with infectious lesions of the mucous membranes or abraded skin. Oral unprotected sex is an important way of sexual transmission. Syphilis can also be transmitted via blood transfusion, or transplacentally from a pregnant woman to her fetus. Mother-to-child transmission of syphilis (congenital syphilis) is usually devastating to the fetus if maternal infection is not detected and treated sufficiently early in pregnancy. The burden of morbidity and mortality due to congenital syphilis responsible for adverse pregnancy outcomes (early fetal deaths/stillbirths, neonatal deaths, preterm/low-birth-weight babies, and infected infants) is still high.[130] Most untreated primary and secondary syphilis infections in pregnancy result in severe adverse pregnancy outcomes. Latent (asymptomatic) syphilis infections in pregnancy also cause serious adverse pregnancy outcomes in more than half of cases. Mother-to-child transmission of syphilis is declining globally due to increased efforts to screen and treat pregnant women for syphilis,[131] while a resurgence of syphilis, which primarily affects men who have sex with men, has clearly been identified in rich-resource countries,[132] but not exclusively.[133,134] Neurosyphilis is a rare but severe form of the disease.[130]

19.4.1.2 Physiopathology/Microbiology

Syphilis is a chronic disease, strictly encountered in humans. This is a multistage disease with diverse manifestations. The etiological agent is *Treponema pallidum* subsp. *pallidum*, a bacterium belonging to the *Spirochaetaceae* family, spiral-shaped bacteria. It is able to invade a wide variety of tissues from the initial site of infection with a rapid dissemination rate. *Treponema pallidum* is able to recognize different cell types allowing the attachment to various tissues through a wide variety of adhesins able to attach to the extracellular matrix components (fibronectin, laminin). The high motility of *T. pallidum* is also a virulence factor enabling the rapid dissemination of the bacteria. Due to these different characteristics, *T. pallidum* gains access to deeper tissues and the bloodstream. Then *T. pallidum* is able to activate endothelial cells, macrophages and dendritic cells via TLR2 receptors, and to induce migration of inflammatory cells, leading to cytokine production.[135] The bacterium is able to evade the immune system to survive. The organisms can survive in different tissues (CNS, eye, placenta) where they grow slowly and are able to reseed other tissues.

19.4.1.3 Clinical Features

The primary chancre, which usually develops 14–21 days after sexual contact, may go unnoticed by patients. If untreated, the disease progresses to the secondary stage, characterized by generalized mucocutaneous lesions affecting both skin, mucous membranes and lymph nodes. The rash of secondary syphilis can vary widely and mimic other infectious and noninfectious conditions, but characteristically affects the palms and soles. The multisystemic involvement of secondary syphilis can affect the CNS: meningitis, uveitis, cranial nerve palsies, and otitis. Early neurosyphilis can be the only symptoms of infection. This is why all patients with unexplained meningitis, uveitis, and sudden deafness should be screened for syphilis.[136] Symptoms and signs of secondary syphilis spontaneously resolve, even without treatment, and if left untreated, the patient enters the latent stage.

Tertiary (or late) syphilis may appear at any time between 1 and 30 years after primary infection and includes neurosyphilis, cardiovascular syphilis, and gummatous syphilis (among which intracranial forms mimicking meningiomas have been described). Late neurosyphilis includes meningovascular (cerebral and medullar) and parenchymatous (general paresis, tabes dorsalis) forms of the disease, as well as ocular and auricular manifestations.

19.4.1.4 Diagnosis

There is no consensus for diagnosis of neurosyphilis. The invasion of the CNS by *T. pallidum* occurs relatively early in infection. However, not all patients will have a symptomatic neurosyphilis. It is thus not recommended to perform LPs in all patients with syphilis. The value of exploring asymptomatic neurosyphilis in HIV-co-infected patients remains controversial. Although these patients are more likely to develop neurosyphilis, the Center for Disease Control does not recommend performing LP in these patients, since the management based on CSF examination does not improve the clinical outcome.[137] The usual parameters taken into account are a high white cell count and a high protein level in CSF, and non-treponemal (venereal disease research laboratory, VDRL) and treponemal tests performed in both the CSF and serum, together with clinical symptoms (neurological or ophtamological). The diagnosis will be based on a non-treponema test (VDRL,) performed in CSF. This test is highly specific but has a poor sensitivity (around 50%). A negative result should not rule out the diagnosis of neurosyphilis. Moreover, as rare false-positives occur, a higher-sensitivity test (fluorescent *Treponema* antibody absorption, FTA-Abs), a treponemal test, could be used as a confirmatory test. Besides the CSF analysis, a serological assay must be performed.

The usefulness of PCR in the diagnosis of neurosyphilis was controversial. Different studies found diverse sensitivity rates, probably due to a transient presence of *T. pallidum* in CSF. As for serological tests, the PCR should be used only when clinical symptoms are present. A recent study showed that a nested PCR targeting the *tpp47* gene (encoding a major lipoprotein of *T. pallidum*) seemed more sensitive in CSF from patients with clinical symptoms of neurosyphilis than CSF VDRL.[138]

A chest radiograph should be performed to exclude asymptomatic cardiovascular syphilis.

When a repeat treatment of syphilis is considered (recurrence of symptoms/signs of active syphilis, increase in the titer of a non-treponemal test), the WHO recommends a CSF analysis in order to evaluate the possible presence of asymptomatic neurosyphilis before repeat treatment, unless reinfection and a diagnosis of early syphilis can be established.[139]

19.4.1.5 Treatment

The recommended regimen of neurosyphilis is aqueous benzylpenicillin, 12–24 million IU by intravenous injection, administered daily in doses of 2–4 million IU, every 4 hours for 14 days. Procaine benzylpenicillin, 1.2 million IU by intramuscular injection, once daily, and probenecid, 500 mg orally, four times daily, both for 10–14 days have been proposed as an alternative regimen; in this case, the regimen should be used only for patients whose outpatient compliance can be assured.

For penicillin-allergic non-pregnant patients, doxycycline, 200 mg orally, twice daily for 30 days or tetracycline, 500 mg orally, four times daily for 30 days have been proposed, but for some experts, however, there is no alternative to penicillin, and penicillin-allergic patients should undergo penicillin desenzitisation before intravenous penicillin therapy.

Although their efficacy is not yet well documented, third-generation cephalosporins may be useful in the treatment of neurosyphilis. The European guidelines on the management of syphilis advise 1–2 g intravenous ceftriaxone/day for 10–14 days.[140]

Corticosteroids (20–60 mg/day) as prevention of Jarisch–Herxheimer reaction, are advised the first 3 days of treatment.[136]

19.4.1.6 Outcome

CSF analysis should be checked 6–12 months after therapy.[136] A decrease in protein level, white cells count and VDRL test should be seen. However, if the initial VDRL titer was high, the return to negative can take years.

19.4.1.7 Prevention[136]

A routine test for syphilis should be taken in all pregnant women or those people who are donating blood, and the following groups are at higher risk of syphilis: all patients who are newly diagnosed with sexually transmitted infection (STI); persons with HIV; patients with hepatitis B; patients with hepatitis C; patients suspected of early neurosyphilis (i.e., unexplained sudden visual loss (uveitis), unexplained sudden deafness (otitis) or meningitis); patients who engage in sexual behavior that puts them at risk (e.g., men who have sex with men, sex workers) and all those individuals at higher risk of acquiring STIs.

19.4.2 Zoonotic BM

19.4.2.1 Leptospirosis, Brucellosis and Q Fever

The zoonotic BM most prevalent in tropical areas and less-developed countries are leptospirosis, brucellosis, and Q fever. However, they represent less than 1% of episodes of BM, and have a strong regional distribution.[141]

19.4.2.1.1 EPIDEMIOLOGY

Infection by direct contact with infected animals is most likely with brucellosis, to which farmers and veterinarians are highly exposed. Leptospirosis occurs predominantly in tropical and sub-tropical parts of the world. Leptospirosis meningitis often follows exposition to fresh water sources contaminated with animal urine (mainly rodents and marsupials). Dairy unpasteurized products are a source of infection for brucellosis and Q fever. Outbreaks of Q fever have also been related to airborne dispersion.[142]

19.4.2.1.2 PHYSIOPATHOLOGY

Leptospira is a spirochete, living as a saprophyte in freshwater and soil. There are 20 species and hundreds of serovars. *Leptospira interrogans* serovar *Icterohaemorragiae* is mainly found in severe human disease. Portals of entry include skin (cuts and abrasions) and mucous membranes (conjunctival or oral). The bacteria then cross the tissue barriers and reach the bloodstream. The level of bacteremia is very high ($>10^4$ CFU/mL) due to the persistence of the *Leptospira* in the blood during the first week of the acute illness. This could be explained by the inability of TLR4 to recognize leptospiral lipopolysaccharide (LPS). This high level of bacteremia leads to a cytokinic storm and severe sepsis. Severe leptospirosis is characterized by dysfunction of multiple organs including the liver, kidneys, lungs, and brain.[143]

Brucella are gram-negative rods. Among the 10 species described, *Brucella melitensis* followed by *Brucella suis* are the main species involved in human diseases.

Q fever is due to *Coxiella burnetii*, a coccobacillus non-stainable with the Gram technique. It grows very slowly and is an intracellular pathogen in eukaryotic cells. It is able to survive in the phagolysosome through production of basic proteins and transporters for osmoprotectants, leading to a persistent infection.[144] Different virulence factors have been described (secretion systems, adhesins, LPS, detoxification mechnisms, etc.). The pathogenicity depends on the route of infection, the *C. burnetii* strain and the inoculum size.

19.4.2.1.3 CLINICAL FEATURES

About 5% of patients with brucellosis have CNS involvement.[145,146] The most common presentation of neurobrucellosis is acute or chronic meningitis or meningoencephalitis.

In leptospirosis, 20% of patients with fever and headache have meningitis.[147] In a study performed in Laos, leptospirosis was diagnosed in 3% of patients whose clinical signs lead to an LP.[148] A variety of other neurologic complications may also occur including hemiplegia, transverse myelitis, and Guillain–Barré syndrome.

Only 1% of acute Q fever patients have meningitis/meningoencephalitis.[149] Myelitis and peripheral neuropathy have occasionally been described in acute Q fever.[150]

19.4.2.1.4 DIAGNOSIS

In neurological leptospirosis, CSF examination includes a white cell count up to 500 mm^{-3} with predominance of lymphocytes, an increased protein level and a normal glucose level. Bacterial culture is possible but needs specialized media, takes time and could be falsely negative. CSF must be cultured the first week of the disease, urine from the beginning of the second week and blood cultures as soon as possible. PCR methods have been developed and increase the sensitivity of the diagnosis. A serological test must also be performed systematically by microscopy agglutination test (MAT). Immunoglobulin M (IgM) appears during the first week of infection. The diagnosis of neurobrucellosis includes bacterial culture of the CSF and a serological test. A PCR can also be performed.[143]

In Q fever with neurological signs (mainly meningoencephalitis), CSF examination shows a predominance of lymphocytes. As *C. burnetii* does not grow on culture media used in routine, the detection of the bacteria is performed by PCR techniques. Besides the molecular detection, a serological test should be performed.

19.4.2.1.5 TREATMENT

Leptospirosis meningitis should be treated like any case of leptospirosis. Among the antibiotics studied in clinical trials, penicillin, ceftriaxone, doxycycline, and azithromycin are effective.[151,152]

Q fever meningitis should be treated with doxycycline.

In brucellosis, a combination of at least two drugs which cross the blood–brain barrier has been recommended.[153] The primary drugs of choice are doxycycline, rifampin, trimethoprim-sulfamethoxazole (SXT), ciprofloxacin, and ceftriaxone. Data from a large retrospective study favored 1 month of parenteral ceftriaxone treatment in combination with doxycycline and rifampin for treating neurobrucellosis.[154]

19.4.2.1.6 OUTCOME

The Istanbul study found sequelae in 19% of patients with neurobrucellosis: (in decreasing order) walking difficulty, hearing loss, urinary incontinence, visual disturbance, and amnesia.[154]

19.4.2.1.7 PREVENTION

Doxycycline might be effective as a pre-exposure or post-exposure prophylaxis in leptospirosis, but evidence of this is still lacking.[155]

Human vaccines are available for infection with Leptospira and Q fever but none of these vaccines has been widely implemented.

19.4.2.2 Rickettsiosis

The proportion of patients with meningitis is estimated at 5% among patients with scrub typhus,[156] with a mortality comparable to patients without meningitis. In scrub typhus, the presence of pneumonitis is associated with the occurrence of scrub typhus meningitis/encephalitis.[157]

Scrub typhus and *Rickettsia typhi* infections are prevalent in parts of South-East Asia.[158] A study performed in Laos showed that the prevalence of these infections in patients whose clinical signs lead to an LP, was 6% (3% for each etiology).[148] The mortality was 27% among the 28 patients with *R. typhi* meningitis, and 14% among the 31 patients with scrub typhus.

The recommended treatment for both infections is doxycycline. However, observations have been made of meningitis developing in the course of scrub typhus while doxycycline had been started early in the management of infection,[157] making it an issue to add rifampicin to doxycycline for treating rickettsiosis meningitis, for which penetration in the CSF is good, in order to treat scrub typhus meningitis more effectively.

19.4.2.3 Relapsing Fever

Among 11 travelers diagnosed with *Borellia crocidurae* relapsing fever, four had a lymphocytic meningitis.[159] The earliest neurologic signs occurred during the second febrile episode, confirming previous studies reporting the onset of neurologic complications after the first episode.[160]

All four patients had traveled to Senegal and had a lymphocytic meningitis. The diagnosis was established on Giemsa-stained blood smear; molecular diagnosis was positive for all patients in blood and for three patients in CSF. All were cured with doxycycline or ceftriaxone.

19.4.3 Mycoplasma pneumoniae

While only 0.1% would complicate *Mycoplasma pneumoniae* infections,[161] it is one of the major causes (5%–10%) of encephalitis in children in some areas.

Meningitis, meningoencephalitis, cerebellitis, myelitis, polyneuropathy, acute disseminated encephalomyelitis, and Guillain-Barré syndrome are the most common neurological manifestations.[162] Myasthenia gravis has also been described.[163]

19.4.3.1 Physiopathology[162,164] and Diagnosis

It seems that there are two ways for CSF infection, one of early onset via a direct invasion of the CSF and one of late onset via an immune-mediated mechanism. In this latter case, the PCR in CSF is rarely positive. *Mycoplasma pneumoniae* expresses adhesion proteins and glycolipids that share structural homology with host cells, leading to autoimmune responses. The CSF shows a pleocytosis and an increased protein level. A serological test in the blood will be also performed. However, the serology has some limitations. Indeed, IgM can be detectable for several months and their detection has poor sensitivity. The full interpretation of a serological test needs a fourfold increase in antibody titer between two sera. However, a convalescent serum is not very useful in clinical practice as it is time-consuming and does not allow initiation of early treatment. The diagnosis of a neurological infection will be based on a positive serology (IgM detection) and molecular detection of *M. pneumoniae* in CSF and/ or in respiratory tract. However, the results have to be interpreted with caution as asymptomatic carriage is observed in children, inducing a positive serological test and even positive PCR in the respiratory tract.

19.4.3.2 Treatment

Macrolides, doxycycline, and fluoroquinolones are the antibiotics of choice. However, resistance to macrolides seems highly prevalent in Asia,[165] and prescriptions of alternatives to macrolides in children is an unresolved issue in these areas. Methylprednisolone usage along with antibiotics has been proposed on the basis of sporadic observations.[166]

Sequels of neurological manifestations of *M. pneumoniae* infections can be observed several years after the disease.[167]

19.4.4 Tetanus

Tetanus is a nervous system disorder characterized by muscle spasms that is caused by the toxin-producing anaerobe *Clostridium tetani*, which is found in the soil.

Neonatal tetanus reporting to the WHO notifiable surveillance system has very low notification efficiency, ranging from 3% to 11%[168] and non-neonatal tetanus is not a reportable condition.

Along with the Maternal and Neonatal Tetanus Elimination Initiative, global deaths from neonatal tetanus was reduced by more than 90% between 1987 and 2010.[169]

The highest reported number of non-neonatal tetanus cases is in the African Region, at 4.0 per million population, followed by the South-East Asia Region at 1.9 per million population.[170]

19.4.4.1 Physiopathology

Tetanus is due to a gram-positive anaerobic sporulating bacteria, *C. tetani*, whose spores persist in the soil. Infection occurs after inoculation of the spores from the soil into a soft tissue wound. The spores will then germinate in vivo and the bacteria will produce a neurotoxin that can be disseminated via the bloodstream. This neurotoxin binds the neuromuscular junctions of the peripheral motor neurons and is then transported to the CNS.[171]

19.4.4.2 Diagnosis

Diagnosis relies mainly on the clinical symptoms. A PCR can be used to detect the neurotoxin.

19.4.4.3 Mortality

Mortality from maternal and neonatal tetanus is high. Neonatal mortality ranges from 3% to 88% in Asia and Africa.[169]

Maternal tetanus arising from internal entry sites (postpartum, postabortional, or intramuscular injection) are associated with higher mortality than tetanus arising from other entry sites.[172]

Adult mortality rates up to 52% are reported in Asia and Africa.[169,173]

19.4.4.4 Outcome in Survivors

In neonatal tetanus, complications reflecting brain damages (microcephaly and mild neurological, developmental, or behavioral problems) have been observed in about one-third of neonates.[174] Cerebral palsy, cognitive delay, and deafness have also been reported.[169]

In non-neonatal tetanus, serious sequelae following unequivocally linked to the disease are observed in less than 10% of survivors.[175,176] Persistent vegetative state, limb amputations, and gait abnormality have been described,[176] as well as residual muscle rigidity.[177]

19.4.4.5 Treatment

Management of maternal and neonatal tetanus involves toxin neutralization (parenteral, intrathecal), bacterial elimination (penicillin, metronidazole), and symptomatic control.[169]

19.4.4.6 Prevention

Neonatal tetanus can be prevented by immunizing pregnant women and women of childbearing age with tetanus toxoid.

The WHO advocates the use of six clean measures to improve birth hygiene: clean birth surface, clean hands, clean perineum, cord cutting, cord tying, and cord care.

Non-neonatal tetanus can be prevented with universal immunization. More than two-thirds of adults hospitalized for tetanus across Africa and elsewhere are men, which suggests that there is an underlying burden of tetanus among adolescent and adult men who have been largely missed by vaccination programs.[170]

Tetanus toxoid does not induce immunity immediately; after the second dose, it takes approximately 2−4 weeks to exceed the minimum level required for short-term protection. For post-exposure prophylaxis, human antitetanus immunoglobulin (TIG) is indicated for all persons at risk of contracting tetanus after being wounded if they are not (fully) vaccinated or are immunocompromised.

References

1. Liu L, Johnson HL, Cousens S, et al. Global, regional, and national causes of child mortality: an updated systematic analysis for 2010 with time trends since 2000. *Lancet Lond Engl.* 2012;379 (9832):2151−2161. <http://dx.doi.org/10.1016/S0140-6736(12) 60560-1>.
2. Ku LC, Boggess KA, Cohen-Wolkowiez M. Bacterial meningitis in infants. *Clin Perinatol.* 2015;42(1):29−45:vii−viii. <http://dx.doi.org/10.1016/j.clp.2014.10.004>.
3. Seale AC, Blencowe H, Zaidi A, et al. Neonatal severe bacterial infection impairment estimates in South Asia, sub-Saharan Africa, and Latin America for 2010. *Pediatr Res.* 2013;74(Suppl 1):73−85. Available from: http://dx.doi.org/10.1038/pr.2013.207.

4. Polin RA, Harris MC. Neonatal bacterial meningitis. *Semin Neonatol*. 2001;6(2):157–172. Available from: http://dx.doi.org/10.1053/siny.2001.0045.

5. Gaschignard J, Levy C, Romain O, et al. Neonatal bacterial meningitis: 444 cases in 7 years. *Pediatr Infect Dis J*. 2011;30(3):212–217.

6. Furyk JS, Swann O, Molyneux E. Systematic review: neonatal meningitis in the developing world. *Trop Med Int Health*. 2011;16(6):672–679. Available from: http://dx.doi.org/10.1111/j.1365-3156.2011.02750.x.

7. Swann O, Everett DB, Furyk JS, et al. Bacterial meningitis in Malawian infants <2 months of age: etiology and susceptibility to World Health Organization first-line antibiotics. *Pediatr Infect Dis J*. 2014;33(6):560–565. Available from: http://dx.doi.org/10.1097/INF.0000000000000210.

8. Keddy KH, Sooka A, Musekiwa A, et al. Clinical and microbiological features of salmonella meningitis in a South African population, 2003–2013. *Clin Infect Dis Off Publ Infect Dis Soc Am*. 2015;61(Suppl 4):S272–S282. Available from: http://dx.doi.org/10.1093/cid/civ685.

9. Kwatra G, Cunnington MC, Merrall E, et al. Prevalence of maternal colonisation with group B streptococcus: a systematic review and meta-analysis. *Lancet Infect Dis*. 2016;16(9):1076–1084. :< http://dx.doi.org/10.1016/S1473-3099(16)30055-X>.

10. Seale AC, Koech AC, Sheppard AE, et al. Maternal colonization with *Streptococcus agalactiae* and associated stillbirth and neonatal disease in coastal Kenya. *Nat Microbiol*. 2016;1(7):16067. Available from: http://dx.doi.org/10.1038/nmicrobiol.2016.67.

11. Nishihara Y, Dangor Z, French N, Madhi S, Heyderman R. Challenges in reducing group B *Streptococcus* disease in African settings. *Arch Dis Child*. 2017;102(1):72–77. <http://dx.doi.org/10.1136/archdischild-2016-311419>.

12. Dangor Z, Lala SG, Cutland CL, et al. Burden of invasive group B *Streptococcus* disease and early neurological sequelae in South African infants. *PLoS One*. 2015;10(4):e0123014. <http://dx.doi.org/10.1371/journal.pone.0123014>.

13. Simonsen KA, Anderson-Berry AL, Delair SF, Davies HD. Early-onset neonatal sepsis. *Clin Microbiol Rev*. 2014;27(1):21–47. Available from: http://dx.doi.org/10.1128/CMR.00031-13.

14. Ka AS. Néonatologie en milieu tropical. In: Cochat P, ed. *Pédiatrie Tropicale et des voyages*. *Progrès en pédiatrie*. Rueil-Malmaison: Doin; 2012:23–33.

15. Pong A, Bradley JS. Bacterial meningitis and the newborn infant. *Infect Dis Clin North Am*. 1999;13(3):711–733, viii.

16. Garges HP, Moody MA, Cotten CM, et al. Neonatal meningitis: what is the correlation among cerebrospinal fluid cultures, blood cultures, and cerebrospinal fluid parameters? *Pediatrics*. 2006;117(4):1094–1100. Available from: http://dx.doi.org/10.1542/peds.2005-1132.

17. Committee on Infectious Diseases, Committee on Fetus and Newborn, Baker CJ, Byington CL, Polin RA. Policy statement—recommendations for the prevention of perinatal group B streptococcal (GBS) disease. *Pediatrics*. 2011;128(3):611–616. Available from: http://dx.doi.org/10.1542/peds.2011-1466.

18. Nhantumbo AA, Cantarelli VV, Caireão J, et al. Frequency of pathogenic paediatric bacterial meningitis in Mozambique: the critical role of multiplex real-time polymerase chain reaction to estimate the burden of disease. *PLoS One*. 2015;10(9):e0138249. Available from: http://dx.doi.org/10.1371/journal.pone.0138249.

19. Leber AL, Everhart K, Balada-Llasat J-M, et al. Multicenter evaluation of BioFire FilmArray meningitis/encephalitis panel for detection of bacteria, viruses, and yeast in cerebrospinal fluid specimens. *J Clin Microbiol*. 2016;54(9):2251–2261. <http://dx.doi.org/10.1128/JCM.00730-16>.

20. Messacar K, Breazeale G, Robinson CC, Dominguez SR. Potential clinical impact of the film array meningitis encephalitis panel in children with suspected central nervous system infections. *Diagn Microbiol Infect Dis*. 2016;86(1):118–120. Available from: http://dx.doi.org/10.1016/j.diagmicrobio.2016.05.020.

21. Weisfelt M, de Gans J, van de Beek D. Bacterial meningitis: a review of effective pharmacotherapy. *Expert Opin Pharmacother*. 2007;8(10):1493–1504. Available from: http://dx.doi.org/10.1517/14656566.8.10.1493.

22. Sáez-Llorens X, McCracken GH. Bacterial meningitis in children. *Lancet Lond Engl*. 2003;361(9375):2139–2148. <http://dx.doi.org/10.1016/S0140-6736(03)13693-8>.

23. World Health Organization. *Pocket Book of Hospital Care for Children: Guidelines for the Management of Common Childhood Illnesses*. 3rd ed. Geneva, Switzerland: WHO; 2013.

24. Laxminarayan R, Matsoso P, Pant S, et al. Access to effective antimicrobials: a worldwide challenge. *Lancet Lond Engl*. 2016;387(10014):168–175. <http://dx.doi.org/10.1016/S0140-6736(15)00474-2>.

25. World Health Organization. *Antimicrobial Resistance: Global Report on Surveillance 2014*. Geneva, Switzerland: WHO; 2014.

26. Landre-Peigne C, Ka AS, Peigne V, Bougere J, Seye MN, Imbert P. Efficacy of an infection control programme in reducing nosocomial bloodstream infections in a Senegalese neonatal unit. *J Hosp Infect*. 2011;79(2):161–165. Available from: http://dx.doi.org/10.1016/j.jhin.2011.04.007.

27. Thaver D, Zaidi AKM. Burden of neonatal infections in developing countries: a review of evidence from community-based studies. *Pediatr Infect Dis J*. 2009;28(1 Suppl):S3–S9. Available from: http://dx.doi.org/10.1097/INF.0b013e3181958755.

28. Bassler D, Stoll BJ, Schmidt B, et al. Using a count of neonatal morbidities to predict poor outcome in extremely low birth weight infants: added role of neonatal infection. *Pediatrics*. 2009;123(1):313–318. Available from: http://dx.doi.org/10.1542/peds.2008-0377.

29. Bedford H, de Louvois J, Halket S, Peckham C, Hurley R, Harvey D. Meningitis in infancy in England and Wales: follow up at age 5 years. *Br Med J*. 2001;323(7312):533–536.

30. Milner KM, Neal EFG, Roberts G, Steer AC, Duke T. Long-term neurodevelopmental outcome in high-risk newborns in resource-limited settings: a systematic review of the literature. *Paediatr Int Child Health*. 2015;35(3):227–242. Available from: http://dx.doi.org/10.1179/2046905515Y.0000000043.

31. Schrag SJ, Verani JR. Intrapartum antibiotic prophylaxis for the prevention of perinatal group B streptococcal disease: experience in the United States and implications for a potential group B streptococcal vaccine. *Vaccine*. 2013;31(Suppl 4):D20–D26. Available from: http://dx.doi.org/10.1016/j.vaccine.2012.11.056.

32. Bomela HN, Ballot DE, Cooper PA. Is prophylaxis of early-onset group B streptococcal disease appropriate for South Africa? *South Afr Med J Suid-Afr Tydskr Vir Geneeskd*. 2001;91(10):858–860.

33. Edmond KM, Kortsalioudaki C, Scott S, et al. Group B streptococcal disease in infants aged younger than 3 months: systematic review and meta-analysis. *Lancet Lond Engl*. 2012;379(9815):547–556. <http://dx.doi.org/10.1016/S0140-6736(11)61651-6>.

34. Rivera L, Sáez-Llorens X, Feris-Iglesias J, et al. Incidence and serotype distribution of invasive group B streptococcal disease in young infants: a multi-country observational study. *BMC Pediatr*. 2015;15:143. Available from: http://dx.doi.org/10.1186/s12887-015-0460-2.

35. Chan GJ, Stuart EA, Zaman M, Mahmud AA, Baqui AH, Black RE. The effect of intrapartum antibiotics on early-onset neonatal sepsis in Dhaka, Bangladesh: a propensity score matched

analysis. *BMC Pediatr.* 2014;14:104. Available from: http://dx.doi.org/10.1186/1471-2431-14-104.

36. Oster G, Edelsberg J, Hennegan K, et al. Prevention of group B streptococcal disease in the first 3 months of life: would routine maternal immunization during pregnancy be cost-effective? *Vaccine.* 2014;32(37):4778–4785. Available from: http://dx.doi.org/10.1016/j.vaccine.2014.06.003.

37. Nuccitelli A, Rinaudo CD, Maione D. Group B *Streptococcus* vaccine: state of the art. *Ther Adv Vaccines.* 2015;3(3):76–90. Available from: http://dx.doi.org/10.1177/2051013615579869.

38. MenAfriCar Consortium. Household transmission of *Neisseria meningitidis* in the African meningitis belt: a longitudinal cohort study. *Lancet Glob Health.* 2016;4(12):e989–e995, Available from: <http://dx.doi.org/10.1016/S2214-109X(16)30244-3>.

39. Borrow R, Alarcón P, Carlos J, et al. The Global Meningococcal Initiative: global epidemiology, the impact of vaccines on meningococcal disease and the importance of herd protection. *Expert Rev Vaccines.* 2017;16(4):313–328. Available from: http://dx.doi.org/10.1080/14760584.2017.1258308.

40. World Health Organization. Meningitis control in countries of the African meningitis belt, 2015. *Releve Epidemiol Hebd.* 2016;91 (16):209–216.

41. Nicolas Lefebvre et Pierre Nicolas. *Méningites bactériennes. Médecine Tropicale.* 6ème éd. Paris: Lavoisier; 2012:593–602.

42. Gessner BD, Mueller JE, Yaro S. African meningitis belt pneumococcal disease epidemiology indicates a need for an effective serotype 1 containing vaccine, including for older children and adults. *BMC Infect Dis.* 2010;10:22. Available from: http://dx.doi.org/10.1186/1471-2334-10-22.

43. Blumental S, Granger-Farbos A, Moïsi JC, et al. Virulence factors of *Streptococcus pneumoniae.* Comparison between African and French invasive isolates and implication for future vaccines. *PLoS One.* 2015;10(7):e0133885. <http://dx.doi.org/10.1371/journal.pone.0133885>.

44. Nhantumbo AA, Gudo ES, Caierão J, et al. Serotype distribution and antimicrobial resistance of *Streptococcus pneumoniae* in children with acute bacterial meningitis in Mozambique: implications for a national immunization strategy. *BMC Microbiol.* 2016;16(1):134. Available from: http://dx.doi.org/10.1186/s12866-016-0747-y.

45. Cohen C, Naidoo N, Meiring S, et al. *Streptococcus pneumoniae* serotypes and mortality in adults and adolescents in South Africa: Analysis of National Surveillance Data, 2003–2008. *PLoS One.* 2015;10(10):e0140185. Available from: http://dx.doi.org/10.1371/journal.pone.0140185.

46. McGill F, Heyderman RS, Panagiotou S, Tunkel AR, Solomon T. Acute bacterial meningitis in adults. *Lancet.* 2016;388:3036–3047. <http://dx.doi.org/10.1016/S0140-6736(16)30654-7>.

47. Pneumococcal meningitis outbreaks in sub-Saharan Africa. *Releve Epidemiol Hebd.* 2016;91(23):298–302:

48. Ngo TH, Tran TBC, Tran TTN, et al. Slaughterhouse pigs are a major reservoir of *Streptococcus suis* serotype 2 capable of causing human infection in southern Vietnam. *PLoS One.* 2011;6(3):e17943. Available from: http://dx.doi.org/10.1371/journal.pone.0017943.

49. Nguyen THM, Tran THC, Thwaites G, et al. Dexamethasone in Vietnamese adolescents and adults with bacterial meningitis. *N Engl J Med.* 2007;357(24):2431–2440. Available from: http://dx.doi.org/10.1056/NEJMoa070852.

50. Chau TTH, Campbell JI, Schultsz C, et al. Three adult cases of *Listeria monocytogenes* meningitis in Vietnam. *PLoS Med.* 2010;7 (7):e1000306. Available from: http://dx.doi.org/10.1371/journal.pmed.1000306.

51. Iliyasu G, Lawal H, Habib AG, et al. Response to the meningococcal meningitis epidemic (MME) at Aminu Kano Teaching Hospital, Kano (2008-2009). *Niger J Med J Natl Assoc Resid Dr Niger.* 2009;18(4):428–430.

52. Kariuki SM, Gitau E, Gwer S, et al. Value of *Plasmodium falciparum* histidine-rich protein 2 level and malaria retinopathy in distinguishing cerebral malaria from other acute encephalopathies in Kenyan children. *J Infect Dis.* 2014;209(4):600–609. Available from: http://dx.doi.org/10.1093/infdis/jit500.

53. Mikita K, Thakur K, Anstey NM, et al. Quantification of *Plasmodium falciparum* histidine-rich protein-2 in cerebrospinal spinal fluid from cerebral malaria patients. *Am J Trop Med Hyg.* 2014;91(3):486–492. Available from: http://dx.doi.org/10.4269/ajtmh.14-0210.

54. Rubach MP, Mukemba J, Florence S, et al. Plasma *Plasmodium falciparum* histidine-rich protein-2 concentrations are associated with malaria severity and mortality in Tanzanian children. *PLoS One.* 2012;7(5):e35985. Available from: http://dx.doi.org/10.1371/journal.pone.0035985.

55. Ploy M-C, Garnier F, Languepin J, Fermeaux V, Martin C, Denis F. Interest of postmortem-collected specimens in the diagnosis of fulminant meningococcal sepsis. *Diagn Microbiol Infect Dis.* 2005;52(1):65–66. Available from: http://dx.doi.org/10.1016/j.diagmicrobio.2004.12.012.

56. Radetsky M. Duration of symptoms and outcome in bacterial meningitis: an analysis of causation and the implications of a delay in diagnosis. *Pediatr Infect Dis J.* 1992;11(9):694–698. discussion 698-701.

57. Auburtin M, Wolff M, Charpentier J, et al. Detrimental role of delayed antibiotic administration and penicillin-nonsusceptible strains in adult intensive care unit patients with pneumococcal meningitis: the PNEUMOREA prospective multicenter study. *Crit Care Med.* 2006;34(11):2758–2765. Available from: http://dx.doi.org/10.1097/01.CCM.0000239434.26669.65.

58. World Health Organization. Contrôle des épidémies de méningite en Afrique Guide de référence rapide à l'intention des autorités sanitaires et des soignants; 2015.

59. Brouwer MC, McIntyre P, Prasad K, van de Beek D. Corticosteroids for acute bacterial meningitis. *Cochrane Database Syst Rev.* 2015;9:CD004405. Available from: http://dx.doi.org/10.1002/14651858.CD004405.pub5.

60. Peltola H, Roine I, Fernández J, et al. Hearing impairment in childhood bacterial meningitis is little relieved by dexamethasone or glycerol. *Pediatrics.* 2010;125(1):e1–e8. Available from: http://dx.doi.org/10.1542/peds.2009-0395.

61. Bonsu BK, Harper MB. Fever interval before diagnosis, prior antibiotic treatment, and clinical outcome for young children with bacterial meningitis. *Clin Infect Dis Off Publ Infect Dis Soc Am.* 2001;32(4):566–572. Available from: http://dx.doi.org/10.1086/318700.

62. de Gans J, van de Beek D, European Dexamethasone in Adulthood Bacterial Meningitis Study Investigators. Dexamethasone in adults with bacterial meningitis. *N Engl J Med.* 2002;347(20):1549–1556. Available from: http://dx.doi.org/10.1056/NEJMoa021334.

63. Scarborough M, Gordon SB, Whitty CJM, et al. Corticosteroids for bacterial meningitis in adults in sub-Saharan Africa. *N Engl J Med.* 2007;357(24):2441–2450. Available from: http://dx.doi.org/10.1056/NEJMoa065711.

64. Bijlsma MW, Brouwer MC, Kasanmoentalib ES, et al. Community-acquired bacterial meningitis in adults in the Netherlands, 2006-14: a prospective cohort study. *Lancet Infect Dis.* 2016;16(3):339–347. <http://dx.doi.org/10.1016/S1473-3099(15)00430-2>.

65. Roine I, Weisstaub G, Peltola H, LatAm Bacterial Meningitis Study Group. Influence of malnutrition on the course of childhood bacterial meningitis. *Pediatr Infect Dis J.* 2010;29

(2):122–125. Available from: http://dx.doi.org/10.1097/INF.0b013e3181b6e7d3.

66. Arditi M, Mason EO, Bradley JS, et al. Three-year multicenter surveillance of pneumococcal meningitis in children: clinical characteristics, and outcome related to penicillin susceptibility and dexamethasone use. *Pediatrics.* 1998;102(5):1087–1097.

67. Roine I, Peltola H, Fernández J, et al. Influence of admission findings on death and neurological outcome from childhood bacterial meningitis. *Clin Infect Dis Off Publ Infect Dis Soc Am.* 2008;46(8):1248–1252. Available from: http://dx.doi.org/10.1086/533448.

68. Wall EC, Cartwright K, Scarborough M, et al. High mortality amongst adolescents and adults with bacterial meningitis in sub-Saharan Africa: an analysis of 715 cases from Malawi. *PLoS One.* 2013;8(7):e69783. Available from: http://dx.doi.org/10.1371/journal.pone.0069783.

69. McCormick DW, Wilson ML, Mankhambo L, et al. Risk factors for death and severe sequelae in Malawian children with bacterial meningitis, 1997-2010. *Pediatr Infect Dis J.* 2013;32(2):e54–e61. Available from: http://dx.doi.org/10.1097/INF.0b013e31826faf5a.

70. Ramakrishnan M, Ulland AJ, Steinhardt LC, Moïsi JC, Were F, Levine OS. Sequelae due to bacterial meningitis among African children: a systematic literature review. *BMC Med.* 2009;7:47. Available from: http://dx.doi.org/10.1186/1741-7015-7-47.

71. Mai NTH, Hoa NT, Nga TVT, et al. *Streptococcus suis* meningitis in adults in Vietnam. *Clin Infect Dis Off Publ Infect Dis Soc Am.* 2008;46(5):659–667. Available from: http://dx.doi.org/10.1086/527385.

72. Edmond K, Clark A, Korczak VS, Sanderson C, Griffiths UK, Rudan I. Global and regional risk of disabling sequelae from bacterial meningitis: a systematic review and meta-analysis. *Lancet Infect Dis.* 2010;10(5):317–328. <http://dx.doi.org/10.1016/S1473-3099(10)70048-7>.

73. van Samkar A, Brouwer MC, Schultsz C, van der Ende A, van de Beek D. *Streptococcus suis* meningitis: a systematic review and meta-analysis. *PLoS Negl Trop Dis.* 2015;9(10):e0004191. Available from: http://dx.doi.org/10.1371/journal.pntd.0004191.

74. Mulwafu W, Kuper H, Ensink RJH. Prevalence and causes of hearing impairment in Africa. *Trop Med Int Health.* 2016;21(2):158–165. Available from: http://dx.doi.org/10.1111/tmi.12640.

75. Adnams CM. Perspectives of intellectual disability in South Africa: epidemiology, policy, services for children and adults. *Curr Opin Psychiatry.* 2010;23(5):436–440. Available from: http://dx.doi.org/10.1097/YCO.0b013e32833cfc2d.

76. Telisinghe L, Waite TD, Gobin M, et al. Chemoprophylaxis and vaccination in preventing subsequent cases of meningococcal disease in household contacts of a case of meningococcal disease: a systematic review. *Epidemiol Infect.* 2015;143(11):2259–2268. Available from: http://dx.doi.org/10.1017/S0950268815000849.

77. Cohn AC, MacNeil JR, Clark TA, et al. Prevention and control of meningococcal disease: recommendations of the Advisory Committee on Immunization Practices (ACIP). *MMWR Recomm Rep Morb Mortal Wkly Rep Recomm Rep Cent Dis Control.* 2013;62(RR-2):1–28.

78. MenAfriCar Consortium. Meningococcal carriage in the African meningitis belt. *Trop Med Int Health.* 2013;18(8):968–978. Available from: http://dx.doi.org/10.1111/tmi.12125.

79. Bonner K, Welch E, Elder K, Cohn J. Impact of pneumococcal conjugate vaccine administration in pediatric older age groups in low and middle income countries: a systematic review. *PLoS One.* 2015;10(8):e0135270. Available from: http://dx.doi.org/10.1371/journal.pone.0135270.

80. Brouwer MC, Coutinho JM, van de Beek D. Clinical characteristics and outcome of brain abscess: systematic review and meta-analysis. *Neurology.* 2014;82(9):806–813. Available from: http://dx.doi.org/10.1212/WNL.0000000000000172.

81. Kabré A, Zabsonré S, Diallo O, Cissé R. [Management of brain abscesses in era of computed tomography in sub-Saharan Africa: a review of 112 cases]. *Neurochirurgie.* 2014;60(5):249–253. Available from: http://dx.doi.org/10.1016/j.neuchi.2014.06.011.

82. Anwary MA. Intracranial suppuration: review of an 8-year experience at Umtata General Hospital and Nelson Mandela Academic Hospital, Eastern Cape, South Africa. *South Afr Med J Suid-Afr Tydskr Vir Geneeskd.* 2015;105(7):584–588.

83. Brouwer MC, Tunkel AR, McKhann GM, van de Beek D. Brain abscess. *N Engl J Med.* 2014;371(5):447–456. Available from: http://dx.doi.org/10.1056/NEJMra1301635.

84. Ambrosioni J, Lew D, Garbino J. Nocardiosis: updated clinical review and experience at a tertiary center. *Infection.* 2010;38(2):89–97. Available from: http://dx.doi.org/10.1007/s15010-009-9193-9.

85. Jones N, Khoosal M, Louw M, Karstaedt A. Nocardial infection as a complication of HIV in South Africa. *J Infect.* 2000;41(3):232–239. Available from: http://dx.doi.org/10.1053/jinf.2000.0729.

86. Zhang C, Hu L, Wu X, Hu G, Ding X, Lu Y. A retrospective study on the aetiology, management, and outcome of brain abscess in an 11-year, single-centre study from China. *BMC Infect Dis.* 2014;14:311. Available from: http://dx.doi.org/10.1186/1471-2334-14-311.

87. Garg RK, Malhotra HS, Gupta R. Spinal cord involvement in tuberculous meningitis. *Spinal Cord.* 2015;53(9):649–657. Available from: http://dx.doi.org/10.1038/sc.2015.58.

88. Chiang SS, Khan FA, Milstein MB, et al. Treatment outcomes of childhood tuberculous meningitis: a systematic review and meta-analysis. *Lancet Infect Dis.* 2014;14(10):947–957. <http://dx.doi.org/10.1016/S1473-3099(14)70852-7>.

89. Rieder HL, Snider DE, Cauthen GM. Extrapulmonary tuberculosis in the United States. *Am Rev Respir Dis.* 1990;141(2):347–351. Available from: http://dx.doi.org/10.1164/ajrccm/141.2.347.

90. Chatterjee S. Brain tuberculomas, tubercular meningitis, and post-tubercular hydrocephalus in children. *J Pediatr Neurosci.* 2011;6(Suppl 1):S96–S100. Available from: http://dx.doi.org/10.4103/1817-1745.85725.

91. Veltman JA, Bristow CC, Klausner JD. Meningitis in HIV-positive patients in sub-Saharan Africa: a review. *J Int AIDS Soc.* 2014;17:19184.

92. Efsen AMW, Panteleev AM, Grint D, et al. TB meningitis in HIV-positive patients in Europe and Argentina: clinical outcome and factors associated with mortality. *Biomed Res Int.* 2013;2013:373601. Available from: http://dx.doi.org/10.1155/2013/373601.

93. Bañuls A-L, Sanou A, Anh NTV, Godreuil S. *Mycobacterium tuberculosis*: ecology and evolution of a human bacterium. *J Med Microbiol.* 2015;64(11):1261–1269. Available from: http://dx.doi.org/10.1099/jmm.0.000171.

94. Thwaites GE, van Toorn R, Schoeman J. Tuberculous meningitis: more questions, still too few answers. *Lancet Neurol.* 2013;12(10):999–1010. <http://dx.doi.org/10.1016/S1474-4422(13)70168-6>.

95. Medical Research Council. Streptomycin in Tuberculosis Trials Committee. *STREPTOMYCIN* treatment of tuberculous meningitis. *Lancet Lond Engl.* 1948;1(6503):582–596.

96. DeLance AR, Safaee M, Oh MC, et al. Tuberculoma of the central nervous system. *J Clin Neurosci Off J Neurosurg Soc Australas.* 2013;20(10):1333–1341. Available from: http://dx.doi.org/10.1016/j.jocn.2013.01.008.

97. Sonawane DV, Jagtap SA, Patil HG, Biraris SR, Chandanwale AS. Intramedullary tuberculoma of dorsal spinal cord: a case report with review of literature. *J Orthop Case Rep.* 2015;5 (2):44−46. Available from: http://dx.doi.org/10.13107/jocr.2250-0685.271.

98. du Plessis J, Andronikou S, Theron S, Wieselthaler N, Hayes M. Unusual forms of spinal tuberculosis. *Childs Nerv Syst ChNS Off J Int Soc Pediatr Neurosurg.* 2008;24(4):453−457. Available from: http://dx.doi.org/10.1007/s00381-007-0525-0.

99. Nussbaum ES, Rockswold GL, Bergman TA, Erickson DL, Seljeskog EL. Spinal tuberculosis: a diagnostic and management challenge. *J Neurosurg.* 1995;83(2):243−247. Available from: http://dx.doi.org/10.3171/jns.1995.83.2.0243.

100. Kamara E, Mehta S, Brust JCM, Jain AK. Effect of delayed diagnosis on severity of Pott's disease. *Int Orthop.* 2012;36 (2):245−254. Available from: http://dx.doi.org/10.1007/s00264-011-1432-2.

101. Zignol M, Dean AS, Falzon D, et al. Twenty years of global surveillance of antituberculosis-drug resistance. *N Engl J Med.* 2016;375(11):1081−1089. Available from: http://dx.doi.org/10.1056/NEJMsr1512438.

102. Sanei Taheri M, Karimi MA, Haghighatkhah H, Pourghorban R, Samadian M, Delavar Kasmaei H. Central nervous system tuberculosis: an imaging-focused review of a reemerging disease. *Radiol Res Pract.* 2015;2015:202806. Available from: http://dx.doi.org/10.1155/2015/202806.

103. Donald PR. Cerebrospinal fluid concentrations of antituberculosis agents in adults and children. *Tuberc Edinb Scotl.* 2010;90 (5):279−292. Available from: http://dx.doi.org/10.1016/j.tube.2010.07.002.

104. Patel VB, Padayatchi N, Bhigjee AI, et al. Multidrug-resistant tuberculous meningitis in KwaZulu-Natal, South Africa. *Clin Infect Dis Off Publ Infect Dis Soc Am.* 2004;38(6):851−856. Available from: http://dx.doi.org/10.1086/381973.

105. Thwaites GE, Nguyen DB, Nguyen HD, et al. Dexamethasone for the treatment of tuberculous meningitis in adolescents and adults. *N Engl J Med.* 2004;351(17):1741−1751. Available from: http://dx.doi.org/10.1056/NEJMoa040573.

106. Prasad K, Singh MB, Ryan H. Corticosteroids for managing tuberculous meningitis. *Cochrane Database Syst Rev.* 2016;4: CD002244. Available from: http://dx.doi.org/10.1002/14651858.CD002244.pub4.

107. Török ME, Yen NTB, Chau TTH, et al. Timing of initiation of antiretroviral therapy in human immunodeficiency virus (HIV)-associated tuberculous meningitis. *Clin Infect Dis Off Publ Infect Dis Soc Am.* 2011;52(11):1374−1383. Available from: http://dx.doi.org/10.1093/cid/cir230.

108. Infections chez l'adulte: prophylaxies et traitements curatifs. Prise En Charge Médicale Des Personnes Vivant Avec Le VIH Recommandations Du Groupe D'experts. *La documentation française.* Paris: DILA; 2013:251−295.

109. Nahid P, Dorman SE, Alipanah N, et al. Executive Summary: Official American Thoracic Society/Centers for Disease Control and Prevention/Infectious Diseases Society of America Clinical Practice Guidelines: treatment of drug-susceptible tuberculosis. *Clin Infect Dis Off Publ Infect Dis Soc Am.* 2016;63(7):853−867. Available from: http://dx.doi.org/10.1093/cid/ciw566.

110. Rajshekhar V. Surgery for brain tuberculosis: a review. *Acta Neurochir (Wien).* 2015;157(10):1665−1678. Available from: http://dx.doi.org/10.1007/s00701-015-2501-x.

111. Seddon JA, Visser DH, Bartens M, et al. Impact of drug resistance on clinical outcome in children with tuberculous meningitis. *Pediatr Infect Dis J.* 2012;31(7):711−716. Available from: http://dx.doi.org/10.1097/INF.0b013e318253acf8.

112. Tho DQ, Török ME, Yen NTB, et al. Influence of antituberculosis drug resistance and *Mycobacterium tuberculosis* lineage on outcome in HIV-associated tuberculous meningitis. *Antimicrob Agents Chemother.* 2012;56(6):3074−3079. Available from: http://dx.doi.org/10.1128/AAC.00319-12.

113. Thwaites GE, Duc Bang N, Huy Dung N, et al. The influence of HIV infection on clinical presentation, response to treatment, and outcome in adults with Tuberculous meningitis. *J Infect Dis.* 2005;192(12):2134−2141. Available from: http://dx.doi.org/10.1086/498220.

114. Woldeamanuel YW, Girma B. A 43-year systematic review and meta-analysis: case-fatality and risk of death among adults with tuberculous meningitis in Africa. *J Neurol.* 2014;261 (5):851−865. Available from: http://dx.doi.org/10.1007/s00415-013-7060-6.

115. Lammie GA, Hewlett RH, Schoeman JF, Donald PR. Tuberculous cerebrovascular disease: a review. *J Infect.* 2009;59 (3):156−166. Available from: http://dx.doi.org/10.1016/j.jinf.2009.07.012.

116. Anuradha HK, Garg RK, Agarwal A, et al. Predictors of stroke in patients of tuberculous meningitis and its effect on the outcome. *QJM Mon J Assoc Physicians.* 2010;103(9):671−678. Available from: http://dx.doi.org/10.1093/qjmed/hcq103.

117. Misra UK, Kalita J, Nair PP. Role of aspirin in tuberculous meningitis: a randomized open label placebo controlled trial. *J Neurol Sci.* 2010;293(1−2):12−17. Available from: http://dx.doi.org/10.1016/j.jns.2010.03.025.

118. Brancusi F, Farrar J, Heemskerk D. Tuberculous meningitis in adults: a review of a decade of developments focusing on prognostic factors for outcome. *Future Microbiol.* 2012;7 (9):1101−1116. Available from: http://dx.doi.org/10.2217/fmb.12.86.

119. Getahun H, Matteelli A, Abubakar I, et al. Management of latent *Mycobacterium tuberculosis* infection: WHO guidelines for low tuberculosis burden countries. *Eur Respir J.* 2015;46 (6):1563−1576. Available from: http://dx.doi.org/10.1183/13993003.01245-2015.

120. TEMPRANO ANRS 12136 Study Group, Danel C, Moh R, et al. A trial of early antiretrovirals and isoniazid preventive therapy in Africa. *N Engl J Med.* 2015;373(9):808−822. Available from: http://dx.doi.org/10.1056/NEJMoa1507198.

121. Roy A, Eisenhut M, Harris RJ, et al. Effect of BCG vaccination against *Mycobacterium tuberculosis* infection in children: systematic review and meta-analysis. *Br Med J.* 2014;349:g4643.

122. O'Brien KL, Ruff AJ, Louis MA, et al. Bacillus Calmette-Guérin complications in children born to HIV-1-infected women with a review of the literature. *Pediatrics.* 1995;95(3):414−418.

123. Araujo S, Freitas LO, Goulart LR, Goulart IMB. Molecular evidence for the aerial route of infection of *Mycobacterium leprae* and the role of asymptomatic carriers in the persistence of leprosy. *Clin Infect Dis Off Publ Infect Dis Soc Am.* 2016;63 (11):1412−1420. Available from: http://dx.doi.org/10.1093/cid/ciw570.

124. Moet FJ, Pahan D, Schuring RP, Oskam L, Richardus JH. Physical distance, genetic relationship, age, and leprosy classification are independent risk factors for leprosy in contacts of patients with leprosy. *J Infect Dis.* 2006;193(3):346−353. Available from: http://dx.doi.org/10.1086/499278.

125. World Health Organization. Global leprosy update, 2014: need for early case detection. *Wkly Epidemiol Rec.* 2015;90 (36):461−474.

126. Reibel F, Cambau E, Aubry A. Update on the epidemiology, diagnosis, and treatment of leprosy. *Méd Mal Infect.* 2015;45 (9):383−393. Available from: http://dx.doi.org/10.1016/j.medmal.2015.09.002.

127. Flageul B. [Diagnosis and treatment of leprous neuropathy in practice]. *Rev Neurol (Paris)*. 2012;168(12):960–966. Available from: http://dx.doi.org/10.1016/j.neurol.2012.09.005.

128. World Health Organization. *WHO Recommended MDT Regimens*. http://www.who.int/lep/mdt/regimens/en/. Accessed 22.10.16.

129. Britton WJ, Lockwood DNJ. Leprosy. *Lancet Lond Engl*. 2004;363 (9416):1209–1219. <http://dx.doi.org/10.1016/S0140-6736(04) 15952-7>.

130. World Health Organization. *WHO Guidelines for the Treatment of Treponema pallidum (Syphilis)*. Geneva, Switzerland: WHO; 2016.

131. Wijesooriya NS, Rochat RW, Kamb ML, et al. Global burden of maternal and congenital syphilis in 2008 and 2012: a health systems modelling study.. *Lancet Glob Health*. 2016;4(8):e525–e533. <http://dx.doi.org/10.1016/S2214-109X(16)30135-8>.

132. Mattei PL, Beachkofsky TM, Gilson RT, Wisco OJ. Syphilis: a reemerging infection. *Am Fam Physician*. 2012;86(5):433–440.

133. Hesketh T, Ye X, Zhu W. Syphilis in China: the great comeback. *Emerg Health Threats J*. 2008;1:e6. Available from: http://dx.doi.org/10.3134/ehtj.08.006.

134. Shah BJ, Karia DR, Pawara CL. Syphilis: is it making resurgence? *Indian J Sex Transm Dis*. 2015;36(2):178–181. Available from: http://dx.doi.org/10.4103/0253-7184.167170.

135. Lafond RE, Lukehart SA. Biological basis for syphilis. *Clin Microbiol Rev*. 2006;19(1):29–49. Available from: http://dx.doi.org/10.1128/CMR.19.1.29-49.2006.

136. French P, Gomberg M, Janier M, et al. IUSTI: 2008 European guidelines on the management of syphilis. *Int J STD AIDS*. 2009;20(5):300–309. Available from: http://dx.doi.org/10.1258/ijsa.2008.008510.

137. Morshed MG, Singh AE. Recent trends in the serologic diagnosis of syphilis. *Clin Vaccine Immunol*. 2015;22(2):137–147. Available from: http://dx.doi.org/10.1128/CVI.00681-14.

138. Vanhaecke C, Grange P, Benhaddou N, et al. Clinical and biological characteristics of 40 patients with neurosyphilis and evaluation of *Treponema pallidum* nested polymerase chain reaction in cerebrospinal fluid samples. *Clin Infect Dis Off Publ Infect Dis Soc Am*. 2016;63(9):1180–1186. Available from: http://dx.doi.org/10.1093/cid/ciw499.

139. World Health Organization. *Guidelines for the Management of Sexually Transmitted Infections*. Geneva, Switzerland: WHO; 2003.

140. Janier M, Hegyi V, Dupin N, et al. 2014 European guideline on the management of syphilis. *J Eur Acad Dermatol Venereol*. 2014;28(12):1581–1593. Available from: http://dx.doi.org/10.1111/jdv.12734.

141. van Samkar A, Brouwer MC, van der Ende A, van de Beek D. Zoonotic bacterial meningitis in human adults. *Neurology*. 2016;87(11):1171–1179. Available from: http://dx.doi.org/10.1212/WNL.0000000000003101.

142. de Rooij MMT, Borlée F, Smit LAM, et al. Detection of *Coxiella burnetii* in ambient air after a large Q fever outbreak. *PLoS One*. 2016;11(3):e0151281. Available from: http://dx.doi.org/10.1371/journal.pone.0151281.

143. Haake DA, Levett PN. Leptospirosis in humans. In: Adler B, ed. *Leptospira and Leptospirosis*. Vol. 387. Berlin, Heidelberg: Springer Berlin Heidelberg; 2015:65–97. <http://link.springer.com/10.1007/978-3-662-45059-8_5>. Accessed 07.12.16.

144. Eldin C, Mélenotte C, Mediannikov O, et al. From Q fever to *Coxiella burnetii* infection: a paradigm change. *Clin Microbiol Rev*. 2017;30(1):115–190. Available from: http://dx.doi.org/10.1128/CMR.00045-16.

145. Riabi HRA, Ahmadi R, Rezaei MS, Atarodi AR. Brucella meningitis. *Med J Islam Repub Iran*. 2013;27(2):99–100.

146. Gündeş S, Meriç M, Willke A, Erdenliğ S, Koç K. A case of intracranial abscess due to *Brucella melitensis*. *Int J Infect Dis IJID Off Publ Int Soc Infect Dis*. 2004;8(6):379–381. Available from: http://dx.doi.org/10.1016/j.ijid.2004.05.003.

147. van Samkar A, van de Beek D, Stijnis C, Goris M, Brouwer MC. Suspected leptospiral meningitis in adults: report of four cases and review of the literature. *Neth J Med*. 2015;73(10):464–470.

148. Dittrich S, Rattanavong S, Lee SJ, et al. Orientia, rickettsia, and leptospira pathogens as causes of CNS infections in Laos: a prospective study. *Lancet Glob Health*. 2015;3(2):e104–e112.:<http://dx.doi.org/10.1016/S2214-109X(14)70289-X>.

149. Reimer LG. Q fever. *Clin Microbiol Rev*. 1993;6(3):193–198.

150. Bernit E, Pouget J, Janbon F, et al. Neurological involvement in acute Q fever: a report of 29 cases and review of the literature. *Arch Intern Med*. 2002;162(6):693–700.

151. Phimda K, Hoontrakul S, Suttinont C, et al. Doxycycline versus azithromycin for treatment of leptospirosis and scrub typhus. *Antimicrob Agents Chemother*. 2007;51(9):3259–3263. Available from: http://dx.doi.org/10.1128/AAC.00508-07.

152. Panaphut T, Domrongkitchaiporn S, Vibhagool A, Thinkamrop B, Susaengrat W. Ceftriaxone compared with sodium penicillin g for treatment of severe leptospirosis. *Clin Infect Dis Off Publ Infect Dis Soc Am*. 2003;36(12):1507–1513. Available from: http://dx.doi.org/10.1086/375226.

153. Pappas G, Papadimitriou P, Akritidis N, Christou L, Tsianos EV. The new global map of human brucellosis. *Lancet Infect Dis*. 2006;6(2):91–99. <http://dx.doi.org/10.1016/S1473-3099(06) 70382-6>.

154. Erdem H, Ulu-Kilic A, Kilic S, et al. Efficacy and tolerability of antibiotic combinations in neurobrucellosis: results of the Istanbul study. *Antimicrob Agents Chemother*. 2012;56 (3):1523–1528. Available from: http://dx.doi.org/10.1128/AAC.05974-11.

155. Chusri S, McNeil EB, Hortiwakul T, et al. Single dosage of doxycycline for prophylaxis against leptospiral infection and leptospirosis during urban flooding in southern Thailand: a non-randomized controlled trial. *J Infect Chemother Off J Jpn Soc Chemother*. 2014;20(11):709–715. Available from: http://dx.doi.org/10.1016/j.jiac.2014.07.016.

156. Taylor AJ, Paris DH, Newton PN. A systematic review of mortality from untreated scrub typhus (*Orientia tsutsugamushi*). *PLoS Negl Trop Dis*. 2015;9(8):e0003971. Available from: http://dx.doi.org/10.1371/journal.pntd.0003971.

157. Kim D-M, Chung J-H, Yun N-R, et al. Scrub typhus meningitis or meningoencephalitis. *Am J Trop Med Hyg*. 2013;89 (6):1206–1211. Available from: http://dx.doi.org/10.4269/ajtmh.13-0224.

158. Lee H-C, Ko W-C, Lee H-L, Chen H-Y. Clinical manifestations and complications of rickettsiosis in southern Taiwan. *J Formos Med Assoc Taiwan Yi Zhi*. 2002;101(6):385–392.

159. Goutier S, Ferquel E, Pinel C, et al. *Borrelia crocidurae* meningoencephalitis, West Africa. *Emerg Infect Dis*. 2013;19(2):301–304. Available from: http://dx.doi.org/10.3201/eid1902.121325.

160. Cadavid D, Barbour AG. Neuroborreliosis during relapsing fever: review of the clinical manifestations, pathology, and treatment of infections in humans and experimental animals. *Clin Infect Dis Off Publ Infect Dis Soc Am*. 1998;26(1):151–164.

161. Koskiniemi M. CNS manifestations associated with *Mycoplasma pneumoniae* infections: summary of cases at the University of Helsinki and review. *Clin Infect Dis Off Publ Infect Dis Soc Am*. 1993;17(Suppl 1):S52–S57.

162. Al-Zaidy SA, MacGregor D, Mahant S, Richardson SE, Bitnun A. Neurological complications of PCR-proven *M. pneumoniae* infections in children: prodromal illness duration may reflect pathogenetic mechanism. *Clin Infect Dis Off Publ Infect Dis Soc Am*.

2015;61(7):1092−1098. Available from: http://dx.doi.org/10.1093/cid/civ473.

163. Yiş U, Kurul SH, Cakmakçi H, Dirik E. *Mycoplasma pneumoniae*: nervous system complications in childhood and review of the literature. *Eur J Pediatr.* 2008;167(9):973−978. Available from: http://dx.doi.org/10.1007/s00431-008-0714-1.

164. Meyer Sauteur PM, van Rossum AMC, Vink C. *Mycoplasma pneumoniae* in children: carriage, pathogenesis, and antibiotic resistance. *Curr Opin Infect Dis.* 2014;27(3):220−227. Available from: http://dx.doi.org/10.1097/QCO.0000000000000063.

165. Kawai Y, Miyashita N, Kubo M, et al. Nationwide surveillance of macrolide-resistant *Mycoplasma pneumoniae* infection in pediatric patients. *Antimicrob Agents Chemother.* 2013;57(8):4046−4049. Available from: http://dx.doi.org/10.1128/AAC.00663-13.

166. Gücüyener K, Simşek F, Yilmaz O, Serdaroğlu A. Methylprednisolone in neurologic complications of Mycoplasma pneumonia. *Indian J Pediatr.* 2000;67(6):467−469.

167. Kammer J, Ziesing S, Davila LA, et al. Neurological manifestations of *Mycoplasma pneumoniae* infection in hospitalized children and their long-term follow-up. *Neuropediatrics.* 2016;47(5):308−317. Available from: http://dx.doi.org/10.1055/s-0036-1584325.

168. Khan R, Vandelaer J, Yakubu A, Raza AA, Zulu F. Maternal and neonatal tetanus elimination: from protecting women and newborns to protecting all. *Int J Womens Health.* 2015;7:171−180. Available from: http://dx.doi.org/10.2147/IJWH.S50539.

169. Thwaites CL, Beeching NJ, Newton CR. Maternal and neonatal tetanus. *Lancet Lond Engl.* 2015;385(9965):362−370. <http://dx.doi.org.10.1016/S0140-6736(14)60236-1>.

170. Dalal S, Samuelson J, Reed J, Yakubu A, Ncube B, Baggaley R. Tetanus disease and deaths in men reveal need for vaccination. *Bull World Health Organ.* 2016;94(8):613−621. Available from: http://dx.doi.org/10.2471/BLT.15.166777.

171. Aronoff DM. *Clostridium novyi, sordellii,* and *tetani*: mechanisms of disease. *Anaerobe.* 2013;24:98−101. Available from: http://dx.doi.org/10.1016/j.anaerobe.2013.08.009.

172. Manga NM, Dia NM, Ndour CT, et al. [Maternal tetanus in Dakar from 2000 to 2007]. *Med Trop Rev Corps Sante Colon.* 2010;70(1):97−98.

173. Woldeamanuel YW, Andemeskel AT, Kyei K, Woldeamanuel MW, Woldeamanuel W. Case fatality of adult tetanus in Africa: systematic review and meta-analysis. *J Neurol Sci.* 2016;368:292−299. Available from: http://dx.doi.org/10.1016/j.jns.2016.07.025.

174. Barlow JL, Mung'Ala-Odera V, Gona J, Newton CR. Brain damage after neonatal tetanus in a rural Kenyan hospital. *Trop Med Int Health.* 2001;6(4):305−308.

175. Flowers MW, Edmondson RS. Long-term recovery from tetanus: a study of 50 survivors. *Br Med J.* 1980;280(6210):303−305.

176. Chalya PL, Mabula JB, Dass RM, Mbelenge N, Mshana SE, Gilyoma JM. Ten-year experiences with tetanus at a tertiary hospital in Northwestern Tanzania: a retrospective review of 102 cases. *World J Emerg Surg.* 2011;6:20. Available from: http://dx.doi.org/10.1186/1749-7922-6-20.

177. Sathirapanya P, Sathirapanya C, Limapichat K, Setthawacharawanich S, Phabphal K. Tetanus: a retrospective study of clinical presentations and outcomes in a medical teaching hospital. *J Med Assoc Thail Chotmaihet Thangphaet.* 2009;92(3):315−319.

20

Viral Diseases

François Denis[1,2] *and Sébastien Hantz*[1,2]

[1]INSERM UMR 1092, Limoges, France [2]Laboratory of Bacteriology-Virology, CHU Limoges, Limoges, France

OUTLINE

20.1 INTRODUCTION

Viruses can be considered according to their route of entry into the body (Table 20.1) and the majority of central nervous system (CNS) infections in the tropics are due to viruses transmitted by arthropod vectors. For others, which are found worldwide, there are specific issues in the tropics. Whatever the route of entry of viruses into CNS (hematogenous, olfactive or neural), clinical signs are dependent on the subsequent spread of virus within the tissues. Viruses can affect any part of the nervous system, but the tropism of viruses for different cells determines the characteristic clinical signs and several manifestations.

Several neurologic manifestations were associated with specific viruses. For example, acute or chronic meningitis and enterovirus, Guillain–Barré syndrome and enterovirus or Zika virus, paralysis and poliovirus, tropical spastic paraparesis and human T-lymphotropic virus (HTLV-1), progressive multifocal leucoencephalopathy and JC virus (JCV) or BK virus (BKV), etc. But the broad spectrum of clinical manifestations in the CNS poses for the clinician not only a problem of prompt diagnosis but also of treatment to allow recovery with a minimum of sequelae.

The diagnosis of viral families and species (Fig. 20.1) may be important in reducing the spread or perhaps eradicating the disease using specific measures and vaccination, if available (measles, poliomyelitis, dengue, yellow fever, Japanese encephalitis, etc.) (Table 20.2).

TABLE 20.1   Routes of Entry of Neurotropic Viruses

Route of entry	Example
Inoculation	
Arthropod bite	Arboviruses
Animal bite	Rabies
Blood transfusion	CMV
Transplantation	Rabies, CMV, HSV, EBV
Respiratory	Influenza
Enteric	Poliovirus
Venereal, blood contamination	HIV, HSV, CMV
Placenta	CMV
Saliva	CMV, EBV, VZV

CMV, cytomegalovirus; EBV, Epstein–Barr virus; HIV, human immunodeficiency virus; HSV, herpes simplex virus; VZV, varicella zoster virus.

20.2 HERPES SIMPLEX 1 AND 2

20.2.1 Epidemiology

Herpes simplex viruses (HSVs) have a worldwide distribution with a variable distribution according to the type and the geographical area, the prevalence of HSV1 being generally higher than that of HSV2. Man is the only known natural host and source of the virus. Rare cases of primary infection have been reported in adults. Clinically, HSV1 is dominant in the orofacial lesions and typically sets its latency in the trigeminal ganglion, while HSV2 is more often found in genital lesions establishing its latency in the sacral ganglia. However, both viruses can infect both anatomical sites. HSV is transmitted by direct contact with a subject excreting the virus during a primary infection, a

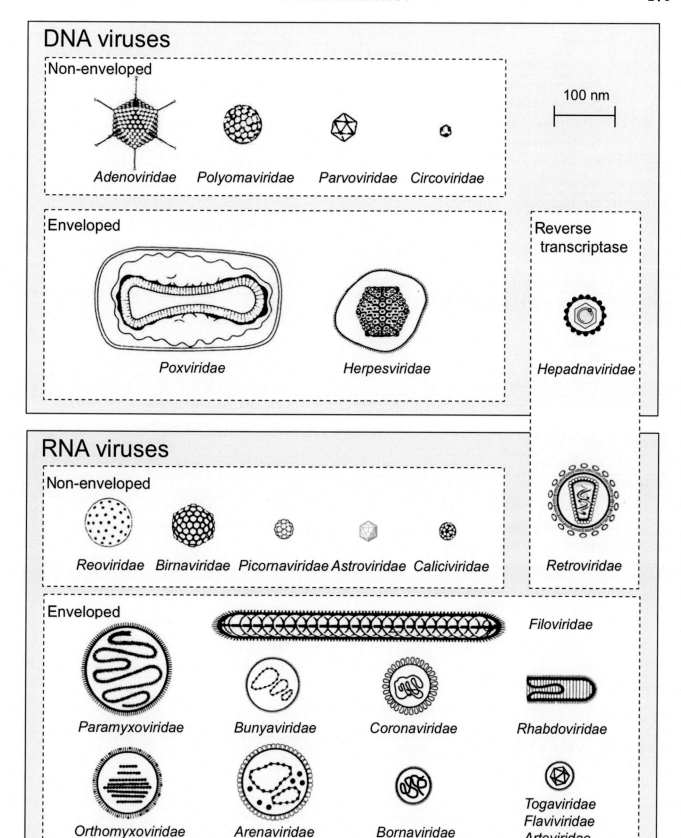

FIGURE 20.1 Approximate size and shape of vertebrate viruses; surface views and cross-sections.

TABLE 20.2   Viral CNS Infections

Groups	Viruses	Geography
HERPESVIRUS (*FAMILY:* **HERPESVIRIDAE**)		
	Herpes simplex virus type 1	Sporadic, global
	Herpes simplex virus type 2	Global, important in people with HIV
	Varicella zoster virus	Increased importance in areas with high HIV prevalence
	Epstein–Barr virus	Increased importance in areas with high HIV prevalence
	Cytomegalovirus	Increased importance in areas with high HIV prevalence
	Human herpes virus 6	Global
ENTEROVIRUS (*FAMILY:* **PICORNAVIRIDAE**)		
	Enterovirus 70	Outbreaks in India in the past
	Enterovirus 71	Especially important in southeast Asia
	Poliovirus	Persists in a few countries of Asia and Africa
	Coxsackieviruses, Echoviruses, Parechovirus	Global
PARAMYXOVIRUS (*FAMILY:* **PARAMYXOVIRIDAE**)		
	Measles virus	Important in areas with low vaccination rates
	Mumps virus	Important in areas with low vaccination rates
	Nipah virus	Malaysia, Bangladesh
RETROVIRUS (*FAMILY:* **RETROVIRIDAE**)		
	Human immunodeficiency virus	Global, but especially important in Africa and Asia
	Human T lymphotropic virus type 1	Especially important in Caribbean, Japan
POLYOMAVIRUS		
	JC/BK viruses	Global
LYSSAVIRUS		
	Rabies, other Lyssa viruses	Africa, Asia, parts of Americas
	Other Lyssa viruses	Localized to different geographical areas
ARBOVIRUSES (MOST ARE ALSO ZOONOTIC)		
FLAVIVIRUS (*FAMILY:* **FLAVIVIRIDAE**)		
	West Nile virus	Africa, parts of Asia, Southern Europe, Americas, Australia
	Japanese encephalitis virus	Asia, Pacific
	Zika	Pacific parts of Americas
	Tick-borne encephalitis	Northern Europe, Northern Asia
	Dengue	All countries between tropics of Cancer and Capricorn
ALPHAVIRUS (*FAMILY:* **TOGAVIRIDAE**)		
	Western, Eastern and Venezuelan equine encephalitis virus	Americas
	Chikungunya	Africa, Asia, Pacific, Australia, Southern Europe
BUNYAVIRUS (*FAMILY:* **BUNYAVIRIDAE**)		
	Lacrosse virus	America
COLTIVIRUS		
	Colorado tick fever	America
VESICULOVIRUS		
	Chandipura virus	India
OTHERS (RARER CAUSES)		
	Influenza viruses, adenovirus, parvovirus B19, lymphocytic choriomeningitis virus, rubella virus	Global

recurrence, or an asymptomatic viral shedding. Vertical transmission is also possible, but rare. Contact sports players (wrestling, rugby) can be contaminated by HSV1, but only occasionally. After oral primary infection, the duration of excretion is an average of 8 days but can reach 20 days. It is shorter in case of recurrence. The seroprevalence of HSV1 infection in adults is 70%–90% in almost all areas with low socio-economic level, where acquisition of virus occurs early in childhood. It contrasts with the European situation where the acquisition is later: the prevalence is closer to 60%–70% in France[1] and Germany.[2] Alongside oro-facial forms that remain the most common, a significant increase in genital herpes associated with HSV1 was described (15%–40% depending on the study), especially in women, related to oral–genital sex.[3] This epidemiological evolution increases the risk of genital primary infection with HSV1 during pregnancy and therefore theoretically that of neonatal infection HSV1. In addition, genital primary infection with HSV1 is more symptomatic than that associated with HSV2. In the general population, 60%–85% of genital herpes (recurrences + primary infections) are due to HSV2. More than 80% of primary genital HSV infections are asymptomatic or unrecognized. HSV2 infections occur at the start of sexual activity and mostly continue during the first two decades of sexual life. In France, the prevalence was estimated at 15.5% in the adult population (35–60 years), which is slightly less than the 21% found in an American study of the 1990s.[4] It may be higher in certain risk groups such as homosexuals or on other continents where it can reach 30%–40% in sub-Saharan Africa, or even up to 80% among sex workers. Prevalence in the general population in developing Asian countries appears to be lower (10%–30%).[5] HSV2 infection risk factors are female gender, early first sexual intercourse, number of sexual partners, history of sexually transmitted diseases, human immunodeficiency virus (HIV) infection, and low-socioeconomic level. Genital primary infection with HSV-2 is most commonly asymptomatic or unrecognized. It occurs 2–20 days (average 6–7 days) after the transmission. Symptoms of primary infection are more severe in women. The frequency of recurrences is higher in cases of HSV2 genital herpes that HSV1.

20.2.2 CNS Diseases Caused by HSV

Herpetic encephalitis can occur at any age but a peak incidence at age 50 is generally described. Its incidence of one to four cases per year per million inhabitants is low compared to the seroprevalence of herpes infection in the general population.[6] However, HSV remains the leading cause of meningoencephalitis in adults.[7] It is usually due to HSV1 (90%) whereas HSV2 is found more frequently in the neonatal forms. HSV encephalitis has been reported from Africa to Asia, where the incidence is thought to be similar to the rest of the globe. HSV2 is a frequent cause of viral meningitis (5%–20%), previously called Mollaret meningitis. The evolution is usually benign but recurrences may occur (20% over a lifetime) as well as complications such as meningo-radiculitis or more exceptionally ascending myelitis, rhombencephalitis, or encephalitis.[8,9]

Vertical transmission of HSV from mother to child causes serious morbidity and mortality in neonates. HSV2 is more often concerned but prevalence of HSV1 encephalitis of the neonates increases. Neonatal HSV can be acquired intra-uterine, perinatally or postnatally, but perinatal infection accounts for about 85% of infections.

20.2.3 Pathogenesis

The entrance door in the CNS classically described is reactivation, with migration from the trigeminal ganglion but more recently an entrance through the olfactory route direct migration to the temporal cortex was suggested. HSV1 encephalitis can also be observed in young children but more frequently during the primary infection. In this case, a deficiency of intrinsic/innate immunity, including the toll-like receptor 3–interferon (TLR3-IFN) pathway is questioned.[10] This is an extremely serious form of herpes infection resulting in necrotizing and hemorrhagic acute encephalitis, typically unilateral, predominantly temporal to frontal or temporoparietal. Symptoms often begin in adults and older children with a prodromal phase of a few days with fever, headache, ear, nose and throat (ENT) and digestive symptoms, whereas convulsions with fever are most often found in infants. At phase state, there is fever, meningeal syndrome, accompanied by various signs of brain injury (headache, altered consciousness, behavioral disorders, hallucinations). Evolution without treatment leads to death in 80% of cases with significant neurological sequels in survivors. When treatment with aciclovir IV is started early, the prognosis is more favorable, even if sequels can be observed in almost 30% of cases. Personality or behavioral impairments are found in 45% of adult survivors.

HSV2 meningitis occurs most often during a genital primary infection but is also observed during reactivation, more frequently in women than men and is likely associated with intense viral replication and viremia.

20.2.4 Diagnosis

In the case of neurological symptoms, a lumbar puncture should be performed as an emergency. HSV infection in cerebrospinal fluid (CSF) is diagnosed by polymerase chain reaction (PCR), with sensitivity above 90% and specificity near 100%. In the case of a negative result when the sample was taken in the first few days of illness, it should be repeated. If the patient has presented late and PCR is negative, then testing the CSF for HSV-specific antibodies can be helpful. In neonatal HSV infection, neonates may have only mucocutaneous or systemic symptoms. However, there may also be CNS infection even without symptoms. Thus, CSF analysis by PCR has to be performed for all suspicions of neonatal HSV infection. Detection of HSV in the blood can also be useful in such cases. In CNS HSV infection, CSF highlights a variable cellularity (10−100 cells/mm^3, predominantly lymphocytes), normal glucose ratio and moderately high protein level. In children, especially young children, there may be no pleiocytosis.[11,12] An electroencephalogram (EEG) is usually abnormal with slow waves of short periodicity and high amplitude in temporal or pseudoperiodic in frontotemporal regions. Brain scans usually find a hypodensity in the temporal region suggestive of edema but this aspect may be delayed (>48 hours) while magnetic resonance imaging (MRI) is a more sensitive technique showing high signal changes in temporal regions on T2-weighted images.[13] In the early phase, an intrathecal secretion of interferon can be detected; although not specific, this confirms the diagnosis.

20.2.5 Prevention and Treatment

HSV encephalitis is a therapeutic emergency, which requires intravenous acyclovir without waiting for laboratory testing. The recommended dose is 10 mg/kg per 8 hours in adults and 20 mg/kg per 8 hours among newborns for 14−21 days because therpaeutic failures at 10 days were observed.[14] Any newborn with a suspicion of neonatal herpes with or without mucocutaneous lesions should be treated as an emergency. The currently recommended initial treatment is intravenous acyclovir (60 mg/kg per day divided into three doses every 8 hours), 14 days in the case of a localized form and 21 days for a disseminated form.[15,16] The child must be isolated because of the risk of contamination of other newborns. Following treatment with oral acyclovir at 300 mg/m^2 per day, three times a day for 6 months showed improved neurological development versus placebo.[17] The delay in the implementation of treatment has a significant impact on mortality: in all age groups, early treatment (<24 hours after admission) is associated with lower death rate (6.6%) versus 9.5% for delayed treatment (>24 hours up to 7 days after admission), this difference is even more pronounced among infants of less than 7 days of age (8.8% vs 16.1%).[18]

To date, no vaccine is available against HSV infection. Use of a condom can protect from genital HSV contamination but protection against HSV1 infection is futile because the prevalence of the virus is very high and many people excrete viral particles without symptoms, leading to a rise in contaminations.

20.3 VARICELLA ZOSTER VIRUS

20.3.1 Epidemiology

Varicella zoster virus (VZV) is the most widespread virus of the *Herpesviridae* family. Primary infection happens most of the time during childhood and causes diffuse vesicular rash without healthy skin intervals named varicella or chickenpox. Secondarily, the virus becomes latent in the sensory dorsal root of the CNS and may reactivate, most often in elderly (>60 years old) or immunocompromised patients. This reactivation called zoster is a local painful eruption of vesicles localized in a dermatome. Seroprevalence of VZV is nearly 95% in the world. In tropical climates, VZV seroprevalence reflects a higher mean age of infection than in temperate climates, with higher susceptibility among adults. Seroprevalence of adolescents and young adults has ranged from 10% to 20% in St Lucia to 80%−100% in Taiwan and Calcutta. Among tribal adults in Eritrea, 44% were seropositive. CNS syndromes may happen after primary infection or reactivation. Transmission of VZV occurs essentially from children with chickenpox via respiratory droplets and skin lesions until complete cure. However, virus shed from zoster lesions (shingles) is also infectious. Although predominantly a febrile rash illness, the overall incidence of CNS disease in chickenpox is estimated to be 1/10,000 to 3/10,000 clinical cases.[19] The most commonly reported CNS manifestations are acute encephalitis and cerebellar ataxia. Encephalitis may occur in approximately 1/10,000 to 2/10,000 cases of clinical chickenpox.[20] Although most total chickenpox cases occur in children, incidence of neurologic disease associated with chickenpox is highest in adults over age 20, and infants <1 year.[19] Cerebellar ataxia is somewhat more common, affecting approximately 1/4000 cases.[19] Case fatality from varicella encephalitis varies considerably, ranging from 5% to 35%, and long-term sequelae may be seen in 10%−20% of survivors.[19]

20.3.2 Pathogenesis

Neurological complications may be observed after chickenpox as well as zoster. However, frequency and clinical manifestation of these complications vary between both forms of the infection.

CNS complications of chickenpox are rare, occurring in less than 1% of children.[21] They include acute cerebellar syndrome with ataxia, headache, vomiting and lethargy, encephalitis with fever, vomiting, reduced consciousness and seizures and other complications such as myelitis, aseptic meningitis, stroke and rarely choreoathetosis, facial nerve palsy and Guillain—Barré syndrome.[22]

The most common neurologic disease produced by any of the herpes viruses is zoster (shingles). A dermatomal distribution, vesicular, erythematous rash, and pain characterize reactivation of VZV from cranial nerve ganglia (usually trigeminal or geniculate), or dorsal root ganglia (usually thoracic level). Frequency of reactivation and neurological complications depend on age, HIV status, and level of immunosuppression. Interestingly, these complications may occur without any cutaneous evidence of zoster.

Postherpetic neuralgia (PHN) is defined as dermatomal-distribution pain persisting for more than 3 months after zoster. Age is the most important factor in predicting its development. Among persons >50 years of age, the incidence of PHN in zoster patients is 18%; in 80-year-old individuals, the incidence is 33%.[23]

Cranial neuropathies are also described. Ramsay Hunt syndrome (herpes zoster oticus) is defined as an acute peripheral facial neuropathy associated with erythematous vesicular rash of the skin of the ear canal, auricle, and/or mucous membrane of the oropharynx.[24] In addition to the painful shingles rash, Ramsay Hunt syndrome may cause facial paralysis and hearing loss in the affected ear.[25]

VZV encephalitis occurs essentially in elderly people and patients with HIV, 1—3 weeks after rash but may be also observed in the absence of rash. Cytology of CSF shows often a pleiocytosis with lymphocytes (but low number of cells is possible), normal glucose ratio and moderately high protein level. Interestingly, VZV stroke-related syndromes have been reported in immunocompetent elderly people following herpes zoster ophtalmica.[26] A similar VZV vaculopathy syndrome seems to be an important cause of stroke in children. Both manifestations are due to affection of major arteries. Small arteries may also be affected inducing vasculatis most often in immunocompromised patients (HIV or organ recipients), and these forms can be observed weeks or months after rash.

20.3.3 Diagnosis

When neurological symptoms are contemporary to dermatological lesions, the virus may be isolated from the vesicles. VZV-antigen detection may be performed by direct immunofluorescence assays whereas liquid of vesicles may be cultured on fibroblast cells to detect cytopathic effects after nearly 7 days.

CSF has to be sampled for PCR assays, which is the most sensible test for detecting VZV during CNS complications. In a study of 662 individuals, VZV DNA was detected in the CSF of six immunocompetent patients.[27] Four manifested meningitis (three had associated zoster), and two had clinical features of brainstem encephalitis. Overall, large survey studies using CSF PCR generated from several countries demonstrate that most patients with detectable VZV DNA in CSF are immunocompetent. Most manifest aseptic meningitis or encephalitis (cerebral or brainstem), but infections are not always associated with zoster rash. CSF may be cultivated but VZV may have some difficulties in cultivation. Detection of intrathecal synthesis of VZV specific antibodies allows confirmation of the diagnosis when VZV DNA is not detectable. Diagnosis of VZV-induced stroke syndrome relies on association of MRI showing segmental restriction of flow and analysis of CSF showing pleiocytosis with elevated protein. PCR on CSF is often negative.

20.3.4 Treatment

Treatment for zoster in people under the age of 50 years is based on symptoms. Analgesics are used to relieve discomfort. Antiviral drugs such as famciclovir 500 mg orally three times daily or valacyclovir 1 g orally three times daily during 7 days speed healing of the rash but are not essential. At any age, zoster in the distribution of the trigeminal nerve should be treated with valacyclovir, 1 g orally three times daily. In immunocompetent patients aged 50 years and older, treatment with both analgesics and antiviral drugs is recommended to prevent PHN. For immunosuppressed patients, antiviral treatment is administered intravenously (acyclovir) for severe forms for at least 7—10 days. PHN is difficult to manage and no universal treatment exists. The same medications used to treat zoster pain are also used for PHN. These include gabapentin, pregabalin, divalproex sodium, opioid analgesics, tramadol, tricyclic antidepressants, antiepileptics, and topical lidocaine patches. Immunocompetent patients with VZV vasculopathy should be treated with intravenous acyclovir, 10—15 mg/kg three times daily for 14 days. Immunocompromised patients or those with recurrent VZV vasculopathies may need a longer course. Since virus-infected arteries typically contain inflammatory cells, oral prednisone, 1 mg/kg daily for 5 days without taper, in conjunction with intravenous acyclovir is advised. Treatment of VZV meningoencephalitis, meningoradiculitis and cerebellitis is intravenous acyclovir as for VZV vasculopathy.

20.3.5 Prevention

A live attenuated vaccine against varicella has been available since 1995 in the USA based on the strain

Oka/Merck produced in Japan.[28] The vaccination schedule is two doses with the first dose administered at 12−15 months of age and the second dose at 4−6 years of age or at any age 3 months after the first dose. It is essentially used in developed countries and prevents severe or moderate forms of varicella but is not highly effective in preventing all varicella.[29]

More recently, a vaccine against zoster and PHN was developed with the same strain 14-times more concentrated than the varicella vaccine. The use of the zoster vaccine reduced the burden of illness due to herpes zoster by 61.1% ($p < 0.001$), reduced the incidence of postherpetic neuralgia by 66.5% ($p < 0.001$), and reduced the incidence of herpes zoster by 51.3% ($p < 0.001$).[30]

20.4 CYTOMEGALOVIRUS

20.4.1 Epidemiology

Human cytomegalovirus (CMV) is a strictly human virus, which infects 40%−70% of the population in industrialized countries.[31,32] The prevalence increases with population density (e.g., Finland: Helsinki 70% vs rural 56%) and low socioeconomic level (Fig. 20.2). It can reach 90 or even 100% in some populations, especially in developing countries, or highly populated areas of Asia (Manila: 100%, Sendai, Japan 86%, Entebbe, Uganda: 100%).[33,34] Transmission of CMV occurs through biological fluids such as saliva, urine, blood, and sperm. There are two main periods of contamination: childhood in nurseries and teens with first kisses.[35] But those not infected in childhood or teens can also become infected as adults through kissing, sexual intercourse, contact with urine of infected toddlers, or transplantation. Infection occurs earlier in poor socioeconomic backgrounds and in developing countries, owing to breast milk transmission and crowded living conditions. Clinically important infections occur in two major groups, those infected congenitally and those that are immunocompromised. The prevalence of CMV in congenitally infected infants is approximately 0.7% worldwide.[31] In developed countries, major complications are observed following

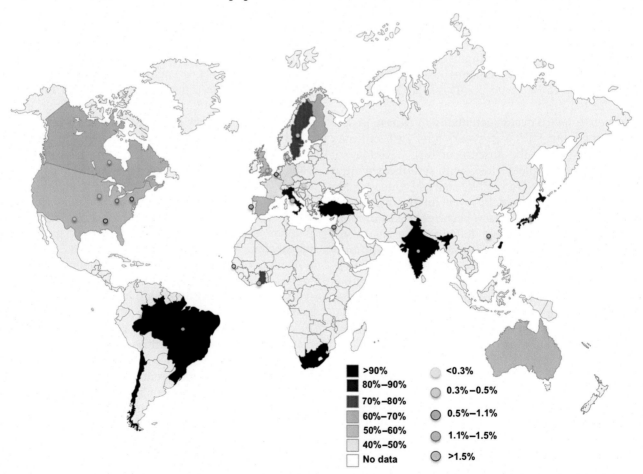

FIGURE 20.2 Worldwide human cytomegalovirus (CMV) seroprevalence rates among women of reproductive age (colored squares) and birth prevalence of CMV congenital CMV infection (colored circles). *Source: Adapted from Manicklal S, Emery VC, Lazzarotto T, Boppana SB, Gupta RK. The "silent" global burden of congenital cytomegalovirus. Clin Microbiol Rev. 2013;26(1):86−102 [33].*

primary infection in CMV-seronegative pregnant women (so that could target nearly 50% of pregnant women). But the incidence of in utero CMV infection is highly population dependent and parallels maternal seroprevalence, probably due to the fact that seroprevalence rates serve as a marker for the size of the reservoir of viruses. Thus, higher seroprevalence rates observed in developing countries lead to an increased chance of either reactivation within a host, reinfection of seropositive hosts (together these constitute non primary infection), or primary infection of seronegative hosts within the population. Immunocompromised patients are at risk of developing a CMV disease. It is a particularly important problem in HIV-seropositive people and transplant recipients (solid organ transplant (SOT) and hematopoietic stem cell transplant (HSCT) recipients). Even if the use of highly active antiretroviral therapy (HAART) was largely extended, most HIV patients in Africa or Asia do not have access to these treatments. When HIV patients arrive in the acquired immune deficiency syndrome (AIDS) state with low levels of CD4+ lymphocytes, opportunistic infections such CMV infection can occur. Transplant recipients can develop CMV disease in two ways: reactivation of CMV after latency in receiver cells following immunosuppressive therapy or contamination with the strain of the donor through the transplant.

20.4.2 Pathogenesis

In immunocompetent patients, CMV infection is asymptomatic most of the time or results in an influenza-like syndrome, particularly in children. In adults, there may be an infectious mononucleosis-like syndrome with abnormal lymphocytes, thrombopenia splenomegaly and impaired liver function. However, CNS diseases (encephalitis, myelitis, myeloradiculitis) may sometimes arise in immunocompetent patients, more often in middle-aged or elderly people.[27] Concerning congenital CMV, only one in 10 newborns infected in utero have obvious clinical igns of congenital infection.[36] But 10%−15% of those without clinical findings develop long-term neurological sequelae, such as lower IQs, behavioral disorders, minor incoordination, defects in perceptual skills and neural deafness.[37] The most serious clinical form is the cytomegalic inclusion disease that affects many organs and leads to death. Sensorineural hearing loss (SNHL) occurs in about 35%, cognitive deficits in up to two-thirds, and death in around 4% of children with symptomatic infection. Visual impairment is thought to occur in 22%−58% of symptomatic infants.[38,39] Far lower rates of sensory and cognitive sequelae have been reported in asymptomatic children. Hearing impairment has been reported in 7%−10% of such

infants,[37] while the risk for cognitive deficits has not been studied systematically and the risk for visual impairment appears to be negligible.[39] Overall (symptomatic and asymptomatic infections), permanent childhood hearing impairment is the most common complication. Primary infection or reactivation are most often more clinically symptomatic in immunocompromised than in immunocompetent patients. CMV disease has many different clinical presentations. Sub-acute or chronic encephalitis with confusion, disorientation, or lethargy have been described among 676 patients in the literature, but mostly in HIV infected patients (85%) whereas only 12% were observed in other immunocompromised patients.[40] Among HIV patients, CMV retinitis is the most frequent neurological manifestation followed by encephalitis, polyradiculitis and multifocal neuropathy.[41]

20.4.3 Diagnosis

Serological diagnosis is based on the dosage of the immunoglobulin G (IgG) and immunoglobulin M (IgM). Interpretation of serology results may be difficult when there is no prior serum to compare IgG levels. In case of primary infection, IgG and IgM are both detected in most cases, but the best proof of primary infection is the seroconversion between the two sera. Indeed, IgM can be detected during a reactivation.

Actually, detection of CMV DNA has to be performed by real-time PCR in blood when there is a suspicion of CMV infection and in CSF when signs of CNS infection are present. Quantification of CMV DNA allows the efficacy of antiviral treatment to be followed. Ventriculoencephalitis caused by CMV in advanced HIV disease is characterized by periventricular enhancement on imaging, with CSF pleiocytosis (characteristically neutrophil rather than lymphocytic).

Congenital CMV infection is suspected after ultrasound imaging in front of nonspecific abnormality such as periventricular calcification, ventricular dilatation, microcephaly and occipital horn anomalies,[42] and non-cerebral abnormalities such as echogenic bowel, intrauterine growth restriction, hepatomegaly, ascites, and cardiomegaly.[43] Ultrasound and, to a lesser extent, MRI are valuable tools to assess fetal structural and growth abnormalities, although the absence of fetal abnormalities does not exclude fetal damage. Amniocentesis with CMV-PCR performed on amniotic fluid, undertaken after 21−22 weeks' gestation, may determine whether maternofetal virus transmission has occurred. But in case of maternal infection without abnormality in imaging, detection of CMV DNA in amniotic fluid is not an indication of symptomatic fetal infection. Thus, only the detection of CMV (by culture or PCR) in urine or saliva of the neonates after birth proves the congenital infection.[44] Finally, CMV-PCR testing of newborn screening cards is particularly useful in retrospective determination of congenital CMV infection in previously undiagnosed infants presenting with late-onset symptoms and sequelae.[45]

20.4.4 Treatment and Prevention

The first therapeutic line against CMV is ganciclovir, administered intravenously, or its prodrug, valganciclovir, given orally. Treatment at full doses should be continued until symptom resolution and negativation of the DNAemia. Patients receiving treatment must be closely monitored for side effects to the drugs (neutropenia with ganciclovir), as well as for response. Drug-resistant CMV is a therapeutic challenge. The first alternative is foscarnet but a high level of renal toxicity is associated with the use of this drug. The nucleoside analog cidofovir may be used in case of foscarnet toxicity or resistance. Immune reconstitution, through reduction in immunosuppression contributes to viral control.

Despite a lot of vaccine trials, no CMV vaccine is available to prevent infection.

20.5 EPSTEIN—BARR VIRUS

20.5.1 Epidemiology

Epstein—Barr virus (EBV) is essentially known as the main cause of infectious mononucleosis. Seroprevalence of the virus is 90%—95% in adults worldwide. The acquisition of EBV begins early in childhood through saliva in daycare centers with a majority of asymptomatic infections and continues among teenagers with the first kisses. Teenagers and adults can develop infectious mononucleosis (symptomatic form of the primary infection). Neurological complications after symptomatic primary infection are reported in 1%—18% of patients. Studies have described several syndromes such as encephalitis, meningitis, transverse myelitis, cranial and peripheral neuropathies and radiculopathies.[46,47] EBV has an important pathogenic role in Burkitt's lymphoma, which is endemic in Africa (Fig. 20.3). The virus is involved in 97% of cases of Burkitt's lymphoma in Africa and Papua New Guinea.

In immunocompromised patients, EBV is responsible for the development of CNS lymphomas. Malignant lymphoma is the second most common neoplasm that occurs in association with AIDS. Approximately 80% of lymphomas arise systemically and 20% arise as primary CNS lymphomas.[48] The incidence of primary CNS lymphomas is increased several 1000-fold in AIDS patients.

EBV is an important pathogen in recipients of transplants (SOT/HSCT). Post-transplant lymphoproliferative disorders (PTLDs) show some similarities to classic lymphomas in the non-immunosuppressed general population. The overall prevalence of EBV-associated PTLD following SOT ranges from 1% to 20%, with rates varying according to the type of

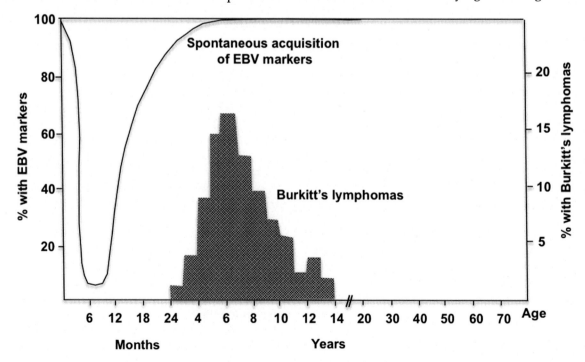

FIGURE 20.3 Correlation between kinetics of apparition of Epstein—Barr virus (EBV) serological markers and Burkitt's lymphoma.

organ transplanted, pretransplant EBV serostatus, and the age of the recipient.[49] Similar rates of incidence of EBV-PTLD are described after HSCT.[50] PTLD localized to the CNS is a rare but potentially fatal complication of immunosuppression for organ transplantation.[51]

20.5.2 Pathogenesis

Meningitis and encephalitis occurring during EBV primary infection have the same presentations as those from other viruses. Although encephalitis most often happens 1–3 weeks after infectious mononucleosis, it can occur before or during the acute febrile illness and occasionally occurs without any features of infectious mononucleosis.[52] Symptoms observed in the case of EBV CNS lymphoma are various and depend on the localization inside the CNS with a variety of focal or non-focal signs.

20.5.3 Diagnosis

Infectious mononucleosis diagnosis relies on serological assays: the Paul–Bunnel–Davidson test (detection of heterophilic antibodies) or specific EBV antibodies (viral capsid antigen IgM). For neurological complications of infectious mononucleosis, quantitative measurement of EBV DNA copies might be useful in assessing the cause of symptoms, intensity of infection, and in the follow-up of the disease. Low or undetectable EBV loads might indicate post-infectious complications rather than acute infection, since EBV loads are higher in patients with encephalitis compared to post-infectious neurological complications.[53] Sometimes, low viral loads in CSF are difficult to interpret and comparison with blood viral load may be helpful if the EBV detected is likely to be significant. Indeed, many patients with HIV have low viral loads due to the immunosuppression without true pathogenesis in CNS. However, when a CNS lymphoma is suspected, CSF exhibits high viral loads. CSF profile is similar to that from other causes of meningitis or encephalitis. In lymphomas, malignant cells can be found in CSF. MRI and tomodensitometry are essential for the diagnosis of lymphomas.

20.5.4 Treatment and Prevention

EBV primary infection does not need any treatment in most cases. If acute CNS complications occur, ganciclovir can be used even if the efficacy against EBV is limited. Treatment of EBV-induced lymphomas includes methotrexate, rituximab, corticoids, and whole-brain irradiation but prognosis is often bad. There is no licensed vaccine for prevention.

20.6 HUMAN HERPES VIRUS 6

Human herpes virus 6 (HHV6) infection is widespread and occurs in early childhood. By 2 years of age, most children have acquired primary HHV6 infection (median age: 8 months). Seroprevalence in adults is 70%–100% worldwide with considerable geographic variation, even within relatively close areas and in similar climates.[54] Transmission occurs essentially via the asymptomatic shedding of HHV6 in the secretions of family members and close contacts. Bone marrow transplantation is another way of possible transmission. A particularity of HHV6 is the ability to integrate its genome in human chromosome leading to a vertical transmission from parents to children.[55]

Diseases occurring after primary infection are rare and appear to be 'an accident' rather than the rule (80% of asymptomatic forms).[56] Clinical symptoms resemble an influenza-like disease with fever, sweats and chills, fatigue, malaise, occasional convulsions and the typical exanthema (roseola). Other diseases are observed in immunocompromised patients: hepatitis, myocarditis, and encephalitis. As with other herpes viruses, the encephalopathy associated with HHV6 is often attributable to the reactivation of a virus previously latent in human brain tissue. Previous reports on HHV6 encephalopathy dealt mainly with virus reactivation in immune-depressed older children. But HHV6 meningoencephalitis was already described in previously healthy children. Among three cases, two of the patients presented invalidating sequelae. In detail, one patient developed speech disturbance and the other persistent hemiplegia and bilateral visual deficit.[57] Complications are rare in healthy children. HHV6 is also suspected to play a role in pathogenesis of multiple sclerosis, but physiopathology of multiple sclerosis is complex and a combination of multiple factors is necessarily required.[58]

Diagnosis of neurological complications of HHV6 infection needs real-time PCR. If the genome is integrated, viral loads are often very high. Therefore, it is recommended to realize an HHV6 PCR on hair follicles to interpret a high viral load in biological fluids (PCR on hair follicles is always positive when virus is integrated). For clinical situations requiring treatment, drugs are those used for CMV infection (ganciclovir (GCV), foscarnet (FOS), and cidofovir (CDV)). No prevention is available.

20.7 ENTEROVIRUS

Human enteroviruses belong to the Enterovirus genus (which is a single-stranded RNA virus) in the

TABLE 20.3　Human Enteroviruses

Enterovirus genus	
Species	**Strains (old nomenclature)**
Human enterovirus A (>21 serotypes)	*Human coxsackievirus* A (CV-A2), 3–8, 10, 12, 14, and 16
	Human enterovirus 71 (EV-71), 76, 89, 90, 91, 92
Human enterovirus B (>59 serotypes)	*Human coxsackievirus* B1 (CV-B1), 2–6
	Human coxsackievirus A9 (CV-A9)
	Human echovirus 1 (EV-1), 2–27, 29–33
	Human enterovirus 69 (EV-69), 73–75, 77–88, 93, 97, 98, 100, 101, 106, 107
Human enterovirus C (>19 sérotypes)	*Human coxsackievirus* A1 (CV-A1), 11, 13, 17, 19–22, 24, 95, 96, 99, 102, 104, 105, 109
	Human poliovirus 1 (PV-1), 2 and 3
Human enterovirus D (>2 sérotypes)	*Human enterovirus* 68 (EV-68), 70, 94

TABLE 20.4　Neurologic Manifestations Associated With Human Enteroviruses

	Poliovirus	**CV-A**	**CV-B**	**EV**	**EV 68–71**
Encephalitis			+	+	+
			2	2, 6, 11, 19	71
Paralysis	+			+	+
					70–71
Meningitis	+	+	+	+	+
		7, 9	2, 5	4, 6, 7, 9, 11, 13, 30, 31	70–71

CV-A, *Human coxsackievirus* A; CV-B, *Human coxsackievirus* B; EV, enterovirus.

Picornaviridae family. The genus includes polioviruses, coxsackie viruses, echoviruses and newer enteroviruses (Table 20.3). Echoviruses 22 and 23 newly reclassified are no longer considered to be enteroviruses. Several neurologic manifestations can be associated with enterovirus from aseptic meningitis to encephalitis and acute flaccid paralysis (Table 20.4).

20.8 POLIOMYELITIS

20.8.1 Epidemiology

Poliomyelitis is caused by one of the three serotypes of poliovirus. Type 1 is epidemiologically the most important. Polioviruses are carried in the human gastrointestinal tract and transmitted via the fecal–oral route. Humans are the only natural hosts and reservoirs. In 1988, when the World Health Organization (WHO) announced that target for polio eradication was the year 2000, the number of cases was approximately 350,000 annually, across 25 countries. Polio will be eradicated, but in 2016 the global eradication had not been achieved. Pakistan, Afghanistan, and Nigeria are countries that have never been polio-free (Fig. 20.4).

20.8.2 Pathogenesis

Following ingestion, poliovirus replicates in the gut wall (pharynx, intestines) and adjacent deep lymph nodes, which leads to a viremia. The virus infects the CNS via the blood. The anterior horn cells of the spinal cord are susceptible to infection with poliovirus and are damaged or in several cases completely destroyed. The lesions may extend to the hypothalamus and thalamus. People remain most infectious immediately before and 1–2 weeks after onset of paralytic disease. Poliovirus is excreted for 3–6 weeks in feces. Paralytic poliomyelitis occurs in 0.1% of all poliovirus infections. Asymptomatic infections or infection with mild febrile illnesses are common only in young children.

Paralytic poliomyelitis is the classic flaccid paralytic syndrome, which may be spinal or bulbar and rarely encephalitic. The paralysis is flaccid and deep-tendon reflexes may be absent. The paralysis is characterized

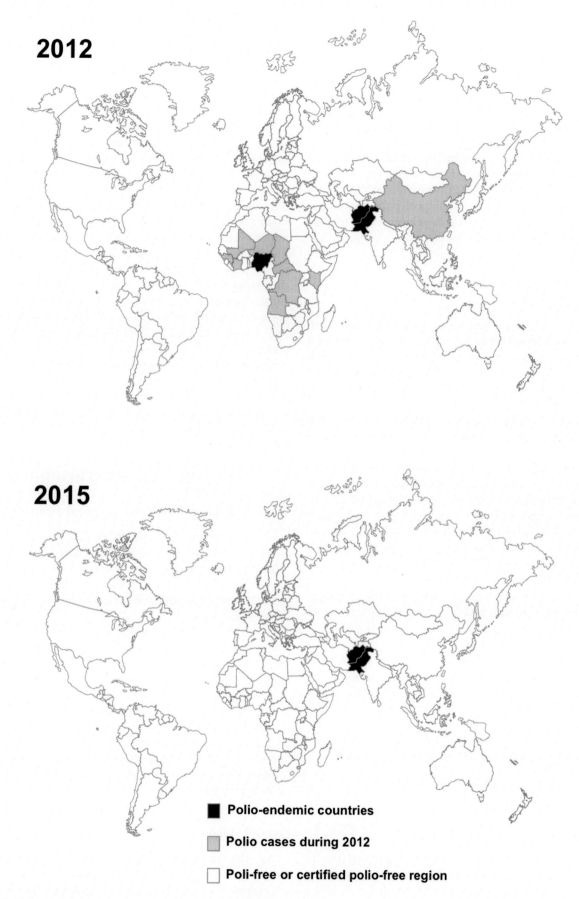

2012

2015

■ **Polio-endemic countries**

▨ **Polio cases during 2012**

□ **Poli-free or certified polio-free region**

FIGURE 20.4 Evolution of poliomyelitis eradication in the world between 2012 and 2015. *Source: World Health Organization/Global Polio Eradication Initiative (for 2012).*

by its asymmetrical distribution, which may range from a single portion of one muscle to quadriplegia.

The evolution of the paralysis is unforeseeable, inconstant and incomplete; an evaluation must be realized after 9 months. Mortality of spinal poliomyelitis ranges between 4% and 6% but in bulbar forms it ranges from 20% to 60%, depending of the quality of intensive care given. Residual motor deficit persists in 75% of patients ranging from minor disability to permanent flaccid paralysis. Poliovirus persistence is involved in the pathogenesis of post-polio syndrome (PPS). PPS is characterized by development of muscle weakness and atrophy at least 10 years after stabilization of motor functions.

20.8.3 Diagnosis

Typically a CSF pleocytosis is observed, often with elevated polymorphonuclear cells, slightly elevated protein and glucose ratio. But CSF punction can precipitate paralysis in the needed muscles and the exacerbation of the disease. Laboratory diagnosis of poliomyelitis is based on the research of poliovirus genome (reverse transcriptase (RT)-PCR), the isolation and culture of the poliovirus (using Hep-2C or other cells) from throat secretions during the first week of illness and from feces subsequently. However, the presence of virus alone in the stool may not be sufficient to establish a diagnosis in endemic areas. Poliovirus strains isolated were serotyped (1, 2, or 3) and studied by intratypic differentiation (ITD) between wild strains, oral vaccine strains and vaccine-derived poliomyelitis strains (VDPR). Molecular biology is likely to further improve the diagnosis of poliomyelitis and the strains study using PCR or genomic sequencing to permit reconstruction of individual chains of transmission.

The WHO has published the standard guide for the isolation and the characterization of polioviruses for laboratories in the WHO's Global Laboratories Network for Poliomyelitis eradication, and it is widely used in other diagnostic laboratories. Sporadic cases of paralysis due to non-polio enteroviruses (as coxsackie A7 and EV70) are clinically indistinguishable from poliomyelitis and virological diagnosis is essential.

20.8.4 Treatment and Prevention

It is well know that there is no specific drug for the treatment of poliomyelitis. Information on physiotherapy and rehabilitation protocols is available in standard textbooks. Orthopedic correction surgery, if required, is performed after 1 or 2 years when maximum spontaneous recovery is established. The mainstay of prevention is adequate vaccination with induction of neutralizing antibodies protecting against disease. Both inactivated polio virus (IPV, Salk) and live polio vaccines (oral poliovirus vaccines (OPV), Sabin) are effective. OPV is almost exclusively used in developing countries because of low cost and ease of administration. Administration of OPV among HIV-infected patients do not appear to be a risk factor for paralytic poliomyelitis caused by wild-type or oral vaccine strains, but HIV patients could be possible long-term excretory. IPV has been combined with DTaP, hepatitis B and Hib vaccines and those combinations are now widely used in countries including the United States for priming use in infants and toddlers and recommended for patients with immunodeficiency including HIV infection. Sabin strains can regain both neurovirulence and capacity to circulate; such circulating vaccine-derived poliovirus (cVDPVs) has caused around 20 outbreaks since 2000 and sequential vaccination strategies with OPV and IPV were studied. Difficulties of immunization in areas of conflict, poor compliance with immunization, emergency of VDPR strains and existence of long-term excretory delayed ultimately the wild-type virus eradication.

20.9 NON-POLIO ENTEROVIRUSES

20.9.1 Epidemiology and Clinical Features

Aseptic meningitis is the most common neurological manifestation of the non-polio enteroviruses (EVs). The etiology particularly among children, with the fecal—oral route and seasonality are dominated by only a few serotypes of echovirus (Enteric Cytopathic Human Orphan).

EV-70 infection is characterized by acute hemorrhagic conjunctivitis followed 2—5 weeks later by acute flaccid paralysis and by indirect transmission (fingers, fomites, objects).

EV-71 caused hand, foot and mouth diseases with associated aseptic meningitis and encephalitis. Large outbreaks across the Asia-Pacific region were observed (1.5 million people affected in Taiwan in 1988) but a circulation of the EV-71 was observed at a low level in Africa, Europe and America. A respiratory transmission is possible.

20.9.2 Diagnosis

In the majority of cases, the CSF examination reveals a lymphocytic pleiocytosis, though polymorphonuclear cells may predominate in the first day of the disease. Traditionally, EV meningitis has been diagnosed by isolating from the CSF, throat or stools. Clinical specimens are inoculated onto cell cultures. Cells are monitored daily for cytopathic effects and identified by serotype in neutralization tests using pools of specific antiserum. Currently, real-time PCR with commercial kits have begun to replace culture.

20.9.3 Treatment and Prevention

Pleconaril has broad activity against most EVs, but not against EV-71, which must be treated with intravenous immunoglobulin in severe infections. The prognosis of children with EV is generally good, but neurologic sequelae were observed in a quarter of EV-71 cases. There is no vaccine against any of the non-polio EVs.

20.10 PARAMYXOVIRUSES (MEASLES, MUMPS, AND NIPAH VIRUSES)

The *Paramyxoviridae* family (RNA virus) contains two sub-families: *Paramyxovirinae* and *Pneumovirinae*. *Paramyxovirinae* are divided into three genera implied in human neuronal infections: Morbillivirus (Measles—MeV), Rubulavirus (Mumps-MuV), and Henipavirus (Nipahvirus).

20.10.1 Measles

20.10.1.1 Epidemiology

Measles is a highly contagious disease (basic reproduction number $R°$ 12—18) whose virus is spread in respiratory secretions. Measles virus does not have a reservoir in nature other than humans. Before the development of an effective vaccine, 99% of people of 20 years of age have been affected. Worldwide, measles is a significant cause of morbidity and mortality. Thus, in 2000 there were an estimated 31—39.9 million cases worldwide, with an estimated 733,000—777,000 deaths, making measles the fifth most common cause of death in children under 5 years old (Fig. 20.5).

Measles virus replicates initially in the respiratory mucosa and regional lymphatic tissues. Virus may also disseminate to distant lymphoid tissue thanks to a brief primary viremia. Viral multiplication in the upper respiratory tract and conjunctives causes, after an incubation period of 10—12 days, the prodromal symptoms and Koplik spots. Viremia occurs towards the end of the incubation period, leading to further widespread dissemination of the virus to the lymphoid tissue and the skin. With the diffuse secondary

multiplication of virus, the prodromal symptoms are intensified and the typical red, maculopapular rash appears, first on the head and the face and then on the body extremities. Bronchopneumonia and otitis, with or without a bacterial component, are frequent complications of the disease. Giant cell pneumonia is a rare disease of debilitated children or those with immunodeficiency. Encephalomyelitis is the most serious complication, appearing about 5—7 days after the rash. Its incidence in most epidemics is about 1/2000 cases (higher in children over 10 years of age and in malnourished infants). The mortality rate of encephalomyelitis is about 10%—20% and permanent mental and physical sequelae have been reported in 15%—65% of survivors (some authors described two distinct presentations: acute and sub-acute encephalitis). Sub-acute sclerosing panencephalitis (SSPE), a progressive fatal complication of measles, is an example of a single virus inducing an acute disease and chronic illness separated by a long interval (many years) with only restricted synthesis and expression of viral genes. Brains from patients with SSPE display a degeneration of the cortex and especially the underlying white matter. Progression is inexorable to a vegetative state with dementia, decortication and decerebration. SSPE typically occurs in 8- to 10-year-old children. The incidence of SSPE is about 1 per million cases.

20.10.1.2 Diagnosis

Virus isolation, direct cytologic examination of clinical material, or demonstration of virus antigen can be used to diagnose measles but more recently RT-PCR can be used to identify measles virus RNA in urine, blood and saliva. Routinely, diagnosis is usually confirmed by IgM (detectable during about 4 weeks after the onset of the eruption) with or without IgG antibodiy detection in serum using enzyme-linked immunosorbent assay (ELISA). Serological tests are effective in identifying cases of SSPE. Patients with this disease have increased serum titers which are 10- to 100-times higher than those seen in the late convalescent-phase sera; there is also a pronounced local production of oligoclonal measles virus antibodies in the CNS.

20.10.1.3 Prevention and Treatment

No treatment is presently available. Live attenuated vaccines are widely used (strains Edmonston B or Schwarz). The rate of seroconversion after vaccination exceeds 90%. Two doses are required. The WHO recommends for most developing countries administration of measles vaccine at 9 months because of the high level of measles morbidity and mortality that occur in the first year of life. Many developed and

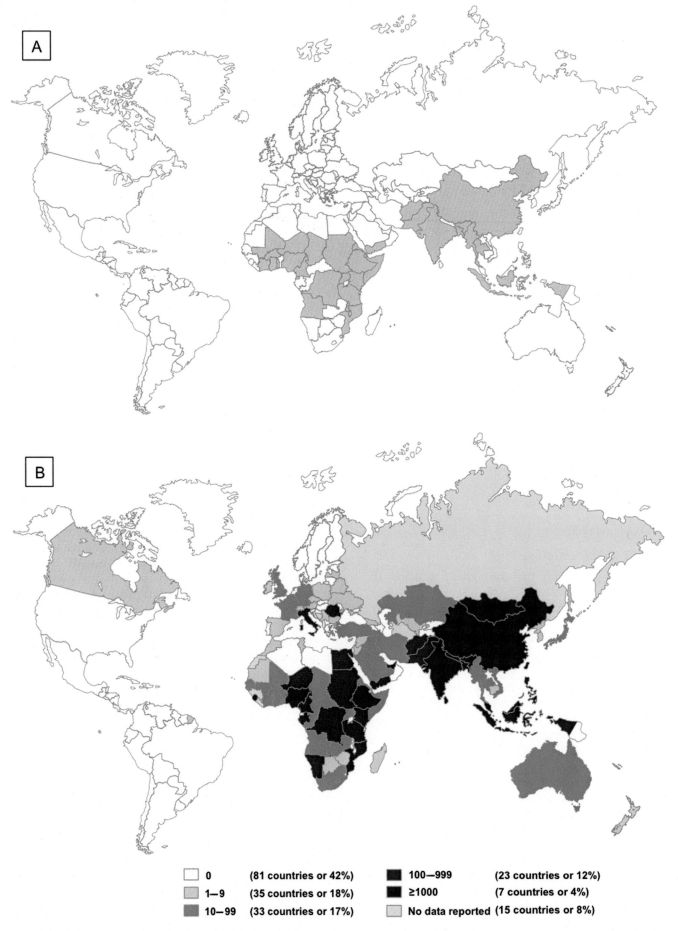

☐	0	(81 countries or 42%)	■	100—999	(23 countries or 12%)
▨	1—9	(35 countries or 18%)	■	≥1000	(7 countries or 4%)
▨	10—99	(33 countries or 17%)	☐	No data reported	(15 countries or 8%)

FIGURE 20.5   (A) Countries with the highest measles incidence rate. (B) Number of reported measles cases with onset date from October 2015 to March 2016. *Source: (A) Adapted from CEPED, French Center on Population and Development (2000).*

some developing countries now have schedules for two routine doses of measles vaccine. The frequency of occurrence of SSPE is reduced by a factor of at least 10 in vaccinated persons.

20.10.2 Mumps

20.10.2.1 Epidemiology

The natural infection is initiated by droplet spread. Mumps is usually acquired between 5 and 14 years of age. Mortality from mumps is related primarily to the complications of meningitis/encephalitis and orchitis.

20.10.2.2 Pathogenesis

After viral replication in nasal or upper-respiratory mucosal epithelium, the virus spreads to lymph nodes with subsequent transient plasma viremia. The parotid glands are often involved. A wide variety of organs have been involved including the testicles, ovaries, liver, and CNS. The onset of mumps meningitis is marked by fever, with vomiting, neck stiffness, headache, and lethargy. Seizures occur in 21%–30% of patients with CNS symptoms. Most cases of mumps meningitis resolve without sequelae. Viral invasion of the CNS occurs across the choroid plexus, encephalitis is observed in 0.02%–0.3% of cases. Although rare, mumps meningo-encephalitis may be fatal.

20.10.2.3 Diagnosis

The clinical diagnostic of mumps is seldom problematic in the presence of parotiditis. Laboratory diagnosis includes determination of virus-specific IgM and IgG levels. Mumps meningitis can be confirmed on the basis of CSF cytology and increase in protein levels. Direct detection of mumps virus in clinical specimens (oropharyngeal swabs or CSF) by RT-PCR has been reported.

20.10.2.4 Prevention and Treatment

No treatment is presently available. Live attenuated vaccines are widely used with the combination measles–mumps–rubella (MMR).

20.10.3 Nipah Virus

20.10.3.1 Epidemiology

Nipah virus first appeared in Malaysia and Singapore in 1998–99, when it caused disease in humans and pigs. The natural reservoir of Nipah virus is fruit bats (genus Pteropus) and serological evidence of Nipah infection has been demonstrated in several species of bats in the world (China, Cambodia, Thailand, India, Madagascar, and Ghana). Pigs became infected possibly through eating fruit, then the virus was excreted in the pigs' urine and respiratory secretions; humans became infected by contact with animals. More than 250 people were affected with more than 100 deaths. Nipah virus has caused subsequent outbreaks in Bangladesh from 2001 to 2007 and Bengal, India (2001).

20.10.3.2 Pathogenesis and Pathology

Humans become infected with Nipah virus by inhalation of respiratory secretions of pigs or bat excreta or ingestion of contaminated juice. After an incubation period ranging from 7 to 40 days, a smaller number of patients may present with atypical pneumonia, but one-third of patients have meningismus.

Encephalitis is characterized by reduced levels of consciousness, myoclonus, areflexia and hypotonia. Several reports have shown that patients who initially have mild symptoms or are asymptomatic can present encephalitis several months after exposure. The mortality has been from 35% to 70% according to the outbreaks.

20.10.3.3 Diagnosis

Nipah virus is a biosafety level 4 pathogen. Direct detection of Nipah virus in clinical specimens (CSF) by RT-PCR or serological detection of IgM by ELISA confirm the diagnosis.

20.10.3.4 Prevention and Treatment

Education of humans about the risk factor is important. There is no vaccine, but ribavirin can reduce mortality.

20.11 RETROVIRUS

Among the family of the *Retroviridae* two genuses were implicated in human diseases: the Deltaretrovirus with the HTLV-1 and -2 and the Lentivirus with HIV-1 and -2.

All the retroviruses are RNA viruses but after reverse transcription, viral DNA becomes incorporated into the host cell DNA and is called provirus. Whereas HIV eventually destroys the cells that it infects, HTLV induces their proliferation.

20.11.1 Human T–Lymphotropic Virus

20.11.1.1 Epidemiology

HTLV-1 infects 15–25 million people worldwide as estimated by seroprevalence. The virus is endemic in southern Japan, the Caribbean, Central and South America, the Melanesian islands, and sub-Saharan Africa (Fig. 20.6). In these areas seroprevalence ranges from 0.2% to 50%, increasing with age, sex (women more than 30 years of age), and ethnic groups.[59]

The virus is primarily transmitted by breast-feeding (between 20% and 25% of children with sero-positive mothers are believed to be infected), spread via blood transfusion or blood exposition and sexual intercourse (sexual transmission is 150-times more efficient from men to women). The same routes as HTLV-1 transmit the HTLV-2 but the incidence is very low.

20.11.1.2 Pathogenesis

HTLV-1 is associated with diseases in only approximately 2% of those infected. The clinical syndrome adult T-cell leukemia/lymphoma (ATLL) was first described in Japan. In the HTLV-1 endemic areas of Japan or Jamaica, the annual incidence of ATLL has been reported to be 3.5/100,000 and 2/100,000, respectively. The acute form of ATLL is rapidly fatal. HTLV-1 associated myelopathy (HAM) and HTLV-1-positive tropical spastic paraparesis (TSP) are clinically and pathologically identical and it is currently recommended that the disease entity be known under the acronym HAM/TSP.

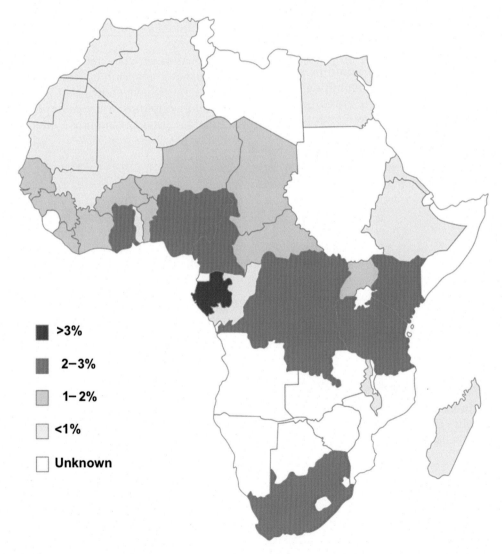

FIGURE 20.6 Human T-lymphotropic virus (HTLV)-1 seropositivity rate in Africa (established after analysis of bibliographic data using confirmed results).

The term "TSP" was introduced in 1969 but the association of TSP with HTLV-1 was first reported in Martinique. In Africa, only a small proportion of TSPs are HTLV-1 positive (10%–20%, personal data). The main neurological manifestation was chronic spastic paraparesis, which usually progresses slowly, remaining static at times after the initial progression, with weakness and hyper-reflexia in the lower limbs, often with clonus and a positive Babinski's sign and bladder disturbance. Minimal sensory impairment was also frequent. The onset of HAM is insidious, a median 3 years after infection. In a large series in Martinique, the median time from the disease onset to use of one walking stick was 6 years and use of a wheelchair was 21 years. HAM is more likely acquired via blood transfusion than breast-feeding.

Neuronal damage is caused by release of inflammatory cytokines, rather than direct viral invasion of neurons. TSP is characterized by a chronic myelitis with degeneration of the posterior column and pyramidal tracts and with loss of myelin.

20.11.1.3 Diagnosis

ELISA is the most frequently used screening test, but false positivity is frequent and western blot must be imperatively realized with strict criteria of positivity, and possibility to distinguish HTLV-1 and 2. In HAM, anti-HTLV-1 antibodies are detectable in the CSF with a high CSF/serum ratio. Cell culture techniques for HTLV are laborious and rarely performed, PCR-based testing to detect proviral DNA in lymphocytes is a diagnostic test that can also differentiate HTLV-1 and 2 and provide quantification of proviral DNA load (prognostic factor).

20.11.1.4 Prevention and Treatment

Screening of blood and organ donors, reduction of needle use, use of condoms, and discouragement of breast-feeding for infants of HTLV-positive mothers reduce the spread of HTLV. Vaccine assays on simians were encouraging but limited for humans by socioeconomic situations. Several treatments have been tested in a small number of patients with HAM: transcriptase inhibitors, interferon alpha and beta with some benefit in clinical neurologic parameters without randomized studies.

20.11.2 Human Immunodeficiency Virus

AIDS was first described in 1981 in the USA. The disease rapidly spread to almost all countries in the world.

20.11.2.1 Epidemiology

HIV is transmitted through contaminated blood or blood products, by use of contaminated needles or surgical instruments, via sexual contact and from mother-to-child.

In 2012, an estimated 35.3 millions people were living with HIV worldwide and new HIV infections (2.3 millions) were down 20% compared to 2001. The regions most affected by HIV, sub-Saharan Africa and the Caribbean, have seen a significant increase in antiretroviral therapy (ART) and a decrease in the number of new HIV infections, while for other regions, such as the Middle East, North Africa, Eastern Europe, and Central Asia, the number of new HIV infections has become worrisome (Fig. 20.7).

20.11.2.2 Pathogenesis

AIDS, a clinical syndrome associated with a progressive deterioration of the immune status of the individual, is characterized by the progressive depletion of the CD4+ T-lymphocytes population. A mild seroconversion illness, with fever, headache, malaise and lymphadenopathy is thought to occur when most people acquire HIV, though in the majority, the cause is not recognized at the time. Aseptic meningitis is less common (10% of patients) and in a small minority of these, there is also encephalitis. HIV enters the brain at the time of the primary HIV infection. Crossing the barrier, monocytes may become activated perivascular macrophages. HIV replicates in these cells, as well as in microglia and astrocytes. At the clinical level, this is characterized by a wide spectrum of clinical symptoms (lymphadenopathy, weight loss, chronic diarrhea, fever, nephropathy, neuropathies) and the occurrence of opportunistic infections such as bacterial infections (tuberculosis, etc.), parasitic infections (toxoplasmosis, candidiasis, cryptococcosis) and severe viral infections (CMV, VZV, HSV, EBV, papovavirus, etc.) and cancers. Numerous neurologic disorders are underdiagnosed or never definitively diagnosed. Neurologic disorders associated with HIV infection are schematically split into three main categories of diseases: (1) neurologic disorders that affect the CNS and are caused by primary infection; (2) neurologic disorders caused by secondary—mainly opportunitic—processes; (3) neurologic disorders affecting the peripheral nervous system, including muscle disorders.

20.11.2.3 Diagnosis

Screening assays for HIV must recognize antibodies for both HIV-1 and HIV-2. The sensitivity of commercial ELISA is excellent, varying from 97.5% and 100%

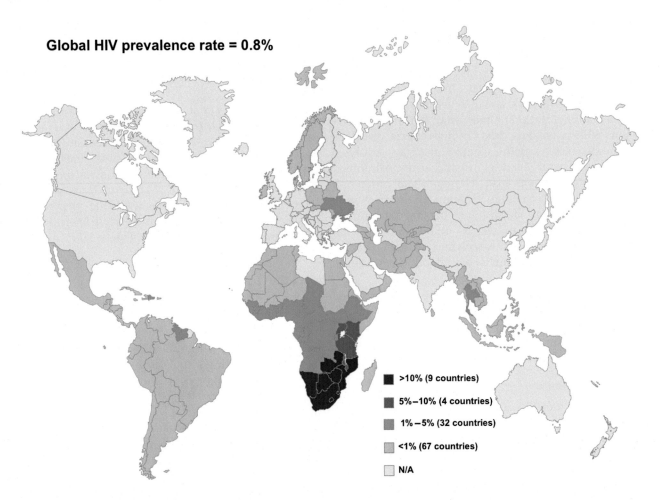

Global HIV prevalence rate = 0.8%

Legend:
- >10% (9 countries)
- 5%–10% (4 countries)
- 1%–5% (32 countries)
- <1% (67 countries)
- N/A

FIGURE 20.7   Adult human immunodeficiency virus (HIV) prevalence rate, 2014. *Source: Kaiser Family foundation, www.globalHealthFacts.org.*

but specificity varied from 74% to 99.5%. Therefore, ELISA-positive sera must be confirmed by western blot. The WHO technical working group recommended criteria for interpreting western blots for HIV-1 and HIV-2. ELISAs are often inappropriate for use in developing countries and simple rapid assays offer the best prospect for routine HIV screening in developing countries.

Plasma HIV RNA level is associated with HIV disease severity.

20.11.2.4 Prevention and Treatment

The prevention of HIV infection remains relevant considering the dynamics of the epidemic. The strategies associate initiative of universal prevention: information, education, communication, and screening. The condom remains the reference method. Global, rapid scale-up of ART programs significantly decreased HIV/AIDS-related morbidity and mortality. Reducing the time interval between HIV infection, diagnosis and starting of ART remains one of the

biggest challenges to reduce the chances of developing severe HIV disease evolution and associated neurological disorders.

20.12 POLYOMAVIRUS AND PROGRESSIVE MULTIFOCAL LEUCOENCEPHALOPATHY

The family of human polyomavirus (DNA viruses) is divided into two genuses. The species BK polyomavirus (BKPyV or BKV) and JC polyomavirus (JCPyV or JCV) were classified in the genus *Orthopolyomavirus*.

The progressive multifocal leucoencephalopathy (PML) is observed in immunodeficient individuals and was suggested, in 1961, to be due to a common virus, which in the immunocompromised host runs an atypical course of infection. The JCV and BKV are named with the initials of the first patient from whom they were isolated.

20.12.1 Epidemiology

The mode of BKV and JCV transmission is unknown, although the rapid acquisition of antibodies has been suggested to be consistent with respiratory transmission. BKV and JCV are widely distributed among healthy individuals (seroprevalence 80%). Antibodies against JCV are acquired by 50% of children at 6 years of age. Reactivation of both JCV and BKV is known to occur in pregnancy, diabetes, chronic disorders and old age.

PML was a very rare disease, associated with a range of immunocompromised conditions. However it has become more common in the area of HIV. Before treatments with protease inhibitors, 3%–5% of all AIDS cases examined had PML.

20.12.2 Pathogenesis

The PML brain is characterized by foci of demyelination that are widespread and vary in size. In advanced cases the areas may be necrotic. Nuclear changes in the oligodendrocytes at the edge of the demyelinated plaques are associated with the presence of JCV. Neurons are unaffected. Symptoms of a multifocal brain disease without signs of raised intracranial pressure in an immunocompromised host suggest the diagnosis of PML.

20.12.3 Diagnosis

MRI of the brain will detect lesions of demyelination. The detection in CSF (or cerebral biopsy) of JCV DNA using PCR (to differenciate JCV and BKV) is in favor with PML diagnostics. High viral load in the CSF may correlate with low survival rate. The confirmation of PML diagnosis requires histocytologic examination on cerebral biopsy or post-mortem brain samples.

20.12.4 Prevention and Treatment

The mortality rate of PML is now between 30% and 50% of patients during the first 3 months, and without treatment the mean survival is around 9 months. The treatment of PML remains very difficult; PML survivors are often left with devastating neurological sequelae.

20.13 *RHABDOVIRIDAE/LYSSAVIRUS*

The *Rhabdoviridae* family (ARN viruses) is divided into the genus *Lyssavirus*, which contains rabies virus and *Vesiculovirus*. The *Lyssavirus* genus contains at least seven members. Type 1 is the rabies virus itself as type species; type 2 is the Lagos bat virus, originally isolated from bats in Nigeria; type 3 is the Mokola virus, isolated from a shrew in Nigeria; type 4 is the Duvenhage virus, isolated from a human bitten by a bat in South Africa. Types 5, 6, and 7 are, respectively, European bat Lyssa virus 1 and 2 and Australian bat Lyssa virus. Actually, no human cases have yet been associated with type 2 virus.

20.13.1 Epidemiology

The epidemiology of human rabies follows closely the epizootiology of animal rabies. The dog is the major reservoir of rabies. Human rabies has been reported from all continents except Antarctica, but the majority of cases occur in countries where canine rabies is not controlled. The WHO estimated between 35,000 and 50,000 human rabies cases in 1997, mostly in India. The annual incidence of rabies deaths per 100,000 was calculated as 2 in India, 0.01–0.2 in Latin America and uncertain 0.0001–13 in Africa (Fig. 20.8). Human rabies is more common in people younger than 15 years, but all age groups are susceptible with a majority of males. An increasing source of rabies is observed in bat populations (approximately 10% of rabies-infected animals in the USA). The epidemiology of human rabies follows that of animals. Recent transmission to transplant recipients from organ donors is particularly worrying.

20.13.2 Pathogenesis

The major route of infection is in the majority via the bite from a rabid animal. Once introduced, rabies virus is quickly sequestrated. It was thought that the virus stayed in the nervous tissue close to the wound site, although studies indicated that the virus replicates in muscle before progressing to the peripheral nervous tissue via neuromuscular connections. But importance of replication in no-nerve cells in the pathogenesis of rabies remains controversial. Rabies virus travels to central nervous tissue via the nerves. Once in the neurons, the virus travels rapidly within the axons at a rate of 15–100 mm/day in humans.

After establishment in the neurons of the brain, it starts to move in the opposite direction, most notably in the nerve plexus and acinar cells of the salivary glands, from which excretion in saliva permits transmission by bite. However, at the end-stage of infection, other extra-neural tissues are also infected. The incubation period of the disease varies and may be as short as 2 weeks but is more commonly 1–3 months, and in a few cases more than 1 year. Development of infection depends on the severity of the exposure, the site of the

FIGURE 20.8   Global distribution of risk to humans of contracting rabies in 2013. *Source: Adapted from WHO (2014) www.who.int/rabies.*

bite and possibly other factors. Neurologic findings may be classified as either "furious" or "paralytic" and are not exclusive. Furious rabies is far more common and is characterized by spasm in response to tactile, auditory, visual and olfactory stimuli (aerophobia and hydrophobia). Such symptoms alternate with periods of lucidity, agitation, confusion and autonomic dysfunction. The alternative form of paralytic rabies ranges from paralysis of one limb to quadriplegia. Disease progresses to coma with neurological complications. The pathophysiology of the fatal outcome is not completely understood. Although encephalitis is widespread, no neuronal destruction is observed. Death probably results from the involvement and dysfunction in brain centers controlling the cardiorespiratory system.

20.13.3 Diagnosis

Especially in the absence of a documented exposure source, clinical diagnosis of rabies requires differentiation from a wide variety of diseases that can cause neurologic symptoms: other forms of encephalitis, for furious rabies (tetanus, drug intoxication, porphyria); for paralytic rabies Guillain-Barré syndrome, poliomyelitis and Japanese encephalitis but Nipah or West Nile viruses or EV-71 should enter into the differential diagnosis.

Definitive diagnosis of rabies infection of humans and suspected animal vectors depends on the detection and identification in infected brain tissue of rabies virus antigen, or suspected intracytoplasmic neural inclusions (Negri bodies); or viral nucleic acid by RT-PCR, on the presence of rabies virus-specific antibodies in the CSF, or in serum of unvaccinated patients; and on the isolation of the virus from brain tissues, saliva or other infected substances. A rapid rabies enzyme immunodiagnostic assay allows the antigen to be visualized by the naked eye and is thus a test that can be carried out in the field.

20.13.4 Prevention and Treatment

The control of rabies is through oral immunization of domestic and, more recently, wild animals. The proper post-exposure prevention of human rabies includes combined vaccination and passive antibody administration (human rabies immune globulins, HRIG) in the naïve patient. The solution to the problem of safety of rabies vaccines lies in the development of vaccines prepared from rabies virus grown in tissue culture free of neuronal tissue.

The post-exposure regimen recommended by the WHO is HRIG on day 0 and a cell culture vaccine on days 0, 3, 7, 14, and 28 administered intramuscularly in the deltoid area only. Other regimens were used in some countries with HRIG indicated on day 0 (2 doses), 7, 21 or 28 administered intramuscularly, or intradermal use in Thailand and other developing countries. The ability of vaccines formulated with classical rabies virus to protect against bat *Lyssavirus* remains questionable. Until recently, there had been no successful human rabies therapy.

20.14 ARBOVIRUSES

Arthropod-borne viruses (arboviruses) replicate in both vertebrates and arthropods. The virus is propagated in the arthropod's gut and if it reaches a high titer in salivary glands, it can be transmitted when a new host is bitten. The viruses often cause diseases in humans and other vertebrate hosts.

Increasing knowledge of viral characteristics has revealed great heterogeneity between the arthropod-borne virus and more than 350 arthropods borne viruses have now been isolated.

Arthropod—borne viruses, because of their vectors, depend strongly on climatic conditions and we relate in detail only arboviruses with tropical geographic distributions and with the CNS as the principal target (encephalitis essentially).

20.14.1 *Bunyaviridae* Family

Rift Valley fever (RVF) is the most notable virus in the genus *Phlebovirus*. Outbreaks in which a few people developed encephalitis have been reported in Nigeria, Egypt, Sudan, and Kenya. More than 20 species of mosquitoes have been implicated as possible vectors. Recently an inactivated RVF vaccine has been developed and a live vaccine is in progress.

20.14.2 Flaviviridae Family, Genus *Flavivirus*

20.14.2.1 *Epidemiology*

Japanese encephalitis virus continues to be the major type of encephalitis in Japan, Malaysia, India and western Pacific Island areas. West Nile virus is distributed in Africa, Europe and Asia and America.

St Louis encephalitis virus is present in the Americas. Murray Valley encephalitis virus (and Kunjin virus) is found in Australia and Papua New Guinea. Rocio virus affects several countries in Brazil. The five viruses have the mosquito as vector (Fig. 20.9).

20.14.2.2 *Clinical Features*

Flaviviruses give three types of clinical syndromes: CNS diseases, mainly encephalitis (St Louis, Japanese B, West Nile, Murray Valley, Ilheus, Kunjin, etc.); severe systemic diseases involving liver and kidneys; and milder systemic diseases characterized by severe muscle pains and a rash that may be hemorrhagic (West Nile, and some of the tick-borne viruses). The severity and extent of encephalitis due to flaviviruses vary with the etiologic agent. For example, St Louis encephalitis virus produces mild lesions, low mortality, and few residua compared with Japanese B encephalitis virus, which has a mortality rate of about 8% and causes neurologic sequelae in more than 30% and persistent mental disturbances in about 10% of clinically diagnosed infections. Although infection with Zika virus often leads to mild disease, its emergence in the Americas, Caribbean and Polynesia has coincided with a steep increase in patients developing Guillain—Barré syndrome.[60] The syndrome is an autoimmune disorder that causes acute or subacute flaccid paralysis with a 5% death rate and up to 20% of patients are left with a significant disability. Several other flaviviruses have been found to be rare triggers of Guillain—Barré syndrome, including Japanese encephalitis virus, West Nile virus and the live-attenuated yellow fever vaccine. Infection with Zika virus among pregnant women coincided with increased numbers of neonates with neurological complications, such as congenital microcephaly.[61] Congenital microcephaly cases have also been reported after intrauterine infection with West Nile virus and Chikungunya virus (Togaviridae/Alphavirus) and other arboviruses, and proved historically after prenatal infections with rubella or cytomegalovirus. The risk of microcephaly associated with Zika virus infection was around 100 cases per 10,000 women infected in the first trimester. In 1 year (2015), 1.5 million Zika cases and more than 4500 cases of microcephaly were observed.

20.14.2.3 *Diagnosis*

The laboratory diagnosis of flaviviruses is based on: (1) isolation or positive RT-PCR from tissue, blood, CSF or other body fluid; (2) antigen detection (immunofluorescence, monoclonal antibody, etc.); (3) fourfold or greater rise in serum antibody titer or positive IgM with ELISA immunocapture in serum or CSF. The results of various tests depend on, besides technical factors, the timing of sample collection and specimen

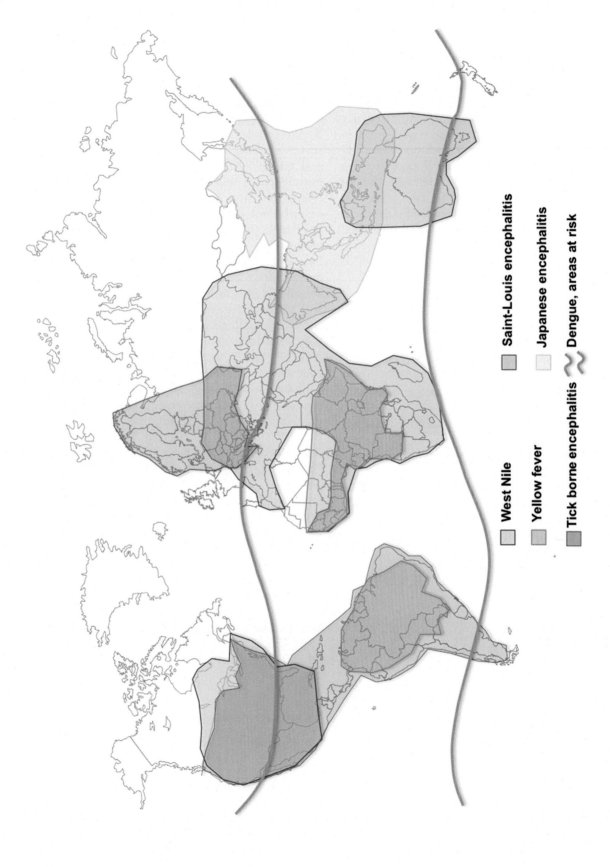

West Nile   **Saint-Louis encephalitis**

Yellow fever   **Japanese encephalitis**

Tick borne encephalitis   ≈ **Dengue, areas at risk**

FIGURE 20.9   Global distribution of flaviviruses. *Source: Adapted from WHO (2011) www.who.int, CDC (2010) www.cdc.gov.*

examined. The research must be large with screening of several arbovirus antibodies, and previous history of vaccination in the endemic area.

20.14.2.4 Prevention and Treatment

Arboviruses can be prevented by vector control. Prevention of epidemics depends on eradication of mosquitos by elimination of breeding sites, larviciding measures and spraying of insecticides. Actually two types of inactivated and one live attenuated Japanese Encephalitis was available; universal primary immunization is indicated for children in endemic areas.

No vaccine is available for West Nile or St Louis encephalitis viruses. Formalin-treated extract of infected mouse brain is used as a vaccine. An effective vaccine against tick-borne encephalitis is derived from chick embryo cell culture-grown virus, which is highly purified and inactivated by formalin. Several vaccines against Chikungunya virus are tested using virus-like particles. Yellow fever is a live vaccine, indicated for 3 reasons:[62] to protect populations living in areas subject to endemic and epidemic disease; to protect travelers visiting these areas; and to prevent international spread by minimizing the risk of importation of the virus by viremic travelers. A single dose of Yellow Fever vaccine is sufficient to confer sustained life-long protective immunity against Yellow Fever disease. Recently, a Dengue vaccine has become available and has been approved in Mexico, Brazil, and Philippines; it is a live attenuated tetravalent vaccine using recombinant DNA technology.

20.15 TOGAVIRIDAE/RUBIVIRUS

Unlike most other togaviruses, rubella has no known vertebrate host and the only natural reservoir is humans.

20.15.1 Epidemiology

Rubella is a worldwide infection, as may be inferred from serologic surveys conducted in many different countries. Some rubella patients excrete large amounts of virus in respiratory secretions and are highly infectious spreaders. The basic reproductive rate before vaccination was estimated to be 6−7 in developed countries and up to 12 in crowded developing countries. Many childhood infections are asymptomatic or without rash and are therefore unrecognized. The large variation seen in seroprevalence in developing countries suggests that rubella occurs in sporadic epidemics except where population density is high. Rubella susceptibility data revealed remarkable differences in developing countries, not correlated with geography.

For example, Malaysia, Peru, and Nigeria were among the countries where more than 25% of women were found to be seronegative, low seroprevalence is also confirmed in Thailand and Turkey.

20.15.2 Pathogenesis

Large particle aerosols spread rubella; primary implantation and replication occurs in the nasopharynx. The incubation period of rubella is 14−21 days, with most patients developing a rash 14−17 days after exposure. During the second week, lymphadenopathy may be observed. Later in the second week, the virus replicates in the blood. At the end of the incubation period, a maculopapular erythematous rash may appear on the face and neck. Although rubella is considered as a benign disease, encephalitis is reported to occur in approximately 1/6000 cases, but in the South Pacific islands, the incidence was estimated as between 1/500 and 1/1000 cases. The mortality rate of rubella encephalitis is approximately 20%; death usually happens within 8 years of onset. The pathology resulting from rubella infection is dependent on the mode of infection: whether it is due to maternal−fetal transmission. Congenital infection may lead to a constellation of abnormalities in neonates of infected mothers.[63] These include by decreasing frequency: bilateral sensoneural deafness, mental retardation, cardiac anomalies, ocular anomalies (i.e., cataracts, retinopathy and microphtalmy), and intrauterine growth restriction. The risk of maternal−fetal viral transmission is related to the gestational age at the time of maternal infection.

Congenital infection occurred in 70%−90% of neonates whose mothers were infected before 10 weeks of gestation, in 25%−50% during the second trimester, 35%−60% in the third trimester and 100% at the end of gestation. But risks of congenital malformations depend on the stage of the pregnancy: 70%−90% before 10 weeks of gestation, 15%−80% between 10 and 16 weeks of gestation, and no risk after 16 weeks of gestation. The epidemiology of Congenital Rubella Syndrome (CRS) is known only for a few countries in the world. Little information is available for the countries in South America, Africa, and most of Asia, although such data suggest that CRS is common even in the developing world. CRS incidences between 0.1 and 0.2 per 1000 births were observed in the absence of an outbreak in Myanmar or Maldives but increased to 1−4/1000 births in the same areas during epidemic periods.

20.15.3 Diagnosis

The common symptoms of rubella, such as low-grade fever and maculopapular rash, should not be confused with other similar infections. Clinical diagnosis of acquired infection is so inaccurate as to be useless without laboratory support. Confirmation of rubella may be made by RT-PCR from throat, urine, blood (during first week of illness) or by specific serological assays such as ELISA. IgM antibody may be positive for up to 6 weeks after acute infection. Congenital rubella infection in the infant is associated with the detection of viral genome, IgM antibodies by immunocapture techniques and low-avidity of IgG antibodies. In severe cases, virus excretion may persist for several years. There is only one serotype of rubella virus; however, there are two clades and within the clades, at least seven genotypes. Worldwide distribution reveals differences between geographic areas: isolates of North America, Europe, and Japan form clade I whereas clade II includes some strains from China, Korea and India. Isolates from CRS cases are not genetically distinct from isolates from acquired rubella. The genomic and amino acid variations between worldwide strains do not affect the efficacy of the vaccines.

20.15.4 Prevention

Rubella virus and particularly the congenital syndrome are legitimate targets for eradication by vaccination. The RA27/3, a live virus vaccine, is produced on human diploid cells. At least 95% of vaccinated people of more than 12 months of age develop protective antibody titers. Vaccine-induced rubella antibodies persisted in more than 90% of vaccinated people at least 15 years after receipt of RA27/3 vaccine. The combination with measles or measles and mumps reduces the cost and avoids any additional expenses of administration. Routine universal vaccination of infants would eliminate CRS in 10–20 years.[64]

20.16 POXVIRUSES/SMALLPOX AND VACCINIA

Smallpox eradication was certified by the World Health Assembly on May 8, 1980.

20.16.1 Epidemiology

Smallpox, an exanthematous viral disease, was once prevalent throughout the world. There is no animal reservoir of smallpox and no long-term human carrier state, but the virus must spread continuously from human to human to survive. Among those not previously vaccinated, the majority of severe variola cases (~90%) carried a fatality ratio of approximately 30%. In India the case fatality rate (from 1974 to 1975) was decreasing with age: 43.5 before 1 year to 17.5 after

20 years. In 1967, the year of the beginning of the Intensified Global Eradication Program, an estimated 10–15 million smallpox cases occurred in 31 countries where the disease was endemic. In the years that followed eradication, the WHO recommended that all diagnostic and research laboratories destroy their stocks of smallpox virus. But research programs using the virus were carried on at the Center for Disease Control and at the VECTOR Institute in Novosibirsk (and probably in secret laboratories of other countries in the context of biological weapons). Variola virus may be lyophilized and is relatively stable as an aerosol. An outbreak of smallpox (induced through aerosol dissemination) in a non-immunized population would be catastrophic.

20.16.2 Prevention and Safety of the Vaccine[65]

In 1796, Jenner showed that the material from the pustule lesions induced by cowpox virus could be inoculated into the skin of an individual and protects this person from smallpox infection. Two live vaccine strains were used: the Lister strain and the New York City Board of Health strain (NYCBH). Thereafter, the vaccine was inoculated intradermally with a bifurcated needle. Initial vaccination was successful (NYCBH strain) in 97.8% of subjects. Historical data revealed that even 70 years after primary vaccination, approximately 78% of cases were still protected against severe and fatal disease. The most frequent symptoms postvaccination were pain at the vaccination site, muscle aches and fatigue. Slight elevations in temperature were common. Vaccines may develop a variety of skin rashes after vaccination. Most are benign but three can be life threatening: progressive vaccinia, eczema vaccinatum and generalized vaccinia (respectively, 0.9, 10.4, and 23.4 cases per million in primary vaccinated patients and 0.7, 0.9, and 1.2 after revaccination). Postvaccinal encephalopathy and encephalitis are the most serious adverse events,[66] although these were quite rare in the contemporary American program, with 16 cases—four deaths in 5,594,000 vaccinees more frequent before 1 year of age (four cases—three deaths in 614,000 vaccinees), but no cases in the revaccination group (14,184,000 vaccinations). The incidence of these complications was substantially higher in Europe with especially reactogenic strains used before the 1960s. Postvaccinal encephalopathy is characterized by general hyperemia of the brain, lymphocytic infiltration of the meninges, and perivascular hemorrhage. Symptoms begin abruptly within 6–10 days after vaccination. Death, when it occurs, follows within a few days. Postvaccinal encephalitis is characterized by perivenous demyelination and microglial proliferation.

Illness usually begins between 8 and 15 days after vaccination. After postvaccinal encephalitis, death occurs in 10%–35% of cases. Some survivors after postvaccinal encephalopathy or encephalitis have residual paralysis or mental impairment. In the context of bioterrorism or biological weapons, the decision of mass primovaccination seems difficult considering the frequency and the severity of adverse events.

References

1. Malkin JE, Morand P, Malvy D, et al. Seroprevalence of HSV-1 and HSV-2 infection in the general French population. *Sex Transm Infect*. 2002;78(3):201–203.

2. Sauerbrei A, Schmitt S, Scheper T, et al. Seroprevalence of herpes simplex virus type 1 and type 2 in Thuringia, Germany, 1999 to 2006. *Euro Surveill*. 2011;16(44):pii: 20005.

3. Xu F, Sternberg MR, Kottiri BJ, et al. Trends in herpes simplex virus type 1 and type 2 seroprevalence in the United States. *JAMA*. 2006;296(8):964–973.

4. Malvy D, Halioua B, Lancon F, et al. Epidemiology of genital herpes simplex virus infections in a community-based sample in France: results of the HERPIMAX study. *Sex Transm Dis*. 2005;32(8):499–505.

5. Weiss H. Epidemiology of herpes simplex virus type 2 infection in the developing world. *Herpes*. 2004;11(Suppl 1):24A–35A.

6. Whitley RJ. Herpes simplex encephalitis: adolescents and adults. *Antiviral Res*. 2006;71(2-3):141–148.

7. Mailles A, Stahl JP, Steering C, Investigators G. Infectious encephalitis in France in 2007: a national prospective study. *Clin Infect Dis*. 2009;49(12):1838–1847.

8. Davis LE, Guerre J, Gerstein WH. Recurrent herpes simplex virus type 2 meningitis in elderly persons. *Arch Neurol*. 2010;67(6):759–760.

9. Noska A, Kyrillos R, Hansen G, Hirigoyen D, Williams DN. The role of antiviral therapy in immunocompromised patients with Herpes simplex virus meningitis. *Clin Infect Dis*. 2015;60(2):237–242.

10. Lim HK, Seppanen M, Hautala T, et al. TLR3 deficiency in herpes simplex encephalitis: high allelic heterogeneity and recurrence risk. *Neurology*. 2014;83(21):1888–1897.

11. Kneen R, Michael BD, Menson E, et al. Management of suspected viral encephalitis in children: Association of British Neurologists and British Paediatric Allergy, Immunology and Infection Group national guidelines. *J Infect*. 2012;64(5):449–477.

12. Solomon T, Michael BD, Smith PE, et al. Management of suspected viral encephalitis in adults: Association of British Neurologists and British Infection Association National Guidelines. *J Infect*. 2012;64(4):347–373.

13. Solomon T, Hart IJ, Beeching NJ. Viral encephalitis: a clinician's guide. *Pract Neurol*. 2007;7(5):288–305.

14. Rozenberg F. Données actuelles sur l'encéphalite herpétique. *Rev Francophone Lab*. 2012;447:27–31.

15. Aujard Y. [Modalities of treatment local and general, medicamentous or not, controlling neonate suspected to be infected/contaminated by HSV1 or HSV2]. *Ann Dermatol Venereol*. 2002;129(4 Pt 2):655–661.

16. Kimberlin DW, Lin CY, Jacobs RF, et al. Safety and efficacy of high-dose intravenous acyclovir in the management of neonatal herpes simplex virus infections. *Pediatrics*. 2001;108(2):230–238.

17. Kimberlin DW, Whitley RJ, Wan W, et al. Oral acyclovir suppression and neurodevelopment after neonatal herpes. *N Engl J Med*. 2011;365(14):1284–1292.

18. Shah SS, Aronson PL, Mohamad Z, Lorch SA. Delayed acyclovir therapy and death among neonates with herpes simplex virus infection. *Pediatrics*. 2011;128(6):1153–1160.

19. Guess HA, Broughton DD, Melton 3rd LJ, Kurland LT. Population-based studies of varicella complications. *Pediatrics*. 1986;78(4 Pt 2):723–727.

20. Choo PW, Donahue JG, Manson JE, Platt R. The epidemiology of varicella and its complications. *J Infect Dis*. 1995;172(3):706–712.

21. Barnes DW, Whitley RJ. CNS diseases associated with varicella zoster virus and herpes simplex virus infection. Pathogenesis and current therapy. *Neurol Clin*. 1986;4(1):265–283.

22. Gilden D. Varicella zoster virus and central nervous system syndromes. *Herpes*. 2004;11(Suppl 2):89A–94A.

23. Yawn BP, Gilden D. The global epidemiology of herpes zoster. *Neurology*. 2013;81(10):928–930.

24. Bhupal HK. Ramsay Hunt syndrome presenting in primary care. *Practitioner*. 2010;254(1727):33–35.

25. Ryu EW, Lee HY, Lee SY, Park MS, Yeo SG. Clinical manifestations and prognosis of patients with Ramsay Hunt syndrome. *Am J Otolaryngol*. 2012;33(3):313–318.

26. Verghese A, Sugar AM. Herpes zoster ophthalmicus and granulomatous angiitis. An ill-appreciated cause of stroke. *J Am Geriatr Soc*. 1986;34(4):309–312.

27. Studahl M, Hagberg L, Rekabdar E, Bergstrom T. Herpesvirus DNA detection in cerebral spinal fluid: differences in clinical presentation between alpha-, beta-, and gamma-herpesviruses. *Scand J Infect Dis*. 2000;32(3):237–248.

28. Takahashi M, Otsuka T, Okuno Y, Asano Y, Yazaki T. Live vaccine used to prevent the spread of varicella in children in hospital. *Lancet*. 1974;2(7892):1288–1290.

29. Weibel RE, Neff BJ, Kuter BJ, et al. Live attenuated varicella virus vaccine. Efficacy trial in healthy children. *N Engl J Med*. 1984;310(22):1409–1415.

30. Oxman MN, Levin MJ, Johnson GR, et al. A vaccine to prevent herpes zoster and postherpetic neuralgia in older adults. *N Engl J Med*. 2005;352(22):2271–2284.

31. Kenneson A, Cannon MJ. Review and meta-analysis of the epidemiology of congenital cytomegalovirus (CMV) infection. *Rev Med Virol*. 2007;17(4):253–276.

32. Ludwig A, Hengel H. Epidemiological impact and disease burden of congenital cytomegalovirus infection in Europe. *Euro Surveill*. 2009;14(9):26–32.

33. Manicklal S, Emery VC, Lazzarotto T, Boppana SB, Gupta RK. The "silent" global burden of congenital cytomegalovirus. *Clin Microbiol Rev*. 2013;26(1):86–102.

34. Cannon MJ, Schmid DS, Hyde TB. Review of cytomegalovirus seroprevalence and demographic characteristics associated with infection. *Rev Med Virol*. 2010;20(4):202–213.

35. Grosjean J, Trapes L, Hantz S, et al. Human cytomegalovirus quantification in toddlers saliva from day care centers and emergency unit: a feasibility study. *J Clin Virol*. 2014;61(3):371–377.

36. Grosse SD, Ross DS, Dollard SC. Congenital cytomegalovirus (CMV) infection as a cause of permanent bilateral hearing loss: a quantitative assessment. *J Clin Virol*. 2008;41(2):57–62.

37. Dollard SC, Grosse SD, Ross DS. New estimates of the prevalence of neurological and sensory sequelae and mortality associated with congenital cytomegalovirus infection. *Rev Med Virol*. 2007;17(5):355–363.

38. Anderson KS, Amos CS, Boppana S, Pass R. Ocular abnormalities in congenital cytomegalovirus infection. *J Am Optom Assoc*. 1996;67(5):273–278.

39. Coats DK, Demmler GJ, Paysse EA, Du LT, Libby C. Ophthalmologic findings in children with congenital cytomegalovirus infection. *J AAPOS*. 2000;4(2):110–116.

40. Arribas JR, Storch GA, Clifford DB, Tselis AC. Cytomegalovirus encephalitis. *Ann Intern Med*. 1996;125(7):577–587.

41. Anders HJ, Goebel FD. Neurological manifestations of cytomegalovirus infection in the acquired immunodeficiency syndrome. *Int J STD AIDS*. 1999;10(3):151–159; quiz 160–151.

42. Teissier N, Fallet-Bianco C, Delezoide AL, et al. Cytomegalovirus-induced brain malformations in fetuses. *J Neuropathol Exp Neurol*. 2014;73(2):143–158.

43. Picone O, Teissier N, Cordier AG, et al. Detailed in utero ultrasound description of 30 cases of congenital cytomegalovirus infection. *Prenat Diagn*. 2014;34(6):518–524.

44. Ross SA, Ahmed A, Palmer AL, et al. Detection of congenital cytomegalovirus infection by real-time polymerase chain reaction analysis of saliva or urine specimens. *J Infect Dis*. 2014;210 (9):1415–1418.

45. Barbi M, Binda S, Caroppo S. Diagnosis of congenital CMV infection via dried blood spots. *Rev Med Virol*. 2006;16(6):385–392.

46. Connelly KP, DeWitt LD. Neurologic complications of infectious mononucleosis. *Pediatr Neurol*. 1994;10(3):181–184.

47. Tselis A, Duman R, Storch GA, Lisak RP. Epstein–Barr virus encephalomyelitis diagnosed by polymerase chain reaction: detection of the genome in the CSF. *Neurology*. 1997;48(5):1351–1355.

48. Knowles DM. Etiology and pathogenesis of AIDS-related non-Hodgkin's lymphoma. *Hematol Oncol Clin North Am*. 2003;17(3):785–820.

49. Allen U, Preiksaitis J. Practice ASTIDCo. Epstein–Barr virus and posttransplant lymphoproliferative disorder in solid organ transplant recipients. *Am J Transplant*. 2009;9(Suppl 4):S87–S96.

50. Uhlin M, Wikell H, Sundin M, et al. Risk factors for Epstein–Barr virus-related post-transplant lymphoproliferative disease after allogeneic hematopoietic stem cell transplantation. *Haematologica*. 2014;99(2):346–352.

51. Castellano-Sanchez AA, Li S, Qian J, Lagoo A, Weir E, Brat DJ. Primary central nervous system posttransplant lymphoproliferative disorders. *Am J Clin Pathol*. 2004;121(2):246–253.

52. Hausler M, Ramaekers VT, Doenges M, Schweizer K, Ritter K, Schaade L. Neurological complications of acute and persistent Epstein–Barr virus infection in paediatric patients. *J Med Virol*. 2002;68(2):253–263.

53. Weinberg A, Li S, Palmer M, Tyler KL. Quantitative CSF PCR in Epstein–Barr virus infections of the central nervous system. *Ann Neurol*. 2002;52(5):543–548.

54. Ranger S, Patillaud S, Denis F, et al. Seroepidemiology of human herpesvirus-6 in pregnant women from different parts of the world. *J Med Virol*. 1991;34(3):194–198.

55. Daibata M, Taguchi T, Sawada T, Taguchi H, Miyoshi I. Chromosomal transmission of human herpesvirus 6 DNA in acute lymphoblastic leukaemia. *Lancet*. 1998;352(9127):543–544.

56. Krueger GR, Koch B, Ramon A, et al. Antibody prevalence to HBLV (human herpesvirus-6, HHV-6) and suggestive pathogenicity in the general population and in patients with immune deficiency syndromes. *J Virol Methods*. 1988;21(1-4):125–131.

57. Bozzola E, Krzysztofiak A, Bozzola M, et al. HHV6 meningoencephalitis sequelae in previously healthy children. *Infection*. 2012;40(5):563–566.

58. Challoner PB, Smith KT, Parker JD, et al. Plaque-associated expression of human herpesvirus 6 in multiple sclerosis. *Proc Natl Acad Sci USA*. 1995;92(16):7440–7444.

59. Verdier M, Bonis J, Denis F. The prevalence and incidence of HTLSs in Africa. In: Essex M, M'Boup S, Kanki PJ, Kalengayi MR, eds. *AIDS in Africa*. New York: Raven Press; 1994:173–193.

60. Cao-Lormeau VM, Blake A, Mons S, et al. Guillain–Barré Syndrome outbreak associated with Zika virus infection in French Polynesia: a case-control study. *Lancet*. 2016;387 (10027):1531–1539.

61. Driggers RW, Ho CY, Korhonen EM, et al. Zika virus infection with prolonged maternal viremia and fetal brain abnormalities. *N Engl J Med*. 2016;374(22):2142–2151.

62. WHO. Vaccines and vaccination against yellow fever: WHO position paper, June 2013: recommandations. *Vaccine*. 2015;33 (1):76–77.

63. Atreya CD, Mohan KV, Kulkarni S. Rubella virus and birth defects: molecular insights into the viral teratogenesis at the cellular level. *Birth Defects Res A Clin Mol Teratol*. 2004;70 (7):431–437.

64. Plotkin SA. Rubella vaccine. In: Plotkin SA, Orenstein W, Offit PA, eds. *Vaccines*. 5th ed Saunders Elsevier; 2008:735–771.

65. Henderson DA, Borio LL, Grabenstein JD. Smallpox and vaccinia. In: Plotkin SA, Orenstein W, Offit PA, eds. *Vaccines*. 5th ed. Saunders Elsevier; 2008:773–803.

66. Lane JM, Ruben FL, Neff JM, Millar JD. Complications of smallpox vaccination, 1968. *N Engl J Med*. 1969;281(22):1201–1208.

Further Reading

Denis F. *Les virus transmissibles de la mère à l'enfant*. Paris: John Libbey Eurotext; 1999.

Farrar J, Hotez P, Junghanss T, Kang G, Lalloo D, White NJ. *Manson's Tropical Diseases*. 23rd ed. Oxford: Elsevier Saunders Ltd; 2013.

Gentilini M, Caumes E, Danis M, et al. *Médecine Tropicale*. 6th ed. Paris: Médecine Sciences Publications, Lavoisier; 2012.

Pasquier C, Bertagnoli S, Dunia D, Izopet J. *Virologie Humaine et Zoonoses*. Paris: Dunod; 2013.

Plotkin SA, Orenstein W, Offit PA. *Vaccines*. 5th ed Philadelphia, PA: Saunders Elsevier; 2013.

Solomon T. Virus infections of central nervous system. In: Farrar J, Hotez P, Junghanss T, Kang G, Lalloo D, White NJ, eds. *Manson's Tropical Diseases*. 23rd ed. Oxford: Elsevier Saunders Ltd; 2013:242–272.

Solomon T, Hart IJ, Beeching NJ. Viral encephalitis: a clinician's guide. *Pract Neurol*. 2007;7(5):288–305.

Wadia NH, Khadilkar SV. *Neurological Pratice*. 2nd ed. New Delhi: Elsevier India; 2005.

21

Other Diseases: Traumatic Brain Injuries, Tumors, and Multiple Sclerosis

Mouhamadou Diagana[1,2,3] *and Michel Dumas*[2,3]

[1]Faculté de Médecine de Nouakchott, Nouakchott, Mauritania [2]University of Limoges, Limoges, France
[3]INSERM UMR 1094 NET, Limoges, France

21.1 TRAUMATIC BRAIN INJURIES

Traumatic brain injuries (TBIs) are most often associated with road traffic accidents. They are one of the factors of gravity, quickly involved in the prognosis of polytrauma patients. Their socioeconomic impact is important, because of the functional consequences they engender among survivors. In tropical areas, both in Africa and Asia, accidents are a real public health problem in many countries due to the development of the fleet, the poor condition of the road network, the noncompliance with traffic laws and the level of organization of support services for medical and surgical emergencies.

Epidemiological data in the general population are rare. However, studies conducted in hospitals and particularly in the emergency wards make it possible to estimate the extent of TBI in the tropics.

21.1.1 Frequency

The incidence rate of TBI is difficult to establish generally in tropical areas, because not only there are few studies but their methodologies differ, particularly regarding the inclusion criteria.

In India, the impact of TBI is estimated at 160/100,000 per year.[1] In Burkina Faso in 2008, in 6 months TBIs concerned 30% of hospitalized patients.[2] This proportion is similar to that of Benin the same year: 31.9%,[3] while in Kisangani (Democratic Republic of Congo), TBIs covered only 17.8% of patients in the surgical department.[4] In Tunisia, the incidence of TBI in children may exceed 200/100,000 per year.[5]

21.1.2 Distribution by Age and Sex

The majority of tropical studies indicate that TBIs mainly involve young adults. The average age was 24 years in India,[1] 28.6 ($\pm$15.4) in Burkina Faso;[2] it was 32 years in Benin[3] and 30.4 years ($\pm$16.1) in Kisangani, Democratic Republic of Congo,[4] where 61.4% of patients were aged from 15 to 44 years.

Male dominance exists in all studies. The male/female ratio of 2:2.5 can rise to 6.7 depending on the series.[1–4] Men are simply more exposed to TBI simply because they use more motorized means of locomotion. In some areas, such as Nepal, this male/female ratio can reach 5;[5] it was also seen in children in Tunisia at a ratio of 2.[6]

21.1.3 The Circumstances of Occurrence

Road traffic accidents are the leading causes of TBIs and, in this case, are a factor in poor prognosis; but there are also TBIs secondary to assault, which are generally benign.

Several occupational groups are particularly exposed to TBIs: cyclists, bikers (in particular moto-taxis), in cities where this mode of transport has grown (64% in the Congolese series[4]). In Guinea Conakry, pedestrians (71.3%) are most exposed to road traffic accidents.[7]

The rapid development of public transport in recent years in almost all countries in Africa, with a fleet consisting mainly of used vehicles imported from Europe, exposes the general population to accidents and TBIs. This mode of transport is involved in 72.6% of accidents in Ivory Coast.[8]

21.1.4 Morbidity and Mortality

TBI is a major cause of morbidity and mortality after an accident. Indeed, the more severe head injuries, defined by a Glasgow score ≤ 8, are often the result of a road traffic accidents with a high rate of mortality or severe physical and/or mental disabilities in survivors.

Deaths unfortunately happen very often before the completion of preliminary investigations, especially computed tomography (CT)-scans that would assess encephalic lesions.

Prognosis is strongly related on the one hand to the quality of care, pre-hospital management, distance and time between the location of trauma and the structures of medical and surgical emergencies and, on another hand, to the extent of brain damage, associated injuries and hospital technical capacities.

TBI will be an increasing issue in the tropics in the coming years due to the development of the transportation network, which is necessary but often inadequate. Therefore, the prevention strategy must include several components:

- sensitization of the population;
- control of the import of used vehicles, which are often too old;
- the strengthening of repression in breaches of regulations and traffic laws, particularly regarding speeding, non-compliance priorities, lack of seat belt or helmet wearing;

- the development of an adequate system of transportation of injured patients to emergency structures; and
- the strengthening of capacity of emergency care units especially in paramedical and medical personnel: emergency physicians, anesthetists, surgeons, radiologists, and neurosurgeons.

21.2 TUMORS

Primary brain tumors (PBTs) are divided into benign and malignant tumors. Neuroepithelial tumors, meningiomas, and neuromas are the most frequently observed tumors in the world.[9]

Most epidemiological studies are devoted to PBTs observed in children and adolescents. Brain tumors in adults are divided into primitive and secondary tumors: brain metastases of a primary tumor in another location, these will not be discussed in this chapter.

The collection of epidemiological data on PBTs in tropical environments encounters several obstacles including: the lack of studies on risk factors of development of specific brain tumors in the tropics; the lack of a registry of cancer in most developing countries; the lack of investigations including magnetic resonance imaging, only recently introduced in many countries, particularly in Africa; and the unavailability of a histological confirmation in some countries.

21.2.1 PBT Risk Factors

Several endogenous and exogenous risk factors are implicated in the development of brain tumors. These factors have been mostly studied in the development of gliomas.

21.2.1.1 Endogenous Factors

21.2.1.1.1 AGE AND SEX

Certain types of brain tumors are strongly linked to age; this is the case for medulloblastomas and pilocytic astrocytomas which are only observed in children.[10] The predominance of the male sex is reported in meningiomas and gliomas.

21.2.1.1.2 GENETIC FACTORS

There are family forms in 5% of cases of certain cerebral tumors.[11] The development of certain types of tumors suggests the role of a genetic factor in the genesis of the malignancy, including the frequency of gliomas in patients with neurofibromatosis type 1 and 2 or tuberous sclerosis; this is linked to a high penetrance of a P53 mutation. A recent meta-analysis including four studies showed that a polymorphism in DNA repair genes (ERCC1, ERCC2, XRCC1) is associated with an increase of cerebral gliomas. The genes encoding certain interleukins (IL-4, IL-4RA, and IL-13) are also associated with a risk of glioma development.[9,11]

21.2.1.1.3 ALLERGIC CONDITIONS

Some studies point out an apparent protective effect of allergic conditions (asthma and eczema),[12] while other studies show that certain ethnicity, sharing the same allergic condition, could instead promote the development of gliomas.[13]

21.2.1.1.4 ANTHROPOMETRIC FACTORS

The weight and size at birth, which reflect maternal nutritional status, associated with exposure to growth hormones could be factors stimulating the proliferation of malignant cells.[9] A meta-analysis of the role of anthropometric parameters of the child including two cohort studies and six case–control studies showed a high risk of development of medulloblastomas and astrocytomas in children with a high birth weight.[14] A cohort of 320 children followed for 40 years showed an increased risk of developing gliomas in adulthood in those who had a high birth weight, and the opposite trend in those who had a low birth weight; a body mass index (BMI) $<18.5 \text{ kg/m}^2$ in adulthood is associated with a lower risk of subsequent development of gliomas.[15]

21.2.1.2 Exogenous Factors

Epidemiological studies show that exposure to environmental toxins increases the risk of developing tumors of the central nervous system.

21.2.1.2.1 IONIZING RADIATIONS

The risk of developing tumors related to radiation exposure was mentioned early in the last century with the use of X-ray medical imaging.[9] An increased incidence of brain tumors in the population of survivors of the nuclear disasters of Hiroshima and Nagasaki was observed. Several studies have demonstrated the increase in the occurrence of some tumors (gliomas and meningiomas) after therapeutic or accidental radiation exposure.[9]

21.2.1.2.2 MAGNETIC FIELDS

Although a positive association seems to exist between prolonged exposure to a magnetic field and the development of brain tumors, case–control studies have not concluded with any certainty the harmfulness of using mobile phones with regard to the occurrence of cerebral tumors.[16]

21.2.1.2.3 VIRAL INFECTIONS

Cytomegalovirus infections have been associated with the occurrence of glioblastomas. The DNA of different viruses has been found in the tissues of cerebral tumors.[16] However, the pathophysiological mechanism of the tumor process is not yet well understood.

Although several cases of brain tumors in patients seropositive for human immunodeficiency virus (HIV) were observed, the carcinogenic risk of HIV is not explicitly stated. Its role in the occurrence of cerebral primitive related lymphoma virus Creutzfeld–Jacob is, however, well known.

The intrauterine infection by influenza virus and chickenpox was also cited as a risk factor for brain tumor development, but in fact their association is low. The role of the Creutzfeld–Jacob virus has also been cited in the occurrence of ependymomas or papilloma of the choroid plexus.[17]

21.2.1.2.4 NITRATES

In Namibia, consumption by pregnant women of water from boreholes with a high nitrate content has been implicated in the relative increase in brain tumors in children in a rural area (Herero).[18]

21.2.2 Frequency

In West Africa, the incidence of brain tumors is estimated to be between 3.7/100,00 and 4.6/100,000 inhabitants per year,[19] which is lower than in developed countries. In low- and middle-income countries, epidemiological studies of brain tumors are rare in the general population. Those that deal with PBT children and adolescents are more common, but studies were essentially conducted in hospitals. In Nigeria, brain tumors represent 3.9% of all cancers.[20] In Cameroon brain tumors account for 8.2% of hospitalizations in neurosurgery in a period of 10 years.[21] In Ivory Coast, 362 brain tumors were observed in children at the University Hospital of Yopougon within 11 years.[22] In Ghana, 30 cases of brain tumors were observed during 2 years[23] and 3.1% of child hydrocephalus in Niger was related to cerebral tumors.[24] In Johannesburg, 142 brain tumor cases were hospitalized in 12 months in two University Hospitals.[25] In Kenya, brain tumors were the least represented among all cancers in children with a proportion of only 0.9%.[26]

It is difficult to determine the age at onset of a brain tumor. In most African series, the average age of diagnosis is between 36 and 42 years,[21,23,26,27] whereas brain tumors in developed countries are seen on average between 60 and 70 years.[28] Could this be explained by life expectancy at birth in developing countries?

A female predominance is observed in some studies with a sex ratio male/female of 0.8[19,21] while the opposite is true in others (e.g., 1.4 in Kenya).[29] In a Moroccan study of 542 cases compiled in Rabat and Casablanca in public institutions,[30] 51.8% of patients were male with an average age of 9.3 years. In Kuwait, a study of 163 cases showed an almost even split between boys and girls (50.8% and 49.2%), while in adolescents a slight female predominance was found.[31] In Asia, in the Taiwanese series[32] of 986 cases collected over 30 years, the average age was 7.8 years and the boy/girl ratio was 1.4, about the same as found in Japan (1.3).[33]

21.2.3 Breakdown by Location and Nature of Tumors

The distribution of the location of brain tumors depends in part on the age but also on the histological nature of the tumor. Tables 21.1 and 21.2 present some results of series in children and in adults. In Taiwan,[32] children's tumors were supratentorial in 58.3% of the cases, infratentorial in 41.1%, and 0.6% had a double localization.

In Morocco,[30] medulloblastoma is the most encountered tumor in children between 0 and 14 years (30%–40%). In this series, certain tumors such as gangliogliomas were over-represented among males, with a boy/girl sex ratio of 4, but the small number (only five patients) did not allow a conclusion to be drawn. Meningiomas were mostly observed in girls with a ratio of 0.7. The most frequent tumors (medulloblastoma, pilocytic astrocytoma, low-grade astrocytomas

TABLE 21.2 Distribution of Some Histological Aspects of Brain Tumors in Adults in Cameroon and in South Africa

	Eyanga et al. [21], Cameroon (%)	Ibebuike and Ouma [25], South Africa (%)
Meningioma	38.5	33.8
Astrocytoma	26.5	
Hypophyseal adenoma	9.1	24.6
Craniopharyngioma	0.8	2.8
Medulloblastoma	0.8	2.8
Lymphoma	5.3	1.4
Hemangioblastoma	1.5	0.7

or craniopharyngiomas) had almost an even split between boys and girls.

21.3 MULTIPLE SCLEROSIS

Since the mid-20th century, the scarcity of multiple sclerosis (MS) in people born and living in a tropical country, has been noticed by physicians working in these countries.[34] The work of Dean and Kurtzke[35] in South Africa has reported these facts and they have also stated that individuals born and who have always lived in this area until adolescence had a low MS risk when they later migrated to a region with high prevalence of MS. Such findings have subsequently been confirmed in all tropical countries.

This double recognition, the low prevalence of MS in the tropics, and the conservation of the low risk acquired in a subsequent migration after adolescence, has been the subject of controversy. The controversy was due to the fact that, firstly, there were, nevertheless, a few cases of MS among indigenous peoples who lived in a tropical region and, on the other hand, the diagnosis could be missed because the number of doctors (and more so, neurologists) practicing in these areas was low.

Since then, multiple epidemiological studies have been well conducted and have concluded that there is indeed a prevalence gradient, showing a decrease from the Polar Regions to the tropics. Moreover, in tropical regions, the prevalence is lower among indigenous populations, (sedentary), regardless of their ethnicity, compared to European populations that migrated from the temperate regions to the tropics. Quite a number of works, including a critical analysis from Coo and Aronson,[36] have also shown that populations migrating after adolescence, regardless of the

TABLE 21.1 Distribution of Some Histological Features in Children in Taiwan, Japan, Morocco, and Cameroon

	Wong et al. [32], Taiwan (%)	JCBTR[a] [33], Japan (%)	Karkouri et al. [30], Moroccco (%)	Eyanga et al. [21], Cameroon (%)
Gliomas	40.9		5.5	
Astrocytomas	31.1	28.2	20.4	20.0
Craniopharyngiomas	8.3	28.2	29.8	17.1
Medulloblastomas	13.2	11.9	28.9	20.0
Seminal tumors	14.0	15.3	0.9	
Epandymomas	5.7	4.5	7.6	
Oligodendriogliomas	1.2	0.9	1.7	
Choroïdal tumors	0.2	1.9		2.2
Pineal tumors	0.9	0.9	0.7	
Meningiomas	1.3		2.2	14.2

[a]Japan Committee of Brain Tumors Registry.

ethnic group, retained the risk of their origin, meaning that ethnicity probably played no role.[37,38] These studies therefore suggested that there is very likely in the tropical regions, one or more protective factors against the disease, if populations have been exposed to these factors during childhood, or, conversely, that there are, in cold and temperate regions, one or more factors favoring the occurrence of the condition.

These facts and assumptions are well accepted now; it therefore seems unnecessary to continue to carry out more prevalence studies. The specific issue to be resolved is the identification of eventual protective or predisposing factors that could explain this difference in prevalence according to latitude, and at the same time conserving the risk acquired during childhood.

It is well established that in this condition, contrary to what was believed for a long time, *genetic factors* are not common; they constitute only 25% of the risk of MS;[39] besides the haplotype HLA-DRB1*15:01, more than 100 variants associated with MS have been identified.[40] These genetic factors have not been well studied among populations in tropical countries. But to explain the low prevalence in the tropics, it is difficult to imagine that there can be in each of the many different ethnic groups living in the many tropical regions of the globe, a single protective genetic factor which could explain the low prevalence of MS in areas as different as equatorial Africa, equatorial America, the tropical part of Australia, or tropical Asia.

Without minimizing the importance of genetic factors, the role of environmental factors therefore seems essential to explain in a tropical region both the low prevalence of the disease, and the risk of an individual after migration.

Among environmental factors, multiple studies have involved the search for infectious factors that may cause this inflammatory disease. Numerous viral, bacterial or parasitic agents were suspected, including *Helicobacter pylori*, the intestinal microbiota, and retroviruses, to name only a few candidates that have aroused a lot of interest. But to date, and despite numerous epidemiological studies, we must recognize that none of these germs, nor prolonged contact with domestic animals or livestock, tobacco, obesity, or many toxic heavy metals and pesticides, could be incriminated with certainty. Only serological tests for Epstein–Barr and/or infectious mononucleosis are found to be consistently positive in MS cases,[41,42] with an odds ratio of 4.5 in a meta-analysis.[43] Of course, this cannot, in any case, account for the low prevalence of MS in the tropics, on the contrary, since the Epstein–Barr virus has a high prevalence in the tropics, particularly in tropical Africa where almost 100% of children aged 5 years are infected, and where this virus of the herpes family, is at the origin of very

serious conditions, such as Burkitt's lymphoma diagnosed in children in its endemic form.

Various studies[44,45] could provide an explanation for this apparent contradiction and led to formulate the concept of the "hygiene hypothesis" that low childhood exposure to infection, so to a small number of antigens, predisposes to the occurrence of autoimmune diseases when infections appear later after adolescence. This leads to the hypothesis that in the tropics, the frequency and variety of infections to repeated antigenic stimuli during childhood, is a protective factor against autoimmune diseases, including MS. If this hypothesis is true, with the gradual improvement of hygiene conditions in tropical countries, and the reduction of the parasitic diseases that appear to play an important role,[46] we should progressively see an increased prevalence of MS in the tropics. Due to the expected slow improvement of socioeconomic conditions, in particular the easy and free access to water, essential for better hygiene, this process may be very slow.

Another hypothesis that could explain the differences in prevalence in relation to latitude, is the predominance of sun in the tropics. Since the middle of the last century, few publications have highlighted the possible role of radiation,[47,48] and finally focused on ultraviolet B (UVB) radiation. This hypothesis, initially controversial, was the subject of several studies, including those by Simpson et al.[49] and Dalmay et al.,[50] and is now almost universally accepted. The regions of high prevalence of MS in the world are less sunny, and vice versa. The sun would be the, or one of the, protective environmental factors against the onset of MS. It is a candidate of choice because, ultimately, it is the only common denominator present in all tropical regions, regardless of the region, place or altitude where an individual may live: the closer to the tropics, the more UVB radiation increases.

Very satisfactorily, this hypothesis can account for a low prevalence in a given place, but apparently cannot explain conservation by individuals of this protection acquired during childhood, where a later migration in a high-risk area occurs. Protection, existing in sunny areas at first glance, cannot explain everything, unless this protective memory continues throughout the life of the individual. Which mechanisms could explain this protection and its maintenance?

Several studies have shown that this protection may be the result of increased secretion of vitamin D (cholecalciferol) induced by the action of UVB on the skin of individuals,[51] which results in increased vitamin D2, the circulating form of vitamin D in the body.[52,53] The biological effects of vitamin D depend on receptors expressed in Langerhans cells and in different cell types of the central nervous system. Among the immune cells, Langerhans cells, immunocompetent dendritic cells, antigen-presenting, form an abundant network in the epidermis and mucous membranes of humans. Responsible for the immune protection of the host, they capture antigens for presentation to CD4T lymphocyte Th1, which then become activated lymphocytes, memory T specific for the presented antigen. It is likely that there is a multiplicity of antigens, all having the common denominator of the ability to present an epitope they share with myelin autoantigens.

The action of the sun, and so UVB and vitamin D3 which is one of the consequences, is carried out in particular by the action of inhibiting the recruitment of Langerhans cells, reducing their number to decrease the expression by these cells of adhesion molecules and co-stimulatory protein (B7, CD40), which is essential for the activation of T lymphocytes; and at the same time, decreasing the production of activating cytokines of Th1 lymphocytes by these cells.[54,55]

It is partly through these inhibitory actions on Langerhans cells that vitamin D3 has an anti-inflammatory action, which is the intended purpose when prescribing such a treatment in patients with MS. This protective role of UVB would be more effective if it occurred early during childhood, limiting from a very early age the recruitment of T-cell memory to a given antigen, thus reducing the possibility of further activation of these lymphocytes. This protective effect of UVB could even occur earlier in the fetus during its intrauterine life.[56,57]

Among the radiation of the solar spectrum, ultraviolet A (UVA) rays, known for their harmful effects in the development of skin cancer, have likewise immunosuppressive action that also occurs on dendritic cells, independently of the UVB.[58,59]

It thus appears that exposure, from a young age, to solar radiation, thanks to the immunosuppressive or immunomodulating properties of UVA and UVB directly on Langerhans cells and indirectly through the secretion of vitamin D, has an immediate and maintained protective action thereby preventing the occurrence of MS.

The low prevalence of MS in the tropics thus appears as the consequence of two environmental factors occurring during childhood:

1. the repetition of antigenic stimuli thus infections, which is a protective factor against autoimmune diseases,
2. a strong and constant sun exposure in these regions.

References

1. Shekhar C, Gupta LN, Premsagar IC, et al. An epidemiological study of traumatic brain injury cases in a trauma center of New Delhi (India). *J Emerg Trauma Shock.* 2015;8:131—139.

2. Sanou J, Bonkoungou PZ, Kinda B, et al. Traumatismes crâniens graves au Centre Hospitalier Universitaire Yalgado Ouedraogo: aspects épidémiologiques, cliniques et facteurs limitant de la réalisation du scanner cérébral. *Rev Afr Anesth Med Urg.* 2012;17:7P (http://saranf.net/Traumatismes-craniens-graves-au.html) (Consulté le 25 /02/2016).

3. Auguémon AR, Padonou JL, Kounkpè PC, et al. Traumatismes crâniens graves en réanimation au Bénin de 1998 à 2002. *Ann Fr Anesth Réanim.* 2005;24:36—39.

4. Talona L, Maoneo A, Baonga L, Munyapara S, Wami W. Profil épidémiologique des traumatisés par accidents de trafic routier aux cliniques universitaires de Kissangani. *Kis Med.* 2014;5:51—57.

5. Bardrinarayan M, Nidhi DS, Sukkla SK, Sinha AK. Epidemiological study of road traffic accidents cases from Western Nepal. *Indian J Community Med.* 2010;35:115—121.

6. Fekik HA, Zayani MC, Trifa M, Ben Khalifa S. Épidémiologie du traumatisme crânien à l'hôpital d'enfants de Tunis au cours de l'année 2007. *Tunis Med.* 2012;90:25—30.

7. Diakité AK, Anzilania, Diaby, Camara ND. Mortalité par accident de la voie publique au CHU Donka. *Mali Méd.* 2005;10:17—19.

8. Konan KJ, Assohoun KT, Kouassi F, et al. Profil épidémiologique des traumatisés de la voie publique aux Urgences du CHU de Yopougon. *Rev Int Soc Med.* 2006;8:44—48.

9. Pouchieu C, Baldi I, Gruber A, et al. Descriptive epidemiology and risk factors of primary central nervous system tumors: current knowledge. *Rev Neurol.* 2016;172:46—55.

10. Johnson KJ, Culten J, Barnhortz-Sloan JS, et al. Childhood brain tumor epidemiology a brain tumor epidemiology consortium review. *Cancer Emidemiol Biomarkers Prev.* 2014;23:2716—2736.

11. Malmer B, Adatto P, Amstrong G, et al. Gliogene: an international consortium to understand familial glioma. *Cancer Epidemiol Biomarkers Prev.* 2007;16:1730—1734.

12. Gu J, Liu Y, Kyritsis AP, Bondy ML. Molecular epidemiology of primary brain tumor. *Neurotherapeutics.* 2009;6:427—435.

13. Zaho H, Cai W, Su S, et al. Allergic conditions reduce the risk of glioma: a meta-analysis based on 128 936 subjects. *Tumour Biol.* 2014;35:3875—3880.

14. Krishnamachari B, Il'yasova D, Scheurer ME, et al. A pooled multisite analysis of effects of atopic medical conditions in glioma risk in different ethnics groups. *Ann Epidemiol.* 2015;25:270—274.

15. Harder T, Plagemann A, Harder A. Birth weight and subsequent risk of childhood primary brain tumors: a meta-analysis. *Am J Epidemiol.* 2008;168:366—373.

16. Little RB, Madden MH, Thompson RC, et al. Anthropometric factors in relation to risk of glioma. *Cancer Cases Control.* 2013;24:1025—1031.

17. Idowu OE, Idowu MA. Environmental causes of childhood brain tumor. *Afr Health Sci.* 2008;8:1—4.

18. Okomoto H, Mineta T, Ueda S, et al. Detection of JC virus DNA sequences in brain tumors in pediatric patients. *J Neurosurg.* 2005;102(3 Sup):294—298.

19. Preston-Martin S, Wessels G, Hecht S, Hesseling PB. Follow-up of a suspected excess of brain tumors among Namibian children. *Sam J.* 2005;95:776—780.

20. Ngulde SI, Fezeu F, Ramesh A. Improving brain tumor research in resource-limited countries: a review of the literature focusing on West Africa. *Cureus.* 2015;7(11):1—8.

21. Eyanga VC, Ngah JE, Atangana R, et al. Les tumeurs du système nerveux central au Cameroun: histopathologie, démographie. *Cahier Santé.* 2008;18:39—42.

22. Broalet E, Haidara A, Zunon-Kipre Y, et al. Approche diagnostique et des tumeurs cérébrales de l'enfant. Expérience du

service de neurochirurgie du CHU de Yopougon Abidjan. *Afr J Neurol Sci.* 2007;26:27−38.

23. Andews NB, Ramesh R, Odjidja T. A preliminary survey of central nervous system tumors in Tema, Ghana. *W Afr J Med.* 2003;22:167−172.

24. Sanoussi S, Bawa M, Kelani A, et al. Using catheter a fentes' for managment of childhood hydrocephalus: a prospective study of nine-six cases. *J Surg Tech Case Rep.* 2009;21:13−16.

25. Ibebuike K, Ouma J. Demographic profile of patients diagnosed with intracranial meningioma in two academic hospitals in Johannesburg, South Africa: a 12 month prospective study. *Afr Health Sci.* 2014;14:939−944.

26. Laurence F, Levy MB, Auchterlonie W. Primary cerebral neoplasia in Rhodesia. *Int Surg.* 1975;60:286−292.

27. Adeloye A. *Neurosururgery in Africa*. Ibadan, Nigeria: University Press; 1989.

28. Lovaste MG, Ferrari G, Rossi G. Epidemiology of primary intracranial neoplasm. Experiment in the province of Trento (Italy) 1977−1984. *Neuroepidemiology.* 1986;5:220−232.

29. Mostert S, Njuguna F, Kemps L, et al. Epidemiology of diagnosed childhood cancer in western Kenya. *Arch Dis Child.* 2012;97:508−512.

30. Karkouri M, Zafed S, Khattab M, et al. Epidemiologic profile of pediatric brain tumors in Marocco. *Childs Nev Syst.* 2010;26:1021−1027.

31. Katchy KC, Alexander S, Al-Nashmi NM, Al-Ramadan A. Epidemiology of primary brain in childhood and adolescence in Kuwait. *Springer Plus.* 2013;2:58−66.

32. Wong TT, Ho DM, Chang KP. Primary pediatric brain tumors. *Cancer.* 2015;15:2156−2167.

33. Japan Committee of Brain Tumor Registry. Report of Brain Tumor Registry of Japan (1969−1996). *Neurol Med Chir (Tokyo).* 2003;43:1−111.

34. Dean G. Annual incidence, prevalence, and mortality of multiple sclerosis in white South-African-born and in white immigrants to South Africa. *BMJ.* 1967;2:724−730.

35. Dean G, Kurtzke JF. On the risk of multiple sclerosis according to age at immigration to South Africa. *BMJ.* 1971;3:725−729.

36. Coo H, Arson KJ. A systematic review of several potential non-genetic risk factors for multiple sclerosis. *Neuroepidemiology.* 2004;23:1−12.

37. Cabre P. Environmental changes and epidemiology of multiple sclerosis in the French West Indies. *J Neurol Sci.* 2009;286:58−61.

38. Cabre P, Signaté A, Olindo S, et al. Role of return migration in the emergence of multiple sclerosis in the French West Indies. *Brain.* 2005;128:2899−2910.

39. Taylor BV. The major cause of multiple sclerosis is environmental: genetics has a minor role—yes. *Mult Scler.* 2016;17:1171−1172.

40. Sawcer S, Franklin RJM, Ban M. Multiple sclerosis genetics. *Lancet Neurol.* 2014;13:700−709.

41. Ascherio A. Environmental factors in multiple sclerosis. *Expert Rev Neurother.* 2013;13:3−9.

42. Leray E, Moreau T, Fromont A, Edan G. Epidemiology of multiple sclerosis. *Rev Neurol.* 2016;172:3−13.

43. Belbasis L, Bellou V, Evangelou E, et al. Environmental risk factors and multiple sclerosis: an umbrella review of systematic reviews and meta-analysis. *Lancet Neurol.* 2015;14:263−273.

44. Leibowitz U, Antonovsky A, Medalie JM, et al. Epidemiological study of multiple sclerosis in Israel. II. Multiple sclerosis and level of sanitation. *J Neurol Neurosurg Psychiatry.* 1966;29:60−68.

45. Fleming J, Fabry Z. The hygiene hypothesis and multiple sclerosis. *Ann Neurol.* 2007;61:85−89.

46. Corralie J, Farez M. Parasite infections modulate the immune response in multiple sclerosis. *Ann Neurol.* 2007;61:97−108.

47. Barlow JS. Correlation of the geographic distribution of multiple sclerosis with cosmic-ray intensities. *Acta Psychiat Neurol Scand.* 1960;35:108−131.

48 Barlow JS. Multiple sclerosis, geomagnetic latitudes and cosmic rays. *Trans Am Neurol Assoc.* 1960;85:189−191.

49. Simpson S, Blizzard L, Otahal P, et al. Latitude is significantly associated with the prevalence of multiple sclerosis: a meta-analysis. *J Neurol Neurosurg Psychiatry.* 2011;82:1132−1141.

50. Dalmay F, Bhalla D, Nicoletti A, et al. Multiple sclerosis and solar exposure before the age of 15 years: case-control study in Cuba, Martinique and Sicily. *Mult Scler.* 2010;16:899−908.

51. Beretich BD, Beretich TM. Explaining multiple sclerosis prevalence by ultraviolet exposure: a geospatial analysis. *Mult Scler.* 2009;15:891−898.

52. Munger KL, Levin LI, Hollis BW, et al. Serum 25-hydroxyvitamin D levels and risk of multiple sclerosis. *JAMA.* 2006;296:2832−2838.

53. Ascherio A, Munger KL, Simon KC. Vitamin D and multiple sclerosis. *Lancet Neurol.* 2010;9:599−612.

54. Dumas M, Jauberteau-Marchan MO. The protective role of Langerhans' cells and sunlight in multiple sclerosis. *Med Hypotheses.* 2000;55:517−520.

55. Bartosik-Psujek H, Tabarkiewicz J, Poscinska K, Stelmasiak Z, Rolinski J. Immunomodulatory effects of vitamin D on monocyte-derived dendritic cells in multiple sclerosis. *Mult Scler.* 2010;16:1513−1516.

56. Willer CJ, Dyment DA, Sadovnik AD, Rothwell PM, Murray TJ, Ebers GC. Timing of birth and risk of multiple sclerosis: population based study. *BMJ.* 2005;330:120.

57. McDowell TY, Amr S, Langeberg P, et al. Time of birth, residential solar radiation and age at onset of multiple sclerosis. *Neuroepidemiology.* 2010;34:238−244.

58. Furio L, Berther-Vergnes O, Ducarre B, Schmitt D, Peguet-Navarro J. UVA radiation impairs phenotypic and functional maturation of human dermal dendritic cells. *J Invest Dermatol.* 2005;125:1032−1038.

59. Clement-Lacroix P, Michel L, Moysan A, Morliere P, Dubertret L. UVA-induced immune suppression in human skin: protective effect of vitamin E in human epidermal cells in vitro. *Br J Dermatol.* 1996;134:77−84.

Glossary

Acculturation for an individual immersed in a new society, it is the loss of the benchmarks of his own culture. The result is the difficulty of being oneself, of locating oneself within this new society, hence the loss of its harmony and vitality.

Addiction a condition involving use of a substance, such as a drug or alcohol, or engagement in a behavior, such as gambling, in which a person has strong cravings, is unable to stop or limit the activity, continues the activity despite harmful consequences, and experiences distress upon discontinuance.

Allele any alternative form of a gene that can occupy a particular chromosomal locus. In humans, there are two alleles, one on each chromosome of a homologous pair.

Amyotrophic lateral sclerosis (ALS) a rare disease, related to upper and lower motor neuron degenerations. It is characterized by extensive paralysis and leads to death generally by respiratory failure in 50% of cases, within 15−20 months after diagnosis.

Amyotrophic lateral sclerosis−parkinsonism−dementia complex refers to the clinical manifestation of a neurodegeneration that combined these three phenotypes and that was identified in Guam island, Kii peninsula of Japan, and West New Guinea in the 1950s.

Analytic epidemiological studies studies investigating risk factors of diseases, sometimes referred to as etiological studies. They are of several types: cohort study or case−control study.

Anthropology the science that deals with the origins, physical and cultural development, biological characteristics, and social customs and beliefs of humankind.

Anxiety an abnormal and overwhelming sense of apprehension and fear often marked by physical signs (such as tension, sweating, and increased pulse rate), by doubt concerning the reality and nature of the threat, and by self-doubt about capacity to cope with it; an uncomfortable feeling of nervousness or worry about something that is happening or might happen in the future.

Bipolar (disorder) also known as manic depression, a mental disorder that causes periods of depression and periods of elevated mood. The elevated mood is significant and is known as mania or hypomania, depending on its severity, or whether symptoms of psychosis are present. During mania, an individual behaves or feels abnormally energetic, happy, or irritable. Individuals often make poorly thought-out decisions with little regard to the consequences. The need for sleep is usually reduced during manic phases. During periods of depression, there may be crying, a negative outlook on life, and poor eye contact with others.

Blantyre coma scale a specific scale to assess malaria-related coma in children. This scale is a modification of the Pediatric Glasgow Scale. Motor response, verbal response, and eye movements are evaluated, and score ranges between 0 and 5.

Blood culture a test that checks for foreign invaders like bacteria, yeast, and other microorganisms in the blood.

Capture−recapture method statistical method first developed to track wild animal populations, based on the use of multiple sources of case ascertainment of cases of a well-defined disease from a temporally and geographically well-defined and closed population. Thanks to the cross-matching of sources (identification for each patient of the type and number of sources which identified ("captured") him), this method allows researchers to estimate the number of cases not identified through multiple sources of case ascertainment, then the total number of cases of a specific disorder, produce adjusted estimates (incidence, prevalence, mortality), and evaluate the exhaustiveness of case ascertainment.

Case−control studies a type of analytical epidemiological study consisting of comparing two groups of subjects: (1) well-defined cases of a specific disease and (2) controls not presenting the studied disease.

Cat scratch disease caused by *Bartonella henselae*, following a bite, lick, or scratch by a cat, less commonly a dog. Acute encephalitis is the commonest neurological manifestation, but cranial nerve palsies, peripheral nerve palsies, sensory neuropathy, and meningoradiculitis have been reported.

Central nervous system the brain, brainstem, cerebellum, and spinal cord.

Cerebrospinal fluid (CSF) a watery fluid that is continuously produced and absorbed and that flows in the ventricles within the brain and around the surface of the brain and spinal cord.

Cerebrovascular diseases a group of conditions that affect the supply of blood to the brain, causing limited or no blood flow to the affected areas. They include cerebral infarction (arterial and venous), cerebral hemorrhage (parenchyma hematoma and subarachnoid hemorrhage). This definition excludes the conditions related to trauma.

Cohort studies a type of analytical epidemiological study that consists of comparing the occurrence of a disease between exposed and non-exposed groups to a defined risk factor.

Culture the set of activities, customs, intellectual productions, art, religious beliefs that define and distinguish a group. It is expressed in customs that are preserved and perpetuated within the group.

Dementia a neurological chronic disease which affects the brain, induces cognitive impairments, behavioral disturbance and social distress, and impacts the activities of daily living.

Depression a mental illness in which a person is very unhappy and more or less anxious (worried and nervous) for a long period and cannot have a normal life during this period. The used term is characterized depressive disorder.

Dermatomyositis polymyositis with characteristic inflammatory skin changes.

Descriptive epidemiological study study of the distribution of diseases and health phenomena in a population; they are conducted either as cross-sectional (prevalence study where subjects are observed only once) or longitudinal (incidence study where subjects are followed-up).

Disability according to the World Health Organization, considered as an umbrella term that include impairments, activity limitations, and participation restrictions. It is a complex phenomenon that reflects the interaction between features of a person's body and the society in which the person lives.

Disability-adjusted-life-year (DALY) a measure combining years of life lost due to premature mortality and years of life lost due to time lived in states of less than full health. The DALY metric is used in the Global Burden of Disease studies to assess the burden of disease consistently across diseases, risk factors, and regions.

DNA extraction a process of purification of deoxyribonucleic acid from a sample using a combination of physical and chemical methods.

Door-to-door screening a screening method conducted in the general population, consisting of interviewing all the individuals residing in the study area to identify suspected cases of one or several diseases.

DSM-IV the abbreviation for the *Diagnostic and Statistical Manual of Mental Disorders* (Fourth Edition, 2000), prepared by the Task Force on Nomenclature and Statistics of the American Psychiatric Association. It is the Association's official manual of mental disorders and provides detailed descriptions of categories of disorders as well as diagnostic criteria. Disorders are placed on one of five axes: axis I includes all the clinical syndromes and V codes except for personality disorders and mental retardation; axis II includes the personality disorders and mental retardation; axis III lists any coexisting physical disorders or conditions; axis IV assesses the severity of psychosocial and environmental stressors; and axis V consists of a global assessment of functioning, using a 100-point scale assessing the highest level of functioning during the past year and the current level of functioning. A new version was released in 2013: DSM-V.

Epidemiology a discipline which studies the distribution of diseases in a population and analyzes the factors that condition their frequency.

Epilepsy a neurological chronic disease characterized by recurrence of seizures.

Active epilepsy *a case of active epilepsy is defined as a person with epilepsy who has had at least one epileptic seizure in the previous 5 years, regardless of antiepileptic drug (AED) treatment.*

Non-active epilepsy *epilepsy with no seizure for ≥ 5 years.*

Ethnic group a group of individuals sharing the same culture, customs, and language, and belonging to an identical human group with the same ascendants and descendants.

Ethnicity which contains specific characteristics of an ethnic group, ethnic quality, or affiliation. Stand for any character or any demonstration (appearance) appropriate (peculiar) to the cultural grouping of a population, by opposition to the characters of the individuals.

Ethnocentrism belief in the intrinsic superiority of the nation, culture, or group to which one belongs, often accompanied by feelings of dislike for other groups.

Ethnology the branch of anthropology that compares and analyses the characteristics of different peoples and the relationships between them.

Food consumption usually the daily alimentary intakes in proteins and energy.

Gene the biological unit of heredity, self-reproducing, and transmitted from parent to progeny. Each gene has a specific position (locus) on the chromosome map.

Global burden of disease study (GBD) the most comprehensive worldwide observational epidemiological study to date. It describes mortality and morbidity from major diseases, injuries, and risk factors to health at global, national, and regional levels. Examining trends from 1990 to the present and making comparisons across populations enables understanding of the changing health challenges facing people across the world in the 21st century.

Global health the health of populations in the global context, the area of study, research and practice that places a priority on improving health and achieving equity in health for all people worldwide.

Guillain-Barré syndrome an acute form of inflammatory polyradiculoneuropathy. It was first described by the French neurologists Georges Guillain, Jean Alexandre Barré, and André Strohl in 1916.

Health spending measures the final consumption of healthcare goods and services (i.e., current health expenditure) including personal healthcare (curative care, rehabilitative care, long-term care, ancillary services, and medical goods) and collective services (prevention and public health services as well as health administration), but excluding spending on investments.

Hirayama disease a disease characterized by unilateral or bilateral muscular atrophy and weakness of the forearms and hands, without sensory or pyramidal signs.

Hypertension high blood pressure, defined as a repeatedly elevated blood pressure exceeding 140 mmHg for systolic pressure and/or over 90 mmHg for diastolic pressure.

Incidence (cumulative) an epidemiological index that can be estimated from a population-based survey, calculating the number of new cases of a well-defined specific disease among a population, divided by the number of people in the population at the beginning of the survey. Cumulative incidence is not a rate.

Incidence rate (density) an epidemiological index that can be estimated from a population-based survey, calculating number of new cases of a well-defined specific disease among a population, divided by the number of people at risk in the population. Density of incidence is a rate.

Institute for Health Metrics and Evaluation (IHME) an independent global health research center at the University of Washington. IHME's policy reports put innovative research into action by translating scientific findings in ways that highlight specific policy implications. IHME researchers produce numerous scientific, peer-reviewed research articles each year on a range of topics.

Interventional studies studies evaluating the effect of a health program such as a new treatment, a screening campaign, a vaccination campaign, etc. They are conducted by experimental or quasi-experimental studies.

K6 forms The K10 and K6 scales were developed with support from the U.S. Government's National Center for Health Statistics for use in the redesigned U.S. National Health Interview Survey (NHIS). The scales were designed to be sensitive around the threshold for the clinically significant range of the distribution of nonspecific distress to maximize the ability to discriminate cases of serious mental illness (SMI) from non-cases. The six-question scale is at least as sensitive as the 10-question scale for discriminating between cases and non-cases of SMI. The K6 is now included in the core of the NHIS as well as in the annual National Household Survey on Drug Abuse.

Kwashiorkor a severe form of malnutrition associating edemas in the parts of the body with cachexia in other parts, and caused by an insufficiency of energy intakes along with an inflammatory status.

Latah a culture-specific startle syndrome found chiefly among Malays and other people of Southeast Asia. It is characterized by exaggerated startle responses and involuntary vocalizations, echolalia, echopraxia, and coprolalia.

Leptospirosis any of a group of febrile illnesses caused by infection with *Leptospira*, specifically one of the serogroups of *Leptospira interrogans*, which occur in worldwide distribution, being transmitted to man by a wide variety of wild and domestic animals which shed the infective organisms in the urine. Human infection is due to direct contact with the urine or tissue of infected animals or to contact with water, soil, or vegetation contaminated by the urine of infected animals.

Low- and middle-income countries refers to the classification of countries by the World Bank based on their per capita gross national income (GNI). The operational classification is reviewed annually by the World Bank on 1 July. In 2016, low-income countries had a per capita GNI of $1025 or less, lower-middle-income countries a GNI per capita of $1026–$4035, while a per capita GNI of $4036–$12,475 defines middle-income countries.

Malnutrition a syndrome induced by a negative imbalance between energy intakes and energy expenditures. It is associated with a number of troubles like an increase of the risk of infections and falls, an increased frailty, a decrease of psychological status, a loss of muscle and autonomy, etc.

Mental health a state of emotional and psychological well-being in which an individual can use his or her cognitive and emotional capabilities, function in society, and meet the ordinary demands of everyday life.

Meta-analysis studies consist of drawing up a synthesis on a scientific issue based on a series of studies already done. They are considered qualitative (always a first step) if they are structured as a systematic review of the literature (including a quality analysis of the studies) and are considered quantitative when a statistical approach is added to estimate a common effect based on the results of the included studies. This approach increases the power because of the higher number of included subjects in the analysis.

Migraine severe, painful headache that can be preceded or accompanied by sensory warning signs such as flashes of light, blind spots, tingling in the arms and legs, nausea, vomiting, and increased sensitivity to light and sound.

Millennium Development Goals basic human rights expressed as quantified targets for addressing extreme poverty in the world. Eight goals were proposed: (1) eradicate hunger and poverty; (2) achieve universal primary education; (3) promote gender equality and empower women; (4) reduce child mortality; (5) improve maternal health; (6) combat HIV/AIDS, malaria, and other diseases; (7) ensure environmental sustainability; and (8) develop a global partnership for development.

Modern medicine western evidence-based medicine.

Monogenic disorder any disease or disorder condition in which only one gene is implicated.

Mononeuritis an inflammatory disorder of the peripheral nervous system that involves only one nerve trunk.

Multigenic disorder any disease or group of diseases in which several genes are involved in its or their pathophysiology.

Multivariate analysis based on the statistical principle of multivariate statistics, which involves observation and analysis of more than one statistical variable at a time. The technique is used to perform studies across multiple dimensions while considering the effects of all variables on the responses of interest. In epidemiology, the most used multivariate analysis is logistic regression in which the dependent variable is categorical (often binary) because of the direct link between the coefficient of the model and the estimation of the odds ratio.

Myopathy any disease involving the muscle itself. It is acquired (inflammatory, toxic, etc.) or hereditary (congenital myopathy, etc.).

Myositis (and polymyositis) inflammatory myopathies.

Native community people who are descended from and identify with the original inhabitants of a given region, in contrast to groups that have settled, occupied, or colonized the area more recently.

Neuromuscular disease any disease that affects the peripheral nervous system (radiculopathy, plexopathy, neuropathy), muscles, and neuromuscular junctions (myasthenic syndrome).

NINCDS-ADRDA National Institute of Neurological and Communicative Diseases and Stroke/Alzheimer's Disease and Related Disorders Association which established diagnostic criteria of dementia, dementia of the Alzheimer type, and vascular dementia.

Nutritional transition transition from traditional modalities of feeding to occidental ones.

Obesity an excess of fat mass, inducing many complications: higher risk of cardiovascular disease, diabetes, and dyslipidemia, increased risk of a large number of cancers and rheumatologic troubles, decrease of psychological status, etc.

Paresthesia a generic name for a painless abnormal sensation that may be characterized by the patients by different words ("tingling," "tickling," "pricking," numbness," or burning") that may be due to central or peripheral nervous sensory injury.

Parkinson's disease a neurodegenerative disorder characterized by progressive slowness or akinesia and rigidity, with or without resting tremor, which responds significantly to dopaminergic treatment.

PCR polymerase chain reaction.

Peripheral nervous system the nervous system outside the central nervous system (cf. definition).

Phenotype the entire physical, biochemical, and physiological makeup of an individual as determined both genetically and environmentally, as opposed to genotype.

Physiopathology pathophysiology or physiopathology is a convergence of pathology with physiology. Pathology is the medical discipline that describes conditions typically observed during a disease state, whereas physiology is the biological discipline that describes processes or mechanisms operating within an organism.

Plexopathy any disease involving the brachial or lumbosacral plexus. It may be due to direct trauma (stretching, stab wound, etc.), infiltration (lymphoma, etc.), compression (local tumor process, etc.), or inflammation (Parsonage–Turner syndrome).

Polyneuropathy a disease involving the peripheral nervous system, which is grossly symmetrical and usually predominates in the lower limbs.

Polyradiculopathy a disease with selective involvement of nerve roots. It may be acute (Guillain–Barré syndrome) or chronic (chronic inflammatory demyelinating polyradiculoneuropathy, CIDP).

Prevalence an epidemiological index that can be estimated from a population-based survey, calculating number of existing cases of a well-defined specific disease among a population, divided by the number of people in the population. Prevalence is not a rate.

Psychiatric disorders a psychiatric disorder is a mental illness diagnosed by a mental health professional, that greatly disturbs your thinking, moods, and/or behavior, and seriously increases your risk of disability, pain, death, or loss of freedom. In addition, your symptoms must be more severe than an expected response to an upsetting event, such as normal grief after the loss of a loved one.

Psychopathology the study of psychological and behavioral dysfunctions occurring in mental illness or in social disorganization; the study of the origin, development, and manifestations of mental or behavioral disorders.

Psychosis a mental illness typically characterized by radical changes in personality, impaired functioning, and a distorted or nonexistent sense of objective reality. Patients suffering from psychosis have impaired reality testing: they are unable to distinguish personal subjective experience from the reality of the external world. They experience hallucinations and/or delusions that they believe are real, and may behave and communicate in an inappropriate and incoherent fashion.

Radiculopathy any disease involving at least one nerve root, usually caused by compression (lumbar disc hernia), infection (meningoradiculitis, such in neuroborreliosis), or neoplasia.

Random-effects model a kind of hierarchical linear model. It assumes that the data being analyzed are drawn from a hierarchy of different populations whose differences relate to that hierarchy.

Remitting-relapsing multiple scleroris (RrMS) the most common multiple sclerosis course characterized by clearly defined relapses of new or increasing neurologic symptoms. These relapses are followed by periods of partial or complete recovery (remissions). During remissions, all symptoms may disappear, or some symptoms may continue and become permanent. However, there is no apparent progression of the disease during the periods of remission. Symptoms of relapses must last at least 24 hours and must occur at least 30 days after the last relapse.

Risk factor any condition which increases the probability of developing a disease.

Schizophrenia a psychotic disorder (or a group of disorders) marked by severely impaired thinking, emotions, and behaviors. Schizophrenic patients are typically unable to filter sensory stimuli and may have enhanced perceptions of sounds, colors, and other features of their environment. Most schizophrenics, if untreated, gradually withdraw from interactions with other people, and lose their ability to take care of personal needs and grooming.

Secondarily progressive multiple sclerosis (SpMS) follows an initial relapsing–remitting course (RrMS) but most people will eventually transition to a secondary progressive course in which there is a progressive worsening of neurologic function (accumulation of disability) over time.

Secondary headaches headaches secondary to neurological disorders or disorders of neck or face structures, or secondary to systemic disorders.

Senility physical and psychological weakness related to old age.

SF-36 the 36-Item Short-Form Functional and Perceived Health Status Questionnaire, an international instrument used to assess multidimensional health-related quality of life, which measures eight health-related parameters, considered in the four previous weeks: (1) limitation of physical activity resulting from health problems; (2) limitation of social activity caused by physical or emotional problems; (3) physical health problems limiting usual activities; (4) bodily pain; (5) general mental health (psychological distress or well-being); (6) limitation of usual activities due to emotional problems; (7) vitality (energy and fatigue); (8) general health perceptions. The SF-36 is designed for use in those over the age of 14 and is useful in comparing the impact of disease and the efficacy of treatments, and identifying those at risk. Each parameter is scored from 0 to 100.

Sickle cell disease an inclusive term for a group disease of related β-hemoglobinopathies characterized by the predominance of sickle hemoglobin (HbS) within erythrocytes.

Snowballing techniques convenient types of samples, snowball samples begin from a core of known elements and are then increased by adding new elements given by members of the original sample. They are so called because of the analogy of the increasing size of a snowball when rolled down a snow-covered slope. Such samples are often used where there is no available sampling frame listing all the elements for the population of interest, for example illicit drug-users. Hence snowball samples are not random and not statistically representative of the population under consideration. They are, therefore, not amenable to inferential statistical techniques.

Sociocultural representation knowledge, beliefs, and attitudes shared by people or social groups about a disease or any specific situation.

Split hand sign a neurological syndrome in which thenar muscles appear disproportionately wasted as compared to the hypothenar muscles.

Stevens-Johnson syndrome a sometimes-fatal form of erythema multiform presenting with a flu-like prodrome, and characterized by systemic as well as more severe muco-cutaneous lesions.

Stigma a social mark of disqualification which makes an individual consider himself as less valuable (perceived stigma).

Stroke rapidly developing signs of focal (or global) disturbance of cerebral function, leading to death or lasting longer than 24 hours, with no apparent cause other than vascular.

Substance use disorder (SUD) also known as drug use disorder, is a condition in which the use of one or more substances leads to a clinically significant impairment or distress. Although the term "substance" can refer to any physical matter, in this context it is limited to psychoactive drugs. SUD refers to the overuse of, or dependence on, a drug leading to effects that are detrimental to the individual's physical and mental health, or the welfare of others. SUD is characterized by a pattern of continued pathological use of a medication, non-medically indicated drug, or toxin, which results in repeated adverse social consequences related to drug use, such as failure to meet work, family, or school obligations, interpersonal conflicts, or legal problems.

Swallowing disorders troubles with swallowing, of which the main consequences are aspirations and malnutrition.

Tailored system perfectly fitted to a condition, preference, or purpose; made or as if made to order; as a system tailored to the needs of individuals.

Tension-type headache attack characterized by lasting from 30 minutes to 7 days, mild to moderate headache with bilateral location, mild or moderate intensity, not aggravated by routine physical activity, and few associated symptoms.

Thermal tolerance the range in which an organism can live and cope with the atmospheric temperature. It is the first step to understand species vulnerability to climate warming.

Traditional medicine different than western modern medicine; based predominantly on magical understanding. The practitioners of this medicine are called traditional healers.

Traditional society society organized in and with the ancestral culture of the group.

Transient ischemic attack a brief episode of neurological dysfunction due to cerebral or retinal focal ischemia, the clinical symptoms of which typically last less than 1 hour without evidence of cerebral acute infarction.

Treatment gap proportion of cases that do not have access to an adequate treatment among all people affected by the pathology. Used mainly in epilepsy.

Tropical neurology stricto-sensu, indicates neurology practice in the countries situated between the Tropic of Cancer and the Tropic of Capricorn. But often, the "tropical" concept also concerns regions situated outside of the tropics, because of environmental, cultural, and socioeconomic similarities.

Tropical pyomyositis tropical myositis (or "myositis tropicans") is a bacterial infection (most frequently caused by *Staphylococcus aureus*) of the skeletal muscles, most commonly observed in tropical areas.

Tropical regions areas between the two tropics, the tropic of Cancer in the northern hemisphere and the tropic of the Capricorn in the southern hemisphere, on both sides of the Equator line.

Tropical spastic paraplegia (TSP) first used to designate spastic paraplegias observed mainly in sub-Saharan Africa in patients after repeated consumption of cassava varieties containing cyanides. Due to clinical and geographical similarity, this term has been misused for chronic spastic paraplegia associated with HTLV-1 retroviral infection.

Westernalization loss at an individual level of traditional ways of life, working habits, cultural patterns, and languages in favor of different attitudes influenced by western culture.

Index

Note: Page numbers followed by "*f*" and "*t*" refer to figures and tables, respectively.

Printed in the United States
By Bookmasters